Darwinism Evolving

Frontispiece Charles Darwin in 1840 by George Richmond. Reproduced with permission of the Charles Darwin Museum, Down House.

Darwinism Evolving

Systems Dynamics and the Genealogy of Natural Selection

David J. Depew and Bruce H. Weber

A Bradford Book
The MIT Press
Cambridge, Massachusetts
London, England

For Mary Depew and Kathy Weber

First MIT Press paperback edition, 1996
© 1995 Massachusetts Institute of Technology

This book was set in Palatino by DEKR Corporation and was printed and bound in the United States of America.

Library of Congress Cataloging-in-Publication Data

Depew, David J., 1942–
  Darwinism evolving : systems dynamics and the genealogy of natural selection /
  David J. Depew, Bruce H. Weber.
    p. cm.
  "A Bradford book."
  Includes bibliographical references (p.   ) and index.
  ISBN 0-262-04145-6 (HB), 0-262-54083-5 (PB)
  1. Evolution (Biology)—Philosophy—History. I. Weber, Bruce H.
II. Title.
QH361.D46 1995
575.01′62—dc20   4-16590
                                                                         CIP

Contents

Preface to the Paperback Edition

This book mainly aims to present an accurate history of the Darwinian tradition conducted from a point of view that is both socially contextualized and conceptually reflective, and to do so as a prolegomenon to estimating the future direction of evolutionary studies. Encouraged by its reception as a history, we commend the book once again. At the same time, those who disagree with our guesses about the future will be pleased to learn that we would make some changes in a putative second edition. For the present, we must content ourselves with listing in this brief preface a number of these contemplated emendations, some of them in response to what we have been fortunate to have: forceful but well-disposed critics. Almost all of these changes concern the three great themes of ontogeny, ecology, and phylogeny that are broached in the final third of the text and that, in our view, are currently teetering uncertainly between Darwinian and non-Darwinian styles of thinking.

We would, in the first instance, situate the molecular revolution in genetics more deeply within the context of the cybernetic and informational paradigm shift of the early fifties, which made it possible (by way of the kind of metaphorical refocusing to which we appeal throughout the book) to think of genes as "containing" information that is "encrypted" in a certain " language," transmitted by "messenger RNA," and "decoded" in ribosomes. Framing the issues raised by the rise of molecular genetics in these terms makes it easier to estimate both the advantages and the limitations of "gene talk." Such estimations are increasingly important in light of the spectacular progress made in recent years by the Human Genome Project. Only a realistic appreciation of the complex epigenesis of organisms can help us here. In this context, we would make more of Susan Oyama's *The Ontogeny of Information*, according to which organisms are self-organizing developmental systems, in which the functional information that guides reproduction cannot be exclusively packed into causally reified genes, or even into complexly interconnected genomes. We would also call attention to the work of Paul Griffiths, Russell Gray, Lenny Moss, and others who have brought this perspective to bear on the units of selection issue in ways that deepen our own conviction that since natural selection evolved out of chemical autocatalysis, its paradigmatic units of selection must be metabolic, developmental, and behavioral cycles (see, for example, P. E. Griffiths and R. D. Gray [1994], "Developmental systems and evolutionary systems," *Journal of Philosophy 91*, 277–304).

We would also acknowledge that, in our haste to bring older traditions in ecology into dialogue with Darwinism, we have insufficiently acknowledged the considerable accomplishments of mathematical population ecology. We remain convinced that ecological systems are coherent enough to have parts that help to preserve wholes, that self-organization underwritten by thermodynamical imperatives plays a role in this process, and that genocentric biases block recognition of these facts. But we would say more loudly than we have that ecological systems cannot form and behave the way they do without a good deal of (potentially conflicting) natural selection working both on individual organisms and on groups, and that in consequence ecological systems do not have trajectories that run in sublime indifference to processes in which organisms and groups, their key nodal points, play decisive roles (see, for example, D. S. Wilson [1992] "Complex interactions in metacommunities with implications for biodiversity and higher units of selection," *Ecology 73*, 1984–2000.) It is odd that we did not say this more forcefully, since it virtually follows from the view we have taken of organisms as informed and integrated ecological systems that ecological communities are precisely *not* as coherent as organisms.

With respect to phylogeny, finally, we note an emerging consensus that although the disparity between phyla during the Cambrian period may be less than previously thought, still the pattern of "sudden" diversification followed by relative quiescence, even stasis, is sufficient to dislodge overly gradualist conceptions of Darwinism that uncritically rely on the dynamical models of an earlier time. If the popular press is to be believed, the overthrow of gradualism, and the uniformitarian view of the cosmos that lies behind it, might be thought sufficient to unhinge Darwinism itself. But that is for us only one more reason to call attention again to the main lesson taught in this book: that Darwinism, far from being trapped behind the dynamical models it has used in the past, has always deepened its explanatory power by switching to new ones.

This book inserted itself into an unstable discursive field in which self-organization and adaptive natural selection were viewed as contending causes of biological order. That our attempt to break the mold by speaking of complex systems dynamics as making possible new conceptions of adaptive natural selection has not (yet) entirely succeeded is evidenced by the fact that those who already hold fixed conceptions of the dynamics of natural selection have interpreted us as less selectionist than we mean to be. It is mostly our fault. We did not state clearly enough that in our view nonlinear dynamics and self-organization provide new models of the paths that natural selection will seek and the entities on which it operates, not alternatives to natural selection. A new edition of this book would tilt more to the selectionist, even adaptationist, side of this difficult relationship by continuing to explore ways in which adaptive natural selection can be liberated from overly reductionistic ontological and epistemological frameworks.

In a book of this size, errors—typographical, grammatical, factual—are inevitable. Nor have readers been slow to bring them to our attention. We hope to correct the errors in a second edition. For the present, however, we content ourselves with noting that Wilberforce was a high-minded Tory, not a Whig (p. 64). Although King Croesus was famously greedy, it was King Midas whose food turned to gold (pp. 117, 119). In the reading guide for Section I, we must surely add Michael Ruse's *The Darwinian Revolution* to any list of essential sources. References to this book on pp. 328 and 334 are, however, actually to another book by Ruse, *The Philosophy of Biology* (1973, reprinted 1989. Old Tappan, NJ: Macmillan), which was unfortunately omitted from the bibliography. Finally, we would give more explicit credit to Bill Wimsatt, along with David Hull, for seeing the philosophical significance of the units of selection controversy and for convincing biologists that philosophy could help solve it.

Finally, we want to draw attention to a short list of important works that have appeared since this book went to press. There are now useful books on Huxley and Owen: Adrian Desmond has published the first volume of his biography of Thomas Huxley (A. Desmond [1994] *Huxley: The Devil's Disciple.* London: Michael Joseph); and Nicolaas Rupke has provided us with the first major biography of Richard Owen (N. A. Rupke [1994] *Richard Owen: Victorian Naturalist.* New Haven: Yale University Press). Also, the first volume of Janet Browne's biography of Charles Darwin has recently appeared (J. Browne [1995] *Charles Darwin: Voyaging.* New York: Knopf). There has also been an interesting debate on Darwin's teleology between Michael Ghiselin and James Lennox in the pages of *Biology and Philosophy.* The first attempt to bring together biographical material on Theodosius Dobzhansky had been undertaken by Mark Adams (M. B. Adams, ed. [1994] *The Evolution of Theodosius Dobzhansky.* Princeton NJ: Princeton University Press). An important book on the concept of evolutionary progress will soon appear by Michael Ruse (M. Ruse [1996] *Monad to Man: The Concept of Progress in Evolutionary Biology.* Cambridge, MA: Harvard University Press). That the range of competing evolutionary theories is still as wide as ever is suggested in four recent books aimed at the general reader: Jonathan Weiner (J. Weiner [1994] *The Beak of the Finch: A Story of Evolution in Our Time.* New York: Knopf), summarizes empirical evidence for the existence of adaptive natural selection in the wild; Niles Eldredge (N. Eldredge [1995] *Reinventing Darwin: The Great Debate at the High Table of Evolutionary Theory.* London: Weidenfeld and Nicolson) argues that population geneticists, such as George Williams, Richard Dawkins, and John Maynard Smith, have missed the importance of the pattern of stasis and punctuated change observed in the paleontological record; Brian Goodwin (B. Goodwin [1994] *How the Leopard Changed Its Spots: The Evolution of Complexity.* London: Weidenfeld and Nicolson) stresses the primacy of laws of morphogenesis; and Stuart Kauffman (S. Kauffman [1995] *At Home in the Universe: The Search for the Laws of Self-Organization and Complexity.* New York: Oxford University Press) stresses the interplay of self-organization and natural selection as causes of biological order. In a much discussed book, Daniel Dennett (D. C. Dennett [1995] *Darwin's Dangerous Idea: Evolution and the Meanings of Life.* New York: Simon and Schuster) gives a philosophical defense of the adaptationist paradigm and boldly extends it to the evolution of mind. Two recent contributions to the philosophy of biology are worth noting: Philip Kitcher (P. Kitcher [1993] *The Advancement of Science.* New York: Oxford University Press) and John Dupré (J. Dupré [1993] *The Disorder of Things: Metaphysical Foundations of the Disunity of Science.* Cambridge, MA: Harvard University Press); Dupré comes to some conclusions similar to ours, though reached from a different starting point.

Preface

This book is a contribution to the history of Darwinian evolutionary theory. The history of evolutionary *theory* is not quite the same thing as the history of evolutionary *biology*. Except by way of illustration, the reader should not expect to find a compilation of discoveries about evolution, numerous and interesting as these are. It is not even a work of popular science, an honorable genre of writing to which Stephen Jay Gould, for example, has contributed exemplary works in recent years. Rather, this book is about the intellectual constructs by which discoveries about evolution are guided, assembled, and justified as contributions to knowledge. That is, it is about evolutionary *theories*.

Scientific theories exist in a strange, interesting, and somewhat insubstantial realm somewhere between concrete inquiry and the social, intellectual, and ideological milieu in which it takes place. It is a realm that historians of science, including historians of evolutionary biology, have explored with increasing success in the past several decades. History of science has, in fact, enjoyed something of a golden age in the last third of this century, and historians of biology have been among the best of the new historians of science. We rely heavily on this body of work to show that there were all sorts of evolutionary theories before and after Darwin that were not Darwinian, that Darwinism was generally supposed to be on its deathbed at the turn of the twentieth century, and that even after a much-amended Darwinian tradition rose to hegemony by integrating itself with genetics it has been riven with controversies, no more so than in recent years. Rather than casting a doubtful shadow over the reality of evolution itself, these raging controversies have, on the contrary, made it clear that evolution, and evolution by means of natural selection, is an undeniably real phenomenon.

As a contribution to the history of evolutionary theory, this work is more than episodic but less than comprehensive. It is more than episodic because it tells a connected story and even projects that story, somewhat tentatively, into the future. It is less than comprehensive, however, because it looks at the history of the Darwinian research tradition from a certain angle. Throughout, we stress that from the outset, Darwinism, while successfully maintaining its autonomy from physics and other basic sciences, has used explanatory models taken from the part of physics called dynamics to articulate, defend, and apply its core idea of natural selection. We stress this point because we want to argue that under the present agitated conditions in evolutionary theory, the Darwinian tradition is deepening and renewing itself by reconceiving natural selection in terms of the nonlinear dynamics of complex systems.

If we look at evolutionary biology from the perspective of evolutionary theories, it is no less true that we look at evolutionary theories from the perspective of philosophy. We do not mean that we rummage through evolutionary theories to find themes and issues that have traditionally interested philosophers or that we are more than passingly interested in the personal worldviews of great evolutionary theorists. Rather, it simply happens to be the case that theories, including evolutionary theories, are the sorts of things whose logical structure philosophers are good at picking apart and reconstructing. Happily, the emergence of a community of historians of biology has been complemented in recent decades by the parallel emergence of a community of philosophers of biology. Their skill in identifying assumptions, testing definitions, reconstructing arguments, and formulating criteria for applying theories to cases have made it abundantly clear that conceptual change has played as crucial a role as new empirical discoveries in the development of evolutionary theory, and continues to do so today.

We believe this book can be read profitably by many audiences. Its narrative form, and the reading guides that follow each part, can help orient general readers and students to the real issues in evolutionary theory. Advanced students and colleagues in other disciplines may find it a useful source of information about scientific, cultural, and ideological themes that are crucial to understanding modernity itself and an invitation to more detailed study of particularly interesting figures, episodes, and ideas. Graduate students and colleagues in evolutionary biology, history of biology, and philosophy of science will find here an explanation sketch of changes in the Darwinian research tradition that may be worth developing and challenging.

The seeds of the book were planted at the Institute on Philosophy of Biology held at Cornell University in 1981 under the direction of Marjorie Grene and Richard Burian. This institute was funded by the National Endowment for the Humanities and the Council for Philosophical Studies. One of its offspring was the International Society for the History, Philosophy, and Social Sciences of Biology. This interdisciplinary group is not only an extraordinarily learned society but a genuinely learning society, whose semiannual meetings have charged us with excitement, provided us with a wealth of material and suggestions, and given us venues to try out some of what is in the book. We are also thankful for cooperation, released time, and financial and staff support from California State University, Fullerton, and from the California State Universities and Colleges. One of us received a Hughes Faculty Grant Program for funding several summers of work and support from a grant from the National Institutes of Health (GM/CA 08258). The other received a CSU Special Research Leave. The Department of Chemistry and Biochemistry at Fullerton has given generously of its material resources. The skilled secretarial assistance of Eileen Simkin, Joyce Calderon, and Shannon Glaab has been even more helpful. We have also been grateful recipients of the hospitality and resources of the Huntington Library, Pasadena, California; the Center for Advanced Studies at the University of Iowa, Jay Semel, director; and the Center for Hellenic Studies, Washington, D.C.

We have many people to thank. This book has emerged from our shared experience of teaching courses in the philosophy of biology for over a decade. Our thanks go first, accordingly, to our students at Fullerton and in the Literature, Science, and Art Program at the University of Iowa, all of whom we remember with affection. Our Fullerton colleagues, especially Jim Hofmann, John Olmsted, and Gloria Rock, have helped and encouraged us in several ways. So has John Lyne, of the Department of Communications Studies, University of Iowa, who co-taught a course with one of us on the material in the book. Among professional colleagues who patiently read part or all of the manuscript, and offered information, advice, and encouragement, are John Beatty, Dick Burian, John Campbell, Chuck Dyke, Marjorie Grene, Jon Hodge, Stuart Kaufmann, Jon Ringen, Krammer Rohlfleish, Stan Salthe, Rod Swenson, Bob Ulanowicz, and Jeff Wicken. Jon Hodge, Stu Kauffman, and Bill Wimsatt graciously made work in progress available to us. Allan Gotthelf pointed us to the correspondance between Darwin and Ogle. David Magnus was helpful on geographical isolation and speciation. Above all, we want to thank Dick Burian, director of the Center for Science in Society at Virginia Tech, Chuck Dyke, of the Philosophy Department at Temple University, and Marjorie Grene, Professor of Philosophy Emeritus, UC Davis. They taught us a lot about genetics, dynamics, and philosophy of science, gave extensively of their time to helping us make this a better book, and bucked us up when we were down. Our publishers, Betty and Harry Stanton, and our editor, Teri Mendelsohn, have been models of cooperation and patience.

The opportunity we have had to work together as co-teachers, conference organizers, and writers was initially facilitated by the deans of our respective schools at CSUF, Donald A. Schweitzer and James Diefenderfer. Both of these fine men have since passed away. We recall their works and days with gratitude, admiration, and affection.

We are better provided with knowledge about perishable plants and animals [than about the immortal heavenly bodies] because we have grown up in their midst. . . . Accordingly, one must not be childishly ashamed when it comes to inquiring into even the least honored animals. For in all natural beings there exists something that is wonderful. What is said about Heraclitus is to the point. Some visitors wanted to meet him, but, upon entering, hesitated to proceed further when they saw him warming himself by the stove. He told them to take heart and come on in. "For there are gods even in here." Similarly, we ought to enter upon inquiry into each of the animal kinds without feeling abashed. For in all of them there is something natural and beautiful. What is not haphazard, but for the sake of something, is especially characteristic of the works of nature. And whatever has been composed for the sake of something, or has become complete, has its share of beauty.

—Aristotle, *Parts of Animals*

There is grandeur in this view of life, with its several powers, having been originally breathed into a few forms or into one; and that, whilst this planet has gone cycling on according to the fixed law of gravity, from so simple a beginning endless forms most beautiful and most wonderful have been, and are being, evolved.

—Charles Darwin, *On the Origin of Species*

1   Introduction: Darwinism as a Research Tradition

This book contains a story, an argument, and some guesses. The story is about the development of the Darwinian research tradition, a family of ways of thinking about living things, looking closely at them, and experimenting with them that began with the work of Charles Darwin in Victorian England, came of age in this century, and may (or may not) be about to enter a period of new growth and explanatory power.

Like any other complex, changing historical subject that has captured the energies, imaginations, and intellects of generations of talented and passionate people, the Darwinian research tradition has had its ups and downs, its crises and triumphs, its periods of decline and of renewed hope. Around the turn of the twentieth century, for example, many biologists intoned that Darwinism was on its deathbed, in part because it seemed to contradict what genetics, recently resurrected after being buried with Father Gregor Mendel's long-forgotten paper on peas, was saying about how the information that guides the transmission of traits is carried from generation to generation. Rather than collapsing, however, Darwinism eventually received a new interpretation, allowing it to rise up with new vigor and to become one of the most fruitful scientific research programs of a century that will be remembered for its spectacular scientific successes.

During most of the twentieth century, the Darwinian tradition has gone under the name of the "modern evolutionary synthesis," which married Darwin's theory of natural selection to the new science of genetics. Under the influence of our rapidly expanding knowledge of molecular biology, however, the modern synthesis has been subjected in recent decades to pressures and puzzlements that have led some to proclaim, once again, that Darwinism is on its deathbed—or at least is due for major surgery. To anyone who is familiar with the history of Darwinism, this can seem "just like déjà vu all over again." Here is where the arguing in this book, and most of the guessing, comes in. We think there is reason to believe that the pressures currently being put on the Darwinian tradition, and on the theory of natural selection in particular, may serve as an occasion for it to transform itself once again into an even more powerful explanatory theory.

In order to make this a plausible guess, we argue for three claims. The first is that the idea of *natural selection is the conceptual core of the Darwinian research tradition.* Natural selection explains adaptedness within species and diversity between them in terms of the higher reproductive rates that can be expected from organisms possessing heritable variant traits that enable them to cope better in a given, and often changing, environment than their competitors. Surprisingly, the notion that Darwin's most salient contribution to evolutionary theory, natural selection, is the essence of Darwinism is not universally accepted. David Hull, for example, has argued that Darwin's substitution of genealogical for essential characteristics as ways of identifying something that changes over time means that it is un-Darwinian to say that there is any stable essence or core to Darwinism—even natural selection (Hull 1985, 1988). What makes a biologist a Darwinian on Hull's view is whether he or she can be located in a particular scholarly lineage, not whether he or she passes some test about "essential" Darwinian beliefs. In support of this view, Hull reminds his readers that the first generation of Darwinians differed widely in their beliefs and that not all of them even believed in natural selection, or at least assigned to it the degree of efficacy Darwin did. Indeed, if there is anything that might seem to be a candidate for the essential defining mark of the Darwinian tradition, it is not natural selection, but commitment to "common descent," the idea that all organism and lineages have descended from a single ancestral form.

We can readily accept these historical points because our claim is not that all and only those people rightly called Darwinians believed in natural selection or even that natural selection is always the preferred mode of explaining an evolutionary phenomenon (Dyke and Depew 1988). Our claim is that the continuity of the Darwinian research tradition over time has been maintained by offering new interpretations of natural selection to accommodate new knowledge. It was because the Darwinian tradition found a new interpretation of natural selection in the early decades of this century that the modern synthesis could be such a productive matrix for evolutionary research for such a long time. Our corresponding guess is that if Darwinism is to revitalize itself, it will do so by finding yet another *conception* (or interpretation) of the *concept* of natural selection.[1]

Our second claim is that the *Darwinian research tradition, while successfully resisting reduction to or incorporation within physics, has from the beginning used explanatory models taken from physics to articulate its core idea of natural selection.* Again, this is not the conventional wisdom. On the contrary, it is now the received view, urged for many years with vigor by Ernst Mayr, one of the founding fathers of the modern synthesis, that Darwinism has come into its own in direct proportion as it has distanced itself from the works and pomps of physicists and has instead reaffirmed its connection with the long tradition of natural history (Mayr 1982, 1985,

1988). With much of this we agree. We hypothesize, however, that too exclusive a concentration on the autonomy of biology from physics can blind one to the fact that ever since Darwin, Darwinians have used dynamical models, most of which have their origin or paradigmatic application in various branches of physics, as ways of envisioning and reconceiving the process of natural selection. Indeed, we think that the Darwinian tradition has maintained its identity and fecundity precisely by doing this (Depew and Weber 1989).

Classically, dynamics is the study of change, rate of change (velocity), and rate of rate of change (acceleration) in systems of all sorts. Dynamics, as distinguished from kinematics, is about more than the path that a system takes over time. It is about how that path can be generated. Dynamics gives a "recipe for producing a mathematical description of the instantaneous state of a physical system, and a rule for transforming the current state description into a description for some future, or per-haps past, time" (Kellert 1993, 2). Galileo was a dynamicist in this sense. So was Newton. We hope to show that Darwin's Darwinism was in-formed, in a metaphorical but far from empty sense, by dynamical mod-els taken from Newtonian physics. Similarly, when the Darwinian tradition arose from its alleged deathbed in the early twentieth century, it used dynamical models taken from statistical mechanics and thermo-dynamics to integrate natural selection with genetics. In a similar spirit, we wonder whether at the turn of the twenty-first century the Darwinian tradition, under pressure from the unexpected complexity of genetic networks and from the perception that major events in the history of life are more subject to chance and less to natural selection than had been assumed, can expand its explanatory power by switching to a family of dynamic models associated with the study of complex systems, nonlinear dynamics, and chaos theory.

Our third claim is about the history of science. It is about how what we call research traditions manage to maintain their integrity over time. Anyone who has dipped into the history of biology cannot help but be amazed at the persistence and resiliency of its various research traditions. Core ideas once declared as dead as doornails have often risen to new prominence by yielding persuasive interpretations of new discoveries. We think that traditions that have fallen on hard times sometimes man-age to get back on their feet by changing what philosophers call their "ontology" (from Greek *on* + *logos*, "the study of being"), that is, the kinds of theoretical entities and processes they recognize. (Whether the world is made of "substances" or "processes" is an example of an ontological issue.) To be sure, a change in ontology does not guarantee that a tradi-tion will maintain its integrity. We will see, for example, that the great nineteenth-century physicists James Clerk Maxwell and Ludwig Boltzmann, in trying to revitalize the Newtonian tradition by shifting to a probabilistic ontology, actually prepared the way for the rise of quantum mechanics, and hence for the fall of the classical physics they both loved too well. Nonetheless, we claim that when in the first half of the twentieth century the Darwinian tradition began to use dynamical models taken from statistical mechanics and thermodynamics, it changed its ontology, and expanded its problem-solving prowess by doing so. Moreover, we think that something like that may well be about to happen again under the impact of nonlinear dynamics. In the final part of this book we will try to say why. Our first task, however, is to provide the reader with a map of the book. We begin with a few basic points about its focal concept, natural selection.

Darwin wanted, among other things, to find a way of explaining the adaptedness of the organisms to their environments and the mutual adaptedness of the traits of organisms—body parts, physiological processes, behaviors—to one another. The pervasive fact of adaptedness long ago induced the otherwise sober father of biology, Aristotle, to break out in lyrical praise of a "nature whose works are everywhere full of purposiveness and beauty" (*Parts of Animals* I.5). Different species of finches, living on islands just a few miles apart, have just the right sorts of beaks to pick up and break open just the sorts of seeds or insects from which each of them lives. Flowering plants have just the right arrangement of colors and designs to attract just the species of insects on which their fertilization depends. The "flower in the crannied nook" over which the poet enthused is a living advertisement, a commercial targeted at a particular insect audience.

Biologists have differed little about whether adaptive traits exist. They have differed considerably, however, about how to explain them. The theory dominant in Darwin's immediate milieu was that of the "natural theologians," and especially of William Paley, an Anglican priest whose works Darwin confesses to having practically memorized during his college days at Cambridge. Paley argued that adapted entities always have at least one property in common. They exhibit *functional organization*, an arrangement of components, processes, and behaviors that when properly combined allow the whole entity to do what it is supposed to do, and do it effectively. Functional organization, Paley's argument goes, never just happens. As in the case of functional contrivances made by humans, such as watches and internal combustion engines, it has to have been put there on purpose and by design. Thus, whereas physical nature, which does not exhibit functional organization, might rest on purely materialistic and mechanical laws, living things do not. Instead, each species shows the work of the Creator, who gave it precisely the right traits for it to flourish in the particular niche *for which it was designed*.

Paley's was not the only explanation of adaptedness current in Darwin's time. Around the turn of the nineteenth century, the French zoologist Jean-Baptiste Lamarck devised an explanation of adaptedness in which the activities of organisms in adjusting to changed circumstances, together with their ability to pass what they have learned to offspring, reflects and advances an inherent tendency of living things to become more complex, and by becoming more complex to stay adapted to environmental change. Lamarck's theory of adaptation led him to differ with Paley about the origin of species. Lamarck asserts that continued adaptation and complexification means that new species are products of a temporally extended process in which new kinds come from old ones rather than directly from the hand of God. Lamarck was a "transmutationist," "transformist," or (later) "evolutionist." Paley, by contrast, was a "creationist."

Darwin, contrary to popular opinion, did not invent the idea of transmutation or evolution. What happened is that he rejected the creationism of his youth and came to agree with Lamarck that adapted lineages do in fact evolve into new species. At the same time, Darwin had a number of reasons from the start for wanting to avoid Lamarck's explanation of adaptation and transformation. In *On the Origin of Species* (1859) he proposed a mechanism of his own, which he called *natural selection*. If adaptation and transformation are the primary phenomena Darwinians try to explain, natural selection is the primary idea they use to do it.

Darwinian natural selection was suggested by the observation that some of the often considerable differences among offspring of the same parents, with which breeders have long been familiar, can be inherited by their offspring. Darwin suspected that it is not purely a matter of chance that heritable variants are passed down through a lineage, any more than it is a matter of chance that a breeder chooses some offspring rather than others. The notion of selection in "natural selection" is, in fact, modeled on the plant and animal breeder's "pick of the litter," work with which Darwin was both practically and theoretically acquainted (Hodge 1985, 1992b). Accordingly, natural selection, as Darwin uses the phrase, is a theory about what causes "differential retention of heritable variation," to use a contemporary formula (Lewontin 1978, 1980). In nature, Darwin says, competition for scarce resources puts a premium on precisely those variant traits that allow the individuals that possess them to do better, metabolically and reproductively, than their immediate competitors. If competitive pressures are constant, severe, and ubiquitous, and if good-making traits are heritable, it follows that over enough generations, only highly adapted traits, and organisms fortunate enough to possess them, will remain. An adaptation, from a modestly reconstructed Darwinian perspective, is a heritable trait that, built up through a selective process over a number of generations, causes differential survival, or relative fitness, of offspring because it offers competitive advantages to its possessors (Lewontin 1978, 1980; Burian 1983; Brandon 1990).[2]

By this route we reach Darwin's rejoinders to Paley and Lamarck. To Paley he says: The living things you see around you are certainly adapted to their environments, but they are not products of a designer. They are the result of a winnowing process, in which what remains after many generations of natural selection looks *as if* it had been designed because the only living things left in a highly competitive world will be those that bear precisely the traits that are needed in order to live adaptively in such a world. All others will have disappeared. To Lamarck Darwin says: Adaptations are not generally the result of the efforts of organisms to solve problems and to pass those solutions to their offspring. It is a matter of chance whether the right sort of variation turns up in organisms living in specific environments. What is not purely a matter of chance is which variations make it through the winnowing to which variation is subjected. The ones that do are favored by natural selection.[3]

Darwin's theory of the origin of adaptations also gave him original theories of transmutation and descent from a common ancestor.[4] The evolution of species does not result from an inherent drive in living matter to become more complex but is the effect of forces impinging on organisms from the environment. Ubiquitous competition places a premium on an organism's ability to extract resources from the environment. Traits that enable an organism to do well in one niche are not precisely those required for doing well in another. Hence, as niches differ and new ones open up, organisms that possess adaptations suitable to those niches will begin to differ from one another. Over time, lineages will depart further and further from each other until they constitute not only new "races" and species but different genera, families, orders, classes, and kingdoms. Hundreds or thousands of generations later, if you had kept track, you would see a branching treelike diagram that provides two pieces of information: a good way of classifying organisms and a record of their descent—what is called phylogeny (from Greek *phylum* + *genos*, "the birth of kinds"). That is Darwin's "branching" or "ramified" (from *rama*, Latin for "branch") theory of evolution. Diversity, for Darwin, is a long-run effect of natural selection as it opens up new niches.

In spite of the power of this idea, neither Darwin's theory of adaptation nor his accounts of transmutation and common descent are obviously true. For one thing, Darwin's theories would fail unless he could show that the variants that are passed on are nonrandomly those that enable their possessors to deal more successfully than their competitors with environmental conditions and contingencies. Darwin could not show this, however, unless he could show first that the competitive stress to which organisms are subjected are constantly and powerfully at work in most populations of most species in most environments and at most times. Otherwise, there would be no motor to push natural selection in a concerted direction, no force to match variation to what is adaptively needed in a certain environment over many generations. Darwin had to prove that the normal situation of organisms is such that environmental stresses are not accidental or occasional but ubiquitous and constant.

Darwin considered himself illuminated on this point by reading the work of the political economist Thomas Malthus. In his *Autobiography*, he says that he got the idea of natural selection when "I happened to read for amusement *Malthus on Population*" (Darwin 1958, 120). In *On the Origin of Species* Darwin says forthrightly that his theory is "the doctrine of Malthus applied with manifold force to the whole animal and vegetable kingdom" (Darwin 1859, 63). Darwin is alluding to the tendency of populations to increase faster than resources. This tendency, which Malthus had postulated as a universal law, provided Darwin with the constant and pervasive pressure that would put all individuals, and not just members of different species, into competition with one another all, or at least most, of the time.

That is not, however, the only influence of "political economy," or what we now call "classical economics," on Darwin. Darwin's theory of diversification, in which new species are created as new niches are opened up, owes something to another economist, Adam Smith, or at least to the pervasive influence that Smith's view of market dynamics exercised in the Britain of Darwin's day (Schweber 1977; Gould 1990). Self-interest, Smith had argued, is generally the motor of economic progress. Each person in a market, or more generally in a society conceived as a market, tries to buy commodities as cheaply as possible and to sell them for as much as he or she can get. The laws of the market, determining what commodities are made, how many are produced, what price they sell for, and what profit is pocketed, are mathematically computable consequences of these psychological and sociological principles of action. Rather than producing economic chaos and social injustice, however, as the traditional view had it, Smith argued that unlimited self-interest, and the unlimited right to buy and sell whatever you want whenever you want it, actually produces what was to be called "the greatest good for the greatest number." Part of the reason is that incentives are created in a market system for dividing labor ever more finely and for finding and exploiting new resources. Everyone benefits from that. In *On the Origin of Species,* Darwin argued in a similar way that "races" and species diversify by acquiring adaptations that enable them to divide labor by entering new niches, just as firms do in a competitive economy.

The influence of the "discourse" of political economy on Darwin is a vexed theme because it raises questions about whether Darwin's biological vision is merely an ideological reflection of Victorian capitalism. Darwin was intimately acquainted with people who argued that Malthus's theory about population always pushing against resources, which its author had devised during the French Revolution to throw cold water on the idea that humanity is perfectible, is actually a motor of progress. Although the "gloomy parson" gave economics its reputation as a "dismal science," and his name has been indelibly linked ever since with the theme of "the limits to growth," the Malthusian enthusiasts of Darwin's day believed that the threat of starvation played a positive role: It stimulated humans to control their reproductive behavior, to become rational economic agents, and to explore new geographic and economic niches (Desmond and Moore 1991). Given this interpretation of Malthus, Darwin was able to derive an Adam Smithean lesson from Malthusian premises (Gould 1990). The conversion of nondirected variation into adaptations offers species a way out of extinction by keeping them tuned to changing environments and by stimulating them to open new niches.

We do not deny these ideological influences on Darwin. We wish to draw more attention, however, to the fact that classical economics was important to Darwin primarily because it gave him dynamical models that made his biological theorizing part of a respectable tradition of British science. For just as the work of political economists stands, for better or worse, behind Darwin's biology, so behind Smith, Malthus, and other British political economists stands the revered British physicist and astronomer Isaac Newton. Political economists metaphorically appropriated the dynamical models Newton had used to explain planetary motion in order to explain the aggregate motion of human production and exchange. By adapting the work of political economists, we claim, Darwin in turn accessed the Newtonian tradition for biology, hoping thereby to gain a hearing for evolutionary theory among the respectable intelligentsia of his country.[5]

Newton's dynamical models are the culmination of physical and astronomical investigations reaching back to Copernicus and Galileo. As such, they represent a sharp break with the assumptions of the ancients and medievals about celestial motion. The old notion was that circular motion is natural motion and that the reason the stars go around in circles is that their motion is more natural than that of terrestrial things. Ever since antiquity, this idea had been contradicted by the known fact that the planets do not move in perfect circles, sometimes even turning back on themselves three or four times in the course of a complete orbit. (The word *planet* means "wanderer" in Greek.) Elaborate mathematical models had been constructed to make the errant ways of planetary motion conform to the assumption of perfect circular motion. Part of the problem dissolved when Copernicus recognized that the planets turn around the sun rather than the earth. But the problem proved more intractable than that. Johannes Kepler had proved that the planets actually move in ellipses, not circles. Newton accounted for this by abandoning the old premise about circular motion being natural motion. Natural, or inertial, motion is rectilinear, as René Descartes had first opined. Thus, the orbits followed by stars, planets, comets, and even by bullets and basketballs, whether they are circles, ellipses, hyperboles, or parabolas, are all carved out by the fact that at each moment the force of gravity exerted on a body by other bodies in its neighborhood deflects it from its natural, inherent straight-line motion and bends it back into the trajectory we see. Moment by moment, *equilibrium* is being restored between the inertial motion of the body and the force of gravity.

Adam Smith's theory is something like that. Each self-interested rational economic agent departs from the circular motion of communal good by "going off on a tangent." The circle of harmony is restored by the forces of the market, which work like gravity. It is self-interest, however, and not altruism or community spirit that does the work. Malthus's notion of an inherent population growth parameter is similarly conceived. The intrinsic rate of population growth is, in effect, an inertial property of organisms. If you postulate, as Malthus did, an inherent tendency in organisms to keep reproducing faster than the productivity of land, and resources more generally, it follows that at some point there will be a crunch. Population growth will be trimmed back by the pressing force of scarcity. Equilibrium between the demand for people, so to speak, and their supply will be reached—on a pile of corpses. If, on the other hand, you think that niche specialization and exploration offers a way to beat the Malthusian devil, as Darwin did, equilibrium between environments and organisms might be reestablished in a happier way. Although the ground will be littered with the corpses of less adapted variants, which are unable to compete for scarce resources, the organisms that remain will be so highly adapted that one might well believe that only a god could have made them.

Darwin sees organic adaptation and the differentiation of lineages in terms of a *Newtonian model*, a more or less abstract picture of how *systems*, whether they be planetary, economic, or biologic, can be expected to behave at each instant and over time. In general, whatever the entities are that conform to this model, they will have an inertial tendency of some sort driving them off on a tangent. This is diverted and shaped by an external force. The result is a system that maintains itself in equilibrium. In Darwin's case, we argue that there is a double appeal to Newtonian models. If individual organisms tend indefinitely to vary, something will be needed to pull them back into the natural kinds, such as species and genera, that we have every reason to think represent real divisions in nature. Because Darwin assumes that parental traits are blended in offspring, sex, like gravity, performs this role: It pulls individual variants back into the circle of kinds (Hodge 1985, 1987). At the same time, the products of sexual blending stay adapted to their environments because natural selection, again like the force of gravity, trims variation to fit circumstances.

Newtonian models had acquired enormous prestige in Darwin's milieu. The first full biography of Newton and a discussion of the importance of the Newtonian approach to science was published in 1831 (Brewster 1831). The philosophers of science whom Darwin most admired dictated not only that Newton's methods were obligatory for every field of inquiry that sought to present itself as scientific but were obligatory because we could reasonably expect that all phenomena governed by natural laws would conform to Newtonian presuppositions about how systems work (Herschel 1830). Darwin's reliance on what he took to be Newton's *methods* has long been recognized (Hull 1973). What is novel in our argument is that Darwin is also faithful to the way that entities are represented in Newtonian systems. Lavoisier and Dalton had already applied Newtonian thinking to the problem of chemical affinities. Charles Lyell had applied it to the hottest science of the day, geology, the field in which Darwin had been trained. Darwin had avidly and thoroughly absorbed Lyell's Newtonian, or "uniformitarian," geology while he was a young man. He thus worked within a very self-conscious scientific culture in which external forces operating on bodies rather than internal drives were expected to drive change over time. Darwin now extended Newtonian model of systems dynamics to questions about biologic origins by taking a detour through economics. That is something Lyell himself hesitated to do. For, as Paley's argument from design suggested, natural law, and Newtonian mechanics, were supposed to stop at the frontier of living things (chapter 4). If Darwin's work unleashed an intense, if short-lived, cultural crisis in Britain, that was not, on our view, because he introduced evolutionary thinking into Britain—in fact, he did nothing of the sort—but because he very nearly succeeded in extending his culture's most respected scientific model to objects and processes from which it was supposed to keep its distance (chapter 6).

In part II of the book, we will see how after a period in which evolutionary theory triumphed, but on generally non-Darwinian terms (chapter 7), the Darwinian research tradition integrated itself with the emerging science of genetics, with which it was originally thought to be in contradiction, to form the modern evolutionary synthesis. This was done by using statistical methods to represent and analyze the distribution of genes in interbreeding populations. We see dynamical models at work here too. Just as Darwin's Darwinism stood on the shoulders of Newton's dynamics, so we believe that this belated triumph of the Darwinian tradition could occur only because latter-day Darwinians, like Ronald A. Fisher, J. B. S. Haldane, and Sewall Wright, represented changing frequencies of genes in populations in terms of models adapted from the statistical dynamics and thermodynamics of nineteenth-century physicists (chapters 9–13).

Throughout the second half of the nineteenth century and the first quarter of the twentieth, techniques were being developed in many sciences for dealing with systems whose elements are too numerous to be tracked by classical Newtonian methods. The general idea was to use statistical averages over populations rather than to follow the movement of each atom, or molecule, or, in economics, the activities of each rational economic agent. Historians of science have done much interesting research of late into this "probability revolution" in the sciences (Hacking 1975, 1990; Gigerenzer et al. 1989; Porter 1986; Krüger et al. 1987). Most concede that an important sea change took place in the whole of science over the period from roughly 1840 to 1920. In the course of this change, whole new fields, many in the social sciences, were brought into the charmed circle of quantitative science by way of the new ability to see statistical pattern in vast arrays of events. The change from Darwin's Newtonian Darwinism to genetic Darwinism and the modern synthesis was part of that larger shift.[6]

The paradigmatic achievements of the probability revolution were the work of Maxwell and Boltzmann in the mid-nineteenth century. Maxwell analyzed the classical gas laws—laws relating the observable or phenomenal characteristics of gases, such as temperature, pressure, volume—in terms of the random motions of millions and millions of molecules of gases colliding with one another in predictable frequencies. Boltzmann used similar techniques to unravel the mystery that had been posed by Sadi Carnot's discovery that there is no such thing as a perfectly efficient steam engine. Boltzmann showed why it was merely a question of the most probable distribution of energy states that not all energy can be converted into work, that is, that "entropy" constantly increases in an isolated system. He provided a statistical and probabilistic analysis of the second law of thermodynamics. Boltzmann's work vividly displayed the idea that certain quantities (entropy, in this case, or workless energy) must increase over time and that wherever there is a difference in energy states, or any analogue of exploitable energy, a system will slide down the "gradient" or energetic hill thus created toward equilibrium, where no more work can be done. This provided a new conception of the concept of equilibrium. It was no longer necessarily or paradigmatically a balance between competing forces or tendencies, but a point to which a system is attracted, when no more changes in the system, considered *as* a system, will occur.

Although others contributed to the foundations of genetic Darwinism, R. A. Fisher, a founder of modern statistics and an ardent Darwinian, explicitly applied the Boltzmannian model to the distribution of genes in a freely interbreeding population (chapter 10). Just as rational economic agents can be presumed to be seeking maximal utility, and physical processes involving energy expenditure can be presumed to be "striving," as the physicist Rudolf Clausius metaphorically put it, toward maximum entropy, so for Fisher genes maximize "fitness,"—reproductive success measured by the genetic contribution of an organism to the next generation. The rate of increase in fitness in a population or array of genes, Fisher proclaimed, is directly proportional to the amount of (additive) variance, that is, to the degree of difference among genotypes, and presumably among the phenotypes that carry them, which serves as a sort of genetic gradient. Just as in a gas, moreover, the laws governing the distribution of these genetical atoms are probabilistic:

The regularity of the [average rate of progress in fitness] is in fact guaranteed by the same circumstance which makes a statistical assemblage of particles, such as a bubble of gas, obey, without appreciable deviation, the law of gases. . . . It will be noticed that the fundamental theorem proved above bears some remarkable resemblances to the Second Law of Thermodynamics. Both are properties of populations, or aggregates, true irrespective of the nature of the units which compose them; both are statistical laws; each requires the constant increase of a measurable quantity, in the one case the entropy of a physical system and in the other the fitness, measured by m, of a biological population. (Fisher 1930, 36)

Fisher thought that among the advantages of his theory was that it gave biology the prestige of having strictly quantitative and general universal natural laws, rather than the myriad of separate adaptive stories that still tied Darwinism to the old tradition of natural history and kept it, in Fisher's view at least, from being as mature a scientific theory as physics. Fisher thought his Boltzmannian model had the additional advantage of freeing Darwinism from its bondage to Malthus. It is simply not the case that natural selection is condemned to work only by way of differential deaths of actual organisms. On the Boltzmannian model, natural selection will take place whenever there is a fitness gradient of any sort to be exploited, that is, whenever there is the slightest degree of difference between the reproductive success of one subpopulation and another.

Appeal to the Boltzmannian model also solved a conceptual problem that had haunted the Darwinian tradition from the beginning. Darwin had been grievously wounded when John Herschel, a high priest of Newtonian science in his day, had contemptuously dismissed his theory on the ground that Darwinian evolution was "a law of higgledy-piggledy." In the Newtonian framework of which Herschel was an anointed keeper, any appeal to probabilities, and any causal role given to chance, looked like a confession of ignorance rather than mature science. In a Boltzmannian framework, however, what had once seemed a defect was now turned into sophisticated wisdom and retrospectively justified Darwin's prophetic insight into role of chance and variation in large-scale natural processes.

Other genetic Darwinians found plenty to object to in Fisher's version of genetic Darwinism (chapters 12–13). It was not enough, they thought, to explain how variation is used up like energy gradients. How the variation on which adaptation depends is created, and even stored, in populations must also be understood. Fisher's critics were no less willing than he, however, to set the entire issue within a statistical framework.

The American geneticist Sewall Wright, for example, exploited a fact about probability distributions that Fisher had dismissed in order to give what he regarded as a better explanation of how populations become adapted. Just as in a game of roulette the ball may land five, ten, or even more times in a row on red, even though the long-range probability of red and black is 50 percent, so, Wright reasoned, genes might be fixed in small populations quite apart from selection pressures by a process Wright called "genetic drift." Natural selection can then sample among the different subpopulations for interestingly novel gene combinations. Wright's "adaptive landscapes," which have become elaborate modeling devises used by genetic Darwinians to this day, are metaphorical extensions of Fisher's genetic gradients.

The Russian geneticist Sergei Chetverikov, meanwhile, passed on to Theodosius Dobzhansky, who emigrated to America, where for a time he worked closely with Wright, a view of genetic natural selection that insisted not only on variation creation but on banking it for a rainy day. That nature stores variation in populations that are chock full of genetic diversity was for Dobzhansky a good argument for resisting the enthusiasm of many earlier Darwinians, including Fisher, for "eugenics," an array of suspiciously antidemocratic ideas about enhancing the fecundity of those who were assumed to be superior and discouraging, sometimes physically preventing, people who were assumed to be less fit from having offspring (chapters 12–13). Dobzhansky's hypothesis about the width of genetic variance in natural populations, and the role of natural selection in maintaining it, has been spectacularly vindicated by several generations of his students, who have dominated recent American evolutionary biology. Indeed, so great is this variation that it has become something of an embarrassment to the Darwinians who discovered it, since it is increasingly difficult to maintain that natural selection is primarily responsible for maintaining that much diversity.

The emergence of the modern evolutionary synthesis on foundations laid down by Fisher, Wright, Chetverikov, and Haldane, and its disengagement from ideological programs with which the Darwinian tradition had long been entangled, was the work not only of Dobzhansky but of Julian Huxley, Ernst Mayr, George Gaylord Simpson, and a number of others. From the outset, the modern synthesis has oscillated between treating natural selection as a deterministic force operating on arrays of genes and a hankering to take advantage of more chancy processes, such as genetic drift, which the new probabilistic explanatory framework made conceptually accessible. What is arguably the greatest achievement of the modern synthesis—its clearer understanding of what species are and how they come into existence than earlier forms of Darwinism—was, for example, the result of combining deterministic and chance aspects of how genes are fixed in populations (chapter 12). It should come as no surprise, then, that doubts being expressed about the explanatory adequacy of the modern synthesis have reopened old divisions within Darwinian ranks about the relative roles of chance and natural selection (chapter 14).

That the modern synthesis has in fact been under pressure and to some extent on the defensive since about 1970 is due in part to the rise of molecular genetics. Francis Crick and James Watson's discovery in 1953 of the structure and role of DNA seemed at first to confirm what genetic Darwinians had been assuming about how genetic information is transmitted and how variation arises from mutation and recombination (Kitcher 1984; cf. chapter 13). Since then, molecular genetics has confronted evolutionists with some surprises. Species are not the only things that evolve. The molecules of life, such as proteins and nucleic acids, evolve too. Much that happens at the level of molecular evolution, however, has not conformed to Darwinian expectations. In particular, it has been shown by the Japanese population geneticist Motoo Kimura and others that the rate of replacement of amino acids in the evolution of structural proteins is too regular or "clocklike" to be consistent with the ups and downs of natural selection (Kimura 1968, 1983). Kimura interpreted these discoveries as supporting the primacy of chance processes like those pioneered by Wright over natural selection in the evolution of the complex molecules on which life depends (chapter 14).

The agitation in contemporary evolutionary theory triggered by Kimura's "neutral theory" of protein evolution has been intensified by analysis of the regulatory genetic mechanisms that guide the development of organisms. This work began with the vindication and expansion of Jacques Monod's and François Jacob's "operon" model of the repressor mechanism for gene expression. (Jacob and Monod 1961). The rapidly expanding field of developmental genetics reveals a much greater degree of complexity in genetic systems than most devotees of modern synthesis have expected, and an increasing sense that a good deal of the order that emerges in the "genome" (the totality of an organism's genes) is the result of the self-ordering properties of large genetic arrays (Kauffman 1985, 1993; Wimsatt 1986). Especially in the evolution of life cycles and the life-history traits on which much of contemporary evolutionary biology concentrates—the number of offspring, the timing of development, reproductive activity, the onset and function of aging and death—it appears that natural selection does not have to do all the work. As Stuart Kauffman remarks, a great deal of order comes "for free" (Kauffman 1991, 1993).

The enhanced roles now being envisioned for chance fixation of genes and self-organizational processes within the genome carry an implicit threat that natural selection is slowly being squeezed between these other "evolutionary forces" and is being assigned a steadily diminishing role in evolutionary dynamics. This threat has been intensified by the suspicions entertained even by some Darwinians that "macroevolution"—the emergence of species and higher taxa—is not a long-range function of gradual change under the control of adaptive natural selection, as the makers of the modern synthesis asserted, but a relatively autonomous process that is "decoupled" from "microevolution," or evolution below the species level. According to advocates of "punctuated equilibrium," for example, macroevolution is not gradual at all but is concentrated in evolutionary time around bursts of speciation (Gould and Eldredge 1977). Between these episodes, things do not change much. Perhaps, then, speciation is caused by sudden reorganizations of largely self-organizing genetic networks, which then remain locked in place for considerable periods of time (Gould 1982a, 1982b). In that case, can we be sure that what makes it through into the phylogenetic tree gets there because of selection (Gould 1989a)? Cut loose from adaptative natural selection, a combination of self-organization and sheer chance might determine phylogeny, severing macroevolution from adaptive microevolution in the same way evolution at the protein level has been partially decoupled from natural selection by Kimura's neutralism.

Generally put, the threat is that the range of processes for which Darwinism, whether old or new, is approximately true might eventually be restricted to the level of organisms and populations and that a more general theory, in which selection and adaptation play markedly diminished roles, may govern evolution in both the grand sense and at very elementary levels. Since these pressures on the modern synthesis concern what happens at the level of the very small (molecular evolution) and the very large (macroevolution), they recall similar pressures exerted on physics earlier in the twentieth century, when quantum mechanics revised received views about particle physics and when relativity revised long-established ideas about space and time. Under appropriate redescriptions, Newton's laws are still predictive for a finite range of processes, including basketballs and planets. That range can be defined by knowing that a more general theory is at work within certain boundary conditions. In a similar vein, Darwinism might have evolved to a point where it is about to be trimmed back by a theory that evolves out of it and then puts it firmly in its place. Just as Darwin's work exploded the Victorians' cozy sense of space and time, so contemporary evolutionary speculation is forcing twentieth-century Darwinians to adjust to the even more expansive, chancy, contingent worldview that is already clearly visible in modern cosmology but that has so far been contained in evolutionary biology by the comforting rationalism of our talk about adaptations, according to which, even if we do not invoke God, we still seem to be able to give good reasons for what we see around us.

It is especially intriguing that the self-organizing properties of genetic systems have attracted attention at a time when dynamics itself has been changing to recognize the self-ordering properties of statistical arrays. In the past several decades, our ability to see mathematical patterns in and construct dynamical models about how complex systems change over time has taken a quantum leap, in part because of the tremendous calculative power of computers and their ability to display the fruits of their computations in vivid graphic displays. Among the startling discoveries is that if you take a simple, deterministic equation and run it over and over again, small differences that sneak in somewhere—a single digit in the twentieth place after the decimal in *pi,* for example—sometimes become wider and wider, less and less predictable, more and more chaotic. That was not supposed to happen. In a Maxwellian or Boltzmannian world, small differences are supposed to wash out, as they do in statistical averages; they are not supposed to amplify. Far less are they supposed to reveal hidden patterns. Yet they do. After many runs through such wobbly, and apparently choatic, iterations, a new, higher-order pattern unexpectedly starts to appear.[7]

"Deterministic chaos," as it is called, is only one aspect of a more general advance in our ability to understand nonlinear dynamics. In nonlinear systems, effects are nonadditive. Mathematically, that means that the solution to two equations will not be equal to the solution of the sum of these equations. More concretely, it means that systems whose dynamics are nonlinear are likely to exhibit what is called "sensitivity to initial conditions," or "the butterfly effect," after Edward Lorenz's surmise that the flapping of the wings of a butterfly in Rio de Janiero might conceivably have consequences for the weather in Texas (Lorenz 1993). Are there parts of nature—previously dismissed as disorderly, and hence "unscientific," or that we have furiously and fruitlessly tried for a long time to cram into the wrong mathematical box—whose logic suddenly reveals itself as like this? In biology, nonlinear dynamics has already been illuminating everything from heartbeat patterns to epidemiology. Many natural systems seem to follow the contours of a deep dynamical background we are only now beginning to understand. In ecology, for example, we see patterns of population rise and fall that had long seemed purely random suddenly reveal themselves as obedient to a higher pattern (May 1974, 1987; Schaffer and Kot 1985a, 1985b). The idea that there is a hidden order now revealed by our massive increase in computative ability produces the kind of awe that Pythagoras once felt when, after knocking his head against the wall trying to find the commensurable unit of a right triangle, he suddenly realized that there is order in the chaos after all but that it was discernible only at a higher level, by drawing squares on the sides of the triangle and measuring their areas. The revolution in dynamics is something like that. It has led us to suspect that we are only nibbling around the edges of deeper dimensions of pattern latent in the natural world.

In this book, the focal question is whether phenomena in evolutionary theory that have so far resisted adequate explanation can be illuminated by nonlinear dynamical modeling. Perhaps the long-sought and often-delayed fusion between genetics and development has had to await the emergence of suitably complex dynamical models. Perhaps the branching pattern of phylogeny reflects sensitivity to initial conditions and self-organizational dynamics within the genome, which produce speciation and the emergence of higher taxa by way of "bifurcation" points in the trajectories of complex genetic networks. These matters are entirely speculative. Our point in mentioning them is not to prejudge their truth but to illustrate our claim that Darwinism has entered a new period of buffeting. It is also to suggest that proper monitoring and evaluation of emerging research programs along these lines will be difficult for those who, in their haste to ensure the autonomy of evolutionary biology from physics, have tended to underestimate the role that dynamical modeling has played in the Darwinian tradition. That is why we devote much of this book to showing how dynamics and Darwinism have been intertwined from the outset.

In spite of their differences the Newtonian and Boltzmannian dynamical frameworks within which the Darwinian tradition has hitherto flourished shared a number of features that facilitated some discoveries while making others less accessible. The Boltzmannian framework differed from the Newtonian insofar as its laws and predictions range probabilistically over arrays of genes in populations. That certainly let twentieth-century Darwinians understand species, speciation, and systematics better than nineteenth-century Darwinians did (chapter 12). But Darwinians of both stripes still shared a commitment to equilibrium models of how systems behave, however differently they conceptualized equilibrium. That is why Fisher assumed that genes are as separate as beads on a rosary, beans in a bag (Mayr 1954), or (more aptly) molecules and atoms in a container and why, like Maxwell and Boltzmann, Fisher thought he could write linear differential equations mirroring the continuous, predictable pathways through which genetic arrays were moving. Now, however, we are seeing something that not even Sewall Wright expected. What nonlinear systems share with Boltzmannian systems is that they are probabilistic. They show what will happen to an array of elements with the passage of time. In this way, both differ from classical Newtonian systems. They differ from each other, however, because the behavior of the statistical arrays in nonlinear systems does not average out. Nonequilibrial small differences in initial conditions take the system on a ride before it settles down to something like order. When it comes, moreover, order arises in part from self-organizing properties that seem to characterize the behavior of large arrays with many possible connections among their elements (Kauffman 1993). While the classical Newtonian framework is restricted, therefore, to organized simplicity, and the Boltzmannian framework to disorganized complexity, the world of nonlinear models now opening up fastens onto organized, indeed self-organized, complexity (Prigogine and Stengers 1984). It might even be said that the complexity revolution completes, or at least extends, the probability revolution.[8]

The relevance of nonlinear dynamics to evolutionary problems becomes especially significant as soon as it is recognized that Darwinism is not the only research tradition in evolutionary biology. By contrast to Darwinism's stress on random variation and selection by external, environmental forces, an older and surprisingly persistent developmentalist tradition has consistently maintained that the evolution of kinds (phylogeny) should be viewed as an inner-driven process like the unfolding of an embryo (ontogeny). Darwinism and developmentalism have been engaged in something of a seesaw battle over evolutionary theory for a long time. The developmentalist tradition displaced Darwinism in the later nineteenth century (chapter 7) but was in turn marginalized by it for most of the twentieth (chapters 11–14).[9] As a rough generalization, we may say that developmentalism, and the internally driven models of change in which it has been encoded, has retained control of evolutionary questions that Darwinism has been unable fully to comprehend. Among these are some of the larger questions that fall within the range of a complete evolutionary theory: the evolution of life cycles and other aspects of ontogeny, ecological succession, and phylogenetic pattern. Maintained by devotees of evolutionary subdisciplines that were never fully integrated into the modern synthesis, often in countries that were never entirely conquered by Darwinism, the developmentalist tradition has patiently been waiting for its chance to rise again (chapter 16). Many of its partisans think that their time has come at last. Emboldened by the resources of nonlinear dynamics for analyzing self-organizing phenomena at various levels of the biological hierarchy, latter-day developmentalists have been asserting of late that natural selection, and so Darwinism, will not survive the transition from the "sciences of simplicity" to the "sciences of complexity" (Brooks and Wiley 1986, 1988; Ho and Saunders 1984; Ho and Fox 1988; Oyama 1988; Wesson 1991; Goodwin and Saunders, 1989; Salthe 1993).

It may be too pessimistic to suggest, as developmental enthusiasts often do, that in the age of complexity, Darwinism is in danger of becoming what Imre Lakatos has called a "degenerating research program," like creationism (cf. Burian 1988; Grene 1990a). Similarly, Darwinians and their philosophical defenders can easily underestimate the significance of nonlinear dynamics for reviving the fortunes of their ancient rival. Darwin's Darwinism was successful in explaining the phenomenon of morphological and behavioral adaptedness. That success was preserved and expanded by genetic Darwinism, which, in addition, could plausibly explain speciation. Darwinism has not fully succeeded, however, in comprehending other phenomena that have remained central for developmentalists. Indeed, genetic Darwinians sometimes even deny that there is much there to explain. Evolutionary direction, for example, is not regarded by many Darwinians as a real phenomenon, in part because the objects of macroevolutionary theory are often viewed merely as accumulated effects of lower-level evolution (Simpson 1944; Dawkins 1992; Ayala 1985). For this reason, Darwinians have more often dismissed than explained putative phenomena that have so far remained within the orbit of the developmentalist tradition. That tradition would gladly shed itself of its semivitalist tendencies if a mathematical and dynamical language became available to explicate and defend its core commitments. This explains why the self-organizational dynamics of complex systems has caused such excitement among Darwinism's traditional rivals (Goodwin and Saunders 1989; Salthe 1993).

In view of these considerations, it is unedifying for Darwinians simply to dismiss the developmentalist tradition as "typological" thinking that betrays irrational resistance to modernity (Mayr 1982, 1988; Hull 1965). For one thing, the modern worldview may itself be changing (Pagels 1988; Toulmin 1990). For another, the developmentalist tradition, as we take pains to show, is more deeply rooted in Aristotelian realism than in neo-Platonic typology. In consequence, it has has been able to express and preserve a number of home truths about living systems that Darwinism has made less central. Orthodox neo-Darwinians are not quite as convincing as they once were, therefore, when they issue reassuring tracts proclaiming that the modern synthesis is as secure and progressive as ever (Stebbins and Ayala 1981). It does not help, for example, to assert that new discoveries in molecular genetics are consistent with known explanatory mechanisms (Stebbins and Ayala 1981; Ayala 1985; cf. Jungck 1983). Only *consistent?* Nor is elegant work by philosophers of biology demonstrating how conceptually coherent the modern synthetic theory is entirely satisfactory (Rosenberg 1985; Sober 1984a). Only *conceptually coherent?* What about empirically adequate and theoretically fecund, ripe with the promise of new explanatory triumphs? Constructing beautiful "rational reconstructions" of a theory and a research tradition that is already assumed to be mature is not an activity calculated to allay suspicions that Darwinism may have drifted into an unstable orbit as physics once did.

We happily acknowledge that some contemporary Darwinians have indeed recognized that the biological world is more complex than their tradition has sometimes been comfortable with. Calling for an "expansion" of the modern synthesis in order to bring Darwinism to bear on the very big and the very small, Stephen Jay Gould, Niles Eldredge, and others would have us recognize that genetic drift, especially at the level of protein evolution, is more pervasive than had been earlier suspected and that natural selection, or a process very like it, can operate on entities at levels and scales other than organisms in populations. There can, it is said, be selection on genes, groups, and even on whole species, in addition to organismic selection (Gould 1980, 1982b; Sober 1984a; Eldredge 1985; Buss 1987; Brandon 1990). Moreover, in contrast to those who incline toward the "pan-adaptationist" tendency within the Darwinian tradition, advocates of an "expanded synthesis" recognize, with their developmentalist colleagues, that natural selection labors under various constraints, including constraints of inherited form. It cannot build any sort of organism it wants (Gould 1982a, 1982b; Gould and Lewontin 1979). "The constraints of inherited form and developmental pathways," writes Gould, "may channel . . . changes so that even though selection induces motion down a permitted path, the channel itself represents the primary determinant of evolutionary direction" (Gould 1982b, 383).

Much of the work along these lines has been done under the auspices of a philosophy of biology that assumes that the complexity of biological systems comes into view in proportion as the law-governed behavior paradigmatically seen in physics is pushed to the background, and attention is instead focused on narrating the manifold, diverse, unpredictable productions of natural history (Gould 1989a; Lewontin 1991). This is to say that the ethos of the expanded synthesis is an extension of the "autonomy of biology" stance promoted most vigorously by Ernst Mayr (1985, 1988). With notable exceptions, accordingly, contemporary Darwinians have tended to underestimate the resources that complex systems dynamics offers to their own tradition, redolent as these ideas are with the habits of mind of mathematical physicists questing after even deeper laws of nature.

The work of the developmental geneticist Stuart Kauffman is one of these exceptions (Kauffman 1993).[10] Whatever the ultimate fate of his particular models, Kauffman has already succeeded in demonstrating that Darwinians as well as developmentalists can take a positive view of complex dynamical models and can extend the explanatory reach of their tradition by using these models to rework developmentalist themes. Like other contemporary evolutionary theorists, Kauffman wants to understand more about the conditions under which significant evolutionary change is likely to take place. One of these conditions is that populations must maintain a large array of variation, as Dobzhansky and his successors have amply demonstrated. In maintaining and utilizing variation, however, Kauffman argues that natural selection does not have to explain as much as the modern synthesis has assumed it does. Indeed, it cannot possibly do so. For, as Kauffman shows, the hand of selection is mathematically constrained by the self-organizing properties that all large arrays display over time, which turn "chaos" into what he calls "antichaos" (Kauffman 1985, 1991, 1993; chapter 16).[11] As developmental biologists have long suspected, it is the antichaotic properties of genetic arrays that keep developmental programs stable and are reflected in phylogenetic order. This does not mean that natural selection is any less important than it has been or that the writ of the Darwinian tradition is contracting. When it is viewed in the light of the new dynamics, Kauffman speculates that natural selection may play its most important, and least understood, role. Natural selection keeps complex genetic systems within the dynamic range between freedom and fixity in which alone significant evolution can take place, and so explains a deep property of living things that might well be called "evolvability":

Living systems are not deeply entrenched in an ordered regime. . . . [They] are actually very close to the edge-of-chaos transition, where things are much looser and more fluid. And natural selection is not the antagonist of self-organization. It more like a law of motion, a force that is constantly pushing emergent, self-organizing systems toward the edge of chaos. (quoted in Waldrop 1992, 302–3; cf. Kauffman 1993)

Kauffman hints at a new conception of natural selection in which it is an integral component of a single process that involves mutually dependent roles for stochastic, selective, and self-organizational components (chapter 18). This has the salutary effect of shifting the emphasis in debates about whether Darwinism is still robust from merely quantitative or bean-counting considerations about the relative extent of selection and other "forces" to deeper questions about the mutual dependence and interpenetration of chance, selection, and self-organization. This mutual dependence and interpenetration is revealed by nonlinear dynamical models of complex systems. If the Darwinian tradition succeeds in transforming developmental biology and macroevolutionary dynamics in these terms, the medium for this transformation will be yet another appeal to the dynamical models that, as we try to show, have always kept the Darwinian tradition fresh.

The argument of this book depends on, and is a case study on behalf of, views about issues in the philosophy of science and the philosophy of biology that we cannot fully articulate or defend here but should at least be made explicit. Our pervasive talk about background assumptions, ontologies, and research traditions suggests that we accord a large role to interpretive frameworks in determining how scientists view data and what they come to believe about how the world works. That puts us on the side of what might be called conceptualists rather than empiricists in the philosophy of science. Conceptualists, to be sure, can approve of the empirical method without being empiric*ists* in philosophy. Indeed, given what modern science actually is and how it works, they had better do so. In contrast to philosophical empiricists, however, who (whether they are inductivists or deductivists, confirmationists or falsificationists) make theory-neutral data decisive in scientific theorizing, conceptualists stress the role of antecedent conceptual schemes in glueing scientific theories together and bringing them to bear on the world.

Conceptualism is no longer as controversial as it once was. What is now controversial is precisely *how* conceptual elements bind scientific theories together and tie them to phenomena and how to keep historically sensitive solutions to these problems from inflating into the notion that scientific theories are merely "social constructions." Thomas Kuhn's famous idea of "scientific paradigms" embodies one of the most influential conceptualist accounts of how science does its work. For Kuhn, paradigms are scientific achievements of known worth, like Newton's account of planetary motion, which for a time serve as models in terms of which other problems can be posed and solved (Kuhn 1970, viii, 10, 23). Kuhn forthrightly, and somewhat scandalously, admitted that change from one paradigm to another is more a matter of conversion than of proof (Kuhn 1962, 1970). On his view, when scientists change their mind about interpretive frameworks they just *see* things differently. And when they persuade others to see things as they do, they do not act in ways sufficiently different from religious converts or political true believers to sustain the key empiricist belief that how science works is all that more rational and progressive than other areas of culture.[12]

In spite of the scandalous nature of these claims, Kuhn is not an irrationalist, even if many Kuhnians are. On the contrary, by setting aside large and intractable aspects of theory change over historical time, Kuhn believes that he is clearing the way for people to see that scientists, once safely embedded inside a paradigm, can devote themselves productively to the detailed problem solving where real progress actually takes place. Nonetheless, Kuhn's concessions about overall progress, and his willingness to say that belief formation among scientists is not as different from belief formation in areas like art, religion, and political ideology as scientists and their philosophical allies usually like to think, have made many people nervous. Not surprisingly, vast tracts of philosophy and history of science in the last third of the twentieth century have been devoted to rebutting and offering alternatives to Kuhn.

Among those who have tried to see more rationality and progress both within and between Kuhn's paradigms, the work of Imre Lakatos has been particularly suggestive (Lakatos 1970, 1978). Lakatos distinguished between the "core" idea of a scientific "research programme," which is constant, and certain "auxiliary assumptions" that attend it, which are not. The core is what the programme is committed to, "come what may." Lakatos called it a "negative heuristic" because the operative rule is: Don't give this idea up. An auxiliary assumption, by contrast, is a proposition that for some contingent, perhaps even irrational, reason is initially regarded as part of the core but proves to be dispensable without destroying the integrity and continuity of the research programme itself. It is a presumed constant that turns out to be a variable. The "positive heuristic" or rule of inquiry is: Throw it overboard if that is what must be done to save the ship. If all goes well, one should be able to buy more time, to get the programme back on track. Thus was Kepler able to preserve the heliocentric core of Copernicanism when it was threatened with difficulties by dispensing with the revered assumption that the planets must go around in perfect circles at uniform speeds. Once freed from that assumption, the way was paved for Newton to comprehend the dynamics that drive the heavenly bodies around in elliptical orbits. The point is that in making these changes we are doing something rational and fostering rational movement within what at first glance looked like arbitrary and fixed interpretive schemes. By analyzing its track record, we should even be able to tell whether a research programme is thriving or what Lakatos called "degenerating."

Our notion that the Darwinian tradition has upon occasion saved itself by changing dynamical models has been influenced by this picture of scientific change. We treat natural selection as the core of the Darwinian research tradition. In telling its story, we will have many occasions to show when auxiliary assumptions are being jettisoned. Fisher's dismissal of Malthusian population pressure is an example. At the same time, our view of research programs has also been influenced by Larry Laudan, who has offered several worthy amendments to Lakatos's account (Laudan 1977, 1984). Scientific research programs can get hung up, Laudan says, by privileging not only empirical assumptions but conceptual ones as well. For this reason, the conflation of auxiliary assumptions with elements of the core often derives from the prestige of revered metaphysical, epistemological, psychological, and even political beliefs. Accordingly, to abandon auxiliary assumptions sometimes requires, and more often leads to, a deep shift in the sensibility of a whole culture, usually led by the sleepwalking into the future of a few gifted individuals. Belief in the magical superiority of circular movement is a case in point. That is what Kepler had to surrender. The result of this emendation is that the historical processes over which Laudan's dynamic ranges are bigger in scope, and more subtly connected to cultural forces, than are Lakatos's "research programmes." Laudan calls these larger patterns "research *traditions*." For Laudan research traditions maintain themselves by solving conceptual as well as the empirical problems in their domain. In talking about research traditions, we are using the phrase in roughly, but not exactly, this sense.

We think Lakatos and Laudan are right that the health of a research tradition is to be gauged by its problem-solving prowess. We are less confident than they seem to be, however, that you can readily count problems and keep score of which competing traditions and programs are generating more and better solutions. We are certainly unconvinced that conceptual problems lend themselves to that way of thinking as easily as empirical problems do. Our view of conceptual change is more holistic and interpretive. In every dimension of culture, we believe, including science, there is an ongoing dialectic between innovation and tradition (Depew and Weber 1985). Cultural evolution thus has an ineliminably narrative dimension, in which traditions and practices change by reinterpreting and reappropriating their own past in the light of new developments and challenges and act to determine the future by trying to see what is authentically, rather than merely conventionally, emergent from their own past. If problem-solving success includes solving conceptual problems, as Laudan rightly says it does, then one must be willing to appreciate the complex, interpretive, and reinterpretive ways in which conceptual issues, as distinct from empirical ones, actually get resolved. Our point is connected to one that Alasdair MacIntyre makes when he holds that a research tradition cannot even be identified or described unless its history comes to be coherently and plausibly rewritten after a period of contestation, for the simple reason that telling a coherent and confirmable story is itself an important part of reconstituting a tradition (MacIntyre 1977). The efforts of successful senior scientists to rewrite the history of their disciplines can appear as one of the weaknesses of old age and too much fame, particularly when the story seems to culminate in their own work. The impulse to do this is not only reasonable, however, but indispensable, for the construction of a history is no less internally important to the continued success of a research tradition than are solid empirical discovery and elegant conceptual articulation.[13]

Our enriched view of research traditions undergirds another significant difference from Laudan. Whereas Laudan at least sometimes thinks that research traditions are bounded by their ontology and that the aim of a tradition is to "reduce as many phenomena as possible to its distinctive ontology," we hold that the continuity of a research tradition through cultural time can sometimes be maintained by passing "ontological amendments."[14] In thinking of dynamical models as "ontologies" we are saying, in fact, that Darwinism as a research tradition has, so far at least, lived under two different ontological regimes, roughly Newtonian and Boltzmannian. Both have allowed it to characterize a process called natural selection and fruitfully to apply it to a range of evolutionary problems. Our suggestion is that in pursuing solutions to evolutionary problems that have thus far resisted Darwinian explanations, the Darwinian tradition can retain its vitality by shifting to the nonlinear dynamics of complex systems, jumping ontological horses yet again.

The mere mention of such extensive flirtation with the truck and produce of physicists cannot fail to elicit prompt and indignant defenses of the "autonomy of biology" by evolutionists and philosophers who are hostile to a physics-based biology of any kind (Mayr 1985, 1988). The impulse to protect evolutionary biology from the clutches of physicists is understandable, for physics-minded philosophers have been quick to point out that evolutionary biology does not fare well when it is compared to physics (Smart 1963; Popper 1972, 1974). Either it has no universal laws or it cannot apply them predictively and deductively to cases.

Its arguments, accordingly, seem little better to physics-oriented philosophers than those of narrative historians. In response to such attacks, advocates of an autonomous evolutionary biology have been prone, like St. Paul, to glory in their infirmity and to proclaim they do not need or want laws anyway. The particular facts of the evolved world cannot be derived from universal lawlike premises. Instead, narrative explanations of particular sequences can be achieved by using key concepts of evolutionary biology, such as natural selection, as heuristic guides (Mayr 1985; Lewontin 1991).

We recognize that from this perspective at least our account of Darwinism's history looks downright reductionistic, seemingly proposing to treat biological processes as cases of deeper physical laws. In responding to this imputation, we note first that our interest in the connection between biology and physics is not in whether the laws of Mendelian genetics can technically be "reduced" to molecular genetics, an issue that has long, perhaps too long, dominated discussions about the putative reducibility of biology to chemistry and physics (Hull 1965; Kitcher 1981; Rosenberg 1985; Ruse 1988a, but see Waters 1990). We are interested instead in what physics says about the temporal aspects of biological systems. The physics we have in mind is diachronic and dynamical rather than synchronic and static. Even so, we are not out to reduce evolutionary biology to thermodynamical physics, or even to "expand" the physics of irreversible processes to encompass biology, although of the two the latter alternative seems preferable to us (Wicken 1987). Rather, we focus on the way in which an admittedly autonomous field called evolutionary biology has appropriated formal models from dynamics. Our concern is that the autonomist position in the philosophy of biology brings with it biases that underestimate the influence of dynamic models on the Darwinian tradition, making it difficult to estimate the prospects of Darwinism in the age of complex systems. We realize that these protestations do not in themselves dispose of the issue of reductionism. Nor do we want them to. On the contrary, we are in effect raising that issue in a new way by asking what reductionism amounts to when scientific inquiry is construed as the use of models to explicate bits and pieces of a presumptively complex world rather than as the search for ever more comprehensive laws governing a world assumed to be essentially simple, predictable, and deducible from a few laws. In this way we hope to steer a course between "physics envy" and "physics allergy."

Part of our case rests on the changing nature of dynamics. As dynamics crosses the complexity barrier, it is becoming a more abstract enterprise than it used to be. Its models range over formally defined entities seen most perspicuously in computer displays. It then becomes a separate problem to determine what the conditions are for physically embodying these models. (In chapter 17 we try to say what those conditions are for living and evolving systems.) So considered, the new dynamics does not propose new laws of nature, at least so far (Kellert 1993). It is a formal science that shows that, due to sensitivity to initial conditions, variables whose trajectories are coordinated by known laws behave in such an unexpected and unpredictable way that they contribute to the "erosion of determinism" that has been a marked cultural phenomenon in this century and that may be leading to radical revisions of our scientific, and commonsense, ontology (Fine 1986). The first lesson to be learned from the new dynamics is that the world contains more novelty, diversity, and complexity than we had assumed. The second is that it is not just physics and biology that must change to accommodate this fact but philosophies of science too, which have generally assumed a simple world and tailored their prescriptions about scientific explanation and prediction to fit such a world.

On the "received view" of theories, a view worked out in elaborate detail by logical empiricist philosophers like C. G. Hempel, the laws that lie at the heart of theories are universally quantified statements about the world. Theories are confirmed or falsified by treating putative generalizations as major premises of hypothetical arguments from which, if the proposed law were true, certain facts would be predicted. Experiments are then devised to test those predictions and by this means to confirm or disconfirm the generalizations that generate the prediction in the first place. A major goal of science, on this view, is to use fewer and fewer laws to explain more and more facts, progressively "reducing" the laws of restricted sciences to those of more basic ones. Explanations of particular facts and events, meanwhile, are achieved on the "received view" by employing confirmed laws as premises from which what is to be explained can be deduced, with the help of "bridge laws" that connect high-level generalizations with data about specific processes. From a "covering law" together with statements about initial and boundary conditions, a statement about the state of the system at a future time, or indeed past times, follows deductively. If these predicted, or retrodicted, statements turn out to conform to fact, what was to be explained (*explanandum*) is treated as having been settled. On this view, prediction and explanation are logically identical, being distinguished merely by the temporal position of the inquirer (Nagel 1961; Hempel 1966).

There is a striking similarity between this model of explanation and Newtonian, or more generally equilibrium, dynamics. What in the material mode appears as laws that determine prior and subsequent states of a system appears here in formal dress as the idea that from lawlike statements and initial conditions one can straightforwardly deduce conclusions (Depew 1986). Doubtless there are many systems that, if they are sufficiently isolated, can be analyzed this way. For centuries, in fact, the thrust of modern science has been to identify those systems and to show that they are "nothing but" cases of even more basic laws. It is increasingly unlikely, however, that the world as a whole consists of anything like a nested set of systems that obey equally well-nested laws. Simple systems certainly exist, but they exist in a sea of complexity, marked off from the surrounding penumbra because very specific boundary conditions hold them in place, at least for a time (Dyke 1988).

What counts as a good explanation looks very different against these two backgrounds. In a simple world composed of simple systems, explanations get deeper as scientists discover more comprehensive and general laws. Since the most basic laws are physical laws, the bias of many methodologists toward a physics-centered philosophy of science, about which Mayr rightly complains, is born here. In a complex world, by contrast, this picture must be abandoned. Explanation depends on knowing just how far a set of boundary conditions holds and how accurately a model applies to what lies inside those boundaries. No less striking than the enshrinement of classical physics in a physics-oriented philosophy of science, therefore, is the fact that in recent decades philosophers of science have been revising the "received" account of theories and of scientific explanation in ways that seem congenial to this conception of inquiry and to the recognition that we live in a presumptively complex world. We mention four revisions in particular that, intentional or not, sort well with the conceptual demands that the complexity revolution is making on the philosophy of science.

First, it is now generally recognized that explanation is an inherently "pragmatic" notion. What calls for explanation (the *explanandum*) as well as what counts as a good one (the *explanans*) is essentially relative to context. Alan Garfinkel's example about Willy Sutton illustrates the context sensitivity of explanation (Garfinkel 1980).[15] A priest, hoping to induce the famous bank robber to reform, asks, "Willy, why do you rob banks?" Astonished at the stupidity of this question, Willy replies, "Because that's where the money is." Garfinkel analyzes the priest's question as occurring against an assumed background in which one might rob or not rob. Sutton, on the other hand, interprets the question against a background in which robbery is taken for granted, and the possibility space that remains contains banks and other sources of money. It should be obvious that banks are the superior targets. But that all depends on background assumptions. Sensitivity to background assumptions is no less prominently at work in scientific explanation. In Aristotle's world, for instance, the fact that heavenly bodies move in circular orbits needs no explanation and gets none. The heavenly bodies are perfect, and everyone knows that circular motion is the most perfect sort of motion. What needs explaining is why they sometimes appear not to turn in circular orbits. By contrast, in modern dynamics, natural motion runs in a straight line. What needs explaining, then, is why things depart from straight-line motion. That is precisely what Newton's dynamics does.

Second, it is increasingly recognized that it is *phenomena*, rather than bare data or isolated facts, that constitute the *explananda* of science (Bogen and Woodward 1988). Phenomena are established over time by sorting through mountains of data and a shifting myriad of facts, no single one of which, or even group, would be sufficient to falsify anything if it failed to be predicted from some law. There is even a longer-run element of wider cultural discourse that goes into establishing a phenomenon as an *explanandum*. On this view, evolution is neither a fact or a theory (*pace* Ruse 1986). It is a complex set of phenomena.

A third new idea about scientific explanation is closely related to the fact that phenomena are the primary *explananda* of scientific inquiry. If this is true, scientific explanation is not so much a matter of connecting laws with data but of connecting phenomena with theories by way of models (Giere 1984, 1988). Data play a role in this enterprise by offering information about how well or badly a given model fits a given phenomenon. A view of scientific explanation conceived along these lines has been defended by those who advocate what has come to be called the "semantic account" of theories (Giere 1979b, 1984; van Fraassen 1980, Suppe 1977; Suppes 1967). On their analyses, the laws that lie at the heart of theories are parts of definitions of kinds of systems, rather than high-level statements about the world directly, as they are on the "received view." Newton's laws, for example, jointly define a particular kind of system, "a Newtonian system." We can then say that a planetary system, or perhaps a market economy, is a Newtonian system if models of these systems prove to be sufficiently isomorphic with Newtonian systems as defined (Giere 1979b, 1984). The beauty of this analysis is that if a certain model does not apply well to a particular phenomenon— something that is bound to happen in a complex world in which phenomena are held in place under definite boundary conditions—the theory is not disconfirmed, as it would be on the received view (since in that case the universality of its major premises would have gone down). Rather, the model is simply judged to be inapplicable to the case at hand. We will make considerable use of this notion in arguing that Darwin's Darwinism embodies Newtonian models, while genetic Darwinism uses models taken from statistical thermodynamics.

Since even tough-minded philosophers of physics now acknowledge that the "laws of physics lie," that is, hold only approximately and intermittently, the idea that theories apply to the world by the degree of similarity between model and some stretch of the world, rather than by way of deductive entailment from a general statement, can be a liberating one (Cartwright 1983).[16] It is perhaps philosophers of biology, however, who have felt most liberated by the semantic account of theories (Beatty 1981; Thompson 1989, Lloyd 1988). John Beatty, for example, has argued that the fact that "Mendel's laws" are contingent products of evolution, and too full of exceptions to count as laws of nature, need not count as an objection to Mendelian genetics or its role in evolutionary theory (Beatty 1981). On a model-based account of theories, Mendel's rules apply where they apply, with no bad consequences where they do not. In a similar spirit, Elizabeth Lloyd argues that the key notion of an "expanded synthesis," according to which selection can occur at a variety of biological levels and on a variety of "units of selection," can much more easily be countenanced by a semantic view of theories than by the "received view" (Lloyd 1988). For our present purposes, however, it is more to the point to note that the greater respect for diversity and complexity expressed by the semantic account of theories reveals an affinity with nonlinear dynamics. "By what method," asks Stephen Kellert,

does chaos theory give understanding? The answer is: By constructing, elaborating, and applying simple dynamical models. . . . Here we see a powerful example of empirical evidence from the sciences working to support a particular position in the philosophy of science. In this case, that position is known as the semantic view of theories, expressed in Ronald Giere's injunction that "When approaching a theory, look first for the models and then for the hypotheses employing the models. Don't look for general principles, axioms, or the like." . . . Chaos theory often bypasses deductive structure by making irreducible appeals to the results of computer simulations. The force of "irreducible" here is that even in principle it would be imposible to deduce rigorously the character of the chaotic behavior of a system from the simple equations which govern it. (Kellert 1993, 85–93)

For those who think of the world as presumptively complex and of understanding as a function of the degree of fit between models and phenomena, explanation is a matter of determining just how well a model fits a phenomenon.

A fourth idea about explanation is relevant to this task. The explanatory power of a model consists in discovering that a system whose workings are puzzling has the same structure as one whose workings are well understood. So far forth, models explain the same way metaphors do. More strongly still, explanation by model *is* explanation by metaphor, since explanation by models works by a process of "seeing as" (Black 1962; Hesse 1966; Bradie 1980, 1984). The work of pigeon breeders, for example, served Darwin not only as psychological inspiration for the idea of natural selection but as an explanatory model that revealed the workings of processes in nature that share the same structure as picking from the litter. The metaphor of picking, in fact, served two functions for Darwin. First, it gathered together a heterogeneous group of facts to constitute a phenomenon to be explained. It then explained these facts by transferring an understandable property of the metaphor to the *explanandum* (Bradie 1980, 1984).

Treating laws as definitions and metaphors as explanations might be thought to lead straight to scientific antirealism and to the related idea that scientific theories are merely "social constructions." We concede that this would be true if the world were composed of simple systems nested within simple systems, for models, metaphors, and approximations do not sort well with the idea that the world is such that someone (or some god) might form a single definitive picture of it, as the received picture invites us to do. In a presumptively complex world, on the other hand, the problem of realism is something of a nonstarter. Both realism and antirealism toil within the contrast between the literal and the "merely" metaphorical that simple science imports into the scene of science. Against a background of presumptive complexity, nothing can be more realistic than a metaphorical explanation with a high degree of fit—and nothing less realistic than a metaphor that lacks this property. Theories may be instruments rather than pictures, but these instruments allow us to see things really.

It is fortunate that a new philosophy of science seems to be emerging just as scientists are beginning to admit that many real world systems are complex and nonlinear. These simultaneous developments reflect not only our increasing technical ability to model complex systems but continued retreat from the cryptotheological worldview that was bequeathed to scientists at the dawn of modernity. It was held then that the world *must* be composed exclusively of simple systems because the Creator's mind must be assumed to be orderly.[17] If one wants to think about how science, evolution and religion are related, it is instructive to think first about this assumption.

I Darwin's Darwinism

2 Evolution and the Crisis of Neoclassical Biology

This chapter deals with evolutionary ideas, and resistance to them, before the appearance of Darwin's *On the Origin of Species.* In the eighteenth and early nineteenth centuries, "transmutation of species" or sometimes "transformism"—the idea was not yet called "evolution"—stood in stark contrast to the classical biology of Aristotle and his Peripatetic school, the West's most persistent tradition of inquiry into living things. Aristotle, who lived in the fourth century B.C.E., maintained not only that species are highly resistant to change but, in a physical world that itself was thought to have no beginning and ending, are themselves eternal. Transformism also contrasts, however, with what we will call the neoclassical biology of the eighteenth and early nineteenth centuries. In spite of a tacitly, and sometimes overtly, Christian conviction that the world was created, along with all the species in it, and that God devises new species to replace any that happen to become extinct, neoclassical biologists as different from one another as Georges Louis de Buffon, Georges Cuvier, Karl Ernst von Baer, and Richard Owen all held, as they believed Aristotle did, that species cannot be turned into other species, or at least that this process could not lead to new lineages of species. The confrontation between neoclassical biology and evolutionary ideas in this period forms the context within which Darwinism was conceived, as well as the body of ideas that it had to work hardest to displace.

It would be a mistake to think that neoclassical biology was a degenerating research program in the decades we are considering. On the contrary, neoclassical biology, often with Aristotle himself as its distant hero, was increasingly successful in putting evolutionary ideas on the defensive almost up to the time of Darwin's *On the Origin of Species.* It did so, moreover, by "fair and square arguments." Resistance to evolutionary thought was not always rooted in irrational factors and interests. The belief that evolution's opponents were blind and their ideas empty springs mostly from the notion, promoted by anticlerical advocates of the Enlightenment, that Aristotelianism of any stripe is intellectually retrograde and that Aristotle himself, in the words of David Hull, was responsible for "two thousand years of stasis" (Hull 1965; cf. Mayr 1982, 1988).

That is certainly what partisans of the new physics, such as Galileo, Hobbes, Descartes, Gassendi, and Newton, thought—and justly so. The new physics abandoned Aristotelianism root and branch when it reappropriated the materialist and atomistic assumptions of people like Democritus, which Aristotle had considered and rejected. The same animus against Aristotle and his works and pomps was not felt, however, by many of those working in the biological sciences. Neoclassical biologists were certainly as eager as their counterparts in the physical sciences to win the imaginary "battle of the books," which, since the seventeenth century, had pitted partisans of antiquity against advocates of modernity and had challenged the moderns to outshine the ancients in every field. They were, moreover, as quick as the physicists to distance themselves from the medieval scholastic philosophies that were called "Aristotelianism." Study of actual Peripatetic works in biology, however, which had been increasingly possible since the rise of Renaissance philology, led many modernizing biologists to distinguish Aristotle himself from the logic-chopping "schoolmen" of the Middle Ages and sometimes to think of themselves as defending and reviving the authentic Aristotle. While admitting, for example, that Aristotle had been mistaken about particular issues, such as the circulation of the blood, William Harvey worked comfortably within Aristotle's general framework. Buffon thought of himself as vindicating the superiority of the moderns by perfecting, rather than rejecting, research programs that the ancient Peripetetics had left in fragmentary stages of development. In this respect, Buffon was like his contemporary Montesquieu, who thought of his *Spirit of the Laws* as completing Aristotle's *Politics,* and like the neoclassical literary critic Boileau, whose canons of taste were based on Aristotle's *Poetics.* The following description of Buffon's aims in his *Natural History* could easily provide a table of contents to Aristotle's own *History of Animals:*

Natural history . . . must center on the relations which natural things have between themselves and with us. The history of an animal must not be the history of the individual, but the whole species. It must treat generation . . . , the number of their young, the care by their parents, their place of habitation, their food, and finally the services they can render us. (Buffon 1954, 16)

It is true that in the early nineteenth century, after the French Revolution had stimulated a break with neoclassicism, Lamarck and his colleague at the National Museum of Natural History in Paris, Etienne Geoffroy Saint-Hilaire, issued challenges to the "fixity of species" or "fixism" of classical and neoclassical biology. In the process they displayed more than a little contempt for Buffon, whom they identified with the *ancien régime.* Moreover, the fact that these advocates of transmutation were associated with revolutionary ideals was explicitly understood, and feared, when transmutation began to spread in France and abroad (Desmond 1989). In defending Lamarck's and Geoffroy's science, Robert

Grant, Darwin's teacher at Edinburgh, was in all probability tacitly defending the democratic side of the French Revolution as well. In responding to the evolutionary challenge, however, comparative anatomists like Cuvier and Owen, as well as embryologists like von Baer, succeeded so well in breathing new life into the neoclassical program that, writing in midcentury, Darwin's defense of transmutation was of necessity tailored to providing an alternative account of the very facts and arguments that Cuvier, Owen, and von Baer had used to smite their evolutionary opponents. Accordingly, while defenders of the neoclassical biological tradition were often politically conservative, it would be wrong to assume that their biological science looked as vulnerable to people like Darwin as Aristotelian physics did to Galileo, Descartes, or Newton. It is not helpful in understanding nineteenth-century evolutionary theory to fail to recognize where the burden of proof fell. It fell on Lamarck, Geoffroy, and Grant, and subsequently on Darwin, whose respect for Cuvier and von Baer was predicated on his level-headed understanding of who and what his own theory had to beat.

In order to understand the neoclassical tradition in biology, we will do well to consider a few of Aristotle's ideas about living things and to see how his modern admirers attempted to develop them. We will focus on four themes in particular: the structural integrity of organisms, classification of kinds, adaptedness, and the link between development and reproduction.

Perhaps the most basic idea of the Aristotelian tradition is that living things can carry out biological functions precisely because they are tightly integrated structural wholes. "The hand separated from the body is not," Aristotle says, "a true hand" (*Politics* I.2.1253a20–21; cf. *De Anima* II.1.412b19–24). Owen captures the flavor of this belief when he says, "A brain, a heart, or a stomach have no independent existence; they have been formed with reference to the whole organized body—remove any one of these and the body becomes a dead mass" (Owen 1992, 213). "The organs of one and the same animal form a single system of which all the parts hold together, act, and react upon each other," Cuvier claims. "There can be no modification in any one of them that will not bring about analogous modification in all" (quoted in Foucault 1971, 265). When Darwin mentions "the mysterious correlation of parts" in *On the Origin of Species* and concedes that it makes trouble for his theory, he is in large measure referring to this holistic cornerstone of classical and neoclassical biology (Darwin 1859, 194–206).

Aristotle insisted on this point because he believed that living things could never exhibit the plasticity of response to environmental contingencies that they do—feeding, responding to light and heat, growing, reproducing, sensing, moving, imagining, thinking, and all the other activities that he collectively calls "soul" (*psychē*)—unless their parts are defined functionally and in relation to the wholes into which they are integrated. Aristotle argued about this point with Plato as well as with Greek materialists like Democritus and Empedocles. Plato, in the beautiful words of the poet Yeats, thought that "nature is a spume that plays upon a ghostly paradigm of things." That is, he argued that the preexistent formal recipes for an animal of this or that kind (Platonic forms) are impressed into evanescent materials, as the form of a cookie is impressed into dough by a cookie cutter, or more generally as an artisan works on materials. Call this *typological essentialism*. Although Aristotle has been described as a typological essentialist (Hull 1965), he was explicitly opposed to it, as well as to many other aspects of Platonism. The result of the "cookie-cutter" view of form, Aristotle thought, would be an entity whose form is external to its material, and so too weakly structured and integrated—literally too insubstantial—to maintain itself and to act coherently in environments demanding differential and plastic responses to contingencies and pressures (*On the Soul* II.1.412a11–413a10). It would be an avatar, not a substantial entity (*ousia*) (Furth 1988).

For the same reason, Aristotle got even more worked up about the ancient materialists, who argued that organisms are collections of independently defined, and even independently existing, parts that happen by chance to come together and to persist that way. In Empedocles' version of "building-block materialism," arms and legs, separately formed by laws of chemical and biochemical compounding, are imagined as happening upon each other and sticking together whenever they form stable structures, tolerated by the environment (*Physics* 198b27–32). This is evolution of a sort, although what Empedocles defends is the survival of the physically stable, not the survival of the biologically fit. Whatever it is, Aristotle could not see how, on this account, organisms could ever acquire the purposiveness, structural integrity, and behavioral plasticity that they clearly exhibit (*On the Soul* II.1.412a11–413a10). What comes together in the way Empedocles suggests can be taken apart as easily.

This obvious point probably mattered not a whit to Empedocles or Democritus, since they may well have doubted the objective reality of functional properties anyway. Functional organization seemed to them little more than a projection of human desires and purposes onto a fundamentally nonpurposive natural world (Sauvé-Meyer 1992). Since Aristotle took the functional adaptedness of parts to organic wholes and of wholes to their environments to be real and pervasive facts about the world, he concluded that biological form, or organic substantial integration, could come neither from "above," as it does for Plato, nor from "below," as it would have for the materialists, if they had believed in organic form at all. Where adaptive organic form does come from, moreover, was perfectly obvious to him: It comes from parents who already possess the same species-specific form (*eidos*). Human gives birth to human, Aristotle says, as horse gives birth to horse, in an eternal linkage between parents and offspring (*Parts of Animals* I.1.640a25).

Since Aristotle assumed as a working hypothesis that "nature does nothing in vain" (*Parts of Animals* IV.10.687a8–18, for example), he also thought it reasonable to conclude that since organic integration and reproduction always accompany each other, the indispensable function of the latter must be to ensure continuity of form from generation to generation. In demonstrating this, Aristotle begins with the fact that organisms have a deep drive to reproduce. "Nothing is more natural," he wrote, "than for a living thing to make another like itself, an animal producing an animal, a plant a plant, in order that, as far as its nature is able, it may partake in the eternal and divine" (*On the Soul* II.4.415a27–29). He also thought that organisms achieve this substitute for immortality with tremendous replicative fidelity, and in sufficient numbers to keep a species line intact even when parts of it are wiped out by environmental contingencies, such as floods, famines, earthquakes, and other "meteorological" catastrophes. (*Politics* 2.8.1269a 4–7; 7.10.1325–31; Plato, *Laws* 67–69) Although he nowhere says so, Aristotle's neoclassical disciples took him also to presume, as they did, that the persistence of species under stable descriptions rests on the fact that replicative error produces developmentally stunted "monsters" (*terata*), and therefore leads to lower rates of successful reproduction among defective offspring. The happy fact that abnormal individuals cannot attract mates or reproduce healthy offspring keeps species stabilized around a viable norm. Asa Gray, an American biologist who was a friend and after his own lights a defender of Darwin, expresses this neoclassical thought well without adopting it himself when he writes: "Although the similarity of progeny to parent is fundamental in the conception of species, yet the likeness is by no means absolute. All species vary more or less, and some vary remarkably. But these variations are supposed to be mere oscillations from a normal state, and in nature to be limited if not transitory" (Gray 1860). We will call this position *constitutive essentialism*.

Aristotle's constitutive essentialism contrasts with the typological essentialism of the Platonists insofar as it denies that organisms have the properties they do because each of them separately exemplifies or embodies the same abstract form. The truth is just the opposite: They are descriptively identical because they inherit their traits from their parents. It is a consequence of the high fidelity of reproduction that kinds have the invariant properties they do.[1] The difference between typological and constitutive essentialism shows up clearly in Plato's and Aristotle's respective treatment of artifacts. Plato's cookie-cutter model is paradigmatically applicable to artifacts, while Aristotle is leery of the analogy between organisms and artifacts (Furth 1988). The latter are not self-moving and self-perpetuating substances at all. That is because external causality at the hands of an artisan can produce only external, and easily shattered, unity. Accordingly, because organic form cannot come from Platonic recipes above or from Empedoclean matter below but only from parents and lineages, Aristotelian essentialism has a temporal dimension that is missing from Plato's typological thinking.[2]

The confusion between Aristotle and medieval Aristotelianism explains why he is so constantly mistaken for a typological essentialist. The medieval Aristotelianism of Ibn-Rushd (Averroes), Moses Maimonides, and Thomas Aquinas, among others, was permeated with Neoplatonic, and hence typological, elements dragged into the Aristotelian tradition by the Greek commentators of late antiquity and Byzantium. This confusion also explains why Aristotle is so often taken to have been fixated on classifying things. The imaginary Aristotle of the Neoplatonizing medieval tradition was indeed a compulsive taxonomist, who imagined a "great chain of being," in which the organic world is composed of a vast series of ranks of animal kinds, in which each kind at the same taxonomic rank is to differ from its nearest neighbor by a single defining characteristic, a "specific difference" (Lovejoy 1936). The diversity of kinds thus defined was thought to reflect a "principle of plenitude," in which lower kinds and ranks support the richer psychic life of higher kinds, and achieve their own purpose or telos in the cosmic household precisely by doing so.

That may be good Neoplatonism, but it is bad Aristotle. For one thing, the real Aristotle was less interested in classifying animals than he has been depicted by those who think of him as a typological essentialist with a hierarchical taxonomic vision (Balme 1962; Pellegrin 1986). In the *History of Animals,* Aristotle postulated eight or sometimes nine "great kinds" (*megista gene*) into which animals could be grouped: insects, testacea, crustacea, cephalopods, fish, cetacea, birds, oviparous quadrupeds, and viviparous quadrupeds (*History of Animals* I.6.490b7–10). Each "great kind" collects species sharing a large number of traits and a common body plan or chassis. These are quantitatively varied in the different species (*eide*) clustered around the *genos* (Lennox 1987). At a generic level the possession of one trait in the cluster sets up the presumption that another will be there too, only a bit modified (*Parts of Animals* I.4.644a13–23). One sort of bird, for example, might have, for adaptive reasons, a shorter beak or a longer leg than another. However, it was clear to Aristotle, as a practicing biologist, that the traits that come together to mark off animal kinds in this way are simply too various, too crisscrossing, and too overlapping to permit a more hierarchical classification scheme than his *genos-eidos* model afforded (*Parts of Animals* I.2–3.642b50–644a12). There is not even a hint at taxonomic levels lying between species and his large genera, or above them (Balme 1962). Aristotle seems simply to have assumed that his two-taxon scheme would continue to be adequate because it recognized complexity and left room for exceptions and borderline cases. Much of the reason for this tolerant attitude lies in

Aristotle's suspicion of Platonism. He cast a lot of cold water on ambitious, aprioristic taxonomic schemes like the ones Plato himself proposed, in which all the animal kinds would be derived by a process of successive dichotomous division of basic traits. By that route one might arrive at what seemed to Aristotle to be a silly, and purely incidental, definition of humans as "featherless bipeds" (*Parts of Animals* I.2–3.642b50–644a12).

Neoclassical biologists were eager to reassert Aristotle's biological realism against the Neoplatonizing typology and hierarchical classification schemes of the late medieval period. That is why Buffon was suspicious of what turned out to be Linnaeus's reasonable proposals for reforming systematics. Since for Buffon, as for Aristotle, "a species is not a collection of similar individuals, but the constant and uninterrupted renewal of those individuals which constitute it," as Philip Sloan puts it, taxonomic categories like genera, classes, orders, and kingdoms would capture real distinctions in nature only if they reflected descent, and therefore only if species degenerated not only into less determinate genera but into even more indeterminate taxa as well (Sloan 1976).[3] Although Buffon conceded that species within a great kind do diversify into genera, as environmental change pounds on them and as replicative fidelity wanes, he could not swallow the idea that the process of degeneration went much further than that. Hence, he regarded Linnaeus's proposals as fictions. By contrast, Buffon praised Aristotle for staying fairly close to the facts. Later neoclassical biologists repeated this praise, even as they hoped that Aristotle could be improved on within his own interpretive framework. This explains why Cuvier's reduction of Aristotle's nine great kinds into four body plans, which he achieved by following the structure of the nervous system rather than the traditional timber of morphology, appeared to be such a triumph for neoclassical biology. As late as 1837, Owen remarked: "It is wonderful, considering that the nervous system, the true key to the primary divisions of the animal kingdom, was wholly unknown to Aristotle, that he would have approximated so nearly in propounding these classes to our modern systems. For his arrangement, although not irreproachable, is more accordant with nature than that of Linnaeus" (Owen 1992, 95).

What interested Aristotle more than classification was trying to explain just why each animal kind has the complex of traits it does. It helps in understanding Aristotle's answer to this question to know that the phrase "*to ti en einai*,"—"the-what-it-is-in something-that-makes-it-be,"—usually translated "essence," is not for Aristotle a classificatory term but an explanatory one. An essential trait is one that is most fecund and salient in understanding how an organism's structure subserves its form of life (*bios*) (*Parts of Animals* V.5.645b1–646a5). Highly explanatory, or essential, traits are in general functional, actional, and psychological rather than morphological. Mentioning the term *rational* in the case of humans, for example, is supposed to illuminate, explain, and justify much else about the dispositions or habits (*ethē*), the movements and actions (*praxeis*), and the body parts (*morphē*) of this interesting species.

Aristotle asserts that inquiry along these lines will show that each kind has precisely the bodily and behavioral traits that enable it to function as well as possible in its particular environment—where "to function" means to be able not only to survive, grow, and reproduce but also to develop and express its unique, defining array of psychological capacities, such as sensing and thinking. In an uncharacteristically lyrical outburst, Aristotle writes:

In all the works of nature there is something marvelous (*thaumaston*). The "not by chance" (*mē tuchantos*), indeed the "for the sake of something" (*heneka tinos*) is especially present in all the works of nature. . . . [In even the meanest of animals] every instrument (*organon*) is for the sake of something, and each bodily part is for the sake of some activity (*praxis*). It is clear, therefore, that the whole body has been constructed for the sake of some complex action. . . . The body is for the sake of the soul, and each of its parts (*moria*) has by nature (*physei*) its own function (*ergon*). (645a17–b20)

We might, on this account, comfortably say that Aristotle was an adaptationist were it not for the fact that contemporary Darwinians tend to use that term to refer exclusively to traits built up over time by interaction with the environment (Brandon 1990). Aristotle did not think that species acquire the traits they have through causal processes like those proposed by Lamarck or Darwin. From a causal point of view, in fact, the reverse is true for Aristotle. Traits are certainly "there" to facilitate action in a certain environment. But species always already have them because of inheritance, that is, because of a highly faithful replicative process that sustains a descriptively invariant lineage (*Generation of Animals* II.1.731b24–732a1), not because they are built up by a separate process. In this restricted sense, Aristotle was not an adaptationist. Call him a functionalist, if you wish, rather than an adaptationist, in order to mark this difference. Aristotle's great successor Cuvier is often called a functionalist precisely for this reason. Do not conclude on this account, however, that either Aristotle or Cuvier was any less interested than Darwin in the adaptedness of organisms. Nor, unlike Plato, does Aristotle's "teleology"—his conviction that explanations of facts about living things refer to ends (*telos* means "end" or "goal" in Greek) or functions—depend on "forward-looking" considerations, as if nature were previewing what an organism would need and, in furnishing it, looking to a disembodied Idea or Plan, like Plato's Demiurge in the *Timeaus*. Aristotle's clearest teleological texts refer for the most part to end-directed, but past-driven, processes like embryological development (*Generation of Animals* II.3.736b2–6) (Gotthelf 1976). Adaptedness depends for Aristotle on the backward-looking fact that one has one's traits because one's ancestors and progenitors had them. It was the Christian creationism of neoclassical Aristotelians that weakened their commitment to this aspect of Aristotle's thought, rendered neoclassical biology vulnerable to evolutionary theories, and gave Aristotelian teleology a bad name.

Given his stress on replicative fidelity as the cause of adaptive response to environments, Aristotle looked to embryology for answers about the mechanism for inheriting adaptations. In *On the Generation of Animals*, perhaps his greatest scientific treatise, he sponsored an early version of the "epigenetic" theory revived by neoclassical biologists like Harvey, Buffon, Caspar Friedrich Wolff, and John Needham (but, oddly, not Cuvier). On the epigenetic view, reproduction is a function of the natural growth and development of an organism. For Aristotle, the "nutritive" or "generative" soul (*psychē threptike, psyche gennetikos*), which all organisms possess, directs metabolism, growth, and reproduction through a single unitary cycle in which the information required for reproduction collects, with maturation, in highly compacted food residues in the male (and to a lesser extent in female) reproductive fluids and is passed to offspring, who begin the cycle over again. (Aristotle did not know that human females had eggs. The fact that some mammals lay eggs was discovered by von Baer only in the 1830s.) The male sperm inscribes successive differentiations and articulations into the less organized and compacted matter provided by the female (*Generation of Animals* II.3.736a19–25). It will do so with perfect fidelity unless some disturbance allows the female materials to "master" it, in which case a "deviation" from the natural course will occur, producing, in the first instance, a female, then, more seriously, a throwback, and finally, in case of radical disturbance, a deformed "sport" or monster (*terata*) (Furth 1988; Cooper 1990). "Folk biology," which is largely the sophisticated theoretical biology of yesteryear, still suggests, in a distant echo of this model, that physical strength is somehow related to sexual drive and reproductive fecundity, although the belief that "potency" is correlated to producing males rather than females has happily waned even further.

The neoclassical convictions that growth and reproduction are integrally related, that the latter is a phase of the former, and that deviation from some sort of norm comes from disturbances in development were opposed, in one of the most intense and divisive quarrels in eighteenth- and early nineteenth-century biology, by "preformationists." In preformationism, advocated in its heyday by figures such as Jan Swammerdam and Hermann Boerhaave, the seed already contains a miniature replica of the parent. The term *evolution* (from Latin *e-volvere*, "to roll out of" or "unfold") originally referred, in fact, to the preformationists' gradual manifestation of what was already there and was transferred later to the notion that kinds unfold through the same kind of inner-driven logic that individuals do (Bowler 1983, 1988; Richards 1992). Preformationism gradually lost ground as neoclassical biology went from strength to strength. However, even neoclassicists like Cuvier sometimes embraced it, perhaps for ideological and religious reasons, since by making it impossible for like to come from anything but like, except full blown from the hand of God himself, it reinforced the doctrine of special creation and species invariance. It is important to note that historians of biology have shown that, in contrast to twentieth-century genetic Darwinism, which in some ways echoes preformationist themes, Darwin subscribed to a generally epigenetic framework (Hodge 1985; Bowler 1989). We will return often to this point. Perhaps he avoided using the term *evolution* to name his version of transmutation and common descent because it had too many echoes of preformationism and of evolutionary theories in which kinds come rolling out automatically the same way individuals do.

An abiding idea of Aristotelian embryology, and of epigenesis more generally, is that all differences are differentiations, progressive articulations of an originally undifferentiated mass (Furth 1988). Aristotle's conviction that he had confirmed this fact by careful embryological observation and experiment was the main evidence he presented against Democritus and Empedocles in *On the Generation of Animals.* Organisms are more tightly constructed than artifacts, he held, and so can act autonomously and flexibly in their environment, because they are not put together from separate parts at all but move toward a final stage of developmental articulation from an originally undifferentiated beginning point.

The idea that embryogenesis is a process of differentiation was crucial for the Estonian embryologist Karl Ernst von Baer. An ardent epigeneticist, von Baer refuted preformationism by experimentally confirming Aristotle's belief that an embryo develops from an originally undifferentiated, homogeneous state to a highly articulated and differentiated one. Like many other German-language biologists of his period, von Baer followed the philosopher Immanuel Kant's ideas about method in the life sciences and so articulated this fundamentally Aristotelian notion by putting a Newtonian spin on it, at least rhetorically (Lenoir 1982). Von Baer thought of epigenetic differentiation as driven by a distinct vital force or life force (*Lebenskraft*), which, like other Newtonian forces, was supposed to propagate from a center after the fashion of an inverse square law. As quantities of this vital force were expended in growth and metabolism, they were at the same time conserved by being passed to offspring through the developmental and reproductive cycle. Although the term *vitalist* has now become little more than a nasty epithet that warring biological theorists hurl at one another, that is exactly what von Baer was, and what he took himself to be. Von Baer and other self-confessed vitalists were convinced that the laws of life, and the vital forces that obeyed them, were irreducible to, and could never emerge from, more basic Newtonian laws and forces, such as gravity, electricity, and magnetism (Lenoir 1982). Like Aristotle, von Baer thought that life could come only from something that already had it. He also thought that each organism has just enough life force to make another like itself, and none left over to perform the complex task of turning itself into a new species. The influence of von Baer in giving neoclassical fixism the edge over evolutionary theories in the decades before *On the Origin of Species* can scarcely be exaggerated.

The conceptual sophistication of Aristotle's biology accounts for the vitality and continuity of the tradition he founded but often comes as a surprise to people who have never taken the trouble to read him, dissuaded, no doubt, by his undeserved image as a scholastic logic chopper and antiempirical dogmatist. Darwin is a case in point. Unlike evolutionists of a Lamarckian and Geoffroyian stripe, Darwin had great respect for neoclassical biology. Thus, it is intriguing to find him in his old age reading Aristotle for the first time, when William Ogle sent him a copy of his Oxford translation of Aristotle's *Parts of Animals*. Rather wittily, Ogle wrote:

I feel some importance in being a kind of formal introducer of the father of naturalists to his great modern successor. Could the meeting occur in the actual flesh, what a curious one it would be. I can fancy the old teleologist looking sideways and with no little suspicion at his successor, and much astounded to find that Democritus, whom he thought to have been effectually and everlastingly squashed, had come to life again in the man he saw before him. (Ogle to Darwin, January 17, 1882)

Darwin responded:

You must let me thank you for the pleasure which the introduction to the Aristotle book has given me. I have rarely read anything which has interested me more, though I have not read as yet more than a quarter of the book proper. From quotations which I had seen I had a high notion of Aristotle's merits, but I had not the most remote notion what a wonderful man he was. Linnaeus and Cuvier have been my two gods, though in very different ways, but they were mere school-boys to Aristotle. I never realized before reading your book to what an enormous summation of labor we owe even our common knowledge. (Darwin to Ogle, February 22, 1882 in F. Darwin 1887 vol. 2, p. 427)

Evolutionary theory, in something like the meaning we now assign the term, first emerged in debates that racked the French Museum of Natural History in Paris in the decades following the French Revolution. The museum had been founded during the revolution in order to produce a rationalized biology that would mirror the rationality of the new France and would distance it from the chaotic medievalism of the *ancien régime*. This mandate stood behind the continuous and intense politicization of biological issues in France throughout the nineteenth century. In particular, it stood behind the notorious disagreements among Cuvier, Lamarck, and Geoffroy, all of whom worked, lectured, and taught at the museum. We now turn to those debates, beginning with Cuvier's role in them.

Cuvier achieved fame and influence in postrevolutionary France from his position as permanent secretary of the French Academy of Sciences and professor of comparative anatomy at the museum. Because of his role in opposing the evolutionary views of his colleagues Lamarck and Geoffroy, as well as his involvement in the conservative power elite of France after the defeat of Napoleon and the restoration of the monarchy, Cuvier has justly come by his reputation as a conservative. Exclusive concentration on Cuvier's conservatism can too easily obscure, however, the fact that he was fully devoted to fulfilling the museum's charge of rationalizing biology and was very successful in doing just that. Cuvier's great achievement was to sustain the contention that neoclassical biology should not be buried with the *ancien régime*.

French biology of the neoclassical age, not unlike French neoclassical tragedy, was bothersome to postrevolutionary intellectuals because it allowed a certain pessimistic note to insinuate itself into its rationalism. Perhaps that is because it carried with it too much medieval Christian baggage about a fallen and disordered world. For example, in order to account for observed diversity and the extent of extinctions, which could no longer escape recognition, Buffon held that species could "degenerate." Buffon could so readily acknowledge and explain transpecific degeneration because he was, in matters geological, a "catastrophist," according to whom pressures caused by violent geological change had split apart natural populations, scrambled the natural correlation of parts, and disrupted growth and faithful reproduction. Species had thus departed from type by way of degeneration. It may be hard for modern evolutionists to appreciate, but theories of "devolution" were common before there were any theories of "evolution." Indeed, adaptationism itself entered biological theory more closely associated with degeneration and disorder than with efficiency. The unnatural environments that catastrophical environmental change had produced occasioned new species whose adaptative features suggested makeshift "tinkering"—what the twentieth-century French geneticist François Jacob, following the French anthropologist Claude Lévi-Strauss, calls *bricolage*—more than sleek rational design (Jacob 1982, 33–35). It is small wonder that the "natural history" that Aristotle had founded, and that Buffon mimed, seemed to more rationalist heads little more than an anecdotal, unsystematic, weird cabinet of curiosities and monsters.

By reducing Aristotle's great kinds to four basic body types, Cuvier was able to rationalize the Aristotelian tradition in ways that Buffon did not, and so protected neoclassical biology by distancing it from prerevolutionary associations. By careful studies of comparative anatomy, Cuvier found that all animals fit into four body plans or *embranchements: radiata*, such as starfish; *articulata*, like insects and crustaceans; *mollusca*, such as snails and clams; and *vertebrata*, the great class of organisms with a central nervous system running down their spine. Cuvier was as much a functionalist as Aristotle. He recognized, therefore, that species differ because they have to live in different environments. Even in a quite varied world, however, Cuvier did not believe that adaptive requirements meant that organisms were bizarre contraptions or products of *bricolage*. Each "body plan" is designed to cope with a range of environments, and each plan can be systematically varied into species, genera, and other taxa by adjusting its basic features to meet environmental circumstances.

A central tenet of Cuvier's system, and the precise point at which he clashed with evolutionists, is that there can be no passage between *embranchements*. The correlation of parts within a plan prevents that. Attempts to cross from one *embranchement* to another, whether by way of devolution or evolution, would so disrupt functioning that the nascent lineage would almost immediately go extinct. As for Aristotle, then, relationships across the great kinds remain analogous rather than homologous. As gill : fish : : lung : quadruped mammal, says Aristotle, although a lung is not a kind of gill, nor a gill a kind of lung (*Parts of Animals* I.V.645b2–32). Cuvier acknowledged, however, as Aristotle did not, that there are manifold extinctions. Indeed, in his day he was perhaps most famous for recognizing that many kinds had gone the way of all flesh, including the recently discovered dinosaurs. Extinction, combined with his strong commitment to the principle of the coadaptedness of functional parts, posed an anomaly for Cuvier's version of the Aristotelian tradition, which he resolved by committing himself all the more fervently to creationism and to preformationism in support of creationism. Cuvier concluded that on occasion new species, based on existing body plans, are inserted by God into a vacated ecological slot. A similar view was held by Cuvier's student Louis Agassiz, a Swiss-born comparative anatomist, paleontologist, and icthyologist who virtually founded scientific biology in the United States and in the process became known as "the American Cuvier." It was this belief in separate creation that, in the eyes of postrevolutionary secularists and materialists, gave the lie to Cuvier's claim to have produced a rationalist biology and that impelled Lamarck and Geoffroy to outrationalize him by holding that *embranchements* are not impassable after all. That claim is precisely what evolutionary theory was originally about and suggests why it was associated from the start with materialism, anticlericalism, and atheism.

Lamarck's efforts at the Museum of National History, where he held the post of professor of invertebrate zoology, were as sustained an attempt as Cuvier's to fulfill the museum's mandate to overturn the old picture of ecological and adaptive disarray. Lamarck achieved this, however, by projecting a biology in which adaptation is rational because organisms take their fate into their own hands. He blunted the dark vision of historical decay that occluded the shining picture of biological order by replacing a fixed world in which things come in preset kinds and then fall from grace when they are contaminated by the mishaps, contingencies, and castastrophes of natural history, with a world in which organisms seem able, on the whole, to meet whatever challenges are thrown at them and, by meeting them, to transform themselves into new species. This process is based on an allegedly inherent tendency of living things to complexify and results in a directional thrust toward higher and higher organisms, culminating (so far) in humans. Lamarck's point was also tacitly political. Just as the revolution was to redeem human history as a site where people and peoples freely determine their own lives, so Lamarck's natural history redeemed nature as a site where autonomous organisms can not only cope with their world but serve as agents of universal progress.

Lamarck did not leave room for a creator even at the outset of this upward march. He held that the microscopic infusorians that formed the object of much of his research had spontaneously emerged from nonliving material, as Empedocles had long ago asserted (and as Aristotle sometimes conceded [Lennox 1982]). Thereafter the inherent tendency of living things to complexify, and by complexifying to increase their own scope for action, meant that once the first step was taken, a progressive tendency was unleashed in organic forms to move up the escalator of morphological articulation and psychological richness. Nonetheless, Lamarck did not hold that all life evolved from a single beginning point. He acknowledged that other invertebrate beginning points than the infusorians (worms, in particular) were possible and actual. Lamarck was not, therefore, committed to descent from a common ancestor, the view of evolution that, if anything does, binds Darwinians of all stripes and Geoffroyians together. Lamarck's theory can be summarized as follows:

1. Organisms adjust their behavior to changes in the (internal and external) environment.

2. "The production of a new organ results from the occurrence of a new, pressing need, as well as from the organic motions occasioned and sustained by that need" (Lamarck 1815).

3. "The frequent use of any organ, when confirmed by habit, increases the functions of that organ, leads to its development, and endows it with a size and power it does not possess in animals which exercise it less" (Lamarck 1809 [1984, 119]).

4. "All that has been acquired, deliminated, or modified in the course of individuals' lives is preserved thanks to reproduction and transmitted to new individuals born from those having undergone changes" (Lamarck 1815).

5. "Nature has produced all species of animals in succession, beginning with the most imperfect or the simplest, and ending her work with the most perfect, so as to create a gradually increasing complexity in her organization" (Lamarck 1809 [1984, 126]).

6. "The plan followed by nature in producing animals clearly comprises a predominating first cause. This endows animal life with the power to make organization gradually more complex. . . . This progressive complexification of organisms was in effect accomplished by the said first cause of all existing animals. *Occasionally, a foreign accidental, and therefore variable cause has interfered with the execution of the plan, without, however, destroying it.* This has created gaps in the series, in the form either of terminal branches that depart from the series in several points, and alter its simplicity, or of anomalies observable in specific apparatus of various organisms" (Lamarck 1815, italics added).

A moment's inspection of the italicized portion of point 6 shows that some of the old degenerative paradigm survives in Lamarck. Lamarck acknowledges that environmental difficulties have to some extent fact battered about the smooth linear series that would otherwise have emerged. Adaptation does sometimes reflect tinkering and introduces some side-branching and sidetracking into lineages, although Lamarck is sure that triumph is inevitable:

If the factor which is incessantly working toward complicating organization were the only one which had any influence on the shape and organs of animals, the growing complexity of organization would everywhere be regular. But it is not. Nature is forced to submit her works to the influence of the environment and this environment everywhere produces variations in them. . . . Progress in complexity exhibits anomalies here and there in the general series of animals, due to the influence of environment and of acquired habits. (Lamarck 1809 [1984, 69–70])

Lamarck could think that a built-in tendency to complexity is consistent with the view that organisms are active beings because his brand of materialism did not rest on a theory of matter that took it to be passive and inert stuff that needs something else, preferably a creator, to set it in motion. Like the Baron d'Holbach and other radical "ideologists," Lamarck thought of the self-creative powers and self-developing energies of matter itself as underlying the spontaneous generation of living things. In his view, matter, as it complexifies, automatically develops emergent capacities for sensing environmental difficulties, actively responding to them, and developing organic structures to deal with them. This process culminates for Lamarck (so far) in human beings. But this does not require the implausible evolution of anything like a spiritually conceived mind or will, for, considered as a biological adaptation to the needs of a fully material being, thought and choice are merely more complex abilities to integrate sensation, imagination, desire, and movement through increased associative and reactive ability. Lamarck could think all this because his materialism was not based on using creaky spring mechanisms as paradigms, as Descartes and Newton had it, but on fluid mechanics. "Subtle fluids," on Lamarck's view, vivify materials and render them responsive to environmental changes. Indeed, for Lamarck life seems to be nothing more than sophisticated fluid dynamics, one of the liveliest French physical sciences of his day.

The crisis to which the title of this chapter refers was precipitated by a highly charged series of debates between Cuvier and Etienne Geoffroy Saint-Hilaire, another professor at the French National Museum of Natural History, debates sometimes waged over the recently dead body of their colleague Lamarck. (Though Cuvier comes off as an old mossback in their debate, Geoffroy had in fact hired him.) These debates took place in 1829–1830, reaching a climax just as Paris was in the throes of a violent revolution that resulted in the replacement of the restored Bourbon monarchy by the constitutional monarchy of Louis Philippe. The Cuvier-Geoffroy debate was not overtly about evolution but about comparative anatomy. Specifically, it was about Geoffroy's contention that Cuvier's *embranchements* are not impassable after all but form a unified and rational series of structural transformations from a single ground plan. Geoffroy maintained that the *embranchements* are not analogies but homologies. This claim spilled over onto the evolutionary territory first opened up by Lamarck because the passage from one *embranchement* to another could easily be interpreted as a real historical process.

The Geoffroy-Cuvier debate was triggered by Geoffroy's contention that the cephalopods share the same body plan with vertebrates, the backbone in the first case being bent back upon itself (Appel 1987). More generally, Geoffroy claimed that Cuvier's functionalist stress on adaptedness had prevented him from seeing a progressively complexifying series across *embranchements* that appears when we look beyond the distractions of function to underlying morphological structure. Geoffroy was saying, in effect, that the prosecution of Cuvier's program of comparative morphology, to which he himself had eagerly contributed (by showing, for example, that the bones of the operculum in fish correspond to parts of the mammalian ear), was so successful that one need not stop at the limits of separate body plans and need no longer concede that biological order is constrained by the batterings of historical contingency. The biological rationality for which the museum was searching would be found, on Geoffroy's view, by discovering the complete sequence of transformations of "rational morphology."

The fact that Geoffroy's approach is usually said to be a "structuralist" rather than a "functionalist" correctly implies that he was not working in the Aristotelian tradition at all. Geoffroy was influenced instead by a group of Romantic philosophers of biology, including Lorenz Oken, Friedrich Tiedemann, Johann Meckel, and Carl Gustav Carus, whose views were informed by the absolute idealism of Friedrich Schelling. The *Naturphilosophen*, as these theorists were called, wanted to see a deeper logic in living nature than Aristotle's maxim that nature does nothing in vain in adapting organs to organisms and organisms to environments.[4] They wanted to see living kinds moving systematically and purposively from simple to complex by ringing ever more complex topographical changes on a basic archetype. Schelling himself was inspired by the poet-scientist Goethe's belief that all the parts of plants are transformations of the leaf and that somewhere there must be an "original plant" (*Urplanz*) in which this basic structure is revealed. *Naturphilosophie* was an extension of this idea to animals as well as plants. Its geometrical, topological way of thinking harks back Plato, and even Pythagoras, more than to Aristotle, even though it is nature itself that does the work, as it develops toward consciousness and, in human beings, expressive self-consciousness, rather than some already realized being or external agent or Demiurge. Since what is represented as unfolding in this way is the ever greater complexity and psychological richness of organic kinds, the *Naturphiloso-phes'* vision can be seen, in fact, as Neoplatonic typological essentialism turned on its side and represented *as if* it were unfolding in time.

We say "as if" because the *Naturphilosophen* were not transmutationists. They merely represented what they regarded as essentially fixed, and temporally coexistent, species as a sequence of temporal moments. Geoffroy himself, speaking as a comparative anatomist, could be understood this way. That is certainly how Goethe, for example, who was far too conservative to be a transmutationist, understood him. Goethe's last works, in fact, were a pair of articles attempting to defend the "deeper truth" of Geoffroy in the face of Cuvier's apparent win on points. When, however, in the course of his debate with Cuvier, Geoffroy began to speak well of the recently deceased Lamarck, in an environment already charged with revolutionary fervor, the great debate about morphology turned into a debate about transmutation and common descent. It was clear that Geoffroy believed not only in transmutation but, unlike Lamarck, in common descent as well, and that his structuralism was supposed to lay bare the inherent tendency to complexification that Lamarck had postulated but obscured by his adaptationism. Thereafter, the galleries of the museum were filled with intense young men and their fainting female companions. The debate was reported dramatically in the press and became a matter of international interest.

What became known as "Geoffroyism" hypothesized common descent from one beginning point, rather than from Lamarck's several or many, pushed along by sudden shifts in organization toward the next accessible level of complexity. Such changes may be induced or triggered by environmental stresses and may enable a new species to cope better with them. But adaptation is a consquence of structural change rather than a reason for it. The resulting combination of Lamarck's evolutionary progressivism with Geoffroy's "philosophical anatomy," as it was called, proved politically explosive, for in associating himself with Lamarck's evolutionism, Geoffroy, whether intentionally or inadvertently, gave a materialist rather than idealist spin to philosophical anatomy, making the doctrine attractive to revolutionaries and radical reformers everywhere, whether they knew much about comparative anatomy or not.

Cuvier was appalled. This was a challenge to the fundamental assumptions of neoclassical biology. Accordingly, he used the solemn occasion of a memorial lecture for Lamarck to ridicule his deceased colleague's ideas. He then took the offensive against Geoffroy's morphological interpretations. It is generally acknowledged that Cuvier won the battle and did a great deal as a result to discredit transformism from the 1830s to the 1850s. At the same time, however, Cuvier's attempt to tar Geoffroy with the brush of Lamarck inadvertently helped give Lamarck posthumous fame that he might not otherwise have enjoyed. Progressive evolutionism came to be associated in French culture thereafter with Lamarck's name and with egalitarian aspirations. Much to Darwin's dismay, for example, his self-appointed French translator, Clemence Royer, subtitled her version of *On the Origin of Species* "Of the Laws of Progress among Organic Beings" and adduced the book as support for Lamarckian and Geoffroyian progressive evolution, as well as for the political progressivism, materialism, and anticlericalism associated with it. Natural selection did not seem to Mlle. Royer a very important aspect of the work.

The potent combination of Lamarck and Geoffroy did not make waves in France alone. It spread immediately to Scotland and England. This could not have happened at a more sensitive time. In the 1820s, England's leaders were no longer terrified by the threat of foreign invasion. They were, however, worried about home-grown revolts. The leaders of the Whig party, by gathering support for a limited number of reforms, were trying to head off the growing resentments of the rising middle and professional classes, who sometimes tended to act as tribunes of the increasingly desperate poor. Having been shut out from privileges long monopolized by the old landowning and clerical elite, young professionals from middle-class backgrounds were often in sympathy with the ideals of the French Revolution, especially with Napoleon's declaration that in a genuinely rational society "careers would be open to talent" rather than reserved for the often-incompetent privileged. We shall examine this agitated social background more fully in the following chapter, for it was Charles Darwin's background as well. At present it suffices to note that, as Adrian Desmond has rather persuasively argued, the new French philosophical anatomy, in its transmutationist interpretation, became an integral part of the pressure that young professionals were putting on the establishment (Desmond 1989). In the years leading up to the First Reform Bill, two Francophile anatomy teachers in particular, Robert Knox and Robert Grant, brought Geoffroy's philosophical anatomy, and evolutionary biology, to Edinburgh. Bruited about in inflammatory medical journals like Thomas Wakley's *Lancet*, these ideas served effectively to encode the aspirations and resentments of the less than privileged medical students to whom Knox and Grant lectured in the private anatomy schools, which had grown up in the shadow of the University of Edinburgh.

It is not surprising that the "Athens of the North" was the site of these efforts or that Knox and Grant found willing audiences there among people who lacked the right background or the right religious beliefs (Anglican in England, Presbyterian in Scotland) to enter the clubby Royal College of Surgeons. Edinburgh had long been the most intellectually progressive city in the realm. In its glory days in the mid-eighteenth century it was, along with Glascow, home to the Scottish Enlightenment. Its medical school was the best in Britain. Erasmus Darwin had attended it, as had Darwin's father, his brother Erasmus, and for a time Darwin himself. Knox and Grant themselves had both graduated from Edinburgh's Medical School in 1814. On their frequent visits to Paris, where they roomed together (sometimes joined by Owen), they were well received by Cuvier, Geoffroy, and Lamarck. Knox, a flamboyant, foppish man, whose father had been a Jacobin and trade unionist, became an advocate of Geoffroy's philosophical anatomy and an opponent of the fatuous "design arguments" into which functionalism had degenerated in England under the clerical influence of Paley's disciples at Oxford and Cambridge. Grant was also an urbane sort, who lectured for profit in Edinburgh as little as possible and spent as much time as he could on the Continent. He was a superb comparative anatomist. He was also a natural historian, who taught invertebrate zoology at the University of Edinburgh, where he served as Charles Darwin's mentor until the younger man transferred to Cambridge. Grant's interest in invertebrate organisms like sponges reflected his Geoffroyian and Lamarckian preoccupations. Sponges stood on the boundary between plants and animals, and thus represented a basic test case for full unity across *embranchements*. He thought that primitive microorganisms living in large mats in the sea might conceivably have come about from spontaneous generation. Grant even had a theory of transmutation of his own. He theorized that new species budded off from old ones when the limited life cycle of the latter reached its natural limit, after the analogy of individual reproduction.

In 1827, Grant was invited to become professor of zoology at the University of London, which had been founded the year before by a fractious coalition of aristocratic Whig patrons, led by Lord Henry Brougham, and middle-class reformers influenced by the utilitarian Jeremy Bentham. As an instrument of projected Whig reforms, the University of London was designed to challenge the educational monopoly enjoyed by Oxford and Cambridge. Its enemies in these high places rejoiced mightily in its early struggles and internal conflicts. Since Grant was a sworn enemy of the clerically dominated biology taught at Oxbridge, it is no surprise that he was invited to become London's first professor of zoology. His radical social views were welcome, or at least tolerated, in the 1820s. They would prove less welcome after the passage of the Reform Bill, when power in the university passed to free-market utilitarians and away from socialist-leaning Lamarckians and when evolutionary thinking had, like democratic radicalism, been put on the defensive.

The most important reason for Grant's eventual eclipse, however, was Owen's successful challenge to Grant in the late 1830s. The conflict between Grant and Owen recapitulated the Geoffroy-Cuvier debate and ended even more clearly than its prototype in the apparent vindication of Cuvier's views. By 1840, it was Owen, professor of anatomy at the Royal College of Surgeons in London, rather than Grant, who was known as "the British Cuvier," an honorific phrase reserved for England's and Scotland's most distinguished comparative anatomist. In Owen's case, the phrase also meant subscription to Cuvier's fixism and creationism, the impassability of *embranchements,* and other tenets of neoclassical biology.

Owen's success depended in large part on his ability to marshal a number of embryological arguments first devised by von Baer against the doctrine known later as recapitulationism, to which most Geoffroyians subscribed, and which had become a sort of test case for their evolutionary ideas. The notion that organisms belonging to higher kinds literally go through or "recapitulate" the adult states of organisms in lower kinds was first advanced by the *Naturphilosophen.* Oken puts it as follows: "The embryo successively adds the organs that characterize the animal classes in the ascending scale. When, for example, the human embryo is but a simple vesicle, it is an infusorian. When it has gained a liver it is a mussel; with the appearance of the bone system it enters the class of fishes" (translated by Ospovat 1976, 4–5, from Oken, *Lehrbuch der Naturphilosophie*).

In its *naturphilosophische* incarnation, recapitulationism was intended to show that there is a wondrous descriptive isomorphism between the "macrocosm" of living kinds and the "microcosm" of individual development. When Geoffroyians adopted this idea, they gave it an evolutionary twist. Recapitulationism so understood implies that at some point in your development, you went through a stage at which you were a fish because at one point you had gill slits. (It also implies that fish are lower in the phylogenetic series than humans and evolved earlier.) This was known in the nineteenth century as the Meckel-Serres law (Meckel having been one of the original *Naturphilosophen,* and Etienne Serres a disciple of Geoffroy who developed and defended it). If true, the Meckel-Serres law would prove a boon to natural historians and comparative anatomists. One could successfully work back and forth between development and evolution, perhaps even hypothesizing about what "missing links" paleontologists should go out and look for. At the same time, recapitulationism was risky business. It might expose evolutionary theory to empirical disconfirmation should its description of individual development prove to be empirically false. Von Baer's attack on evolutionism was built on this challenge. So was Owen's attack on Grant.

Owen's great moment came when he gave the annual Hunterian Lectures at the College of Surgeons in 1837. The Hunterian Lectures originated when the college acquired John Hunter's extensive collection of displays illustrating how the various organ systems compare within and across various *embranchements*. A condition of this bequest was that an annual series of public lectures be given explicating Hunter's displays and praising his genius. Owen's Hunterian Lectures create a mildly comical effect, purely unintended by such a humorless and single-minded young man, when Owen repeatedly turns from his obligatory and slightly condescending bows to Hunter to praise the "all-commanding intellect of Aristotle," whose sober functionalism had immunized Cuvier from the drunken "transcendental speculations" of Geoffroy, and whose comparatively trivial "errors and omissions . . . arose entirely from Aristotle not having had the assistance of magnifying glasses," a defect Aristotelians have not had to endure since Harvey (Owen 1992, 104, 104, 107). To this invocation of the pantheon of classical and neoclassical biology, Owen adds his belief that the college had preserved its sanity in these manners in part because "our library is particularly rich in editions of the works of Aristotle, especially the treatise on the *History of Animals*, which Cuvier justly pronounced to be one of the most admirable works that antiquity has bequeathed to us" (Owen 1992, 113).

All of this authority-creating and tradition-constituting maneuvering was preparatory to Owen's reports of the "truly philosophic inquiry now in progress," in which von Baer and others of his school "are cautiously but steadily laying the foundations of a just and true theory of animal development and organic affinities" (Owen 1922, 191–92). Von Baer had demonstrated that embryos of higher kinds do not in fact pass through adult stages of supposedly lower kinds, as the philosophical anatomists held. What happens is something quite different. Embryos of higher organisms go through the early, undeveloped stages of lower kinds, *not* their terminal stages. It follows that organisms of different kinds resemble each other most during the earliest phases of development; that adult stages of lower kinds more fully resemble their own embryonic states than do the adult states of higher organisms; and that adult stages of lower kinds resemble one another more than adult stages of advanced kinds resemble one another. The fact that you had rudimentary slits that would develop into full-fledged gills in a fish, if you had been in the fish lineage, does not imply, then, that you, a human being, once had gills, and it certainly does not imply that you were once a fish.

All of this demonstrated to von Baer that full developmental articulation and species determination occurs only at the terminal point of growth. Indeed, von Baer took it to be a law of nature that earlier ontogenetic stages are amorphous and homogeneous and arrive at full heterogeneity only in their final developmental stage. For von Baer, as well as for Owen, this strongly epigenetic view of development constitutes an argument against evolution when other assumptions dear to the neoclassical tradition are thrown into the mix, such as the depth of organic integration, and the notion that a limited amount of vital force drives the developmental-reproductive cycle, a theory von Baer and Owen appropriated from Johannes Müller (Sloan 1986). "The individuals of each species," says Owen, "have a characteristic durability of Life—the operation of the organizing energy in them is limited. . . . This organizing energy exists in its state of greatest concentration in the germ" (Owen 1992, 223, 230). Whatever quantum of the stuff is left is devoted to building as much structure as possible. Since the maximal amount of energy will be allocated either to ensuring reproduction or to maximizing complexity, no quantum of living force is left to propel a species further into a new kind. That feat would, in any case, require a massive reorganization of the whole rather than a simple addition to a terminal stage, since "complication of organized structure increases the reciprocal dependence of the different organs," and would demand a much greater quantity of life-force than seems available (Owen 1992, 227). For Owen, the fact that no human was ever an ape, or even an ape stage, and that no ape could ever become a proto-human, was both explained and guaranteed by these considerations. If, moreover, these arguments were sufficient to rule out even simple transmutation of species within single *embranchements*, a fortiori they made it virtually inconceivable that new forms could pass across distinct body plans. Owen's lectures thus supported the tenaciously argued monographs in which he knocked down each of Grant's attempts to see unity of type in strange new organisms like the Australian duck-billed platypus.

Owen's Hunterian Lectures were a terrific success. Soon, however, Owen was talking much more than he had before about species and higher taxa as distinct archetypes in the mind of God. In this turn from Aristotelian neoclassicism to the Neoplatonizing rhetoric of idealism, Owen was following Joseph Green, his mentor and predecessor as Hunterian lecturer. Green had studied German *Naturphilosophie* under Oken and saw organic kinds as a logically continuous series, albeit in a distinctly nonevolutionary mode (Desmond 1989; Sloan 1989, introduction to Owen 1992). Back in England, Green fell under the spell of the Romantic poet, idealist philosopher, and conservative social theorist Samuel Taylor Coleridge. Coleridge had sensed very early in the game that the old arguments of Paley, still taught as gospel at Oxford, Cambridge, and within the closed corporation of the Royal College, were no longer very effective against the new comparative anatomy and evolutionism. He called, accordingly, for a new philosophy of biology in which the creation of species was to be portrayed, as the German idealists had portrayed it, as the creative self-externalization of a divine mind immanent in nature rather than as a collection of highly rigid, if well-adapted, machines produced by a quaint eighteenth-century Deist designer. Owen apparently agreed with Coleridge and Green that the old argument from design had to be replaced with a new argument based on divine self-expression. Moreover, he seems to have felt that he had finally done the job. Why should not the separate creation of each fixed species and impassable type mean the successive appearance in time of an active, emerging Deity's coherent thoughts? Why should not creation be an ongoing, continuous affair, if not an evolutionary one? If, in short, Geoffroy's rational morphology had forsaken its idealist roots and taken a materialist turn when it allowed itself to be mingled with Lamarckism, why should not Owen turn the tables and put an idealistic metaphysical spin on Cuvier's and von Baer's neoclassical biology?

Some scholars have seen Owen as a typological essentialist all along, as well as a willing tool of reactionary Tories, to whom he supposedly owed deliverance from the isolation and poverty of his Lancastershire origins (Desmond and Moore 1991). Owen's idealistic rhetoric, however, driven by his desire to use his newly acquired influence to heed Coleridge's and Green's call for a new argument from design, did not become the most pronounced feature of his thought until the 1840s, becoming ever more strident thereafter as evolutionary theory began to make a comeback (Sloan, introduction to Owen 1992, 71–72). This fact has obscured the roots of Owen's arguments in neoclassical biology. Indeed, Owen's commingling of neoclassical vitalism and idealist metaphysical gestures has been so influential that it has contributed, especially in English-speaking countries, to systematic conflations between the Aristotelian tradition on which we have concentrated in this chapter, and the quite different, idealist tradition of the philosophical anatomists. This conflation more than anything else is responsible for Aristotle's belated, and quite false, reputation as a typological essentialist. It also contributed to a long-standing tendency to think of von Baer's laws of development as themselves versions of recapitulationism.[5]

Advocates of Darwinian evolutionary theory are often complicit in these misunderstandings. Assimilating neoclassicism to the neomedievalizing obscurantism of *Naturphilosophie* makes it all the easier to assert Darwinism's monopoly on biological common sense and to paint its enemies as ideologically motivated reactionaries. This is, however, historically unsound. It fails to recognize that Darwin's version of evolutionism incorporated many elements of Cuvier's, von Baer's, and Owen's neoclassical biology, even as he rejected typological thinking and vitalism.[6] In the years following his return to England from the five-year voyage of the *Beagle*, Darwin and Owen were in close contact. It is even possible that Darwin attended Owen's 1837 lectures (Sloan, introduction to Owen 1992). It became increasingly injudicious in these years, however, for Darwin to appear as a discipline of Lamarck or Geoffroy, or a former student of Grant, even though he acknowledged to himself by 1837 that he had in fact transmuted into a transmutationist. Evolutionism, under the old dispensation, was in full retreat. It would take some working out to find an evolutionary theory that honored insights such as reproduction as a function of growth, disturbances to growth as sources of variation, epigenetic development, and conservation of inherited traits. It would also take more than a little discretion. That, however, is precisely what Darwin discretely set out to do.

3 A Short Look at "One Long Argument": The Origins of *On the Origin of Species*

In October 1836, Charles Darwin disembarked from the great five-year adventure of his youth, the voyage of the *Beagle*. The *Beagle* had been sent out under the command of the irascible and depressive Captain Robert FitzRoy, nephew of the reactionary Tory prime minister Castlereagh, to make a scientific survey of the coast of South America and nearby regions of the South Pacific. The mission was yet another manifestation of the glories made possible by Britain's dominion over the seas. When Darwin shipped out on the *Beagle* from Plymouth on December 27, 1831, he was a twenty-two-year-old youth, recently graduated from Cambridge. Although he was an amateur, whose main job was to serve as a gentlemanly dining companion to FitzRoy, Darwin was fairly well equipped to serve as ship's naturalist. His collections and ruminations during and after the expedition about such things as coral reefs reflected the interest in invertebrate biology he had acquired under Grant at Edinburgh, where he had enrolled when he was thinking of following his father's footsteps into medicine (Sloan 1985, 1986). His fascination with the connection between earthquakes, such as the one he experienced on the coast of Chile, and the fossils of sea creatures he kept finding on the sides of mountains, reflected his training in geology. He had acquired that under John Henslow and Adam Sedgwick at Cambridge, where he had transferred from Edinburgh with the not very fully formed intention of training as an Anglican priest. When Darwin left Plymouth, he was still vaguely assuming that he would wind up a country clergyman, if a botanizing and geologizing one, but was giving his father, a respected provincial physician in Shropshire in the west of England, fits about whether he would ever settle on a career at all. At the time Darwin was, or persuaded himself that he was, a conventional creationist after the manner of Paley, whose natural theology had been proclaimed as gospel at Cambridge.

When he returned to Britain five years later, Darwin was a mature man on the verge of allowing himself to doubt the fixity of species. What had provoked this change was the overwhelming belief he had acquired on his voyage that species (including humans) change their character dramatically over space and time. Through a continuous range, they differ imperceptibly but inexorably. On islands, like the Galapagos, finches and tortoises seemed to differ markedly even if separated by just a few miles. On the Argentine pampas and in the Andes, Darwin had found fossil remains of huge creatures that were oddly like the smaller sloths and llamas still roaming the same places. The suspicion that these were allied but different species had dawned on him. Yet it was only when his surmises were confirmed by the expert systematists to whom he consigned his collection upon his return to London that Darwin began seriously to consider transmutation. In July 1837, while living unhappily in sooty, crowded, and contentious London, he started taking notes on the subject in a chapbook he called *Zoonomia*. It was the title of a book his grandfather, Erasmus Darwin, had written on evolution in the more freethinking days before revolution and counterrevolution had put the lid on.

Darwin had already changed his views about geology while still aboard the *Beagle*. He had seen with his own eyes that earthquakes can suddenly and violently rearrange the furniture of the world. At Concepción, devastated by a quake he had felt two hundred miles away, he observed dead clusters of mussels and shellfish stranded well above the waterline. The ground had been raised by two feet. Such events, he mused, seem catastrophic. They are certainly fatal to many creatures and sometimes to whole species. But on the large and slow-moving scale of things to which Charles Lyell's recently published *Principle of Geology* was at the time converting Darwin from the catastrophism Sedgwick had inculcated in him at Cambridge, these "catastrophes" are the merest of disturbances along a continuous curve. We live, it seems, in the interstices of an order more vast than we had ever suspected.

Only someone who had escaped the insular world of little England and had sailed to what the poet Matthew Arnold was to call "the vast edges drear and naked shingles of the world" could have his sense of space and time pried open to that extent. The encounter with nature on such a large scale left a permanent mark on Darwin, enlarging his sense of what is possible in this world to a degree that set him apart from many of his contemporaries. There were, of course, other Victorians who felt that their tidy world was threatened. As reports from the empire's far corners filtered in, it became clear what vast and grinding processes of creation and destruction were at work in the world, what a sheer and potentially meaningless multiplication and variety of differences existed—differences even in the human world, some of whose peoples scarcely seemed to Europeans to be members of their own kind. These perceptions terrified some. "The lavish profusion of the world," wrote the poet Tennyson, "appals me." Some reacted to this sense of discomfort by compulsively reimposing order on it, constructing hierarchies in which European man stood at the top, master of the universe. Darwin's inclination was to move somewhat in the other direction.

Since childhood, Darwin had had a deep sense of identification with the natural world, a feeling of participating in its life, of being buoyed up by it, of finding himself by losing himself in it. In his first published book, *The Voyage of the Beagle*, half travelogue, half natural history, Darwin wrote, "Epithet after epithet was found too weak to convey to those who have not visited the intertropical regions the sensation of delight which the mind experiences" (Darwin 1839, 591). He experienced an "intense sense arousal" in the presence of nature and a keen "zest for the observable world" (Beer 1983, 34, 41). Darwin's writings convey a strong sense of the kinship of all living things. His was a world order in which revelations of the affinity of humans and animals have the effect of ennobling both rather than of degrading humans to the status of animals, as he was usually taken to imply. He was also convinced of the affinity among all humans. Although he was shocked by the lives of the primitive peoples he encountered on his voyage, he thought that each population would make progress if left pretty much to its own devices and that it was an act of unjustified arrogance for one people to dominate another. Darwin's most trying times with Captain FitzRoy were over the issue of slavery, which the liberal Whig Darwin abominated and the conservative Tory FitzRoy defended.

This sensitivity to nature's ways had been nourished in Darwin's youth by Romantic poetry, especially by Wordsworth and the Romantic reading of Milton, as well as by a peculiarly Romantic conception of science that flourished during his formative years. In the *Autobiography* he wrote late in life for his grandchildren, Darwin says that "During my last years at Cambridge I read with care and profound interest [Alexander von] Humboldt's *Personal Narrative*. This work stirred in me a burning zeal to add even the most humble contribution to the noble structure of natural science" (Darwin 1958, 67). Humboldt had traveled around the world, as Darwin had himself now done, Humboldt's book in hand, practicing a view of science as a process of acquiring knowledge through identification with nature, of learning by forgetting oneself, of undertaking an almost mystical trip into the sublime inner life of the world in order to bring back reports of hidden marvels. The old Greek sense of wonder, thought poet-scientists like Goethe and Humboldt, must be restored to a world whose cognitive instruments had been dangerously narrowed by Enlightenment mechanism.

Such was the man who, soon after his return to England, became engaged to his cousin Emma Wedgwood and set up house in London. Darwins had been marrying Wedgwoods for two generations and would continue to do so. They formed a closely knit clan whose fortunes and social status had been rising steadily. Having been "in trade" and the professions, they were by now proper gentry, equipped with country houses, hunting dogs, and social lives not entirely unlike those admired by Jane Austen. Darwin's mother, Susannah, was the daughter of the patriarch of the clan, Josiah Wedgwood I, whose exquisite pottery is well known to this day. Darwin was thus doubly among the heirs to the considerable Wedgwood fortune, as well as master of the yearly sum his successful physician father settled on him and Emma. He did not have to earn a living in order to support the large family he planned, or at least acquired. Not only could he live like a gentleman, but he was free to engage in scientific activity like a gentleman. That, it was now clear to him, is what he wanted to do with his life.

Darwin's first task was to secure his position in the scientific establishment, even if it condemned him to living for a time in the capitol. Accordingly, he went about soliciting the help of well-known scientists to aid him in describing, classifying, and interpreting the collection of specimens he had gathered as the *Beagle*'s naturalist. John Gould, who did for Britain what John James Audubon did for America, described his birds. Darwin's Cambridge friend, the Reverend Leonard Jenyns, an accomplished parson-naturalist now rotting in a country vicarage, as Darwin himself could easily have been, did the fish. Jenyns's brother-in-law, the Reverend John Henslow, who had tutored Darwin and had secured his appointment on the *Beagle* in the first place, took the plants. Owen undertook the large fossils.

All in all, Darwin's haul was impressive enough for him to be lionized by this distinguished but ideologically diverse company just for assembling it. He was asked to join the prestigious Geological Society, presided over by Lyell, as well as the Linnean and Zoological societies, among whose influential members his letters to Henslow from the *Beagle* had already privately circulated. But Darwin was out to interpret his haul for himself. First he published several scientific papers on geological topics. Lyell, whose work had first weaned him away from geological catastrophism as its three volumes caught up with him in various romantic ports, quickly came to regard Darwin as a protégé, an ally, and a friend. In 1837, Lyell pressed Darwin to serve as secretary of the Geological Society.

In 1839, Darwin's Humboldt-inspired book, *The Voyage of the Beagle*, was published. It was not only a popular success but a vindication of Lyell's uniformitarian geology. Darwin was made a member of the Royal Society. By 1842, however, his intense and growing hatred of the big city got the better of him. He moved his family to a country house in the village of Downe in Kent, about seventeen miles south of London. There he remained for the rest of his life, where in a strange sort of way he became a secular version of a botanizing, entomologizing, and pigeon-fancying country curate after all (Moore 1985). Down House (the "e" is omitted) had been a parsonage. Moreover, in spite of his own increasing disbelief (which became total after the death of his daughter Annie in 1850), Darwin served as vestryman of the parish Anglican church and its rector's close friend, while his gentle, dutiful, religious, and very Victorian wife played Lady Bountiful to the locals.

The England to which Darwin had returned in 1836 was different from the one he had left in 1831. Even as Darwin was founding a household and securing his position in the scientific establishment, the country was changing at dizzying and often dismaying rates. In 1830, the Whigs had managed to push through a Reform Bill, which gave greater political representation to the middle classes, who were pressing not only for greater access to professions like law, medicine, and education, but for more market-oriented economic policies. Throughout the 1830s and 1840s, the consequences of the First Reform Bill worked themselves out. The move to market society increased the economic, social, and political power of the professional and mercantile middle class but did little to improve the lot of industrial workers in the Midlands or of urban and rural poor. Between 1837 and 1842, in fact, England fell into the worst economic depression of the century. Starvation and riots were commonplace. The threat of rebellion increased until in 1848 it finally broke out, as it did nearly everywhere in Europe. But after the defeat of the democratic Chartist movement, which had unsuccessfully petitioned Parliament for universal manhood suffrage, the right to hold office without meeting property qualifications, and freedom to form labor unions and cooperatives, the power of the industrial and commercial middle class slowly consolidated. By the late 1850s, influential sectors of the population began to feel mildly confident that market society would continue to lift the general standard of living and that its discontents were worth bearing. It was only then that Victorian society, and the middle-class morality that sustained it, entered into safe harbor. Spreading opposition to social revolution began to be accompanied by fervent, if distinctly patronizing, efforts of middle-class reformers to alleviate the lot of the workers and the unemployed. Sanitation was improved, popular schools established, "factory acts" passed, "workingmens' protective associations" founded. Charitable organizations of all kinds went to work. Capitalism was tempered by intensified sentimentality about the lot of those less fortunate. That was the lasting deal that England cut with modernity.

During this period, ideological, political, and personal responses to the breakdown of premodern society fell into a complex continuum. At one end of the spectrum was the "squirearchy," whose chauvinistic ideology had dominated England during the Napoleonic wars. Rooted in the interests of the inefficient rural gentry, these conservative Tories were led by the old war hero Wellington. FitzRoy was a member of this class and shared its antimercantile, monarchical, patriarchal, military, rural, religious, and increasingly antiscientific view of the world. There were, however, important reformist Tories as well. Led by Sir Robert Peel, this stratum wanted as badly as their "king-and-country" cousins to preserve the numinous, religious view of the world that had sustained traditional authority and underwritten an ethic of noblesse oblige. Rather than blindly denying that the landed aristocracy and clergy had failed to meet its inherited obligations, reformed Tories proposed to address the social unrest of the day by revitalizing and reforming religion, education, political representation, government ministries, health organizations, and charitable institutions through religious romanticism and philosophical idealism. The strength of this group was in the old universities, Cambridge and Oxford, whose own badly needed reform was to serve as a platform for reforming the whole country. Peelites were as incredulous as socialists that the money-grubbing middle classes would make everyone better off by throwing everything onto the market. It is not surprising, then, that their idea of what was needed was what their acknowleged sage, Coleridge, thought he saw in Germany: a national "clerisy" of high-minded intellectuals and adminstrators who would extend the ideals of the wise and good into politics, administration, and corporate and professional boards. From these positions of power they would effectively and benignly look after society, containing the vulgar free-market fantasies of middle-class industrialists and merchants within their natural, and subordinate, bounds. Owen was sympathetic to this point of view.

The Peelites were trying to head off the threat to Tory power that had been posed by the Whig party, which took power under Earl Grey in 1831. Indeed, they were regularly accused by conservative Tories of almost treasonous collusion with the Whigs. The Whig coalition itself, however, was a fractious lot. Its patrons were Whig aristocrats, landowners so powerful that they did not need to oppose the demand of their industrial and mercantile allies that protective tariffs be lifted, as Tories of all stripes did. On the contrary, by honoring these demands, which would allow cheap food to flow into the country, lower industrial wages, and make English goods more competitive, the Whig grandees saw how they could wield power in alliance with Manchester factory owners and London merchants. The parliamentary representatives of the latter were known as Radicals. They were led by John Bright and Richard Cobden, the brains behind the Anti–Corn Law League. Nonetheless, Whig leaders like Earl Grey and Lord John Russell clearly had a tiger by the tail. At times the fate of their party seemed to be in the hands of Irish Catholics led by Daniel O'Connell, who, not content with newly acquired freedom to practice their religion, to own property, and to run for office in their own country—concessions reluctantly granted by their English conquerers—now pressed for independence for Ireland itself. "Single-issue" pressure was also exerted by Dissenting Whigs who, having successfully pressed for abolition of the slave trade, now sought to outlaw slavery itself in British colonies.

The left wing of the Whig coalition was formed by the "Philosophical Radicals," disciples of Thomas Malthus, Jeremy Bentham, and the French positivist Auguste Comte, who wanted to disestablish and privatize religion and who saw in secularized and humanistic science especially a statistically based "social science," the salvation of society. The Philosophical Radicals were ardent free marketeers. They recognized in Malthus's grim idea that population tends naturally to exceed food production proof that brutal competion need not lead to mass starvation but can be a motor of self-exertion, sexual restraint, and an irreplaceable source of social energy, innovation, expansion, and "progress." The University of London, with its collection of Whig aristocrats sitting uncomfortably at meetings with egghead reformers like Jeremy Bentham, gives an image of the ideological tensions in the Whig coalition, for the support of the Philosophical Radicals for free markets extended to free markets in ideas and power as well, that is, to institutional mechanisms that would allow outsiders like themselves eventually to displace their patrons by sheer talent, energy, and numbers.

Sometimes the idea of a "free market in ideas" led middle-class reformist intellectuals to the heretical, and usually evanescent, thought that support ought to be accorded to nonrevolutionary labor unions and the right to vote extended to the workers. On the whole, however, the Philosophical Radicals distanced themselves from leaders of the working class as much as other members of the Whig coalition. The feeling was mutual. The democratic and socialist leaders of the masses regarded Malthusians as their sworn enemies, who had doubly slandered working people by implying that starvation is the fate poor people deserve for not being capitalists and by imputing to them an animallike inability to control their sexual appetites.

Standing slightly below the Whig aristocrats, but slightly above their industrial, commercial, and intellectual clients, could be found an influential stratum of genteel Whig Dissenters, who often mediated effectively between them. Among these folk were Wedgwoods, Darwins, and Lyells. High-minded and well-off Unitarians, Quakers, and Deists, Whig Dissenters asked above all for disestablishment of the Anglican church (or at least demanded that candidates for positions in professional corporations like law and medicine, or for political office, should no longer have to swear fealty to the Thirty-Nine Articles). Rationalists of an eighteenth-century sort, they projected an image of Enlightenment reasonableness, balance, and moderation. Accordingly, while they despised both Tory anti-intellectualism and the Methodist "enthusiasm" and evangelicalism of the people, Dissenters were nonetheless generally religious, in a rationalist sort of way. Although the Darwin men secretly thought of themselves as freethinkers, or Deists at best, the Wedgwoods were devout Unitarians, who revered Jesus as a moral exemplar rather than a divine incarnation. (When the pious Emma Darwin went to Anglican services in Downe on Sundays, she remained silent when the Creed was recited.) Wedgwoods and Darwins were also creationists, less after the fashion of Genesis I, however, than of deism, which assumed that a reasonable God would not scurry about his universe patching it up with miracles but would govern (as would a rationally reformed political class) from behind the curtain of inviolable natural laws, or "secondary causes," that he had set up.

As advocates of science, Dissenting Whigs searched everywhere for applications, extensions, and analogues of the natural laws that keep the clockwork universe in balance, laws that their (more than slightly reconstructed) hero, the great physicist Isaac Newton, had revealed. The very fact that Darwin "burned with zeal to add even the most humble contribution to the noble structure of natural science," was a reflection of his Dissenting Whig background. Such people were pleased to discern in the mechanisms of supply and demand discovered by "political economists" like Adam Smith, David Ricardo, and Thomas Malthus laws governing society as fully as gravity governed the physical world. They were therefore supporters of free-market reforms. Indeed, while Josiah Wedgwood himself may have started out as a craftsman, employing "artisans," he ended as a factory owner, hiring and firing "hands," and substituting technological innovations for skilled labor as fast as he (and his successor Josiah II, Darwin's uncle and personal counselor) could. Yet, unlike Benthamites, Dissenters did not see every facet of experience or every aspect of society in market terms. They believed, for example, in the call of conscience, and hence in the irreducibility of morality to long-run self-interest. It was Dissenting Whigs, led by Wilberforce, who successfully opposed the slave trade on moral grounds and who almost as soon as the Whigs took power set about banning slavery itself in British colonies. The sentiments about slavery that Charles Darwin could not refrain from expressing to the unreformed Tory FitzRoy were inherited from his family. One of Josiah Wedgwood's pottery designs featured the silhouetted head of an African man, who appears over the words "Am I not a man and a brother?"

Because the role of science in a society struggling with modernization was politically charged, each of the parties contending against the old Tories developed a distinctive philosophy of science. Indeed, it was at this time that the discipline we now know as philosophy of science or methodology was invented. Reformed Tories like Owen, for example, typically followed the Cambridge polymath and master of Trinity College William Whewell, who in his *History of the Inductive Sciences* (1837) used Coleridge's favorite German philosopher, Immanuel Kant, to argue that, in order to have any experience at all, we necessarily must presuppose that the world is highly ordered. Accordingly, Whewell assumed that the work of science goes on within a definite, fixed philosophical framework, and indeed that a good deal of science's work is already done for it by metaphysics. From his high level of confidence that humans know a lot already in a highly knowable world and can fairly easily find out more,

Whewell held that support for a theory rests primarily on its ability to make unitary sense of an interrelated range of phenomena. Whewell called this process a "consilience of inductions." (It might less misleadingly have been called a "fecundity of deductions.") One of Whewell's deepest convictions, which he used a version of Kant's philosophy to defend, was that the vital forces governing living things will never be reduced to the forces that govern mere matter. Accordingly, Whewell, reburnished Aristotle's maxim that "nature does nothing in vain," arguing not only that there is a purpose for nearly everything organic but that finding and describing those purposes and functions is in itself explanatory. Teleological explanation, Whewell says, is "so far from barren that, in the hands of Cuvier and others, it has enabled us to become intimately acquainted with vast departments of zoology to which we have no other mode of access" (Whewell 1837 [1897, 2:489]).

At the other end of the ideological spectrum, Philosophical Radicals eventually found their champion in John Stuart Mill. For Mill, all reasoning, even apparently deductive reasoning, is ultimately inductive. By this Mill meant to say more than that inquiry moves, as empiricists had always said, from particulars to generals. He meant that reason uses generalizations as "inference tickets" that guide us as we move from particulars to particulars. "Universals," for the Utilitarian Mill, are not objects in or facts about the world but instruments that help us move around in it. What pushes science along is the observation of constantly conjoined phenomena, which can be used to extrapolate, but only revisably, predictions of future events that might allow of human intervention. Mill was so insistent on this that he took even mathematical inferences to be inductive: "Two plus two equals four" is a sound inference ticket because it has always turned out that way in the past. The very principle Mill employs to make this argument—that we can expect the world tomorrow to be pretty much what it was today, and therefore can make predictions—was for Mill an inductive truth rather than the presupposed philosophical axiom that Whewell reasonably took it to be. For these reasons and others, Mill opposed Whewell's notion that a mere consilience of inductions can establish a hypothesis just because all sorts of nice consequences flow from it. The fact that incompatible hypotheses can equally well meet this criterion seemed to Mill to be an excuse for hanging on to traditional beliefs and practices, thereby slowing the rate at which new truths might be learned. Mill was particularly suspicious of citing purposes as explanations, in both biology and the infant human sciences, as Aristotle's and Whewell's teleological maxim required. The bias toward intentional and purposive, so deeply embedded in what are now called "folk biology" and "folk psychology," seemed to Mill to hold back the desirable extension of experimental and statistical methods, which had proved so fecund in the natural sciences, to the human sphere.

Among Dissenting Whigs, such as Lyell and Darwin, it was not Whewell or Mill, however, who served as semiofficial philosopher of science. It was the astronomer John Herschel, a man of their own sort, whose *A Preliminary Discourse on the Study of Natural Philosophy* (1830) became their canonical treatise on scientific method. Herschel, whom Darwin made a point of meeting in Cape Town, South Africa, where the astronomer was charting the southern constellations, greatly lauded Newton's discovery that the world is governed by simple, uniform, constant, inviolable, and generally mechanical laws, laid down by God. Herschel's view of scientific method thus stood midway between Whewell's and Mill's. He did not think that God reveals his mind to those who think they see purposes in every nook and cranny of the world. At the same time, he did not approve of abandoning the search for simple uniform laws by contenting oneself with exception-ridden statistical generalizations that have at most a pragmatic status.

The search for uniform universal laws of nature begins, according to Herschel, with the recognition by direct perception of true (by which Herschel means "real" or "actual") causes (*verae causae*). "Whenever any phenomenon presents itself for explanation," he wrote in the *Preliminary Discourse*, "we naturally seek . . . to refer it to some one of those real causes which experience has shown to exist and to be efficacious in producing similar results" (Herschel 1830, sec. 141). Herschel argues that Newton, like Bacon, took the right approach because his natural philosophy begins from the observation of a *vera causa*: "We see a stone whirled round in a sling, describing a circular orbit round the hand, keeping the string stretched and flying away the moment it breaks. We never hesitate to regard it as retained in its orbit by the tension of the string, that is, by a *force* directed to the center. For we feel that we do really exert such a force. We have here *the direct perception* of a cause" (Herschel 1830, sec. 142).

Experiment and observation show how wide the writ of a true cause is. A truly generalized cause is expressed by simple explanatory laws, like Newton's laws of motion. The deductive applicability and explanatory fecundity of these laws depends on their connection to true causes. Thus, Herschel implied that Mill and Whewell each had hold of only half of what was required. A theory missing either a known mechanism or an extension to a wider range of cases testified to by a "consilience of inductions" should not be accepted.

Darwin's Herschelian philosophy of science meant that his secret option in 1837 for transmutation, and even more radically for "common descent" of all species from an original ancestor, committed him to search for a true or actual process that, by suitable generalization, could produce a natural law governing "the origin of species." Preferably that law should work analogously to Newton's laws, for organic change must fit into a world fully governed by Newtonian physical, chemical, and geological forces. That was Darwin's research program. It posed a number of difficulties—some intellectual, many more political, ideological, and personal.

Although the Enlightened Whig view of the world that he shared with Herschel and Lyell was the abiding center of Darwin's thought, he was in other ways a virtual epitome of the diverse currents in British political and scientific thinking we have been describing, having encountered all of them close up at one time or another. Perhaps the tensions among these styles of thought were reflected in the lifelong stomach illness that first appeared when he started to think seriously, and dangerously, about transmutation. Darwin's career as a gentleman naturalist had been promoted by Cambridge clerics like Sedgwick, under whose tutelege he had practically memorized Paley's *Natural Theology.* On his trips to Cambridge after his return to England and in the high-toned scientific societies to which he belonged, he regularly rubbed shoulders with Peelites like Owen and Whewell. Whewell in fact served as president of the Geological Society while Darwin, at Whewell's request, was its secretary. These men would instantly have dismissed a known advocate of transmutation, and even more a transmutationist who believed in common descent, from their lofty company. Even Lyell, his patron and friend, who was an almost perfect mirror of his own Dissenting Whig inheritance, denied transmutation with horror and did not know that Darwin was a closet evolutionist. The constant changes in the geological world that Lyell described in *Principles of Geology* would certainly put pressure on organisms to change, and some, unable to do so, would go extinct. However, Lyell, in apparent disregard of Herschel's canons, regressed to creationism to fill the gaps. The evolutionary alternative seemed too horrible to contemplate. It would degrade human reason to the level of an ape, and, as it had in revolutionary France, would tend to empower irrational democrats. Darwin felt increasingly alone with his thoughts.

Nor was the idea of transmutation yet much bandied about even by Malthusian liberals, although, having already swallowed the paradoxical idea that the threat of starvation was the mother of improvement, they loved to fancy themselves speaking the unspeakable. Darwin had become acquainted with the Philosophical Radicals' leading lights when he dined, as he often did, at his medical brother Erasmus's table in London. It seems that the talkative Harriet Martineau, a leading writer for the radicals' journal, the *Westminster Review,* had become his brother's part-time girlfriend. Their circle included George Eliot (Mary Ann Evans); her husband, George Henry Lewes; Thomas Carlyle; and later Herbert Spencer, and John Stuart Mill and his talented wife, Harriet Taylor. Soon the Malthusian ideas that served as leitmotivs for this group would give Darwin his evolutionary clue. But even after they became evolutionists themselves, their conception of it would be that of Spencer, a member of their own circle, more than Darwin's. At present, though, the idea of evolution was socialist property and was discussed positively only in moods when the group sympathized with the working class. Darwin found no support there, and would probably not have wanted it in any case. It would have been ill regarded by his classier connections.

The plain fact is that when Darwin opted for transmutation and common descent in 1837, the only set of theoretical ideas he had to go on were the heretical Lamarckian and Geoffroyian ones he had learned long before from Grant. Darwin had not forgotten them. A deeply retentive thinker, he forgot or abandoned very little, saving ideas and facts for further thought or good use (Gruber 1985; Hodge 1985; Desmond and Moore 1991). Accordingly, Darwin's transmutation notebooks in the period leading up to his formulation of his theory of natural selection are full of talk about self-moving monads and species budding off from older species whose life cycle is at an end. Since these Grant-inspired ideas were deeply associated with French materialism and with radical politics, they provoked, both when Darwin first encountered them as a young man in Edinburgh and later in London, spasms of guilt. "Oh you materialist," he wrote to himself while reflecting that "mind might be an effect of organization" and that "love of the deity" might be "nothing but" a consequence of slight morphological differences between humans and orangutans. Darwin might easily have talked the issues over with Grant himself and found support there. Teaching by then at the University of London, his old tutor lived only a few blocks away. Significantly Darwin never contacted him, and when Grant offered to describe some of his specimens, Darwin politely gave him the brush. The problem was clear: A man whose family was moving up the social scale and who was himself quickly ascending the scientific hierarchy secretly held a theory whose local habitation was at the other end of that pecking order and was sure to cause scandal for Darwin and those he loved if it were ever known.

The depth of this agony cannot readily be grasped by people living in the secular societies of the late twentieth century. One must learn to appreciate what was connoted by materialism of Lamarck's stripe and how it differed from even the most reductionist forms of British thought. The conception of matter long dominant in Britain was one in which basic stuff has no inherent or spontaneous ability to create itself, to act, to develop, to adapt, to self-organize. This idea of matter was fairly easy to reconcile with theism. Indeed, if matter is inert, something like divine will seems absolutely necessary if order was to be brought out of chaos at all. "When nature underneath a heap of jarring atoms lay," begins a poem by Dryden. It was precisely because Newton construed matter as passive stuff, which fairly cried out for a God to give it an initial shove if the natural laws he had devised were to work like a clock thereafter, that it was so easy for the Royal Society to turn the great physicist into an icon of respectable science and moderate opinions. (Even more charming was that Newton also held that the clock occasionally had to be rewound.) Accordingly, when the French physicist Pierre-Simon de Laplace informed Napoleon that he had "no need for the hypothesis of God" to explain the universe, he was alluding to an aspect of French materialism that would have shocked educated Englishmen. Viewed from within the Newtonian framework, Lamarck's stress on the boot-strapping activities of organisms in bringing about their own complexity sounded as though they were to "will" their own adaptations. It was fairly easy to cover one's tracks by ridiculing such an idea. Darwin did it himself. "Heaven forfend me from Lamarck's nonsense about a 'tendency to progression,' 'adaptations from slow willing of animals, etc.,'" he wrote (Darwin to Hooker, January 11, 1844, in Darwin 1887, 1:384; CCD 3:2). Yet this does not mean that Darwin did not understand the general idea, or how deeply intertwined it was with any extant sort of evolutionary theory, or that he did not feel in his bones how socially unacceptable it was. The Britain in which Darwin grew up, in which threats of revolution and consequent cold and hot wars with France had produced an ideological big chill, had made advocates of French evolutionism seem as dangerous to sound morality as Russian bolsheviks appeared to early-twentieth-century Americans.

In spite of his careful protestations to the contrary, however, Darwin and the men in his family had never been nearly as far from this tradition as they led others to believe. His own grandfather, the freethinking and sexually freewheeling Erasmus Darwin, held radical views about matter and sensation in the insouciant, neopagan era before the French Revolution. (The worldly wise French minister Talleyrand famously remarked that no one who had not lived before the revolution could ever know how sweet life was then.) The elder Darwin's *Zoonomia* was an updated version of Lucretius's *De Rerum Natura*, a neo-Epicurean hymn to sensationalism, sexual pleasure, and relief from the terrors of religious orthodoxy, which projected a view of evolution close to Lamarck's, to which Charles Darwin was exposed at any early age. The antirevolutionary milieu in which Darwin lived precluded any overt identification with this inheritance and led him to circumscribe his life with sober Victorian rectitude. Nonetheless, the bacillus was there and in fact formed part of the creative fire of Darwin's thinking. Clearly, however, any theory of evolution that he would publicly announce could not be yet another warmed-over version of Lamarck and Geoffroy. If he was to do it at all, evolution would have to be fitted to the canons of solid British science, and in particular to Herschel's norms. It would have to be evolution à la Newton.

Such a theory would not be the same old evolution in different clothes in part because Darwin was not, or at least was not for long, a recapitulationist in the strong sense.[1] "It is not true," he wrote in an 1842 sketch of his still secret theory, "that one passes through the forms of a lower group" (DAR 6, 42, quoted by Richards 1992, 116). Humans were not fish before they were people. Darwin, it appears, shared large parts of von Baer's and Owen's picture of ontogenetic differentiation. Only undeveloped stages of growth are analogous to adult stages of lower kinds. Thus the full determinacy that enables you to say "trout" or "human," or even "fish" or "mammal," occurs in kinds that are related in the way cousins are rather than as ancestors to descendants. (At the same time, Darwin was willing to say that not every contemporary group had undergone the same degree of differentiation, adding to the sentence cited above, "No doubt fish [are] more nearly related to [the] fetal state.") What Darwin wanted, in briefest compass, was a picture of phylogeny that paralleled von Baer's ontogeny. He wanted it, moreover, to track real common descent from mostly extinct ancestors, whose shared traits were conserved, rather than recapitulated, as species and even body plans, like the parts of individuals, grew ever more differentiated from one another.

Moreover, Darwin wanted a phylogenetic history that was mirrored in the science of systematics, or classification, in a way that differs from what might be expected from strong recapitulationist principles. Given his conception of development as progressive differentiation, evolutionary branching takes place in a tree that does not point straight up toward humans but in a thick bush whose buds represent nascent species ("races"), whose twigs represent nascent genera, whose thicker branches represent the body plans of larger phyla, and whose trunk represents the oldest strata of conserved, inherited traits. Modern humans are to be found at the end of one of those branches. The ground around the tree, it should be noted, will be littered with broken blossoms, fallen leaves, dead twigs, and rotting branches. These are the fruits of extinction.

What Darwin did not share with von Baer and Owen, accordingly, was the rigid typological essentialism, metaphysical holism, and vitalism that backed up their metaphysical opposition to transformism and common descent. In order to picture evolutionary branching in the way he envisioned, Darwin would have to follow a two-track research program. He would, in the first instance, have to find a theory of inheritance that would allow newly acquired information to pass into the developmental and reproductive cycle in ways that were not stopped dead in their tracks by limited "vital force" or by rigid body plans. Darwin was, therefore, as Hodge has put it, a "life-long generation theorist," whose alleged indifference to these issues has been exaggerated by twentieth-century Darwinians too eager to believe that Darwin left room for Mendelism (Hodge 1985). If being some sort of materialist was the price for this, Darwin was willing to pay it. Darwin speculated and experimented continuously about the little "granules" or "gemmules" that in his view carried reproductive information (Sloan 1976). Still, if Darwin's theory was to pass muster with the likes of Herschel and Lyell, who demanded theories that looked as much as possible like Newton's, in which matter was to be more acted upon than acting, Darwin would have to find an external force, like Newtonian gravity, rather than an internal drive, that impinged on the developmental and reproductive cycle with sufficient force to drive and shape evolutionary diversity. That force must, moreover, be a *vera causa*, a real cause that can be seen at work this very day.

Here we reach the main contention of this chapter, and indeed of this entire section of the book: In calling the external force that drives evolution "natural selection," Darwin would be extending the "artificial selection" of variant organisms by plant and animal breeders, thereby treating the general mechanism they share in common as a *vera causa*. In this way, Darwin proposed to meet Herschel's methodological criteria. But he was doing something else as well. In portraying within- and between-species change as occurring through selection pressure in an overpopulated, competive, force-filled Malthusian world, Darwin was *applying the highly prized Newtonian models that Lyell had already applied to geology to the history of life, bringing evolutionary theory, for the first time, into the conceptual orbit of respectable British thinking.* This was done by portraying the world of nature as very like the world as political economists saw it. What is perhaps most significant about Darwin's use of political economists is not that he inscribed into nature itself the capitalist ideology of his class and circle but that he recognized something suitably Newtonian, and in his context rhetorically powerful, about their dynamical models (Schweber 1977).

Given the importance of this incipient project, it is not odd that Darwin himself would long afterward point to what happened on September 28, 1838, as of the highest significance for the development of his thought. He recalls it this way in his *Autobiography:*

Fifteen months after I had begun my systematic inquiry, I happened to read for amusement *Malthus on Population,* and being well prepared to appreciate the struggle for existence which everywhere goes on, from long continued observation of the habits of animals and plants, it at once struck me that under these circumstances favorable variations would tend to be preserved, and unfavorable ones to be destroyed. The result of this would be the formation of a new species. Here I had at last got a theory by which to work. (Darwin 1958, 120)

We should not imagine that "for amusement" meant anything other than what we call "study" or that Malthusian ideas were new to Darwin. The notion that a clash between population and resources might be an agent of progressive change, was, as we have seen, commonplace among people with whom Darwin associated. In letters to his sister from the *Beagle,* Darwin mentions that the officers had been passing around, and eagerly discussing, one of Harriet Martineau's treatises on the curative powers of Malthusian economic and social principles.[2] What was new in

September 1838 was that Darwin now began to use this idea, and political economy more generally, to devise a theory of evolution free from the ideological burdens borne by Geoffroyian Lamarckism because it shifted the causal accent from inner drives to external forces.

The basic idea is simplicity itself. If it is an inherent, lawlike tendency for organisms to reproduce at rates higher than the means of supporting them, then the resulting competition will be ubiquitous not only between species but among individuals within species. Under these conditions, variant traits that enable their possessors to command more resources, and so to live and reproduce more effectively than their competitors, will, if they are heritable, gradually mold lineages whose adaptedness to their niches is a result of constant reequilibration between organisms and environments. Increasing differences between lineages—races, species, and higher taxa—reflect the fact that organisms under this kind of pressure will tend to explore and exploit new and different resource bases, making them different as well.[3]

Darwin cleaved to this basic model continuously from 1838. He wrote sketches of his hypothesis of transmutation by means of natural selection in 1839, 1842, and 1844; however, he never published any of these, and mentioned his idea to his friends only a few times. The reasons for "Darwin's delay" in publishing his theory have occasioned much speculation. Clearly some of the problem was theoretical and even empirical. It was not until the mid-1850s, for example, that Darwin figured out how to use natural selection to account for diversity in any really coherent way or had gathered sufficient evidence to support his new analysis. It is noteworthy that this breakthrough, as we will see in a later chapter, came by applying a little more economics to the problem. It is also clear, however, that there were psychological aspects of the case. Darwin himself acknowledges in the *Autobiography* that he was affected by the pain that publication of his ideas would cause his pious wife and feared the ill repute that would befall him and, more important, his family (Darwin 1958). It is commonplace to relate Darwin's ill health, whatever its physical cause may have been, to the stress he felt about this issue (Colp 1977; Bowler 1990, 73–75). This psychological stress, however, becomes more real the more it is described in sociological terms (Desmond and Moore 1991). No matter how Newtonian his mechanism, or how Herschelian his methods, Darwin always realized that there was much higher risk that his theory would be assimilated to the heterodox Geoffroyian Lamarckism from which he was trying to break away than that it would be seen as elevating biology to the status of a discipline conforming to approved Newtonian canons.

Darwin might have gone public in 1844. Enthused about his latest draft, he confided his views to his new friend Joseph Hooker in January of that year. "I am almost convinced," he wrote, "(quite contrary to the opinion I started with) that species are not (it is like confessing a murder) immutable" (Darwin to Hooker, January 11, 1844; in Darwin 1887, 1:384; CCD 3:2; cf. Colp 1985, 1987). Hooker replied noncommittally. Later in the year, however, an Edinburgh man named Robert Chambers anonymously published a book entitled *Vestiges of the Natural History of Creation* (1844), in which he portrayed the Geoffroyian recapitulationist transformism that had taken hold in the cut-rate medical schools of his home town as God's way of using secondary causes to bring about his ends (Hodge 1972). This somewhat incoherent attempt to patch together Tory natural theology, Newtonian lawfulness, Lamarckian inheritance, and Geoffroyian progress met with vilification on all sides. Sedgwick in particular lambasted it in a vicious review. The point was not lost on Darwin. He did not kid himself that his theory of adaptation would fare much better, even if it used a different mechanism and avoided Chambers's manifold mistakes.

In April 1856, however, Darwin finally confided his views fully to a few of his peers whom he had invited for a weekend at Down House. They were to act as a sounding board and, by their reactions, to help Darwin test the waters again. Hooker, who always played Dr. Watson to Darwin's Sherlock Holmes, was there. He was the son of the curator of Kew Botanical Gardens and, like his father, whose successor he became, a botanist. He too had been a ship's naturalist, serving in that role on the voyage of the H.M.S. *Erebus* to Antarctica. It was a bond with Darwin. Thomas Henry Huxley, a struggling anatomy lecturer who would soon become Darwin's most vocal champion, was also invited. From the start, Huxley was more interested in defending Darwin because he was an evolutionist, and in defending evolutionists because they were anticlerical scientistic humanists, than in defending Darwin's theory of natural selection. Personally, he thought that species must emerge suddenly, by leaps or "saltation" (from Latin *saltus*, "leap"), rather than gradually, as Darwin was proposing—and promptly told him so. Thomas Vernon Wollaston, an entomologist who had written an interesting, and Darwin thought potentially supportive, article on the great range of variation in insect species, had also been invited that weekend. But Darwin had misjudged him. Like Lyell, he thought that no matter how much variation a species could tolerate, there were essentialist limits to it. He was shocked by Darwin's hypothesis and became its enemy.

Soon after meeting with these men, Darwin consulted with Lyell, who, as Darwin's biographers put it, "went away staggered. As always he saw the starkest implications. . . . Was man only a sort of brute? Was he 'improved out of' some Old World ape?" (Desmond and Moore 1991, 438). Lyell's problem, it seems, was the flip side of Huxley's. He did not worry, as Huxley did, that Darwin's mechanism of gradual adaptive change might fail to produce new species. On the contrary, he thought that it might do just that, and therefore that the Newtonian uniformitarianism that permeated his *Principles of Geology* would be used to ground a conclusion that would refute one of the main contentions of his own book and undermine his entire view of the world.

The occasion for Darwin's risky confidences in the spring of 1856 was the appearance in an agricultural journal the previous September of a paper by Alfred Russel Wallace to which Lyell had alerted Darwin. Wallace was a "muddy boots biologist" who collected species in far-off places not for the glory of England but to make a living. The paper in question had been written when its author was alternately burning up and shivering with malaria in a fascinating area of the Dutch East Indies where the Australian biota comes up against the furthest extension of the biota of Southeast Asia, the biogeographic boundary being marked by a narrow strait. Entitled "On the Law Which Has Regulated the Introduction of New Species," Wallace's paper proposed to elevate into laws of nature the same observations Darwin had made on the *Beagle*. It is a general rule governing the distribution of organic types, Wallace wrote, that new species "come into existence coincident both in space and time with a pre-existing closely allied species." With respect to the distribution of species in space this implied, he said, that "no species or genus occurs in two very distant localities without being also found in intermediate places"; and with respect to temporal distribution that "species of one genus, or genera of one family, occurring in the same geological time, are more closely allied than those separated in time." It also meant that "no group or species has come into existence twice" (Wallace 1855, 184–196). Wallace was alert to the implications of this "biogeographic law" for systematics, for his proposed law suggested to him that the larger taxa (classification categories) simply reflect the persistence over time of similarities among lineages that are accumulating further differences at lower taxonomic levels, yielding, at least in principle, a "true or natural system of classification" based on descent. Wallace's paper came close to asserting that new species might have been descended from those most closely related to them, instead of being separate types, instances of which God might insert into the world wherever and whenever he wanted, but Wallace refrained from saying so directly.

Lyell reacted to Wallace's paper by urging Darwin to get into print as soon as possible. Suppose Wallace published yet another paper in which he proposed an evolutionary explanation for this "true or natural system of classification" like Darwin's? Lyell apparently thought that the climate of opinion, especially among younger biologists like Wallace, was shifting quickly toward transmutation. It would be better, he probably reasoned, for someone respectable and responsible like Darwin rather than someone more radical to seize the day. His shock at finding out what Darwin had told Hooker, Huxley, Wollaston, and now himself probably reflected his sudden recognition that his protégé had long been harboring more heterodox notions than he had assumed. Nonetheless, Lyell continued to urge Darwin to go ahead, for, as he wrote to Hooker, "Whether Darwin persuades you and me to renounce our faith in species, I foresee that many will go over to the infinite modifiability doctrine" (Lyell 1881, 2:213-14). Better this in any case than Chambers's warmed-over Lamarck, Geoffroy, and Grant.

The issue was still moot, however, because Darwin again declined to publish—in part because he underestimated the possible effects of what Wallace was saying. On first reading Wallace's paper, he had noted to himself: "Laws of geographical distribution. Nothing very new." These were, after all, observations with which Darwin himself had long been familiar, which had convinced him of the fact of evolution as early as 1837. They were not new to him at all, even if they were to Lyell, who was so stunned by Wallace's paper that he opened his own set of notebooks on the species question, which were as tortured, and as secret, as Darwin's had been twenty years before. Darwin did, however, begin to work on earnest on a big manuscript now called *Natural Selection*. He also entered into correspondence with Wallace. Thus, on June 18, 1858, when Darwin received a new paper from Wallace, together with a letter from him soliciting help in getting it published, it was far from a bolt out of the blue. Still, the contents of the paper confirmed Lyell's earlier fears, for "On the Tendency of Varieties to Depart Indefinitely from the Original Type" contained a hypothesis to account causally for the biogeographic and systematic claims Wallace had made in his earlier paper that seemed to Darwin substantially identical to the theory of natural selection he had carefully secreted, in every sense of the term, for almost two decades.

Wallace articulated his hypothesis by contrasting it with a not very good, but typically English, interpretation of the theory of transmutation that had been proposed by Lamarck:

The powerful retractile talons of the falcon and cat tribes have not been produced or increased by the volition of these animals; but among the different varieties which occurred in the earlier and less highly organized forms of these groups, those always survived the longest which had the greatest facilities for seizing their prey. Neither did the giraffe acquire its long neck by desiring to reach the foliage of the more lofty shrubs, and constantly stretching its neck for that purpose, but because any varieties which occurred among its antitypes with a longer neck than usual at once secured a fresh range of pasture over the same ground as their shorter necked companions, and on the first scarcity of food were thereby enabled to outlive them. (Wallace 1858, 61)

What Wallace proposed was, in a nutshell, a (if not the) theory of evolution by natural selection. As it turned out there were some important differences between Wallace's and Darwin's theories that went unnoticed at the time. While both men thought that the exigencies of the environment mold the traits of lineages, Darwin stressed competition among individuals *within* a population, treating other members of the same species as problematic features of the individual organism's environment. Wallace, by contrast, stressed competition *between* populations of the same species struggling in relation to the same inorganic environment. Wallace thus tended to think that the environment eliminates only those "varieties" totally unfit to survive in it, whereas Darwin thought that individual competition favors only the very fittest and sacrifices all the others. Wallace, accordingly, regarded natural selection as a pruning mechanism that might leave standing a fair amount of natural growth, whereas Darwin viewed it as a more demanding and creative force. Wallace's group-oriented socialist sympathies showed. So did Darwin's more systematic use of political economy to articulate natural selection. Nonetheless, it was precisely because they shared the same Newtonian explanatory framework that the two men could, and later did, have interesting conversations and precise disagreements about topics now debated under such headings as cooperation versus competition, group versus individual selection, and negative versus positive selection (Kottler 1985).

When Darwin first read Wallace's paper, however, he saw no difference between their views. "If Wallace had my MS sketch written out in 1842," he wrote to Hooker, "he could not have made a better short abstract" (Darwin to Hooker, June 1858, DAR 114, 238). It is generally believed that Darwin's initial failure to recognize his differences with Wallace resulted from a hasty and emotion-driven reading of his paper. If he had been too insouciant about the first essay, the theory goes, he was in a state of panic about the second. Thinking that his thunder had been stolen, he projected his own theory onto Wallace's (Desmond and Moore 1991, 469). It is true that Darwin was under emotional strain at the time; another child died soon after he received Wallace's paper. But even if Darwin had recognized where he and Wallace differed, he might still have felt that Wallace had stolen his thunder in a deeper way. Wallace's own theory of natural selection might differ from Darwin's, but they shared a virtually identical research program. Darwin's assimilation of Lyell had led him to postulate that external, environmental pressures and forces, rather than Francophile inner drives, play the causal role in speciation. The rhetorical beauty of Darwin's theory was that he was increasingly able to operate within the prescribed framework of Newtonian external causation, where real forces exert pressure on real populations, rather than imagining an internal, developmental dynamic that is awakened, steered, or thwarted by external stimuli. Darwin had found in the lawlike mechanisms of economics the connecting thread. Yet this was precisely the framework within which Wallace was also operating. Referring to the culling of adapted variations driven by Malthusian population pressure, Wallace wrote, "The system works like the centrifugal governor of the steam engine, which checks and corrects any irregularities almost before they become evident" (Wallace 1876, 42).

Darwin and Wallace shared the same framework in part because they had had similar adventures. Both had taken long biological expeditions to the outer reaches of the world. (Wallace had carried to the Malay Peninsula not only Humboldt but Darwin's own *Voyage of the Beagle*.) Both had acquired in the course of their travels a sense of the vastness of time and space, and of the wild variety of living things that time and space contained, which contrasted dramatically with the coziness and insularity of England and the English. Both were prepared to live with the sheer overwhelming immensity of things. Both, moreover, were biogeographers, whose studies focused on the distribution of organisms through the continuum of space and time. Like Darwin, Wallace too conducted his biogeographic inquiries against the background of Lyell's uniformitarian geology, which asserted, as Wallace put it in his first paper, that "during an immense, but unknown period, the surface of the earth has undergone successive changes; that these changes have taken place, not once merely, but perhaps hundreds, perhaps thousands of times; and that all the operations have been more or less continuous but unequal in their progress" (Wallace 1855, 184).

Wallace and Darwin both saw that these conditions of constant, grinding geological and environmental change would put continuous pressure on organisms to keep up with changes in the conditions of their own life. Extinction was telling evidence of failure in this struggle. "After a certain interval," Wallace wrote, "not a single species exists which had lived at the commencement of the period" (Wallace 1855, 185). Where, then, did new species come from to fill up the vacant ecological niches? Lyell had no good answer. The theory of evolution by natural selection, whether under Wallace's or Darwin's dispensations, did. Most striking of all is that the idea of natural selection in Wallace's case too had been precipitated by Malthus, a new edition of whose *Essay on Population* he read while he was in Malay.[4]

Personally, and in virtue of his and his family's place in English society, Darwin was a gentleman. He could not very well just burn Wallace's paper and tell him it had never arrived. But he was not about to be left behind, and so at last he decided to take the plunge. With help from Lyell, he arranged for Wallace's paper to be read at a meeting of the Linnean Society in London, together with a dated copy of a letter he had sent to Asa Gray at Harvard in September 1857 and some pages from his 1844 manuscript. Darwin was in effect politely claiming priority. (Darwin's letter to Gray, outlining his theory, had been provoked by Gray's curiosity about why he had been asked to provide certain data. Darwin swore him to silence.) Lyell, Hooker, and Darwin chose the Linnean Society because it was more civil than the Zoological Society. It was so civil, in fact, that both Wallace's paper and Darwin's reports met with shrugs, or at best muttering. This meeting occurred July 1, 1858. After that, Darwin, his cover blown, went home to Downe and set about composing what to us is a big book but was to him a mere summary, taken from the "the big species book" on which he had been working. In it he set forth his reasons for believing that natural selection is the true cause of departure from type and of descent from a common ancestor through a complex pattern of phylogenetic branching. The book was called *On the Origin of Species by Means of Natural Selection, or the Preservation of Favoured Races in the Struggle for Life*. The first edition sold out the day it was published, November 24, 1859, and a second printing did so a month later.

On the Origin of Species is a brief, after the manner of Lyell's lawyerly *Principles of Geology*, in favor of a case. It is, as Darwin remarked near its end, "one long argument" (Darwin 1859, 459). It is marked, accordingly, by admirable attention to the problem of evidence and explanatory adequacy and by rhetoric that courts the reader by flattering his or her intelligence. This literally judicious way of making his case renders Darwin's *On the Origin of Species* one of the least dogmatic, and one of the most attractive and inviting, of all the major works of Western thought. Francis Darwin, Darwin's son and literary executor, rightly speaks of the "courteous and conciliatory tone" that Darwin takes toward his readers. "The reader is never scorned for any amount of doubt which he may be imagined to feel, and his scepticism is treated with patient respect" (Darwin 1887, 1:132).

The book begins by reviewing the ancient art of the breeder, by which over time the characteristics of lineages are shaped by artificially selecting mating pairs from among the slightly differing offspring that nature seldom fails to provide. Pigeons provide the primary case study. This is the first step in the argument: the existence and profusion of variation within species and the effectiveness with which generations of breeders can shape it. That is Darwin's candidate for a Herschelian *vera causa*, like Newton's sling. It is a cause whose reality can literally be perceived, as Herschel demands. Darwin then goes on to question the categorical distinction between such varieties and true species, a distinction Lyell had sought to shore up in his refutation of Lamarck in *Principles of Geology* and that Darwin now calls into question once more, narrowing the difference between what goes on in nature and what pigeon fanciers can achieve. In chapter 3, Darwin then asserts that there is a severe struggle for existence going on among organisms due to Malthusian population pressure:

A struggle for existence inevitably follows from the high rate at which all organic beings tend to increase. Every being, which during its natural lifetime produces several eggs or seeds, must suffer destruction during some period of its life, and during some season or occasional year, otherwise, on the principle of geometrical increase, its numbers would quickly become so inordinately great that no country could support the product. Hence, as more individuals are produced than can possibly survive, there must in every case be a struggle for existence, either one individual with another of the same species, or with individuals of distinct species, or with the physical conditions of life. *It is the doctrine of*

Malthus applied with manifold force to the whole animal and vegetable kingdoms. (Darwin 1859, 63, italics added)

The argument depends on a counterfactual analysis, that is, on saying what would have happened if something had not intervened (like later versions of Malthus's law). There must be a severe competition for limited resources, including competition among individuals of the same species. Otherwise we would see many more organisms of the same kinds than we do, in accord with Malthus's calculation of the natural, geometrical rate of increase of offspring when it is unconstrained by scarce resources. Hence

in looking at nature, it is most necessary . . . never to forget that every single organic being around us may be said to be striving to the utmost to increase in numbers; that each lives by a struggle at some period of its life; that heavy destruction inevitably falls either on the young or old, during each generation or at recurrent intervals. Lighten any check, mitigate the destruction ever so little, and the number of the species will almost instantaneously increase to any amount. The face of Nature may be compared to a yielding surface, with ten thousand sharp wedges packed close together and driven inwards by incessant blows, sometimes one wedge being struck, and then another with greater force. (Darwin 1859, 66–67)

The startling image of the wedge is a bit obscure here. It is made clearer by the fact that it appears early in the course of Darwin's copious theorizing in his early notebooks, where it refers to nature's way of using competition under the hammer blows of scarcity as a way of opening up new resource niches, and in particular to the activities of adventurous, pioneering human populations, driven by want, as they invade and conquer older, more settled people. This notebook entry had in fact been provoked by Darwin's famous reading of Malthus, who recognized that competition for scarce resources drives people into every possible niche and explained European colonizing in these terms. The wedge image, especially in this human context, is rather gruesome. It does at least as much work in the argument as the picking or selection in which the animal breeder engages. For whereas "selection" more vividly calls attention to the fact that what is selected varies than the wedge image, the wedge image picks up the idea that powerful external forces are constantly bearing down on organisms. The Malthusian element is as prominent in *On the Origin of Species* as it is in the notebooks because the scarcity principle, articulated within the quasi-Newtonian framework of political economy that Malthus had inherited from Adam Smith, provides Darwin with the unrelenting and pervasive pressure or force he needs if organisms and their traits are to be literally molded to fit environments. Just how strong the force of population pressure and the consequent "struggle for existence" is is vividly suggested by the following passage from *On the Origin of Species*:

We behold the face of nature bright with gladness, we often see super-abundance of food; we do not see, or we forget, that the birds which are idly singing round us mostly live on insects or seeds, and are thus constantly destroying life. . . .

What war between insect and insect, and between insect, snails and other animals with birds and beasts of prey—all striving to increase, and all feeding on each other or on the trees or their seeds and seedlings, or on the other plants which first clothed the ground and thus checked the growth of the trees. (Darwin 1859, 62, 75)

In the crucial fourth chapter in the book, "Natural Selection," from which this passage comes, Darwin starts to draw conclusions from the principles which he has already established: variation, Malthusian repro-duction, and a consequent "struggle for existence." At the end of the chapter, Darwin himself offers the following superb summary of the argument (we have broken up the text on the page, highlighted the logical structure of the argument, and eliminated some passages in order to make the argument clearer to those of us who, living in a lesser age, no longer have sufficient wit to appreciate the glories of Darwin's fine Victorian prose style):

If
> during the long course of ages and under varying conditions of life organic beings vary at all in the several parts of their organisation [*variation*],
>> And I think this cannot be disputed;

If there be,
> owing to the high geometrical powers of increase of each species [*Malthusian reproduction*]
> at some age, season or year a severe struggle for life [*struggle for existence driven by scarce resources*],
>> And this certainly cannot be disputed;

Then,
> considering the infinite complexity of the relations of all organic beings to each other and to their conditions of existence, causing an infinite diversity of structure, constitution, and habits to be advan-tageous to them [*variation correlated to potential utility*]

I think it would be a most extraordinary fact if no variation ever had occurred useful to each being's own welfare, in the same way as so many variations have occurred useful to man [*probability that some variation is in fact adaptively useful by analogy to breeding or artifical selection*].

But,
> if variations useful to any organic being do occur,

Then
> assuredly individuals thus characterised will have the best chance of being preserved in the struggle for life [*differential survival*];

And
> from the strong principle of inheritance [*principle of inheritance*]
> they will tend to produce offspring similarly characterized. This principle of preservation I have called *Natural Selection*. (Darwin 1859, 126–277)

The first line of argument is taken as having been established in earlier chapters: Variation plus Malthusian reproduction leads to a struggle for existence. The observed number of individuals is relatively stable and far less than the number projected by the intrinsic rate of population growth when it is left to operate on its own, so something must be limiting it. But now the struggle for existence is said to lead to the superior adaptedness (fitness) of some variants to a certain set of ecological conditions. These are the subset that will, on the whole, survive. The argument is hypothetical and probabalistic: Differences in adaptedness (fitness) will occur only if there is a correlation between variation and what is needed for survival (under conditions of struggle) in a particular niche. Nothing guarantees that. But, Darwin argues, given the considerable amount of variation in nature, the large stretches of time available, and the wide diversity of possible niches, as well as the instructive analogy with human breeding practices, this match seems probable. This argument yields, however, only a necessary, and not a sufficient, condition for natural selection, as Darwin uses the term, for it is also required that differential survival in the struggle for life, that is, greater ability to live, to thrive and to reproduce, be combined with ability to transmit one's advantageous traits, so that adaptations can build up over time.

Darwin goes on to anticipate a few implications of natural selection protracted over many generations. Extinction, he remarks, is easy to explain on this theory. When appropriate variation is not forthcoming under changed ecological conditions, that is precisely what one would expect. Nothing guarantees, moreover, that it will be forthcoming. The geological world, as Lyell had shown, is constantly changing; useful variation is not producible on demand; and (contra Lamarck) adaptations cannot be hatched up whenever they are needed. Darwin is more impressed, however, by what nature creates out of these chance conjunctions variation and utility, and out of the apparent waste and carnage of natural selection (Gould 1990), for the good news is that natural selection, under Malthusian conditions, will lead to the exploitation of every possible niche, and hence to maximally diverse and well-adapted forms of life:

Natural selection, also, leads to a divergence of character; for more living beings can be supported on the same area. The more they diverge in structure, habits, and constitution, . . . the more diversified these descendants become, the better will be their chance of succeeding in the battle for life. Thus the small differences distinguishing varieties of the same species, will steadily tend to increase till they come to equal the greater differences between species of the same genus, or even of distinct genera. (Darwin 1859, 127–28)

Populations will become specialized, like firms in a competitive economy, through the progressive division of labor. That is, in effect, how

Darwin thinks new "races" and species are produced. It is also how, as the diversity stacks up over time, the ramified or treelike system of classification into which organic kinds naturally fall comes into view: "On these principles, I believe, the nature of the affinities of all organic beings may be explained" (Darwin 1859, 128).

Darwin's "long argument" thus has four major steps:

Malthusian reproduction + Resource scarcity = Struggle for life

Struggle for life + Variation = Differential adaptedness of variants

Differential adaptedness + Strong inheritance = Adaptive natural selection

Adaptive natural selection + Niche diversification + Many generations = branching taxa.[5]

It was precisely the last steps in this argument that preoccupied Darwin during the silent decades. A key event during this period was Darwin's use of Henri Milne-Edwards's notion of the body's "physiological division of labor" to model diversification of niches and species. It is noteworthy that Milne-Edwards, a Belgian-born disciple of von Baer, acknowledged that he got this idea from Adam Smith's famous principle of the division of labor. The influence, if only indirect, of British political economists was picked up as soon as he opened the book by the quick-witted Karl Marx, who could spot this sort of thing a mile off: "It is remarkable how Darwin recognizes among beasts and plants his English society, with its division of labor, competition, opening up of new markets, 'inventions,' and the Malthusian struggle for existence. It is Hobbes' *bellum omnium contra omnes* [war of all against all]" (Marx to Engels, 1862, in Marx-Engels 1937).

A more subtle preoccupation, however, is Darwin's tendency in *On the Origin of Species* to look over his shoulder at theology. His apparent sang froid in observing the process of natural selection does not mean that Darwin denied that nature, even if not nature's god, is beneficent. In a passage that recalls and revises biblical imagery, Darwin writes:

How fleeting are the wishes and efforts of man! how short his time! and consequently how poor his products will be, compared with those accumulated by nature during whole geological periods. Can we wonder, then, that nature's productions should be far "truer" in character than man's productions, that they should be infinitely better adapted to the most complex conditions of life, and should bear the stamp of far higher workmanship? (Darwin 1859, 84)

A strange permutation of theodicy (a form of argument as old as Hesiod's *Works and Days* and the *Book of Job*, which is intended, as the poet Milton put it, "to justify the ways of God to man") can actually be felt in vivid passages like these, as well as in the following:

Man selects only for his own good; Nature only for that of the being which she tends. . . . Natural selection is daily and hourly scrutinizing, throughout the world, every variation, even the slightest; rejecting that which is bad, preserving and adding up all that is good; silently and insensibly working, whenever and wherever opportunity offers, at the improvement of each organic being in relation to its organic and inorganic conditions of life. (Darwin 1859, 83–84)

Darwin's fascinated immersion in nature and its sublime ways gave him a far less anthropocentric view of these matters than was, or still is, common. He was always warning himself and others against anthropocentrism (Gould 1977, 1990). Others, not heeding this lesson, may have seen in his line of argument the looming shadow of the problem of evil: If God could not find a better way to create than this, he must be less than all powerful or else less than wholly benign. But Darwin's habit of looking at the matter from nature's point of view, *sub specie naturae*, as it were, induces not horror, or even the bland stoicism one might expect, but lyrical wonder at what nature creates out of its apparent indifference to individuals.

In spite of its obvious power, however, and its enthusiastic anticipation of large consequences, the argument of *On the Origin of Species* to the end of chapter 4 is not intended to have proved species transformation and monophyletic descent. On the contrary, it is meant to show only the existence of natural selection as a process and to begin to establish its competence to account for transformism and phylogenetic diversity (Hodge 1977, 1989). The existence of the process as a "true," or actual, cause (*vera causa*), as demanded by Herschel, is predicated on the analogy with artificial selection. It is, as Herschel demands, a process one can observe at work, as Lyell's stress on subsidence and elevation of continents can be observed. The competence argument depends on showing that the characteristics that undergird artificial selection are present in great profusion in nature. By the end of this stage of the argument, Darwin can rightly claim that to have offered some good arguments to the effect that the kind of world we live in, roughly a Malthusian world, is one in which a process like natural selection would be likely to have massive effects, many of which are quite worthy of admiration. The question of whether natural selection is actually responsible for the whole process of phylogenetic diversification is, however, still to be undertaken in chapters 6–14. This is a matter of finding empirical support and of overcoming objections. It is not a job that can be accomplished with a quick, deductive, knock-out punch. The remaining chapters of *On the Origin of Species* read even more like a lawyer's brief. Replies to objections alternate with positive arguments, and the jury is asked to conclude only that there is a preponderance of evidence in favor of the hypothesis.

Upon reading the book, Lyell appreciated the way it had been written, perhaps recognizing that his *Principles of Geology*, and Herschel's norms, had served as Darwin's models. It was not at all like Chambers's ill-fated 1844 work on evolution. After reading the proofs Lyell wrote: "I have just finished your volume, and right glad I am that I did my best with Hooker to persuade you to publish it. . . . It is a splendid case of close reasoning and long sustained argument throughout so many pages, the condensation immense, too great perhaps for the unitiated, but an effective and important preliminary statement" (Lyell to Darwin, October 3, 1859, in Darwin 1887, 2:2).

Even so, Lyell continued to doubt whether Darwin had made his case. What Darwin had done was to extend Newtonian methods and models, as Lyell and Herschel understood them, to biology. He had thereby taken the next logical step in the research tradition to which they all belonged. Why, then, was Lyell resisting? Was his reluctance to follow Darwin simply a case of unprincipled opposition to evolution, which Lyell continued to think reduced humans to the level of beast and in the process undermined rational authority in society? To answer that question, we must review the history of that research tradition and assess what was at stake in Darwin's extension of it.

4 Tory Biology and Whig Geology: Charles Lyell and the Limits of Newtonian Dynamics

In this chapter we begin to show that Darwin is most perspicuously viewed as extending Newtonian dynamical models to biology. If true, it follows that the crisis unleashed by *On the Origin of Species* was rooted in the fact that, far from shilling for something as foreign as French evolutionary theory—which could always be dislodged from respectable culture by ritual denunciations like those that befell Grant and Chambers—Darwin worked within the main line of respectable eighteenth-century British science. It is, we argue, the very possibility that Darwin might succeed in producing a Newtonian biology that caused the cultural crisis surrounding his book. *On the Origin of Species* challenged a long-standing cultural compromise, according to which Newtonian natural science would not be allowed to trespass onto the terrain of historical biology, so that one step up the line, theological narratives about the human condition and philosophical theories about human nature and valuation might be afforded protection. Darwin let the wolf come closer to the door.

By following the spread of the Newtonian research tradition in Britain from field to field, and in particular Lyell's extension of it to earth history, we will learn more about why Darwin's book unleashed the reactions it did. In the following chapter, we will dwell more fully on how Darwin used economics in constructing his Newtonian theory of biological origins. We will then be prepared to consider in chapter 6 the immediate reception of Darwin's book in Britain and how the issues debated at the time were resolved. We begin with a brief review of Newton's explanatory paradigm.

Sir Isaac Newton was the Western European Enlightenment's greatest culture hero and far and away the most influential intellectual of the last half millennium. Disillusioned twentieth-century thinkers, who have in their own time seen the Enlightenment program frustrated, have recognized that there is some irony in this, for the hero of Reason's sweetness and light was himself a crabby and paranoic person, whose worldview made him a better candidate to be the last medieval than the first modern man (Manuel 1968). Newton thought that the spatially extended world in which perceptible objects are displayed is God's sensory threater (*sensorium*). Time is the uniform, unidirectional flow of God's consciousness. The point of doing physics, accordingly, was literally to pick the Creator's brain. The trick was to figure out the language in which God talked to himself and therefore in which he wrote the Book of Nature. Clearly God's native tongue was mathematics. What Newton prided himself on was discovering the right mathematics. It was the calculus of infinitesmals, which enabled both God and Newton to track instantaneous changes in the "system of the world." Having figured out the Book of Nature, Newton went back to work on the Book of History, spending his time trying to figure out the mathematical code behind the Book of Revelation.

If there is anything at all that deserves to be characterized as a "scientific paradigm," it is Newton's model of planetary and stellar dynamics. A paradigm is rooted in a successful example of scientific achievement. The inner workings of the example, or "paradigm" case, are analyzed in terms of a schematic model, which is then extended to a wider range of problems. If the model is markedly fecund, it sometimes is generalized into a worldview (Kuhn 1970). Newtonian physics conforms paradigmatically, as it were, to this definition. It is concretely anchored in examples like simple self-regulating, but not self-creating, machines. When it is said that Newton "mechanized" the world picture, what is meant is that he and his successors viewed the heavens as a machine like a pendulum clock. In Newton's analytical model of the behavior of systems like these, an inertial tendency, rooted in the mass of an object, is countered instant by instant by the gravitational force exerted on it by other massy entities in the system. Forces counter forces. As a result, the system constantly restores itself to a steady state, in which mass remains constant and momentum is conserved through changes.

In a rare display of humility, Newton recognized that in writing the *Philosophiae Naturalis Principia Mathematica* (*Mathematical Principles of Natural Philosophy* [or *Principia* for short], 1687) he had "stood on the shoulders of giants" like Copernicus, Galileo, and Kepler. In turn, the makers of the Western European Enlightenment recognized that they stood on Newton's shoulders. There is no better short summary of the self-proclaimed Enlightened thought of the eighteenth century than this: It meant the spread of Newton's model of dynamics from astrophysics to other fields, a process that reached breakneck speed at the end of the eighteenth and the beginning of the nineteenth century.

What elevates the Newtonian paradigm to the status of a generalized worldview, however, is that the success of Newtonian physics led not only to attempts to see Newtonian dynamics in other kinds of natural systems but in political, social, and economic systems as well. For example, a Newtonian cast was imparted in the eighteenth century to constitutional theory. The Constitution of the United States, designed to be "a machine that goes of itself" because its separate governmental powers are allegedly balanced like a planetary system, is a monument to the Newtonian imagination. It is no coincidence that Benjamin Franklin, who had a hand in writing it, was already internationally famous for extending Newtonian mechanics to electricity. (That is what the story of the kite is all about.) The very idea of what we mean by the Enlightenment is bound up with the Newtonian paradigm. Enlightenment ideologues like Voltaire, who knew little physics, wrote books praising Newton's system of the world. By the same token, Romantic critics of the Enlightenment, who formed what Isaiah Berlin calls the Counter-Enlightenment, got their point of view across by writing against Newton. Goethe tried to refute his theory of colors. Blake prayed to be delivered from "Newton's sleep."

The enormous influence of Newton's model of celestial mechanics in Darwin's milieu requires that we briefly recount a story that has already been told often and well (for example, by Cohen 1960). The story is about how modern physics, culminating in Newton's *Principia*, displaced the old medieval cosmology. It begins this way. Once Platonized, and later Christianized, Aristotelianism had been cut loose from Aristotle's biological realism, there was nothing to stop the purposiveness that Aristotle attributed to living things from being massively overextended to physical reality as a whole. Thus, Aristotle's remarks in the *Physics* about the natural directions in which the four elements move (earth goes down, fire goes up) and his even more fragmentary suggestions that the stars, the sun, the planets, and the eternal species lineages try as best they can to approximate God's unmoved motion by turning in circles, were made central to the later semi-Platonized reconception of Aristotelian teleology (from Greek *telos*, "end" + *logos*, "account," meaning an explanation of something in terms of its end, purpose, goal, or function). This is what Dante meant when he spoke at the end of the *Divine Comedy* of "the love that moves the sun and other stars." Clearly "teleology" or end directedness no longer referred primarily to the developmental trajectories of embryos, as it did for Aristotle. It meant that everything in the world, from the purely physical to the psychological, must serve some purpose in a vast cosmic drama.

In early modernity, Platonized Aristotelian physics of this overly teleological cast was subjected to devastating criticisms by the likes of Galileo and Descartes. It is now acknowledged that the collapse of Aristotelianism was already underway within medieval natural philosophy itself, for Aristotle's own physical and biological theory, with its insistence on the eternality of the world and of species, as people like Ibn-Rushd (Averroes) and Thomas Aquinas realized, was flatly inconsistent with Christian or Islamic creationism. Thus, some of the first and best critiques of the old physics were made by conservative and pious Augustinian and Franciscan theologians who opposed the "secular Aristotelians" at the University of Paris, who, following Averroes, sided with "the

Philosopher" against the theologians. For example, Galileo's speed-distance-time equations had already been worked out by Nicholas of Oresme in the fourteenth century (Grant 1971). Nonetheless, it was not until the seventeenth century that the old cosmology overtly collapsed under the increasing weight of mechanical models, whose projection onto nature was undoubtedly spurred by advances in technology. It was not until the achievement of Newton, however, at the end of the century, that there eventually emerged a new paradigm of order out of the rubble of that collapsed cosmology. It was only then that the pessimism and disoriented cast of thought that led John Donne to write, "All lost, all coherence gone," gave way to the optimism of the British Enlightenment or "Augustan Age," provoking Alexander Pope to proclaim:

Nature, and Nature's Law lay hid in night,
And God said, Let Newton be! and all was light.

Because Newton's laws of "natural philosophy," and his way of picturing the "system of the world," will play an important role in what follows, and in this book as a whole, we will take a moment to review them. Newton's laws of motion are about the calculable forces exerted on entities in a system and what the predictable effects of these forces will be. The Newtonian way of thinking begins with Galileo, who asked himself how a body would behave in the absence of any external forces affecting it. Newton finally answered these insightful questions: Such a body will move at a constant speed in a straight line. If the forces acting on the body balance each other, it will behave in exactly the same way. If no net force is exerted on a stationary body, it will stay still. That is Newton's first law: inertia.

Now the fact is that we usually do not see things working like that. Planets and stars, whose motion certainly seems natural, go around in stable orbits, not off in straight lines. That, Newton says, is because forces in the environment of the body are operating on it to make it deviate from its natural motion. Thus, the natural motion of a body is not, as Aristotle thought it was, what is actually observed to happen "always or for the most part." Rather, natural laws tell us not only what happens under certain conditions but what would happen if something else were not masking it or intervening to deflect it, even most or all of the time. Natural laws, that is, can stand up to what philosophers call counterfactual questions. (We have already seen Darwin using counterfactual reasoning in connection with Malthusian reproduction: Unless something were interfering, there would be many more organisms than we see.) Now consider the planetary orbits again. Aristotle had said that it was natural (by which he meant normal) for them to move in perfect circles. But if natural motion is in straight lines, why do the planets actually move in circular or, as Kepler correctly pointed out, elliptical orbits? The answer is that their circular or elliptical motion can be analyzed as the sum of a series of instantaneous movements that "go off at a tangent" and at the very same instant are pulled back by an external force. It is as if circles or ellipses are actually polygons with an infinite number of sides, natural straight line motion being yanked at each instant from the direction in which it would otherwise move. Planetary orbits are constructed instant by instant. Accordingly, physics is analytical rather than merely observational.

Newton's invention of "fluxions," an early version of calculus, gave him the mathematical tools with which to analyze motion considered this way. It permits treatment of instantaneous changes of many variables and sums these up to see what trajectory a body moves in. We come by this route to Newton's second law, which tells us precisely how a body will deviate from inertial motion or rest when a net external force is exerted on it. It will accelerate, that is, change the amount of space it covers in a given time, by an amount directly proportional to the strength of the force being exerted on it, and inversely proportional to its own mass, which is a drag on its acceleration: $F = ma$. Newton then stipulates in a third law that all bodies in a given system, all entities having mass or "quantity of matter," are to be thought of as exerting force on each other in this way. "The pressure of my finger on a table," Newton says, "is at the same time the pressure of the table on my finger," in spite of the fact that I, and not the table, feel the pressure. For every action there is an equal and opposite reaction. This is to say that if you have a closed system, in which no additional force, matter, or other influence enters or leaves, the total momentum (mv: mass times velocity or speed) is constant or conserved. A change in momentum of one part of the system will be compensated for by a corresponding, opposite change in another part. If you suddenly get up in a rowboat your partner in the stern is bound to appreciate this fact instantly.

With these laws in hand, Newton was able to formulate a fourth general law, applicable to planets, projectiles, and much else, measuring the mutually attractive forces that all bodies exert on one another: "Any two bodies exert forces on each other proportional to the product of their masses divided by the square of the distance between them": $F = Gmm'/D^2$. That is Newton's "inverse square" law of universal gravitation. It explains the force that, for example, keeps the moon orbiting the earth, balancing its inertial tendency to fly off on a straight line with the gravitational pull of the planet. Newton declined to speculate what gravity is. All he knew was that he could measure it.

From these laws, Newton was able to deduce the orbits of the planets, and hence to confirm Kepler's three laws of planetary motion by deriving them from more general, deeper laws. He was also able to deduce the orbits of comets, which had long seemed so erratic that Aristotle regarded them as sublunar phenomena like the weather. (That is the reason for our strange use of the word *meteorology* to designate not the study of meteors but of the weather.) The orbits of the planets and of comets are stable ellipses because the system of bodies of which each of them is a part is in what is now called dynamic equilibrium. The tendency to fly off on a tangent, inherent in bodies by the first law, is just balanced by the gravitational pull of other bodies in its neighborhood. Thus, in a great reversal of the Aristotelian and medieval view, these orbits are not an expression of an inherent tendency to move in a circle (because circles are allegedly more perfect and divine than other figures) but are constructed instant by instant by a balance of opposed forces. Because the orbits of the planets are in equilibrium, any change would provoke the body either to fly off forever into the heavens or to crash into a body that attracts it by exerting gravitational force on it. Both effects can be observed in the behavior of spaceships. These laws suggested to Newton's Enlightenment followers, Kant and Laplace, a compelling hypothesis about the origin of solar systems and galaxies: matter, probably gases at first, collects into planets, stars, and galaxies because, except for small pieces of junk, gravity makes everything congeal into a finite number of massy balls.

At this point, we can see better how Newton stood on the shoulders of giants. The history of astrophysics is most simply represented as a process in which what were formerly taken as constants are treated as variables, while something increasingly less obvious stays the same or is conserved. For Aristotelians, velocity or speed (v) is constant, while distance and time vary. The planets go around in circles at a uniform speed. Kepler's revelation that planetary orbits are ellipses destroyed this idea forever by showing that velocity varies with distance from a focal point. For Kepler, what stays the same is that equal areas are swept out by a planet in equal times. That is Kepler's second law. Galileo too knew that velocity varies under different conditions. He demonstrated that when the distractions of air pressure, friction, and other such interferences are (counterfactually) disregarded, bodies of different weights fall freely at the same speed but that their velocity changes incrementally and uniformly through time: Distance traversed is proportioned not to time directly but to the square of the time. What remains constant in Galileo's world is the rate at which speed changes, or acceleration. Tacitly, however, Galileo was treating gravity as a constant, for his laws of uniform acceleration hold only at the same level of gravitational force. Newton took the next step. By adopting Descartes' hypothesis about natural motion being rectilinear, he was able to unify Galileo's mechanics with Kepler's astronomy by treating gravitational force as a variable: It falls off with distance after the fashion of the "inverse square law." That is why the same object will have different weights at different gravities and why a body (such as an astronaut) free of gravity has no weight at all. What remains constant for Newton is mass, the amount of matter a body contains, regardless of whether it is on the earth or traveling in a spaceship. We can see a bit more clearly at this point why Newton needed his calculus. At any instant, it is not only the speed and acceleration of planets that are changing but gravity itself, as distance from other planets increases and decreases. That is a lot of balls to be keeping in the air at the same time. (Indeed, the calculations cannot be done without introducing approximations for more than two bodies at a time.)

In a spectacular display of what Whewell called "consilience of inductions," Newton was able to explain more than the solar system with his laws. They also accounted for almost everything that was then known about the movement of bodies on earth. They explained, for example, the oceans' tides as an effect of the moon's gravity. And as people began to apply Newton's laws, or laws closely related to them or derived from them, to other fields, laws taking the form of an inverse square became the goal with which to explain chemical affinities, light and color, heat and fluid flow, and other phenomena. It became a particularly important task in the late eighteenth century to link Newton's laws to electricity and magnetism. Everything worked well enough when you articulated the geometry of inverse square laws to allow for repulsive as well as attractive forces and for forces running at right angles to each other, as well as in straight lines between central points. In sum, Newton's model had tremendous explanatory fecundity. Tweaked and kicked a bit, it yielded good explanations not only of the phenomena it had set out to explain but of new phenomena as well. That was the element of Newton's achievement most admired by Whewell.

It was because of its demonstrated power within physics that Newtonian systems turned into a generalized model for describing and explaining phenomena in fields beyond physics, even social systems. It was by way of this further, overtly metaphorical extension that Newtonianism blurred into a generalized Enlightenment worldview. An even more abstract way of describing Newtonian systems comes into view at this point. A Newtonian way of looking at things, we will say, prevails when any system of entities, whether it consists of physical bodies or not, conforms to the pattern paradigmatically seen in Newton's laws of motion: an external force, directed in straight lines, whether between the entities in question or at right angles to them, deflects and shapes the motion of an inertial mass of some sort with a magnitude that falls off exponentially with distance between the centers of force. So construed, Newtonian systems include those that conform to Newton's laws directly, as does classical mechanics; systems that obey laws derived from, reducible to, or isomorphic with Newton's laws, like those of the classical theory electricity; and, more weakly still, systems whose laws are analogous to Newton's laws, such as the "laws of the market" in Adam Smith's economics. No matter what kinds of entities the model ranges over, however, whether they are basketballs or rational economic agents, there remains at least some recognizable analogue of an inner inertial tendency, of conserved momentum, and of a gravity-like force, the intensity of which decays with distance.

Viewed from this high level of abstraction, it is possible to isolate at least four conceptual presuppositions that must hold if a system is to be Newtonian:

1. Newtonian systems are *closed*. If they were open to outside influences, none of the laws governing force-induced changes within the system could be expected to hold. It would be like playing croquet with the Queen in *Alice in Wonderland*. Every time an opponent's ball looks as though it will move through a wicket, the wicket will be moved.

2. Newtonian systems are (in the absence of friction and its analogues), *deterministic*. Given the initial position, mass, and velocity of every entity in the system, a completely specified set of forces operating on it, and stable closure conditions, every subsequent position of each particle or entity in the system is in principle specifiable and predictable. (Call this Laplace's Hypothesis.)

3. Newtonian systems are (again in the absence of friction and its analogues) *reversible*. The laws specifying motion can be calculated in both temporal directions. There is no inherent "arrow of time," or necessary directionality, in a Newtonian system.

4. Newtonian systems are strongly *decomposable* or *atomistic*. Newton committed himself to a "corpuscular" or atomic theory of matter, and even to the corpuscular theory of light, long before any direct evidence for such a theory was forthcoming because he needed an ontology of point masses and additive quantities of matter and force. "It seems probable to me," he wrote in the *Opticks*, "that God in the beginning formed matter in solid, massy, hard, impenetrable, movable particles." That is partly why it seemed so important for the Newtonian research program to vindicate atomism, as Dalton did.

The Newtonian research tradition was not a static affair. Indeed, the extension of what came to be called "classical physics" to new fields eventually led the Newtonian picture of dynamics, as we have modeled it, to the brink of its own transcendence and demise. In the course of bringing new fields under its sway and exploiting its problem-solving potential, the tradition progressively amended its original conceptual and ontological structure out of existence. The process began with Augustin Fresnel and the mathematical development of a wave theory of light. The emphasis shifted at that time from hard little particles and forces to the transmission of a state of motion through a medium. André-Marie Ampère's extension of wave theory to electrodynamics was developed by James Clerk Maxwell into field theory. Under the influence of William Thompson (Lord Kelvin), Maxwell, Helmholtz, and others, this process culminated at midcentury with the displacement of Newton's central concepts of force and mass by the concepts of energy and work as the architectonic concepts in physics. That did not seem necessarily fatal to Newtonianism at the time. Indeed, as we will see in a later chapter, Maxwell thought that he was preserving the Newtonian tradition by giving it a new ontology in order to explain systems in which vast numbers of elements interact in seemingly random ways but nonetheless can be averaged over the aggregate. So did Ludwig Boltzmann, who used analytical techniques like Maxwell's to explain irreversible thermodynamics. Both Maxwell and Boltzmann insisted, however, that these analytical devices were merely substitutes for classical processes lying further down, processes that some demon, but not us, might be able to see and even calculate. It was, in fact, not until the further extension of probabilistic reasoning pioneered by Maxwell and Boltzmann to basic subatomic processes led to quantum mechanics in the early twentieth century that it became clear that Maxwell's and Boltzmann's attempts to preserve the Newtonian tradition had actually led to its demise by a massive erosion of the four presuppositions we have attributed to Newtonian systems. In twentieth-century physics, all four of these presuppositions have lost their claim to universality. Somewhere along the line, the pretensions of classical physics to be, as Newton called it, a "theory of the world" became unsustainable, and the term "Newtonian" or "Newtonian system" (reanalyzed for the most part by Sir William Hamilton) was restricted to the behavior of limited classes of systems under contingent, and approximate, boundary conditions.

The spreading out of the Newtonian paradigm to new fields, and from the natural to the human sciences, was particularly intense and creative in Scotland. There the Glasgow moral philosopher Adam Smith devised a Newtonian economics, and his friend David Hume a Newtonian psychology. The Edinburgh chemist Joseph Black and later John Dalton began to apply Newtonian thinking to the problem of chemical affinities. (Darwin himself studied Dalton's atomic theory of chemical combination during his short-lived days as a medical student in Edinburgh. He and his brother Erasmus were enthusiastic amateur chemists.) James Hutton, another Edinburgh man, was the first to propose a Newtonian geology in his *Theory of the Earth* (1788). Hutton's turgid book was popularized by his Edinburgh colleague John Playfair, a mathematician, in *Illustrations of the Huttonian Theory of the Earth* (1802).

This extension of Newtonian models had been facilitated, since the mid-eighteenth century, by the use of minimization and maximization principles in analyzing the trajectories of Newtonian systems. Maupertuis had said that Newtonian entities move along a path of "least action." At the same time that they minimize some quantities, however, Newtonian systems maximize others, such as the most amount of work for a given amount of effort. (The ultimate maximization principle is entropy, which, as we will see in chapter 10, "strives toward a maximum," in Rudolf Clausius's famous formula.) Maximization and minimization assumptions, now called "extremum principles," can be seen at work in, and used to analyze, power-driven machines. It was at least vaguely by seeing economic and reproductive self-interest in these terms that Newtonian models spread, if metaphorically, to economics and to evolutionary biology. The rational economic agents portrayed by Adam Smith "maximize" their self-interest, a property supposedly rooted in their very nature as rational economic atoms, by getting the most profit for the least expenditure of inputs like rent, labor, and raw materials. The organisms Malthus portrayed "maximize" their fecundity by a similar natural impulse. The "varieties" portrayed by Wallace and Darwin are generated and culled by an environment that works, in Wallace's words, like "the centrifugal governor of a steam machine."

The wonderful clockwork world of eighteenth-century social science, which fired the political imagination of the founding fathers and of classical political economists like Adam Smith, looks quainter today than Newton's physics itself. Something of the radical spirit of the continental Enlightenment was, in its day, at work in the Scottish and American Enlightenments. In Europe, the medieval program of universal teleology was never fully given up by the educational, religious, and political establishment. Accordingly, Enlightenment scientism, when it came, took on a furiously reductionistic, materialistic, neopagan, anticlerical, democratic (and later still communistic) cast. Although the Scottish Enlightenment was never like that, it became noticeably tamer still when it spread south into England under the influence of rationalist Whigs. It was in Scotland that Darwin felt daring. In England he insisted, above all, on his respectability and reputation for being judiciously moderate, like any other good Whig.

This difference in tone is related to England's political history. England was arguably the first nation to undergo, and more or less successfully to emerge from, what we now recognize as a modernization crisis. The English Revolution of 1640 may very well have inaugurated a two-century, worldwide orgy of king killing elsewhere, but at home the Glorious Revolution of 1688, which settled that revolution, initiated a protracted process of political and ideological compromise in Britain and its colonies. Newton was thereupon reconstructed, through the good offices of the Royal Society and Whig ideology, as the cultural icon of a great compromise. Newton's discernment of the balance of opposing forces that binds the physical world into an ordered system gave hope that similarly balanced forces might be found in the social and political world. It was in this spirit that Pope invoked Newton in the couplet we quoted. As part of this ideology of moderation and as an expression of its dignified retreat from the religious "enthusiasm" (from Greek *en* + *theos*, "wrapped up in God") of the Puritans, God was now to be seen at work in physical and social systems through impersonal laws or "secondary causes." In his wisdom, he had doubtless originally set these laws up. They did not, however, require his personal supervision moment by moment, any more than political and economic life required the constant supervision of a constitutional monarch.

This spirit formed the conceptual background of Dissenting Whigs like Herschel, Lyell, and Darwin. Absorbing as much as they could of the more daring Scottish Enlightenment—Herschel's stress on finding real causes (*verae causae*) in science had originally been developed by the Scottish "common sense" philosopher Thomas Reid—they stressed the reliability of natural laws as agents of the divine purpose. Fearing atheistic materialism, however, which they regarded as just as fanantical a violation of the ethic of moderation as that of religious enthusiasts, Dissenting Whigs continued to insist that God hovered in the wings. It was the singular achievement of Charles Lyell, in his *Principles of Geology* (1830–1833), to give a Newtonian account of geological processes that fit the Whig mold much better than Hutton's and Playfair's efforts, and so to extend one step further the Whig program of treating natural laws as divinely appointed "secondary causes." Lyell represents Whig science at its apogee.

At this time, geology included not only earth history but the historical part of biology as well, since the earth's various strata, which geologists had begun accurately to date, contained fossils of animals, indeed of whole biota, that are no longer with us (Rudwick 1972, 1985). Geology, so construed, was the site at which the extendability of Newtonian models was contested most strenuously in Darwin's youth. What divided "catastrophists" from "uniformitarians" was precisely whether geological explanations were to be fully Newtonian. According to catastrophists, the different geological strata, and the biota whose remains are embedded in them, reflect very different geological and biological eras, usually separated by "catastrophes," each of which might be governed by different laws from those on which we now rely. Catastrophism was narrative geology. Its narrative cast gave it an affinity with biblical creationism. Catastrophists, however, were not usually religious creationists, who used Genesis I as a stick with which to beat theories of origins, but scientific creationists, like Buffon, Cuvier, and Agassiz, whose theories threw more light on Genesis than Genesis threw on them. Scientific catastrophists tended to think of the earth as progressively cooling, congealing like pudding, contracting, and cracking. They thought that the general replacement of smaller with larger, homeothermic animals reflects adaptive compensation for this change.

Something like catastrophism, having at last shuffled off its theological coil, has in fact made a modest comeback in the twentieth century. Our view of earth history is in some ways closer to it than to uniformitarianism. The major episodes in the history of life, we now realize, are punctuated with mass extinctions, which, even if they do not rely on different laws than obtain today, as the old catastrophists thought, do at least use catastrophic events to explain biological eras different from ours. In the eighteenth and early nineteenth centuries, however, catastrophism constituted a barrier to the extension of Newtonian methods, models, and mechanisms to an important part of natural science. According to uniformitarians like Herschel, sound empirical scientific method relies on the assumption that the causal processes and natural laws that govern the world have always been uniform in the sense of being "the same laws." To uniformitarians, therefore, the catastrophists' narrative explanations of particular events, without appeal to uniform, presently acting laws, meant that they were unscientific. That is why uniformitarians tended unfairly to assimilate scientific catastrophists to biblical literalists. It was part of this allergy to the dramatic, narrative quality of catastrophism, as well as out of their transcendent respect for Newton's celestial paradigm, however, that uniformitarians tended to think not only that natural laws are uniform across all eras but that such laws work, like Newton's heavens, *uniformly*—that is, gradually and at a fairly constant rate. Much trouble from that conflation lay ahead.

What Lyell accomplished is perhaps best brought out by contrasting his versions of Newtonian geology with that of his predecessor, James Hutton. Hutton was an Edinburgh man in the glory years of the Scottish Enlightenment. He knew Adam Smith, Hume, and, significantly, James Watt, the inventor and theorist of the steam engine. Accordingly, Hutton was heavily invested in the project of spreading Newtonian systems dynamics around. Gould writes that in Hutton, "The light of Newton's triumph continued to shine brightly, and the union of other disciplines with the majesty of this vision remained a dream of science at its best. Hutton yearned to read time as Newton had reconstructed space" (Gould 1987, 78). Not surprisingly, therefore, Hutton's theory of the earth, published in 1785, portrayed geological change as a Newtonian machine that "goes by itself." First, there is decay and erosion. Mountains wash into the sea. Soils are sedimented there in the layers or strata now clearly visible, seashells and all, when mountains are exposed by uplift. That happens because, as strata horizontally pile up, their weight increases, producing heat, which, in the form of magma, forces its way up to the surface, producing new land. That in turn erodes, starting the cycle again. In places where land once stood, then, there is now water, and vice versa. Hutton's world is a dynamic balance of opposing forces.

For all this stress on machinery, however, there remains in Hutton's work more than a touch of narrative drama and of theocentric teleology. The old Stoic-Aristotelian cosmology, in which everything has a purpose, is especially prominent when Hutton talks about the relationship between the earth and the beings that live in it. The earth, he writes, "is a machine of a peculiar construction by which it is adapted to a certain end" (Hutton 1788, 209), and the "theory of the earth is . . . a general view of the means by which which the end or purpose is attained" (Hutton 1795, 270). The purpose is to provide the continuous and sustained creation of "a world contrived in consummate wisdom for the growth and habitation of a great diversity of plants and animals; and a world peculiarly adapted to the purpose of man, who inhabits all its climates, who measures its extent and determines its productions at his pleasure" (Hutton 1788, 294–95). Significantly, Hutton's champion, the Edinburgh mathematician Playfair, removed some of this talk in his *Illustrations of the Huttonian Theory of the Earth* (1802). Some thirty years later Lyell pushed natural theology and teleology even further to the margins of Huttonian geology in his *Principles of Geology* in order to ensure that his geology of secondary causes would contrast as vividly as possible with the interventionism of his catastrophist opponents. Hutton may have been a Newtonian, but he was not a consistent uniformitarian.

Lyell's attack on catastrophism was, in the first instance, methodological. Only forces currently seen to be operating could ever count as objects of scientific inquiry, since only they are empirically accessible. The fact that it does no good to speculate about what might have happened in bygone ages is underlined by the fact that such speculation always takes the form of a narrative. If the aim of science is to see how much one can explain by rules governing observable phenomena, however, a scientist must avoid giving up the search prematurely simply by postulating ad hoc, and question begging, stories that could have accounted for the occurrence of something long ago. Science is not narrative. As in physics, the rules or laws governing changes in the earth must be presumed to be the same throughout time. What goes on in the world, therefore, is an endless series of law-governed instances rather than a concatenation of episodes.

This methodological point about appealing to presently active forces is what is most properly meant by "uniformitarianism." Not surprisingly, Lyell legitimated his methodological strictures by appealing to Newton's example. Just as Newton had built a world around presently observable processes like the whirling of a stone at the end of a string or the way water sloshes up the sides of a spinning bucket, so Lyell took elevation and subsidence to be a *vera causa* in Herschel's sense. So did Herschel himself. "The elevation of the bottom of the sea to become dry land," Herschel wrote, "has really [verily, *vera*] been witnessed so often, and on such a scale, as to qualify it for a *vera causa* available in sound philosophy" (Herschel 1830, sec. 138).

Lyell's uniformitarianism is, however, more than methodological. His scientific imagination was deeply imprinted with Newton's image of forces kept in balance instant by instant. He reasons that any upthrust in the earth's crust at one point will be compensated for by subsidence elsewhere. Properly observed, then, the earth itself, in spite of the vast changes that have occurred through earthquakes, volcanoes, meteor impacts, and a whole host of other "catastrophes," is not only subject to the same uniform laws but is sustained by a uniform process in which a steady state was and is always being maintained. Although rates at which uniform forces and laws work may vary—we may expect, for example, more upheaval and subsidence when volcanism is prominent—change in rates is gradual enough to rule out catastrophes.

Gould has argued that Lyell's means of commending gradualism are less functions of the evidence he marshals than of his lawyerlike ability (he was in fact a lawyer) to deploy what might be called a rhetoric of perspective (Gould 1987). It is obvious that the further you extend time, the less significant seems any event within it. Extend your perspective far enough, and change smoothes itself out into a linear curve. The point can be seen by considering its opposite. It is an effect of the theocentric worldview of biblical creationists to contract time within the limits of a dramatic story with many ups and downs. Many people even in modern societies live within the bounds of religious narratives with strikingly small time frames and action-packed plots. In removing the storylike elements that plague the work of scientific catastrophists Lyell, much more than Hutton, uncovers a "deep time" that moves as uniformly as Newton's absolute time. Indeed, in modeling geology on astrophysics, and in taking the narrative drama out of it, Lyell tended to abolish time's drama by taking directionality out of it altogether. Just as there is a celestial "great year," he thinks, in which the heavens find themselves in exactly the same configuration that they have been in before, so there will be a geological "great year," in which the earth, and everything in it, including species, returns to a previous position. From the perspective of deep time, the clattering cycles and phases of Hutton's world machine disappear into a stately, and for all we know eternal, vision of smooth changes, moving in serene indifference to what we, caught in our false perspective, think of as catastrophic upheavals. By this route, says Lyell, "the mind is slowly and insensibly withdrawn from imaginary pictures of catastrophes and chaotic confusion, such as haunted the imagination of the early cosmogonists. Numerous proofs are discovered of the *tranquil* disposition of sedimentary matter and the *slow* development of organic life" (Lyell 1830, I:84, italics added). When this mood is upon him, Lyell sometimes intimates something that is not strictly entailed by uniformitarianism: that the laws of nature work at a *constant*, rather than merely at a *gradual*, rate.

Lyell's work did not please everyone. The readiness with which Whig intellectuals like Lyell were willing to place the accent on secondary causes, and to push to the far periphery any inferences that might be drawn from nature's laws to their first cause, had long made the more culturally conservative, largely clerical, intellectuals of Cambridge and

Oxford nervous. Tory intellectuals feared that the process of converting primary into secondary causes would erode God's direct revelation in his creation, as well as their own clerical authority in society, as the Newtonian model marched confidently from field to field, for every step in this program, even if it was attended by pious, even sincere, remarks about God's providence in setting up laws, meant that the heavens no longer proclaimed the glories of God quite as immediately as they previously had. In that case, the cosmos would no longer be a system of signs that, when decoded by the proper authorities, reveal what St. Augustine had called God's footprints or vestiges (*vestigia*). (Note that Chambers sought support for his controversial evolutionary theory by rhetorically appropriating this metaphor in the title of his *Vestiges of Creation*.) At best, a clever and well-meaning architect-engineer God would have left a system of impersonal laws to be admired. We who have grown more or less used to this de-divinized world can hardly imagine the sense of unease that "God's recessional" induced in earlier generations. This sense of loss was, in fact, one of the chief sources of the Romantic movement, which longed to restore through intense feelings what was no longer available through rational argument, a world where sermons could still be found in stones and worshipful impulses in vernal woods, and where every fact or event could reasonably be interpreted as witnessing to God's comforting presence to his creation.

The impulse to set limits to the advance of Newtonian thinking was already apparent by the turn of the nineteenth century in the work of the Reverend William Paley and other members of the tribe of "natural theologians." Biologizing and geologizing clerics like Jenyns, Henslow, and other members of Darwin's circle at Cambridge were not oddities. Even today readers of English fiction and fans of English film are familiar with the image of somewhat dotty English clergymen prowling around the countryside with butterfly nets. Well-informed amateur biologists were a minor industry among otherwise underemployed British clergy in the late eighteenth and early nineteenth centuries. It turns out that a serious research project was underway here, aimed at showing, by minute and complete cataloging of the intricate coadaptedness of the parts of organisms to each other, and of whole organisms to the precise ecological niches in which they function, that if in the physical sphere God was present, only through secondary causes and natural laws, at least in the biological world he was closer at hand. Detailed study of the intricate coadaptedness of parts of organisms to one another and of organisms to their environments would reveal a degree of functional order that, on any reasonable account, simply could not have been the product of natural laws of a Newtonian type.

In his *Natural Theology, or Evidences of the Existence and Attributes of the Deity Collected from the Appearances of Nature* (1802), Paley acknowledged, as any other Newtonian would, that organisms are in fact mechanisms.

(The fact that the Aristotelian tradition, and even Kant, would flatly deny this shows how hard the wind of Newtonian physics was blowing in the face of British theologians.) But, he argued, "the contrivances of nature surpass the contrivances of art" to such an extent that, if the former cannot be understood without postulating a skillful artisan, even less so can the world of organisms, which exhibit the same properties to a much higher degree. Such things, accordingly, come to us directly from the hand of God, and not indirectly through the workings of natural forces or laws.

If on a hitherto deserted island, Paley reasoned, one happened upon a watch, "no man in his senses would think the existence of the watch . . . accounted for by being told that it was a [chance assembly] out of possible combinations of material forms." Rather, one would immediately suspect that somewhere in the history of that watch's coming-to-be and to be found in that place there lurked an intelligent watchmaker. That is because this mechanism, with its subordination of different parts to a single purpose, exhibits functional organization. Now, argues Paley, "The difference between an animal and an automatic statue consists in this—that in the animal we trace the mechanism to a certain point, and then we are stopped: either the mechanism becoming too subtle for our discernment, or something else beside the known law of mechanism taking place" (chap. 3). Paley is saying that the more closely we look at myriad cases of this sort, the more we will be convinced that a supernatural power is directly at work. Look carefully, for example, at the fact that moths and butterflies "deposit their eggs in the precise substance— that of cabbage, for example—from which not the butterfly itself, but the caterpillar, which is to issue from her egg, draws its appropriate food. Yet in the cabbage, not by chance, but studiously and electively, she lays her eggs. There are, among many other kinds, the willow caterpillar and the cabbage caterpillar. But we never find upon the willow the caterpillar which eats the cabbage nor the reverse" (chap. 18).

As we have seen, Aristotle too had marveled at such adaptations. But no one until the British "natural theologians" had thought it necessary to pile up case after case of organic adaptedness in order, by a persuasive inductive argument, to block the suspicion that all this had come about by blind natural law. Now, however, it became extremely important, for unless God's immediate creative art was shown to be palpably manifest in his biological world, thereby reassuring us that "life is passed in his constant presence" (chap. 27), we might easily forget that the impersonal, mechanical laws according to which it was now conceded that the physical world runs, and the inherently inert atoms out of which it is made, were created by God as instruments for sustaining organic, and ultimately rational, entities. Thereupon we might slip by degrees into the sort of atheistic materialism (and revolutionary excess) so unfortunately endemic across the channel.

Paley was far from the first person to think of the argument from design. The tradition of "natural theology" goes back a long way. You will find the argument listed among Thomas Aquinas's five proofs for the existence of God, written in the thirteenth century. It came to him from Cicero's *On the Nature of the Gods*, in the first century B.C.E. Cicero had gotten it from the Stoic philosophers of Greece. It is important to recognize, however, that the argument had been shown to be fallacious by the time Paley deployed it in such magnificent profusion. Scandalously, David Hume, the most incisive and skeptical mind produced by the Scottish Enlightenment, had challenged it in a work he never had the courage to publish in his lifetime. Hume showed that the argument was question begging or circular, assuming its own conclusion. You are certainly free to look at nature as a whole, Hume said, or indeed any natural system or individual object, as a product of design. But nothing about these things or processes compels you to do so, for we know so little about the causes of natural things and processes that whatever properties they exhibit, or that we in our ignorance project into them, might just as well have come about in a way of which we are completely ignorant. Matter might, after all, be alive (as various Frenchmen seemed in fact to believe). Hence, no hypothesis about natural origins is powerful enough to rule out an indefinite list of rivals, with the result that no definite conclusion about these matters can, or should, be reached at all (Hume 1779).

Closer inspection of Hume's refutation shows, however, that the versions of the argument from design on which he concentrates his fire are largely based on instances of merely physical, and in particular astrophysical, order, such as the regular, and thus apparently directed, orbits of the stars and planets. That is what you will find in Cicero and Aquinas. Hume's sharpest thrust, for example, concerns the assumption, masquerading as a conclusion, that the physical universe as a whole is like a house, insofar as all of its parts serve some function or purpose. You can certainly see it *as* a house, but there is nothing about it that dictates that way of looking at it. You can see giants in the clouds too. What is significant about Paley's revival of the argument after Hume's attack is that it shifts the ground revealingly. In a physical world now acknowledged to be governed by Newtonian physics, it seems best to take organisms, and their adaptedness to their surroundings, as the best evidence of design. Functional organization may no longer be a property of the physical world, Paley implies, but it certainly is of the biological world. Indeed, Paley flatly proclaims something that would have astonished Thomas Aquinas: "Astronomy . . . is not the best medium with which to prove the agency of an intelligent creator," even though once that hypothesis has been accepted "astronomy shows God's brilliance" (chap. 22). Moreover, Paley held that while plants irreducibly and ineliminably exhibit functional organization and adaptedness and are sufficient to guide an inductive inference to the existence and goodness of God, animals, being both more complex and more behaviorally versatile, exhibit this characteristic much more fully, and so are paradigm cases.

Paley had, in effect, thrown down a challenge. *No natural law comparable to, or derivable from, other genuine natural laws, he tacitly claimed, would ever be found to explain organic function and adaptedness, including the morphological coadaptedness of parts to one another and the ecological fittingness of organisms to their niches.* There never would be what Kant called a "Newton of a blade of grass" (Kant 1790, sec. 67).[1] The line between phenomena governed by Newtonian laws of nature and spheres governed by other forms of meaning and explanation was thus to be drawn at the frontier of biology. A sort of protective shield was to be thrown in front of what Darwin called "the citadel itself," the human soul.

The increasing influence of Scottish Newtonianism in England throughout the 1820s, signaled by the Whigs's role in founding the University of London, caused conservatives to worry anew. The first volume of Lyell's *Principles of Geology* appeared in 1830, the same year as Herschel's *Preliminary Discourse.* Not quite coincidentally, that was also the year that Whig politicians engineered the First Reform Bill. Thus, it was perhaps no accident either that in the following year, 1831, a group of learned writers was commissioned by Francis Henry Egerton, the eccentric eighth earl of Bridgewater, to update Paley's argument for a new generation exposed to the creeping success of Newtonianism as it moved from field to field, as well to as the dangerous materialism of continental evolutionists that could be felt not far behind. The authors of *The Bridgewater Treatises,* as they were called, were chosen by the president of the Royal Society, with the advice of the archbishop of Canterbury and the bishop of London. They were men known for their piety as well as for their scientific reputations, such as Whewell and Peter Mark Roget (of *Thesaurus* fame). For their efforts each was to be given the considerable sum of one thousand pounds, plus royalties.

The Bridgewater Treatises are not as bad as their later reputation, or the motives for commissioning them, might suggest. Their authors, charged with showing that cutting-edge science continued to support Paley's main point, the argument from design, generally carry out their mandate by conceding that Newtonian laws, and the various forms of balance and equilibrium that they sustain, are pervasive throughout nature, including living nature. Sometimes what they say is clever. They invariably go on to argue, however, that these laws cannot have produced the very objects they govern. Explanation by secondary causes, accordingly, fairly cries out for explanations by divine first causes.

William Prout's *Chemistry, Meteorology, and the Function of Digestion Considered with Reference to Natural Theology* (1834) is a good example of the genre. Prout recognizes Newtonian equilibrium in many natural processes, including metabolic chemistry:

Amidst all that endless diversity of property, and all the changes constantly going on in the world, we cannot avoid being struck with the general tendency of those to a state of repose or equilibrium. . . . The formation of this state of equilibrium, and its preservation, may be considered as a result of those wonderful adjustments among the qualities and quantities of bodies . . . the qualities being such as to neutralize each others' activity, while the quantities are so apportioned as to leave one or two only predominant. . . . The state of equilibrium here described is not absolutely fixed. . . . The whole is so adjusted . . . that slight deviations, or oscillations about a neutral point of rest or equilibrium, take place . . . though these changes are bounded within very narrow limits, and greater deviations would prove fatal. (Prout 1834, 161–62)

Prout goes on to argue, however, that for hydrogen, carbon, oxygen, nitrogen, light, heat, and so forth to constitute a homeostatically regulated organism in this way, an external agent or power, perhaps no longer in operation, is required. Since "organized beings at the present time are at least as fixed and permanent in their nature as the state of equilibrium in which they have been placed," their existence depends on their present organization. They could not, accordingly, have spontaneously emerged, Prout concludes, from cruder states of organization (Prout 1834, 164–65). Therefore they point to a creator. Indeed, since Prout was a catastrophist in geology, he thought that the need to find new ecological equilibria after catastrophes requires an analogue for the organic world of Newton's old idea that from time to time God must intervene to set the mechanism straight. Prout was, in effect, trading on the Newtonian notion that matter is inert and needs an external agent to set it into action. The inherent passivity of matter is the crucial line of defense of British science against French materialism.

In general, *The Bridgewater Treatises* are monuments to the philosophy of science displayed in Whewell's *History of the Inductive Sciences*, according to which scientific explanations depend, directly or indirectly, on properties that natural laws sustain but cannot originate. Whewell's own Bridgewater treatise on astronomy and physics is a case in point. Gravity itself, he argues, perhaps the most important source of natural order, is not a necessary property of matter (Whewell 1833, 14). Even in basic physics, then, the mechanistic program, when pursued far enough, limits itself by revealing the need for an external agent and hence gives way once again to a logic of purposes or teleology. "Wherever laws appear," says Whewell, "we have a manifestation of the intelligence by which they were established" (Whewell 1833, 361–62). What goes for physics goes a fortiori for functional processes.

It is especially salient to Whewell that "the mutual adaptation of the organic and the inorganic world" requires us to recognize an "Intelligent Author of the Universe," who not only must create physical, chemical, and geological laws that are "accommodated to the foreseen wants of living things" but must in turn create organisms that can make appropriate and efficient use of what is provided by inorganic nature (Whewell

1833, 20). In general, Whewell claims that the closer we look, the more we see "how unlike chance everything looks. Substances . . . exist exactly in such manner and measure as they should to secure the welfare of other things. . . . The laws are tempered and fitted together in the only way in which the world could have gone in. . . . This must, therefore, be the work of choices, and if so, it cannot be doubted, of a most wise and benevolent chooser" (Whewell 1833, 144). In making these points, Whewell thinks he is repeating what Kant, arguably modernity's greatest philosopher, had said about progress in biology in the *Critique of Judgment* (1790). There are, however, big differences between Kant and Whewell. In fact, on the crucial points, Whewell and Kant are flatly opposed. Whewell is trying to reconstruct the argument from design. Kant follows Hume in rejecting it. Moreover, Whewell, following Paley, thinks that organisms are complex machines but then goes on to argue that machines, being just machines, need something else to bring them into existence. Kant's view is just the opposite. The fact that organisms are so complex, so full of feedback loops in which each part is "cause and effect of itself," means that we cannot possibly describe them as machines in the first place. That in itself is enough to guarantee that there will never be a Newton of a blade of grass. Having said this, Kant thinks that biologists can and should move fearlessly to figure out the physical and chemical means by which these purposive functions are carried out. Overeagerness to find God in this process is more likely to impede biological inquiry than to advance piety.

In many cases, in fact, the prosecution of this kind of analytical inquiry is likely to reduce the scope previously allowed to teleological explanations, for by following this method, processes hitherto taken to be irreducibly purposive can often be shown to be straightforwardly mechanical. If, for example, rain carries soil by way of rivers from mountains to alluvial plains, with the result that trees, or crops for human inhabitants, grow there, we can trace the causal pathway well enough to deny that rain carries soil to the plains *in order to* provide crops for humans, or even to let trees grow, as the old Stoic-Aristotelian cosmology had it (Kant 1790). These are mere effects, not functions. Kant is sure that the collapse of functional processes into mere mechanical effects will never be reached in analyses of organisms. His point, however, is that whenever we see it, we must be willing to reject what he called external teleology, which collapses into mechanical cause and effect, so that the genuine inspiration we can derive from the internal teleology of organic integrity will not be exposed to ridicule.[2]

This is not a lesson anyone would draw from *The Bridgewater Treatises*. Indeed, seen in the sophisticated light of Kant's philosophy of biology, these efforts look remarkably naive. In their haste to refloat the argument from design, their authors run roughshod over the distinction between external and internal teleology, and they ignore the cautionary notes it sounds. The result is that they habitually fall into what Gould and Lewontin, in another context, call "the Panglossian paradigm" (Gould and Lewontin 1979; cf. chap. 14, below). Dr. Pangloss, you may recall, was the philosopher in Voltaire's novella *Candide*, who justified all the rotten things that happen to the other characters in the book, and even to himself, by saying that they serve a purpose in this "best of all possible worlds." (Voltaire is satirizing, probably unfairly, the rationalist philosopher Gottfried Leibniz.) The point is that if you look at the biological world through rose-colored lenses like these, as Paley and his Bridgewater epigones did, you are likely to see purposes and functions where there are only contingent effects. You are likely, for example, to assert confidently that the purpose for which we have noses is to hold up eyeglasses. The result will be to expose genuinely purposive arguments to the ridicule of rationalists and skeptics like Hume. The harmony between the inorganic world and the organic, and the fittingness of the former to sustain the life, pleasure, and virtuous activity of humans, is a particularly vulnerable case. So is the invitation to God to maintain this fittingness by occasionally creating new species to compensate for losses. So too is the physiologist Charles Bell's argument in *The Hand: Its Mechanism and Vital Endowments as Evincing Design* that Malthusian population pressure, by enforcing the deaths of less able members of a kind, is an inspired way of holding species to their essential type, and thus of serving a divine purpose as an equilibrium-maintaining secondary cause of organic design (Bell 1833).

These weaknesses, rooted in the British intellectual tradition's bent toward coupling inert mechanisms with divine intervention—a bent that can be traced to the great eighteenth-century compromise—caused the Bridgewater episode to have effects quite unlike those they were intended to produce. Advocates of evolution à la Geoffroy and Lamarck lambasted the treatises as they rolled off the press. Even Darwin, polite to the core, repeated a fashionable jibe by calling them "the bilgewater treatises." It is arguable, in fact, that the Bridgewater experience, almost more than anything else, galvanized Tory intellectuals like Coleridge, Green, and Owen to take a more processive, expressive, and less mechanistic view of creation by utilizing the conceptual resources of German idealism, vitalism, and romanticism (Desmond 1989, cf. chap. 2 above). God's creative activity was now to be envisioned as as ongoing and unfolding expression of a self-creating immanent divinity rather than as the intermittent interventions of a worried engineer-God into an otherwise self-sustaining world. In general, however, the new idealistic version of Tory philosophy did not impress Dissenting Whigs. On the contrary, Whig scientists of the younger generation, such as Darwin, tended to push the divine, no matter how conceived, even further to the periphery of a law-governed world than their Whig mentors, taking, in the name of scientific professionalism, an "agnostic" view of religious belief and natural theology. (The very word *agnostic* comes from that advocate of professional science, Huxley.) Bathing creation in a Romantic glow, accordingly, as Owen increasingly did, did not prevent Darwin from remarking coolly that this way of looking at the appearance of new species was "just words," imparting nothing new or useful about the actual mechanism of speciation and adaptation (Desmond and Moore 1991, 499).

In one area, Lyell's work was unlikely to raise the hackles of conservatives. What he says about the biological side of geology is virtually indistinguishable from what any Bridgewater writer might have written about it or, in spite of his admiration for Herschel, the methodological dicta of Whewell's teleology-centered philosophy of science. In order for species to adapt to changing environments and to save God the trouble of constantly creating new species to keep the ecology balanced after extinctions, Lyell claims that the divine workman originally built into each species a great deal of variability and adaptability. This meant that God must have foreseen the wide variety of conditions in which species would be forced to operate and "preadapted" each of them with the relevant room for maneuver (Ospovat 1977). Lyell's appeal to God's "prevision" of what is needed for a species "provision" (Hodge 1987), motivated no doubt by his Whiggish desire to keep God from having to scurry about a patchwork universe, ironically results in a theocentric vision of the natural world that would have pleased Whewell as much as any other Tory intellectual:

Many species most hostile to our persons or property multiply in spite of our efforts to repress them; others on the contrary are intentionally augmented many hundred-fold by our exertions. In such instances, we must imagine the relative resources of man and of species . . . to have been prospectively calculated and adjusted. To withhold assent to this supposition would be to refuse what we must grant in respect to the economy of nature in every other part of the organic creation. For the various species of contemporary plants and animals have obviously their relative forces nicely balanced, and their respective tastes, passions, and instincts so constituted that they are all in perfect harmony with each other. (Lyell 1832, 2:42)

Even more striking is the following descent into Panglossism:

It seems fair to presume that the capability in the instinct of the horse to be modified was given to enable the species to render greater services to man. . . . It also seems reasonable to conclude that the power bestowed on the horse, the dog, the ox, the sheep, the cat, and many species of domestic fowls, of supporting almost every climate, was given expressly to enable them to follow man throughout all parts of the globe in order that we might obtain their services and they our protection. (Lyell 1832, 2:44)

The course of reasoning that leads Lyell to these surprising conclusions is not hard to reconstruct. Because he is a Newtonian and thinks that geological and climactic change are products of external forces, Lyell admits that environmental change exerts tremendous force on organisms, populations, and whole species. Those that cannot stay attuned to their resource base, or push others out of theirs, will perish. This sort of pressure will be constant and pervasive, moreover, rather than occasional and catastrophic, as incessant cycles of subsidence and elevation produce changing distributions of land and sea, changing distributions of land and water in turn produce climatic changes, and climatic changes affect the food supply of plants and animals (Ospovat 1977; Hodge 1982). Even very small changes in the ecological equation will, on this view, initiate aggressive reactions on the part of species less badly affected by a specific ecological shift, with the result that the numbers of a stressed population will diminish, or even go extinct, while the numbers of a better-adapted species will rise, as they move in and take the place of those unable successfully to confront ecological disaster. The fact that there are no observable gaps in the economy of nature, that is, in the mutual dependence of species in food chains, suggests to Lyell, however, that it is generally the same species that are conserved through these changes, or at least that the same genera are always around, out of which, on the odd occasion when God's intervention is needed, he can without much effort produce new species closely related to existing ones. Accordingly, God does not have to create a new species to fill gaps very often, just as Newton requires him to reset the clock of the universe only occasionally. (God, like a good engineer and a rational economic agent, works on a principle of minimum expenditure of energy.) In spite of a good deal of stress and consequent biogeographic moving about, then, the overall distribution of species remains constant and mirrors the presumed constancy of geological change.

Lyell's view that species are fixed, even if quite flexible, shows his fidelity to the fundamental principles of classical and neoclassical biology, especially the irreducible coadaptedness of functional parts, the intrinsic limits on variation departing from type, and the negative correlation between departure from type and reproductive fitness. "There are fixed limits," Lyell writes, "beyond which the descendants of common parents can never deviate from a common type" (Lyell 1832, 2:23). Lyell remains unwilling to challenge Paley's claim that functional organization will never be explained by natural law. What this implies, however, is that Lyell's theory embodies very strong internal tensions: *He has Newtonian environments filled with Aristotelian organisms.* Fixed types come up against external forces, forces against which they cannot be presumed to be as well buffered as they would be for an Aristotle or a Cuvier, who do not accord to environmental pressures such a powerful causal role in the first place.

In order to relieve those tensions while at the same time avoiding what his catastrophist opponents posit—massive extinctions requiring equally massive spurts of new creation—Lyell simply increases by fiat the amount of adaptability and behavioral plasticity that fixed kinds are presumed capable of mustering. In doing that, however, Lyell cannot appeal to a presently acting force, or pressures that fall off with distance, since what is in question is future utility. He has to invoke God's prevision and provision. With that, however, a great deal of the old cosmology that Whigs like himself claimed to eschew floods back in, as well as considerable compromise with Lyell's own methodological strictures.

Lyell might conceivably have lowered the tension between external forces and fixed kinds by taking recourse to Lamarck's theory of adaptation. In a world where environmental pressures are assumed to operate on populations, why not concede that organisms can stay tuned to changing environments by passing on traits they have individually acquired in the course of responding to challenges? Isn't that a nice Newtonian mechanism of instant-by-instant balance of forces? If a giraffe needs food from trees whose leaves are increasingly less accessible (through changes in the environment or the consequences of competition from other giraffes), he might simply stretch his neck to reach the higher leaves, until, through habit and the inheritance of characteristics acquired in this way, ever taller necks are favored in successive generations.

Lyell had been attracted to this approach at one time. In volume 2 of his *Principles,* however, written, significantly enough, on the eve of the First Reform Bill and at a time when Grant and Knox were at the height of their influence, and later in the tortured notebooks he began keeping after he had read Wallace's worrisome first paper, Lyell devotes considerable rhetorical and argumentative energy to blocking Lamarck's ideas, redolent as they now are with the abhorred evolutionism of Geoffroy's disciples. Lyell concedes that use inheritance might result in, or enhance, local adaptedness. But, he reasons, if overall evolutionary direction is under the causal control of an autonomous drive toward complexification, as Lamarck says it is, we should expect to see more phylogenetic direction than we do. "If . . . the tendency to progressive development were to exert itself with perfect freedom, it would give rise . . . in the course of ages to a graduated scale of beings" (Lyell 1832, 2:14). Lyell denies that over time species progress in this way on what appear to be empirical grounds. The remains of less complex as well more complex species are found in every geological stratum. Any appearance to the contrary is an artifact of willfully aggregating isolated data points. Nor is this situation likely to change in favor of Lamarck and Geoffroy, since it is highly unlikely that the relevant fossils would have been preserved in any event.

Lyell's deepest objection to Lamarckian adaptationism, however, was conceptual. Lamarckism is not and cannot be a canonically correct theory because it does not conform to the Newtonian model of dynamics. It pushes causality back from the environment into nonempirically verifiable, metaphysical causes assumed to be operating within organisms, indeed within matter itself, and so is not really a case of maintaining balanced external forces at all. For Lamarck, internal development rather than Newtonian inertia is the natural or normal state. When Lamarck admits that external causes interfere with the natural progess of organic lineages toward complexity, he is asserting that natural motion is the expression of an inner tendency rather than the product of external forces. For the Newtonian Lyell, this was to get the matter exactly backward. What happens to organisms in their environment is causally prior to what they "want" to do. This is an uncharitable reading of Lamarck, but it sends a coded message that for very high-level reasons, having to do with preferred models of dynamics, Lamarck's theory is not in the ballpark.

Darwin was deeply influenced by Lyell. Although he had been trained in the catastrophist rather than the uniformitarian tradition by his Cambridge mentors Henslow and (in what amounts to a postgraduate year) the eminent Adam Sedgwick, Darwin was converted to Lyell's uniformitarian geology almost as soon as read *Principles of Geology* on board the *Beagle*. When he returned to England, he promptly became Lyell's loyal supporter and looked to him for help in making a scientific career. Lyell not only gave Darwin the account of geological processes that stayed with him throughout his life but his overall philosophy of nature as well. Lyell's uniformitarian method and his gradualist view of natural processes, absorbed during the most receptive period of Darwin's life, seeped deeply into his imagination. His habit of seeing things *sub specie naturae* was magnified and fixed by it. That is how Darwin came to see the world in terms of enormous spans of time—spans longer even than those of the physicists of his day. Lyell is the source of Darwin's assumption that, viewed from the proper perspective, organic change, both within species and across species boundaries, moves at rates that may speed up here and there but are nonetheless gradual. Darwin also adopted Lyell's perspectivalism about evidence. If we think that geological and biological history are punctuated by discrete, dramatic, catastrophic changes, that is only because, with all our scratching and digging at the earth, we come up only with isolated pieces of data that we falsely aggregate into sudden large changes.

Yet from the start, Darwin sensed the insupportable tensions in Lyell's view of life's history. He knew that while a Newtonian geology was acceptable to Whigs and Tories alike, a Newtonian biology was not. There were limits to how far the writ of Newtonian models ran. They were not supposed to apply to living things. He also knew that Lyell went to considerable lengths not to violate this understanding. The result, however, was an incoherence in Lyell's research program that Darwin resolved to relieve. Relief would come by accepting transmutation and common descent, but it would not be on Lamarck's, Geoffroy's, Grant's, or Chambers's terms. Darwin had a genuine, indeed a burning, desire to find a theory of organic origins that conformed, as far as possible, to Newtonian canons, and that complemented Lyell's uniformitarian geology better than Lyell's own account. The simplest way of putting what Darwin did, secretly at first and publicly after 1859, is to say that he proposed yet one more extension of the Newtonian paradigm, and forthwith of the Scottish Enlightenment and English Dissenting Whig research traditions, applying Newtonian models to the historical origins of organisms themselves, and not just to the physical environments in which they live.

In a high culture in which Newton's model of dynamics was taken to be the norm to which every scientific field must conform, Darwin sought to show that pressures being exerted on organisms, rather than internal, spontaneous processes deep inside them, drive adaptation, speciation, and the diversification of taxa. What was required was a theory in which external rather than internal causes do most of the explanatory work. In characterizing the sort of theory Darwin proposed to meet these criteria, it helps to draw a distinction between what Richard Lewontin calls "transformational" versus "variational" theories, or what Elliott Sober calls "developmental" versus "selectionist" theories (Lewontin 1983; Sober 1985). Lamarck's and Geoffroy's theories fall into the first type, Darwin's into the second. What is important to recognize is that these distinctions instantiate a more basic cleft between internal and external causation. Darwin used variational and selectionist mechanisms to shift causality from the organism to the environment. Indeed, every step in the slow development of Darwin's mature theory, as we will see in the following chapter, thrust more of the causality from inside to outside. Darwin tried systematically to operate within the prescribed Newtonian framework, where real forces impinge on real populations and organisms, rather than positing an internal, developmental dynamic that is awakened, steered, or thwarted by external stimuli.

We suspect, then, that the crisis that followed the publication of *On the Origin of Species* would not have occurred if Darwin had not unexpectedly appeared to people in his own milieu to meet standards that were long assumed to be, and were perhaps designed to be, unmeetable. The British compromise between the naturalistic and theistic viewpoints, which went back to 1688, held that mechanical laws could govern physics, chemistry, and geology but that purposes, and therefore values, must govern biology, psychology, and society. This compromise was enforced by Paleyesque arguments to the effect that no natural law of an approved Newtonian type would ever be able to explain the property of functional organization in organisms. The fate of this claim in a rapidly modernizing, secularizing, and scientizing society was the underlying issue at stake in the competing philosophies of science of Darwin's day. Grant and Chambers posed no real threat to the received agreements. Their naturalism rested on developmental and transformational laws. Nor did Lyell, for he retained a suitably classical view of organisms. It was Darwin who took the next step. He exhibited a suitably Newtonian process, natural selection, that was capable of explaining functional organization, and displayed that process as a certified *vera causa*. In doing so, Darwin shifted the boundaries of the Whig compromise, dragging biology down into a world of impersonal and purposeless natural laws, thereby exposing "the citadel itself" and calling into question the fundamental premises on which the leaders of his still only half-modernized country depended for their legitimacy.

Given such a highly charged background, it is not surprising that Darwin's long-standing, and often fond, connections with the clerical biologists who had nurtured him at Cambridge were among the first casualties of the publication of the *On the Origin of Species*. One might have expected, however, that Lyell, the very model of Whig science and Darwin's personal friend, would have taken a different view. He had, after all, virtually commissioned Darwin's book. Darwin nervously awaited his verdict. Beyond the polite letter we quoted at the end of the last chapter, however, in which Lyell congratulated Darwin on the way he had made his case, and himself for urging Darwin to publish, as well as a good deal of hemming and hawing about transmutation in print, Lyell's support was never forthcoming. Worried about the dignity of humans, the foundations of morality and social order, and the necessity of an afterlife if hope in a just world was to be retained, Lyell in the end sided with the clerical intelligentsia. He belonged, after all, to the old order.

The Newton of a Blade of Grass: Charles Darwin and the Political Economists

A basic idea of the Darwinian tradition is that natural selection is a two-stage process. Variation arises independently of adaptive advantage and selective pressure, which then shapes it and gives it direction (Mayr 1978; Brandon 1990, 4–6). This notion is crucial to distinguishing Darwin's from Lamarck's theory of adaptation, which implies that particular variations arise because of what an organism needs. Darwinism's commitment to this element of chance is strong enough also to rule out, or at least to view dimly, even the idea that variation, while not hatched up to deal with antecedently given problems, is generally biased in an antecedently useful direction.

The role of chance in Darwin's theory is not restricted to the origins of variation. Although Darwin thinks that given enough time, variation, and opportunity there will in all probability be sufficient matches between the variation that is forthcoming and what is needed to drive adaptive and evolutionary change, he thinks there is also an element of contingency, or what Herschel invidiously called "higgledy-piggledy," in the very process of matching of variation to adaptive needs. Nothing says those matchings have to occur except the overwhelming improbability of their not sometimes occurring. "I think it would be a most extraordinary fact," Darwin writes, in a rather probabilistic vein, "if no variation ever had occurred useful to each being's own welfare, in the same way as so many variations have occurred useful to man" (Darwin 1859, 127).

In making this double appeal to chance and chances, Darwin was not saying that variations, or their manner of fixation, are random or chance events in the sense that they are uncaused (Hodge 1987). On the contrary, Darwin never questioned that every event has a cause, or perhaps a myriad of conjoined causes. What we call chance is for Darwin, as well as most of his contemporaries, a matter of our ignorance of causes, as is our reliance on probabilistic judgments to compensate for that ignorance. What is more, Darwin always had definite, if evolving, ideas about what the general causes of undirected variation are. Throughout his life, he thought that variations are caused when environmental stress affects the development of the embryo. In *On the Origin of Species* he writes, "We have reason to believe . . . that a change in the conditions of life, by specially acting on the reproductive system, causes or increases variability" (Darwin 1859, 82; cf. Hodge 1985). While this assumption does not mean that the direction of variation is responsive to stress, it does reflect Darwin's tacit retention of the classical view that reproduction and development are parts of a single cyclical process, one that can be disrupted by impediments and insults to growth. We first found that idea in Aristotle himself, who thought of reproduction as a culmination of nutrition and growth. There are even echoes, especially in the early Darwin, of Aristotle's notion that reproduction is an organism's way of saving itself from extinction. These ideas, as we saw in chapter 2, were commonplace among neoclassical biologists from Buffon to von Baer.

Darwin saw few difficulties in integrating his developmental view of generation with his Newtonian picture of ecological dynamics, in part because he tended to view variation and natural selection, operating over time, as a unitary process that embraces not only the selective downstroke of the variation-retention cycle but the upstroke, the production of variation, as well. Darwin's conception of a cycle suggests one way in which he retained the Newtonian imagination of Hutton, Lyell, and Adam Smith. Just as an economy, or Lyell's earth, is a machine that runs on its own, like a heat engine, Darwin's world is a machine for producing and then selecting variations. It is stress that produces flurries of variation, which are then culled by the force of natural selection.

Within this general picture, Darwin's problem was to determine what keeps populations, in which at least as many maladaptive as adaptive variants are presumably arising, tuned to their environments. The textual problems connected with the development of Darwin's theory, and not least this aspect of it, are tangled and still under debate (Ospovat 1981; Hodge 1985; Hodge and Kohn 1985; Richards 1987, 1992). We do not wish to prejudge them. Our own tendency, however, is to think that in earlier days, when he presumed that variation arises sporadically and only as needed to reequilibriate populations with their environments, and when he thought that natural selection was itself a functional mechanism whose "final cause" is to keep organisms and environments in balance, Darwin placed the causal accent on the conservative effects of continuous crossings and blendings, in which offspring generally come out as a mixture of both parents' traits (Hodge and Kohn 1985; Hodge 1985). Considered in terms of Newtonian thinking, this would be to treat the tendency of offspring to vary as a property like straight-line motion and sexual crossing as an analogue of gravity.[1] Species are conserved because blending preserves adaptive traits as momentum is conserved when bodies interact. Darwin never gave up this way of thinking about inheritance. Later, however, when he came to think that variation is ubiquitous rather than episodic and that the cycle of reproduction, variation, and selection is less teleological than he had earlier assumed, Darwin placed the causal accent more on natural selection, viewed as an external force that impinges like gravity on organisms in highly constrained and competitive conditions. Darwin thus makes a double appeal to the Newtonian model. It trims the tendency of variation to go off on a tangent by using sex to drive offspring back into the circle of natural biological kinds, and it trims variation to fit the adaptive requirements of environments.

This pattern in the development of Darwin's theory conforms with the main conclusion we reached in the last chapter. Darwin's quest to become the Newton of biology depended on his ability to shift the causal and explanatory accent of evolutionary theory from inner drives to outer forces. The aim of this chapter is to show that this transition was facilitated at every step by Darwin's use of explanatory models taken from the discourse called "political economy" (Schweber 1977). In particular, we argue for three influences of economics on Darwin. In addition to enabling him to challenge the dominant evolutionary theory of Geoffroy, Lamarck, Grant, and Chambers by transferring causality from tranformational drives to environmental forces, the discourse of political economy helped Darwin challenge Lyell's essentialism by bringing the individualist ontology of political economy into biology. Finally, models of economic diversification, already introduced into physiology by Milne-Edwards, enabled Darwin to explain the diversity of living things in both space and time. These influences were not felt all at once. On the contrary, they were absorbed very slowly. If this process begins with Darwin's self-described moment of illumination in September 1838 when he reread Malthus, it culminates in a second self-described moment of illumination in 1856, when on another September day Darwin realized that evolutionary diversity is related to the principle of the division of labor first made famous by Adam Smith.

To understand how political economists helped Darwin become the Newton of a blade of grass, we must first review some of the main ideas of the classical political economists, especially the revisionist brand of Malthusianism that was a part of his immediate environment. We have already noted that the great writers of the Scottish Enlightenment were the first to apply something like Newtonian thinking to the human sciences. Hume wanted to produce a Newtonian psychology, with sense impressions serving as analogues of mass points. Out of these units of sensation our complex ideas of things would be formed by the action of associative habit, which thus plays the same role in "mental philosophy" that gravity does in "natural philosophy." Hume also toyed with the idea that a whole social system might be described as a machine cunningly contrived by an invisible but godly hand to turn self-interest into justice (Hume, *Treatise on Human Nature*, bk. 3). Hume's friend Adam Smith, professor of moral philosophy at Glascow, applied this kind of thinking to that part of the social system that was most likely to conform to the general idea of economic activity. (Economics, politics, and law were all part of "moral philosophy" well into the nineteenth century. In writing about economics, Smith was not muscling into somebody else's academic turf.) Smith wanted an economic theory that worked like Newtonian astrophysics, and in the *Wealth of Nations* (1776) he produced one. In this theory, self-interest, rather than being a corrupter of persons and a distorter of natural prices, as tradition held, is both the driving force and the stabilizer of economic activity. "It is not from the benevolence of the butcher, the brewer, or the baker that we expect our dinner," Smith famously wrote, "but from their regard to their self-interest" in buying cheap and selling high.

At first sight, this may sound like a recipe for unconstrained greed and social chaos. The market mechanism, with its stern laws of supply and demand, prevents the self-interested individual from going off too far on a self-aggrandizing tangent. The market does indeed accelerate in proportion as opportunities for trading at a profit present themselves to the individual, as in Newton's second law. But if others are free to cash in on the same opportunity, supply, prices, and wages will be neatly trimmed back down into equilibrium with demand. Just as stable planetary orbits are constructed out of a moment-by-moment balance between inertia and gravity, then, a free market, into which producers and consumers can enter and exit at will, will produce stable patterns of production, exchange, and consumption In this way the economic sphere will be run by laws analogous to those of a Newtonian system.

But for Smith political economy was not just about economics. It was also about politics. To say that there are self-regulating laws governing something called "the economy" was to help establish the hitherto unheard of idea that the economy constitutes a relatively autonomous system within society that might thrive by being liberated from political control. It was also to give a piece of cheeky advice to political authorities: laissez faire, or "leave it alone." If God need not interfere with the stars and planets, neither should governments interfere in the economy. On the contrary, the natural laws of economics should inform and constrain the artifices of public policy.

Today we are so used to Smith's ideas that we must pause to see how novel and counterintuitive they once were, differing as much from received economic wisdom as Copernican heliocentrism differed from Ptolemy's geocentric astronomy, or as Galileo's mechanics differed from Aristotle's. Aristotle himself, for example, was notoriously suspicious of market-oriented societies (Finley 1973). In his view, when market relationships are allowed to dominate a city, the morals of its people and its political institutions are bound to become corrupted (*Politics* VII.1326b39–27a40; 1331a30–31b3). In his *Politics*, Aristotle argued for this view by contrasting the temperance, courage, and political freedom of the Greeks with the hedonism and slavishness of barbarian societies. The Greeks (or so Aristotle liked to think) lived in relatively small city-states, limited their population and their pursuit of wealth, defended themselves in time of war without relying on mercenaries, and in time of peace engaged in worthwhile leisure pursuits. Markets were peripheral appendages to the self-sufficient agrarian households (*oikoi*) of free male citizens, each of which was managed by frugal womenfolk and worked by (barbarian) slaves. Trade was limited to something like barter between the excess products of different households and regions and to imports of farm tools, weapons, pots and pans, and perhaps jewelry. For the most part, external trade was handled by foreign, or at least noncitizen, merchant communities, and no self-respecting Greek ought to try to take their place. There was no market in land or, for the most part, in labor. In theory, it was all very Jeffersonian. By contrast, barbarian societies (or so Aristotle liked to think) spawned vast, uncontrolled urban aggregations like Babylon, whose inhabitants were unreflectively devoted to animal-like production and consumption mediated by markets (*Politics* III.1276a26–31, 1285a19–24). Whoever had the power to make others produce was able to consume most. Thus, wealth was regarded as a horde of money that enabled one to maximize consumption and to buy military and police protection, as the tale of the barbarian king Croesus suggests. (Croesus starved because, in accord with his fondest wish, everything he touched turned to gold.) Greeks, Aristotle sneered, had better things to do with their time than that. Men could to go down to the city square to talk, to the gymnasium to keep themselves fit for battle, to the courts to judge their fellows, to the assembly to decide public policy, or to the groves of Academe or the cool porch of the Stoics or Aristotle's Lyceum to take in some philosophy lectures, in which they could learn how much more noble than the barbarians (and than women) they were.

Given a picture like this, it is easy to see why Aristotle might be nervous about societies in which markets play a large role. If a society is dominated by commercial activity, he thought, it will soon degenerate into a barbarous way of life (Depew 1988). Since there are always tendencies in this direction, it is the chief responsibility of the military and political class to control markets and to educate the young citizens to despise consumerist values. Aristotle, accordingly, would have thought Smith's ideas about market society absurd and the notion that he was a professor of moral philosophy laughable. Indeed, for Aristotle the phrase *political economy* itself would have sounded self-contradictory. Except for some very thin connective tissue between them, the economy (from Greek *oikos* + *nomos*, "rule of the household") was one thing, and politics (from Greek *polis*, "city-state") quite another. Nor did medieval and early modern attitudes about economic activity differ all that much from those of the ancients. It was still assumed that an uncontrolled market was economically and morally a bad idea. That is why there were guilds to control production and officials to control prices. The church sanctioned this notion by proclaiming that greed, once let out of the bag in a world cast down by original sin, would run amok. Accordingly, the clergy disapproved of interest even more than Aristotle, who called it "money breeding money."

Clearly, then, Smith's proposal was as revolutionary as the new astronomy. Aristocracies from antiquity to the *ancien régime* had thought that the willingness of a military man to risk his life, and to despise the animallike consumerism of the many, would facilitate the cultivation of even higher values and thoughts. These values would make the difference between a good and a bad state. Looking at the historical record, however, Smith, Hume, and other luminaries of the Scottish Enlightenment concluded that aristocratic values, and the regimes they legitimated, had actually brought upon humanity an inordinate amount of misery and parasitism. The prickly code of honor, revenge, quickness to violence, and contempt for one's alleged inferiors that characterizes the ruling classes of such societes contrasts badly with the rational self-control of a merchant, who, even though he is vulgar enough to hope to get rich, must use his head to calculate his long-term interest. In the process, he controls his passions and frees himself from their destructive effects (Hirshmann 1977). In this rationally calculative frame of mind, morality, especially a sense of fairness to others grounded in long-term self-interest but internalized into the norms of conscience, will flourish. So will prudent statecraft, in which self-reliance will be demanded of everyone and contempt will be heaped on those who consume without producing. Conspicuous displays of social superiority will be replaced by the cultivation of a well-regulated, sober, and, above all, private inner life. The Victorian middle-class morality of Darwin's milieu was rooted in these "bourgeois" values, the values of city dwellers (from French *bourg*, "town").

It should be clear from these considerations that the notion that markets should be freed from political control was not a merely technical issue of efficiency for Smith. As a moral and political philosopher, he was arguing in effect that a little greed is not a bad thing. A commercial republic, or perhaps a constitutional monarchy, rather than being unstable and ignoble, will actually sow the seeds of moral development, economic progress, and political stability. Only increasing riches keep people from spending most of their time killing each other (Skinner 1978). Views like these signal the emergence of modern European republicanism and liberalism, in the widest sense of these terms. It is small wonder that these theories were worked out in Scotland, a Calvinist country whose secular Enlightened wing was affected by the individualistic and moralistic values that tied Protestantism to the rise of capitalism and republicanism. Rooted in the great revolutionary transformation that by the middle of the nineteenth century had displaced the old aristocracy and left the bourgeoisie in command of society and state, "political economy" was not "value-free science" but a way of encoding a powerful new view of the social world. Not surprisingly, it was a view that made much rhetorical use of the same Newtonian explanatory models that had ushered in a new picture of astrophysics.

In spite of his importance, Smith was not the first political economist. He adopted and transformed the ideas of the French physiocrats, notably those of François Quesnay. The physiocrats were devoted to keeping shortsighted monarchs and parasitic aristocrats from making the mistake of poor old Croesus, who thought of wealth as money. (The long-term consequences of this delusion were plain to see in the decline of the Spanish empire, whose unwillingness to invest its New World gold, rather than alternatively to hoard and waste it, had brought predictable ruin.) Nor was it wise for rulers to overly protect and regulate industry. It was against the internal tolls and tariffs of France that the cry "laissez faire" was first heard. Wealth, thought Quesnay, who was a royal physician, is like the circulation of blood. It keeps the body healthy by circulating freely through the society. Neither should authorities worry that in being spent it will be dissipated. Value (henceforth a central, but contested, concept in economics) is conserved through exchange until it disappears in consumption and is renewed by the inherent productivity of the land. Value courses through society in ever-changing forms in a gigantic cycle. It is easy to see why the main technical aim of the physiocrats was to produce a giant flow chart, a *tableau économique,* showing what these pathways are.

Unlike the physiocrats, it was not Smith's view that the natural productivity of the land is the ultimate source of value. For Smith, value is a conserved quantity because it consists in units of labor of all sorts, which are aggregated, stored, traded, consumed, and created anew by more labor. This being so, profit is made by dividing labor more efficiently and finely than one's competitor, a condition that presupposes a free market not only in commodities but in land and labor as well. The overall result of Smith's self-governing economy will be that a maximum amount of output is produced with a minimum amount of labor and other productive inputs. When the market is free to take its natural dynamical path, not only will it create of its own accord a balance of supply and demand, and spontaneously natural prices, but will progressively divide labor in ever finer ways, creating more general wealth and sustaining in consequence a more stable and advanced society. Smith's economy was a machine that ran by itself, like Hutton's earth, and that improved society in the bargain, like Watt's steam engine. What greater triumph of Enlightenment modernism over ancient conservatism could there be than the discovery that the aim of statesmanship is not to administer scarcity but to break through its limits—and to assert that making everyone richer, rather than insisting fruitlessly that they be virtuous, is the best way to allow republican, and even democratic, governments to at last have their day in the sun.

Because he encoded his theory in the associationist psychology typical of British philosophy, and especially of Hume, rather than in the sociological terms typical of French thought, Smith gave a more individualistic twist to economics than the physiocrats had. What happens to the whole is a function of what the separate parts do. The relation between microeconomics of the individual and firm and the macroeconomics of the society and state is additive. This individual-centered way of putting things bequeathed to Smith's successors a set of issues about what the long-term trajectory of an economy will be when economic decisions are placed entirely in the hands of "rational economic agents," each of whom is presumed to be seeking nothing other than his or her own maximum profit. It was David Ricardo who explicitly cast political economy in these terms in his *Principles of Political Economy* (1817). Although he showed that an economy run this way would be maximally productive and efficient, a certain pessimistic note about long-run tendencies can also be heard in Ricardo's work. A Ricardean economic agent prowls the world looking for ever-new sources of profit. However, the rate of profit tends to fall over time and the economy to move closer to a steady state. That is because there are limits and constraints built into Smith's wonderful machine that its designer seemed to have overlooked.

Ricardo agreed with this friend Malthus, who taught economics at the in-house college of the East India Company, that the improvement of an economy will necessarily mean more new mouths to feed. Malthus took it to be a law of nature that organisms tend to maximize their offspring. It is an inertial tendency of living things to reproduce up to the limit of available resources. Since this continuous production of new bodies will, if unimpeded, always outstrip new increments of sustenance, equilibrium between supply and demand is constantly being restored by the force of scarcity, an analogue of gravity. If populations grow because of increased production, they nonetheless do so under constraint. They are always at or near "Malthusian limits." Not even modern economies, Ricardo thought, can beat the game forever, for staying ahead will depend on bringing new sources of food into the picture. But land, as Malthus had shown, is a limiting quantity, and the productivity of land is relatively inelastic. Ricardo added insult to injury by arguing that the cost of feeding and housing workers is inelastic too. The capitalist's sole source of profit is the difference between the value that a worker creates and what he gets paid. But there is a mimimum standard of living that the workers must be paid if they are to be capable of working. Thus, although its monetary value will vary, wages are fixed by Ricardo's "iron law of wages" at an absolute minimum. As competition ensues, therefore, the capitalist's source of profit, as Marx gleefully pointed out, progressively dries up. Accordingly, an economy, even as it expands, moves closer and closer to a terminal state where not much new happens and where social tensions will undoubtedly rise. Only the division of labor in manufacturing keeps the game going at all.

Economies, like heat engines, need ways to blow off steam. Accordingly, statesmen began to pick up from practitioners of the dismal science advice like this: It might be a good idea to open up the country to the import of cheap food, to pack off excess population to colonies, to urge workers to abstain from having children—and above all to make sure that the unemployed cannot retreat to the dole, which will make things economically and politically worse, whether in the name of charity or of utopian hope. These were the policies of the Reform Bill. None of these proposed remedies is intrinsically inconsistent with the Malthusian principle, for the law does not say that people can do nothing about the gap between population and resources, even if plants and animals cannot, but only that what they do is constrained by a range of possibilities and comes up against limits. It is true that early Malthusians, especially Malthus himself, were more pessimistic about these expedients than later ones, for they assumed that the behavior of the poor would continue to be animallike. However, the neo-Malthusian "Philosophical Radicals," whom Darwin knew through his brother Erasmus, such as Martineau and Spencer, put a new spin on the doctrine and in the process pictured a world much closer to Smith's than to Ricardo's.

In the short run, the new Malthusians argued, society can arrange things so that the resources each person gets do not fully depend on what each is capable of purchasing from the results of one's labor or the labor congealed in rent or profit. That's what socialists (and some liberals, like John Stuart Mill and his wife, Harriet Taylor) wanted to do. However, society will run a lot better, Martineau and Spencer argued, and in the long run make everyone better off, if it chooses to keep a little Malthusian pressure on everybody. This will energize individuals, put a premium on new ways of making a living, encourage thrift and hard work, make birth control and abstinence ever more attractive, and allow the self-discipline required for higher intellectual, aesthetic, and moral pursuits to flourish. Rather than refuting Adam Smith's wonderful world, Malthusianism would, on this view, usher it in.

We are entitled to think that when Darwin reread Malthus on September 28, 1838, he would at least tacitly have been aware much of what we have been reporting. In his *Autobiography* Darwin says that this event was important to him because "I had at last got a theory whereby to work" (Darwin 1958, 120). One might think from this remark that the theory of natural selection sprang forth fully formed from Darwin's head on that very date. Closer inspection of Darwin's notebook entry for September 28, 1838, suggests, however, that this was not the case. Indeed, it has been somewhat of a puzzle to scholars precisely what Darwin did get out of reading of Malthus that day (Mayr 1977; Ospovat 1981; Schweber 1977; Hodge and Kohn 1985). The relevant notebook entry reads as follows:

We ought to be far from wondering of changes in numbers of species from small changes in nature of locality. Even the energetic language of De Candolle does not convey the warring of the species as inference from Malthus—increase of brutes must be prevented solely by positive checks, excepting that famine may stop desire—in nature production does not increase, while no check prevail, but the positive check of famine and consequently death. I do not doubt everyone till he thinks deeply has assumed that increase of animals exactly proportionate to the number that can live—Population is increased at geometrical ratio in far shorter time than twenty five years—yet until the one sentence of Malthus no one clearly perceived the great check amongst men—there is a *spring*, like food used for other purposes as wheat for making brandy—Even a few years plenty makes population in man increase and an ordinary crop cause a dearth. Take Europe: on an average every species must have same number killed year by year by hawks, by cold, etc.—even one species of hawk decreasing in number must affect instantaneously all the rest—the final cause of all this *wedging* must be to sort out proper structure and adapt it to changes—to do that for form, which Malthus shows is the final effect (by means however of volition) of this populousness on the energy of man. One may say there is a *force* like a hundred thousand wedges trying to *force* every kind of adapted structure into the gaps by *forcing* out weaker ones. (Darwin 1987, 374–75, italics added)

It was certainly not the bare notion of superfecundity that Darwin first realized on this occasion. As he does here, Darwin just as often refers to Augustin De Candolle in this connection as to Malthus. Nor was this Darwin's first realization of the consequences Malthus and his followers drew about the effects, whether disastrous or benign, of superfecundity in a world of scarce resources. Darwin, as we know, had been acquainted with the Malthusian line taken by Martineau and others ever since his trip on the *Beagle*. On the other hand, reading Malthus does seem to have affected how Darwin thought about superfecundity. For one thing, he may have been impressed by the idea that there is a stable quantitative measure of the gap between reproduction and resources (Schweber 1977, 1980). Malthus says that population increases geometrically (exponentially) and food supply only arithmetically (additively). He also shows mathematically that it makes little difference whether the species in question is as fast breeding as a fruit fly or as slow breeding as an elephant. Darwin found that interesting.

Whatever chain of influences may have led up to his close encounter with Malthus, however, what most vividly stands out in this passage is that Malthus's stress on external circumstances and quantitative relations leads Darwin to appreciate, with a depth of feeling that is palpable, the tremendous external force that population pressure regularly, and therefore by universal law, exerts on all organisms all the time. In the passage,

Darwin uses Malthus's high intrinsic rate of population growth to explain the apparently stable number of organisms. The explanation is framed not in terms of internal restraints on procreation but as a function of the pervasive competition and death that are necessarily brought about when population increases faster than resources. Given the fact that population always tends to outstrip resources, competition is universal rather than episodic. It is the constancy of this pressure that provokes Darwin's analogy between bioeconomics, as we may call it, and machines, an analogy deeply rooted in Scottish economics that is fleetingly but powerfully present in Darwin's comparison between a "spring" mechanism and a process that trims population growth to fit available resources. By the end of the passage, Darwin is already glimpsing an explanation of adaptedness, if not of speciation or evolution. Surely the organisms that survive this pressure cooker are those that can command scarce resources better than their competitors. There is a "force," Darwin says, "like a hundred thousand wedges trying to force every kind of adapted structure into the gaps by forcing out weaker ones."

This is not yet, however, a theory of natural selection. For one thing, when at the end of the passsage Darwin speaks of "energy" and "will" he is talking about adaptive traits that are built up by use and habituation and are passed to offspring. "Habits give structure," Darwin writes elsewhere in the notebooks. "Therefore habits precede structure, therefore habitual instincts precede structure" (Darwin 1987, 301). He also says that the arms of the children of blacksmiths soon come to resemble those of their fathers (Gruber 1974, 338). It is clear that Darwin first envisioned the feedback relation between stress and variability in a generally Lamarckian way, in which population pressure serves to motivate organisms to move about and experiment. What Darwin was doing on September 28, we conclude, was placing an account of adaptation conceived along these lines into a more force-filled external context, thereby blunting its connection with the inner drive toward complexity that at the time even he found inseparable from evolutionary theory. Darwin might well have remembered this moment as crucial, even though he did not discover natural selection on September 28, for someone who was seeking a theory of evolution that conformed to Newtonian canons would have recollected vividly when he first found the explanatory means to replace inner drives with outer forces.

This interpretation of the significance of his encounter with Malthus is supported by the fact that Darwin's musings on September 28 seem to have helped him in the first instance to deflect an argument of Lyell's against Lamarck about which he had been troubled for some time. If Darwin's notebook entry is to be our guide, it seems likely that "the one sentence of Malthus" that caught his attention on September 28 was not the bare statement about the massive, regular, quantifiable effects of unchecked population growth but a sentence in which Malthus talks about the effect of population pressure in stimulating people to become ruthless colonizers and pioneers. The threat of starvation and extinction is the mother of invention. Accordingly, when Darwin says that population pressure motivates humans to exert themselves, he is alluding to the Malthusians of his own acquaintance, who see population pressure as useful in motivating the unemployed of free-market Britain to emigrate to new colonial territories, where they will displace less energetic indigenous peoples. This was a process, it is worth noting, that Darwin had had occasion to observe. He was distressed to see Spanish colonists in Argentina commiting palpable genocide on the native population. It turns out, however, that in volume 2 of *Principles of Geology* Lyell himself had considered a scenario very like one that Darwin encountered while reading the sixth edition of Malthus's *Essay on Population*. A climatic change triggers invasions of hardier species into the range of settled ones, whose immobility reflects their inability to cope well with changed ecological circumstances. Lyell uses this thought experiment to argue against Lamarck. Since Lamarck's adaptations by use, habituation, and the transmission of acquired characteristics must be assumed to be brought about only gradually, whereas adaptive response to a situation like the one portrayed must be quick, the result must be, Lyell says, that the native population "must perish before they had time to become habituated to such new circumstances. That they would be supplanted by other species at each variation of climate may be inferred from what we have said of the known local exterminations of species which have resulted from the multiplication of others" (Lyell 1832, 2:173–74). Lyell thinks this refutes Lamarck. Darwin sees, however, that Malthusian population pressure will drastically affect this thought experiment. Under Malthusian conditions, only organisms that quickly adjust to circumstances by acquiring new habits and manage to pass these on to their young in a hurry will have a future. It is not the settled population, however, that must somehow change, as Lyell and Lamarck assume. They plainly will not be able to do so in time, as Lyell demonstrates. Instead, it is the invaders who change. Indeed, their very mobility and aggression testifies to the fact that they have already risen to the challenge presented by the scarcity that environmental change has imposed on them. Darwin's underlying thought is that energetic emigrants are, by that very fact, under considerable stress, and hence will tend to adapt faster to changing circumstances.

It not was until some weeks after his encounter with Malthus, when Darwin started playing with an analogy between nature's way of picking and the selective picking of breeders, that natural selection proper makes its first appearance (Hodge 1987; Hodge and Kohn 1985). The emergence of a reasonably clear conception of natural selection was, however, facilitated, we believe, by Darwin's enhanced appreciation of external forces in late September. What led Darwin to defend Lamarck against Lyell led him in turn to criticize Lamarck. According to the metaphor on which the notion of natural selection rests, the plant or animal breeder does not have any effect on what differences turn up in a litter but only on which variants will have a future—if he picks them. (Note that a breeder is an external agent.) That is certainly Darwin's candidate henceforth for a Herschelian *vera causa* (Hodge 1987). We do not think, however, that the selection metaphor, by itself, was sufficient to give Darwin his *vera causa*. The ubiquity and sheer force of population pressure, available in compressed form in the spring and wedging metaphors that are so prominent in the passage just quoted, had to have been there already if a would-be Newton of a blade of grass was to convert a particular causal mechanism of adaptation by way of an external agent (selection) into a law-governed, and ubiquitous, external force. People who tend to dismiss Darwin's theory as mere metaphor significantly tend to think that the only metaphor in question is that of picking (Midgley 1985). The metaphors of wedging and springing are just as important. Nor does selection replace the others. Rather, picking is superimposed on springs and wedges, with the result that the process called natural selection occurs on a field whose underlying structure is that of a schematized Newtonian system, as we explicated that notion in the last chapter. In consequence, not only does selection become a natural force, like gravity, rather than an elective, random, or sporadic activity, like the choices of breeders, but Darwin is able to shift the emphasis from the internal, developmental causes that had long been used to encode theories of transmutation to external forces, thereby conforming his theory even more closely to Herschel's Newtonian canons.

The notion that external forces induce adaptive change by bearing down on organisms was not, however, merely a conceptual or ideological triumph for Darwin's Newtonian project. Natural selection had in addition the problem-solving power that Darwin always took as the test of ideas. The Lamarckian Darwin had for some time been troubled by the conviction he had acquired from William Yarrow, a knowledgeable breeding expert, that older, more deeply embedded traits predominate over newer ones in crosses. What Darwin called "Yarrow's Law" seemed to make it very difficult for traits acquired by adults to pass deeply and quickly enough into the inherited developmental program to produce permanent and significant adaptive reequilibriation. The enhanced stresses to which organisms were exposed in a Malthusian world went a long way toward solving this problem. Constant stress on adults, particularly females carrying young, would doubtless affect embryos. Variation might occur at any time in the developmental cycle and, if it proved adaptive, would recur at the same time in the development of offspring (Hodge 1985, 1987). Rather than having to depend on the heritable actions of adult organisms, then, as those stressing the inheritance of acquired characteristics were forced to do, Darwin concluded that new adaptive traits of far greater significance might be forthcoming, albeit on a more chancy basis, when constant competitive Malthusian pressure triggered developmental errors in embryos and then selected the most highly adapted of them. Darwin's new theory thus showed its problem-solving power by yielding a theory of adaptation, and for the first time of evolutionary change itself, that was consistent with Yarrow's law. Darwin never denied that Lamarckian inheritance and adaptation were possible or that they actually occurred. His new view, however, was that this processs would affect only local and perhaps evanscent adaptive problems, such as those faced by giraffes, blacksmiths, and pioneers, and therefore that its role in significant evolutionary change was subordinate to that of natural selection. In this way, the ubiquitous stresses induced by Malthusian population pressure helped Darwin insert adaptation into an even more intense matrix of external forces than when he was a follower of what was conventionally taken to be Lamarck's position. That, of course, is precisely what a Newton of a blade of grass had to do.

These early advances should not lead us to overlook the fact that Darwin's notebook entry for September 28, 1838, also betrays a number of quite traditional ideas from which Darwin would free himself only gradually. For instance, he interprets the production of adaptedness by competition somewhat teleologically. He says, "The final cause of all this wedging must be to sort out proper structure and adapt it to changes." So deep in fact was Darwin's early tendency to look at secondary causes in a rather Whewellian light that Malthusian population pressure itself is viewed from a functional perspective. Although he does not say that there is a "final *cause*" of population pressure, Darwin does say that it has a "final *effect*," a curious phrase. The "final effect" to which Darwin alludes is probably the Utilitarian identification of the good and the useful, which formed the foundation of the Malthusian enthusiasms of his brother's circle of friends.

Even after Darwin found his way to natural selection, the teleological way of looking at biological equilibrium retained its grip. In the *Essay* of 1844, the production of variation under stress and consequent reequilibriation by natural selection is treated as a functional process, like any other aspect of the physiological machinery of generation and development. Its point, Darwin holds, is to keep populations perfectly, or nearly perfectly, adapted to their changing (Lyellian) environments (Ospovat 1981; Hodge 1985; Schweber 1980). This accords well with Darwin's developmentalist view of the stressful conditions under which variation is assumed to arise in otherwise well-adapted populations, for, on this interpretation, variation, albeit undirected, must be presumed to be scarce, since it occurs only when abnormal conditions provoke developmental disorders, which in turn serve as a materials for reequilibrating organisms and environments. (If variation appears plentiful and diverse in the case of extensively bred species like dogs and pigeons, that is presumably because organisms in this situation have been taken out of their natural environments and have by that very fact been subjected to a great deal of variation-inducing stress [Hodge 1987, 243].)

It was only after Darwin moved toward a more individualist ontology that he began to treat variation, adaptation, and transmutation as effects of contingent processes rather than as goal-directed processes. We see in this shift a second effect of classical economics on Darwin's conception of biological processes. Over time, the habit of looking at bioeconomics after the manner of the political economists gave Darwin not only a way of intensifying the environmental forces being exerted on organisms but induced him to move steadily toward inscribing into his way of talking about organisms and species the more individualistic, antiessentialist conceptions of the economists (Schweber 1977, 1980).

Lyell still assumed, as classical biologists did, that the primary locus of competition is between different species, which make a living by preying on each other. So did de Candolle. There are also traces of this view in the journal entry we have been analyzing. Like de Candolle, Darwin assumes at the beginning of the passage that Malthusian population pressure intensifies the war between species, not yet fully realizing what the later parts of the passage imply: that intraspecific warfare, that is, contestation for the same resources by members of one species, is the primary site of competition and the root of interspecific competition. When they are viewed in the light of Malthusian considerations, however, Lyell's own examples of stress-induced competition suggest that the primary competitors of each organism are actually members of its own kind, which depend on the same resource base, rather than members of different species, which utilize different resources. As in economics, the true analogue of a mass point is not the species but the individual, and the basic relation between conspecifics is competition rather than cooperation.[2]

This shift toward an individualist ontology advances Darwin's Newtonian program considerably. If competing individuals are the primary realities, Darwin was free to abandon the essentialist assumptions of classical and neoclassical biology still retained by Lyell. In their place he could populate Lyell's stress-filled geological and biogeographic world with competing individuals, whose dealings with one another and with their inorganic environment are brought into balance in the same way that free markets bring supply, demand, and price into equilibrium. Indeed, if all causal transactions are reducible to forces exerted by and applied to individuals competing in a closed system, no high-level, conceptual reasons remain for continuing to assume that variation must be confined within a species-specific limit, or that a species necessarily snaps back to some preestablished norm, or that as one departs from type (whatever that may now mean) organisms are necessarily less fit. In a world made up entirely of individuals and their relationships, there is nothing to which this alleged limit on variation could apply. In the changing environments that Lyell's Newtonian geology had postulated, what is fit at T_1 may be unfit at T_2. The very fact that Lyell declares that species tolerate wide amounts of variation in order to remain adaptable suggests that he needs, even if he cannot bring himself to acquire, a much less essentialistic conception of species. Darwin relieves the internal tensions in Lyell's theory by concluding that species are not real units in nature. Individuals are.

This shift away from the most basic assumption of classical and neoclassical biology—a shift that Mayr, Hull, and others rightly take to be the crux of the Darwinian revolution (Mayr 1982, 1988, 1991; Hull 1965, 1980)—was in part, we conclude, an effect of Darwin's slow introjection of the ontology of classical political economy into biology. The inscription of political economy into biology, a discipline that had long been more affected than any other by essentialisms of all stripes, had several additional consequences of great moment, which progressively began to manifest themselves in Darwin's mature thought.

At first, as we have seen, Darwin assumed that variation is triggered only when populations get out of kilter with their environment (Ospovat 1981; Burian 1983). On this view, variation can be part of an explicitly functional process. It can have a final cause. This is even classical and neoclassical biology of a sort, since it makes adaptedness a functional process. It already breaks away from the Aristotelian tradition, however, by characterizing variation as potentially useful, that is, as something more than deviant or monstrous. But Darwin's subtle, and perhaps for the most part unconscious, replacement of classical biological essentialism with the individualist ontology of modern economics soon turned this break with the classical tradition into a presumption that organisms must be assumed to be inherently prone to indefinite variation. Since they are not representatives of constraining types in the first place, why should they be presumed to be ready to snap back to some inner norm? This presumption certainly undermined the classical and neoclassical idea of deviation from an alleged species-specific norm. At the same time, however, it undermined Darwin's own residually teleological view that the function of variation is to maintain perfect adaptation. If only individuals exist, why shouldn't they vary, if only under stressful disturbances of the developmental sequence? And, if so, why shouldn't the match between surviving organisms and environments be a contingent effect rather than a functional process?[3]

Darwin's individualist ontology has profound implications for what a species actually is. A species is not, Darwin slowly and haltingly came to recognize, a type defined by a set of necessary and sufficient properties but a group of reproductively linked individuals in time and space sharing more or less the same degree of distance from common ancestors.

Absent essentialism, whether typological or constitutive, a group's taxonomic status as a species, rather than a variety or genus, is to some extent, then, a function of arbitrary judgments about classificational rank. This is what Darwin means when he writes in *On the Origin of Species* that "we shall have to treat species in the same manner as those naturalists treat genera, who admit that genera are merely artificial combinations made for the sake of convenience" (Darwin 1859, 485). In holding this, Darwin does not mean to say that species are not real things, a claim that would make the title of his book ironic indeed (Beatty 1985). On the contrary, the fact that the entities Darwin is talking about in his book are real things at last becomes unproblematic when they cease being seen as instances of kinds, types, or forms and begin being viewed as populations (Mayr 1988). Darwin did not always see this clearly himself. Clarifying the conception of "species" to which the Darwinian evolutionary theory is committed has involved a major, and largely successful, effort by twentieth-century Darwinians (Ereshefsky 1992; cf. chap. 12). To the extent that Darwin did see into the populational nature of species, however, more was involved than the nominalist epistemology that he inherited from Hume (Richards 1987). Having nominalist proclivities certainly helped. But what really did the work was Darwin's decision to describe biological reality in terms of the individualist ontology that was already a prominent feature of how the political economists saw human reality. In sum, in relieving the tension between Lyell's Newtonian geology and his Aristotelian biology, Darwin rids the organic world of Aristotelian essences altogether by generalizing the individualist ontology of political economy.[4]

A notable aspect of Darwin's individualism and antiessentialism is that they change what needs to be explained about species and what does not. We have already registered our belief that what counts as explanatory is a context-relative affair. What needs to be explained depends on what background assumptions are taken as unproblematic (Garfinkel 1980). We have seen, for example, that Aristotle's conception of natural motion meant that circular orbits did not need explaining while deviation from circularity did. By contrast, Newton's conception of natural motion as rectilinear made it necessary to explain why planets do not go off in straight lines. Similarly, we have seen that classical and neoclassical biologists, from Aristotle to Lyell and Asa Gray, assumed that species, like the perfect circles of ancient astronomy, are faithful cycles of growth and replication. In that case, variation within a species is essentially limited. What needs to be explained is departure from type. We can now see that Darwin's conception of species as groups of interbreeding individuals sharing a common ancestry reverses what can be taken for granted and what needs to be explained. Darwin takes departure from type so far for granted that not much sense can even be assigned to the idea, in spite of his retention of this very phrase in the title of his most famous book. What now needs explaining is not why populations and species change but why they remain as coherent as they do! In a world where Aristotle's assumption of replicative fidelity is not built in from the outset, the stability that both Aristotle and Darwin see in the world must be reconstructed as a dynamical equilibrium between the past and the present of the lineage, just as Newton's dynamics constructed orbits out of instantaneous motions. Once more, causal explanation is transferred from inner tendencies to outer forces.

If novelties, in the form of spasms of undirected variation, can be introduced at the whim of every passing environmental contingency, and if differing individuals are all that really exist, populations might readily fragment so often that the relatively coherent taxonomic hierarchy we have every right to think we observe in nature (even if we do not know how to map it precisely onto the biological continuum) would look like a miracle. Some force is required, thought Darwin in his Newtonian way, to pull interbreeding groups back together without blocking change. Certainly the relatively stable boundaries of the ecological niches within which organisms toil do some of the shaping work. But Malthusian pressure puts such a premium on specialization and differentiation that niches themselves will be multiplying and changing regularly. We agree with Hodge that what primarily ties populations together for Darwin is sex (Hodge 1985). Rather than being treated as a source of variation, an idea that became compelling only after the discovery of independently sorting and recombining genetic elements physically located on chromosomes, sex served Darwin as a source of order. Darwin explains how species, considered nonessentialistically, remain more or less constant across their range by relying on his lifelong conviction that healthy offspring are a blend of parental characteristics. "Blending inheritance" means that "if the offspring does not comes out intermediate between the character of the two parents that is a sign that the normal function of crossing, preserving constancy, is close to frustration" (Hodge 1985, 220). Accordingly, "Crossing, with the blending of parental characters, keeps the species constant as long as conditions are constant overall and changing only temporally and locally" (Hodge and Kohn 1985, 188).

On the other hand, even while interbreeding groups remain relatively coherent entities in space and time in this way, adaptive variations, induced by stress and selected because they provide enhanced fitness, can be moved along in a certain direction, in spite of the homogenizing force of sex, by continued selective pressures adding up and safely passing along small differences, in the process pushing populations apart from one another into new "races" and species. Thus, "a new variety can be formed and adapted to new conditions if the conservative action of crossing is circumvented by the reproductive isolation and inbreeding of a few individuals in the new conditions, whether that reproductive isolation arises with or without geographical separation" (Hodge and Kohn

1985, 188). In consequence, a new, and constantly changing, equilibrium between organisms and environments is maintained: "Blending inheritance, which diminishes these deviations at each reproduction, [is] counteracted by the effect of the conditions of life in stimulating [and culling] fresh variation in each generation" (Olby 1985, 47).

A double analogy to Newtonian dynamics seems at work in Darwin's thought. It is an inertial tendency of organisms both that they maximize their offspring and that they tend by their nature as individuals to vary from parents and ancestors. If new variants have a tendency to go off on a tangent in the second of these ways, the mechanics of blending inheritance will pull them back into a relatively coherent form. That is how the analogue of gravity works at the level of Darwin's "life-long generation theorizing." If in remaining tuned to changing environments, on the other hand, populations are driven further and further apart even while remaining coherent, that is because the inertial tendency to superfecundity is countered by the gravitational force of natural selection in retaining only those variants that are best able to use available recources. Here is dynamic equilibrium indeed!

To underwrite, ensure, and explain blending inheritance, as well as the transmission of useful variation from generation to generation, Darwin invented his theory of "pangenesis," according to which information about traits passes from all the body's cells and tissues to the reproductive system, where they are banked and blended by crossing (Darwin 1868). Because Darwin was wrong about how inheritance actually works, later Darwinians have tried to paint pangenesis as a peripheral and speculative hypothesis designed to account for the secondary influence of acquired characteristics, and hence to be shed with few tears at the first sign of something better (Mayr 1982). Recent historians have shown, however, how central to Darwin's conception of pangenesis is (Hodge 1985). Darwin writes:

On any ordinary view it is unintelligible how changed conditions, whether acting on the embryo, the young or the adult animal, can cause inherited modifications. . . . But on our view we have only to suppose that certain cells become not only functionally but structurally modified; and that these throw off similarly modified gemmules. This may occur at any period of development, and the modification will be inherited at a corresponding period. . . . Each living creature must be looked at as a microcosm—a little universe, formed of a host of self-propagating organisms, inconceivably minute and as numerous as the stars in heaven. (Darwin 1868)

In an individualistic biological theory that needs to explain species continuity, rather than deviation from a norm, something like pangenesis serves as an analogue of Newton's atomism. Additive effects at the populational level, both those that keep a species relatively coherent and those that drive them apart, could only be the result of "inconceivably minute" gemmules that are "as numerous as the stars in heaven."

This picture of dynamics, working on competing but at the same time interbreeding individuals, helps explain Darwin's unbending commitment to gradualism, the component of Lyell's science to which he was most attached. Huxley, we may recall, told Darwin that in clinging to the doctrine that "nature does not make leaps" (*natura non facit saltum*), he had saddled himself with an unnecessary burden (Huxley to Darwin, November, 23, 1859, in Darwin 1887, 2:26–27). For his part, Huxley thought speciation, if not adaptation, must occur by sudden saltational change. Darwin, however, refused to take the easy way out: "If it could be demonstrated that *any* complex organ existed which *could not possibly* have been formed by numerous successive slight modifications, my theory would absolutely break down. . . . We should be extremely cautious in concluding that an organ *could not* have been formed by transitional gradations" (Darwin 1859, 190, italics added).

The fact that Darwin clung so tightly to gradualism registers the fact that in his conceptual model, as in Newton's, instantaneous infinitesmal changes, working on myriads of slightly different individuals, are the sole locus at which force can be applied, and hence at which causal transactions can occur. Darwin shifted the site of causal interaction in biology down into the infinitesmally small differences between individuals, preserved and transmitted by the hereditary units he called "gemmules." Large-scale changes are the result of adding up very small, and very gradual, changes. Working within similar continuationist assumptions, Newton had explained a great deal about the physical world. Darwin now proposed to do the same thing for the biological world. It is no surprise, then, that he clung so tightly to his gradualism. So deeply was it embedded into the conceptual structure of his theory that without it Darwin, in his own view, would have surrendered the explanatory power of his model.

We have been arguing that as Darwin progressively inscribes more of the individualist ontology of political economics into Lyellian environments, Newtonian dynamical models play ever deeper roles in his thinking. Malthusian reproduction is an inertial tendency that is trimmed into equilibrium by competition for scarce resources. Variation, unconstrained by preset limits, is free to accumulate in populations until it is culled and shaped into adaptations by the requirements of the environment. Since on this view it is easier to think that coherent interbreeding populations will fragment and dissipate than to assume that they will also remain "essentially" the same, Darwin invokes the homogenization brought about by sexual interchange to explain why species remain as coherent as they do. Here too there is an analogy between inertia and force. We come now to a final example of advancing the program of a Newtonian biology by inscribing the ontology of political economy into evolutionary biology. Darwin's appeal to competitive economics was crucial in his protracted struggle to map natural selection onto biological diversification, and hence onto classification by way of his branching conception of nature's diversity. This problem was resolved only as the watershed year of 1859 approached. Its resolution depended on appealing to Malthusian considerations to establish what looks like a biological analogue of Adam Smith's division of labor, mediated by the "physiological division of labor" that Darwin found in the work of Henri Milne-Edwards. This was an idea that had affinities with main claims of the Malthusian circle of Darwin's younger days. It comes into its own only as Darwin rids himself, under the increasing influence of economic thinking, of still more residues of biological essentialism and teleological thinking.

We have seen that in the late 1830s and early 1840s Darwin tended to think of the production and culling of variation as a functional phenomenon, as having a "final cause": to keep populations as perfectly adapted to their environments as Paley assumed them to be (Ospovat 1981). Once Darwin had firmly internalized an individualist ontology that made variation natural and ubiquitous, however, he no longer had any reason to think that, even though variation is indeed triggered by stress, its production and culling is part of an adapted or adapting mechanism. If individuals differ naturally and prolifically, variation will occur in any case. As a result, natural selection's use of these variants can be viewed as a mere effect rather than as a function. Whatever remnants of external teleology lurked in Darwin's mind evaporated with this recognition. Relative adaptedness replaced perfect adaptedness produced by a "final cause" (Burian 1983). Natural selection will preserve whatever variants happen to be ever so slightly better than others at thriving and replicating in a given environment, even if they are not perfectly adapted. "Natural selection . . . adapts the inhabitants of each country only in relation to the degree of perfection of their associates," Darwin wrote in On the Origin of Species. "Nor ought we to marvel if all the contrivances in nature be not, as far as we can judge, absolutely perfect" (Darwin 1859, 472). The upshot of these changes was a "view of life" in which Darwin thought there was a certain "grandeur," even though the world was devoid of purpose or overall direction. For "whilst this planet has gone cycling on according to the fixed law of gravity," he wrote in the last sentence of the On the Origin of Species, "endless forms most beautiful and most wonderful have been, and are being, evolved" (Darwin 1859, 490).

Even as these shifts were taking hold, however, Darwin continued to cling to an earlier assumption that for adaptive change to occur and for enough departure from a common ancestor to collect to produce species, isolation of selected variants at a biogeographic periphery would be required. Until the mid-1850s, geographic isolation seemed to Darwin indispensable for the emergence of new species, not only because that is where stress, and hence variation, is strongest but because if novelties were not to be "swamped," according to Darwin's theory of blending inheritance, by mating with members of the same species, isolation in small populations would be required. When this idea is applied to the emergence of new species it is now called "allopatric speciation" (from Greek *allos*, "different," + *patris*, "fatherland": "in a different place of birth"). It is ironic that even though Darwin was about to abandon the idea, it is now the accepted account of the most favorable conditions for speciation. What Darwin would eventually abandon, as he tried to explain how new species arise, was not his assumption that variation is caused by stress or that inheritance is blending but his earlier belief that biogeographic frontiers are the presumptive locus of that stress.

Darwin changed his mind on this point when the facts seemed otherwise. Large genera with wide ranges, he was led to believe, vary more than small genera, and so produce more species, while small genera, with limited or peripheral ranges, speciate less often. All of the field biologists and systematists who responded to Darwin's inquiries, and especially his American correspondent Asa Gray, were telling him so. So was his own painstaking research on classifying barnacles. This is not what his theory of allopatric speciation would have expected. Darwin finally resolved this problem by giving up the assumption that variation-inducing stress occurs only at biogeographic boundaries. What determines the number of species is the range of different resources available in a given unit of space. The pervasiveness of Malthusian competition now meant that since stress, and variation, are ubiquitous, there is always a premium on finding new resources. Even well-adapted populations, well within their range, will feel it, and natural selection will reward those possessing variant traits that enable them to do something about it. The crucial fact, Darwin realized in a flash in September 1856, is that it is in resource-rich environments, which provide a greater number of ways of making a living, that one finds, on average, more species than in resource-poor environments. Thus, in a note to himself in January 30, 1855, which is echoed in a famous passage in *On the Origin of Species*, he compares "a heath thickly clothed by heath, and a fertile meadow, both crowded. Yet one cannot doubt more life supported in [second] than in first, and hence (in part) more animals are supported. This is not a final cause, but mere result from struggle. (I must think out this last proposition.)" (DAR 205.3, 167, quoted in Kohn 1985, 256; cf. Darwin 1859, 489).

By September 1856, Darwin had clearly seen that nature favors a maximum amount of life, and its greatest possible diversity, in a given unit of ecological space. This empirical law, or quantitative correlation under determinate conditions, is a result, he concluded, of an Adam Smith–like division of labor driven by Malthusian competition: "The advantage in each group becoming as different as possible may be compared to the fact that by division of land—[cancel] labour—more people can be supported in each country. . . . Not only do the individuals of each group strive one against the other, but each group itself with all its members, some more numerous, some less struggling against other groups, as indeed follows from individual struggling" (DAR 45, 65, September 23, 1856, quoted in Ospovat 1981, 181).

Darwin attributes this idea to the influence of reading Milne-Edwards's *Introduction à la zoologie générale* (1851) the year after it was published (Darwin 1859, 116, 418). Milne-Edwards had become one of the most eminent physiologists of his generation by applying von Baer's laws of development, according to which embryos grow from the homogeneous to the heterogeneous, to the way in which energy and materials are allocated to different bodily organs and processes. That will happen in such a way that a minimum of resources is used, which occurs when a maximum of diverse functions divides the necessary work. The idea is clearly reminiscent of Adam Smith's division of labor. Milne-Edwards cites the British political economists as a source. Darwin, in any case, probably recognized the original provenance of the idea (Schweber 1980). For his part, Milne-Edwards was attempting to defend Owen's and von Baer's antitransformist idea that a limited amount of energy means that species do not have the wherewithal to develop into new species, and thus to defend Cuvier's fixity against a new generation of Geoffroyians (Hall 1990, 47). By contrast, Darwin's way of taking up Milne-Edwards's idea depends on the assumption that ecologies, like economies, but unlike physiologies, are driven by Malthusian imperatives. When populations move into unoccupied niches and find a new way of making a living, they are taking advantage of the fact that competition can thereby be reduced. Darwin recognizes, accordingly, that given the ubiquity of Malthusian competition and the consequent availability of variation, tendencies toward diversification, and hence new species, can be expected to occur everywhere, and not just in the biogeographic borderlands. It is not quite that Darwin abandons the notion that isolation is required for adaptive change and species differentiation to take hold. What changes is his conception of what constitutes isolation. It is now a function of ever so slight but ever so constant niche diversification, the analogue (as Marx recognized) of new technologies and markets in economics.

The novelty of this idea is that it appeals to grim Malthusian struggles to produce a biological version of Adam Smith's benign economic trajectory, in which the division of labor keeps an economy indefinitely expanding by opening up new resource bases. It appears that by leading Darwin to see that there are no preset, species-specific norms around which variation is constricted, that individuals can indefinitely vary, and that variants will be selected solely in accord with their adaptive value in a given competitive environment, Malthus's competitive individualism eventually led Darwin to the very un-Malthusian, but quite Adam Smithean, idea that competition might just as readily lead to new ways of making a living as to ghastly catastrophes (Gould 1990). That is precisely the way in which the Malthusian liberals of Darwin's day saw economies.

Biogeography conceived along these lines gave Darwin not just spatial distributions but a window on time. The size of genera reflects not only how widespread and varied its member species are but how long lineages have been evolving from a common ancestor. Biogeography thus grounds classification in phylogenetic history through the intensified action of natural selection in rewarding ecological diversification. If it is ecological niches that provide the sites at which natural selection generates phylogenetic diversity, Darwin's intuition that phylogenetic pattern is like a bush rather than a tree proves sound. There is nothing that drives phylogeny upward in a single direction, as Lamarck and Geoffroy seem to think, but much that drives it into every nook and cranny where resources can be found. The roots, trunk, and main branches of the bush of life represent the oldest strata of conserved traits. As primitive species diverged and diversified, they became genera. With the continuation of the same process, conserved traits became even more deeply entrenched, yielding the large phyla and architectural ground plans of Cuvier's old *embranchements*. Along the way much extinction took place. Many lineages went nowhere. New growth, meanwhile, takes place at the ends of twigs, where buds in the form of species are born, and almost as often die (figure 5.1).

It is commonplace to deny that Darwin himself read this as a theory of biological progress (Ruse 1988b; Mayr 1988; Gould 1977). Robert Richards is right to point out, however, that the fact that Darwin's ideas about evolutionary progress are subtle and highly qualified does not mean that he did not believe in evolutionary progress at all (Richards 1992; cf. Nitecki 1988) Although he famously warned himself in his notebooks, "Never say higher or lower," Darwin writes in *On the Origin of Species*: "The inhabitants of each successive period in the world's history have beaten their predecessors in the race for life, and are, insofar, higher in the scale of nature; and this may account for the vague yet ill defined sentiment, felt by many paleontologists, that organization on the whole has progressed" (Darwin 1859, 345).

What Darwin does not have is the idea that progress is an inevitable result of an inner drive toward complexity. In a biosphere that works like a cycling heat engine, the inner drive to reproduce that Malthus ascribes to all organisms is no longer anything like Lamarck's inherent tendency toward complexification. It is simply the undifferentiated inertial tendency of any organism to maximize its reproduction and to vary. If the history of life reflects complexification, accordingly, that will be the result of external forces rather than the expression of an internal drive conceived in terms of developmental or transformational laws. Darwin's transfer of causality to external forces means that whatever increasing

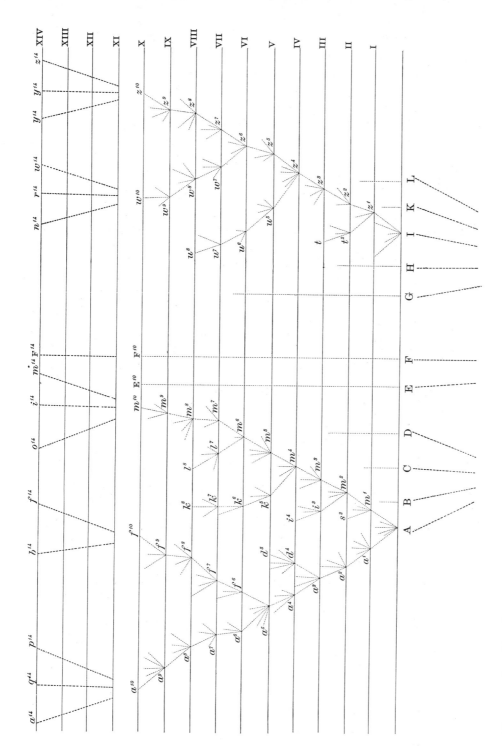

Figure 5.1 Branching diagram of divergence through modified descent. The only diagram, between pages 116 and 117, in *On the Origin of Species* (Darwin 1859). It appears in the sixth edition (Darwin 1872), from which this is reproduced.

complexity is seen is a consequence of the fact that competition generally rewards innovation, just as capitalist competition rewards technological innovation. It also implies that since under Malthusian conditions every conceivable niche will be occupied and utilized, simplification will sometimes predictably be the best adaptive strategy. Indeed, the group he knew most about, barnacles, illustrates this very point. Darwin was able to reclassify them as simple crustraceans rather than as complicated mollusks by remaining open to the adaptive benefits of simplicity (Richards 1992).

These reflections have two additional implications that bear on the issue of progress. The first is that even though Darwin allows one to think vaguely about overall progress, it would seem possible to measure it only along a single branching line. If you consider evolutionary time as an expanding sphere, the center of which represents the first instant and the surface the present, you will find contemporary kinds, including fish and humans, at various points on or near the sphere itself, with different lines leading back into it. On the surface there is maximal diversity and, in that sense, complexity, although any given organism may well be less complex than some of its ancestors. Progress does not appear across the surface of the sphere but comes into view only as one begins to move into the interior along a single line. As Darwin says in a letter to Hooker, "'Highest' usually means that form which has undergone more 'morphological differentiation,' from the common embryo or archetype of *the class*" (Darwin to Hooker, June 27, 1854, Darwin 1887, 1:76, italics added). If this differentiation did not occur, you would not see so much diversity at the surface of the sphere.

Second, although there is admittedly a modest degree of recapitulation in this view, Darwin's way of envisioning phylogeny and classification means that any recapitulation of older traits will have a von Baerian rather than a Geoffroyian stamp. The very sources that Darwin used to construct his theory suggest that. Darwin's aim was to show that the von Baerian pattern can bear an evolutionary rather than solely the antievolutionary interpretation that Owen, Milne-Edwards, and von Baer insisted in putting on them. In *On the Origin of Species,* he remarks that it will be intelligible on his theory, "If it should hereafter be proven that the ancient animals resemble to *a certain extent the embryos* of more recent animals of the same class" (Darwin 1859, 345, italics added). All that was needed was to transfer causality from the inside to the outside. One might choose to see adults of primitive single-celled organisms in the eggs or sperm of humans, or fish in what look like gill slits. But egg and sperm have within them the seeds of what is destined to be a human, and what look like gill slits are on their way to becoming lungs. It might be more productive, therefore, to look at primitive, and generally extinct, species as showing, in their adult state, traits that reveal themselves much earlier in the development of contemporary organisms. Traits viewed in this way are better described as conserved than as recapitulated.

Schweber has argued that "when Darwin undertook to unravel the mechanism of evolution he realized he was attempting to do for biology what Newton had done for physics and astronomy" (Schweber 1979, 83). He also sees that if Darwin was something of a Newton of a blade of grass, it was because he applied to the organic world dynamical models taken from the British tradition of political economy (Schweber 1980). In this chapter we have attempted to sketch several aspects of the case for this view of Darwin's achievement. Economic thinking enabled Darwin to envision a force-filled external environment, making his original teleological functionalism ever more recessive. It pushed him from essentialism to individualism. Finally, it gave him a mechanism for explaining diversity in space and time.

The fact that the economic models in question were themselves appropriations of Newtonian systems dynamics provides, we think, the relevant link between biology and economics. Ideology is involved, but it cannot reasonably be discussed except by way of the scientific ideals that Darwin inherited and sought to advance (Hodge 1987). This project was carried out more in terms of the classical Newtonian ideals of Lyell, Herschel, and other Dissenting Whigs than in anticipation of a new statistical theory of nature in which the Darwinian tradition eventually came of age. This is reflected in the fact that the fractious reaction to the publication of the *On the Origin of Species* was focused, among Darwin's peers, on whether he had in fact met the Newtonian criteria they shared. We will consider this reaction, the debate that it stimulated, and how the issue was resolved in Victorian England in the following chapter.

6 Domesticating Darwinism: The British Reception of *On the Origin of Species*

In the middle third of the nineteenth century, philosophical views about science divided in Britain roughly into the teleological science of the Oxbridge Tory clerics; the uniformitarian, law-governed science of leisured Whig gentlemen like Lyell, Herschel, and Darwin; and the "radical-liberal" science of self-made professionals like Huxley. In spite of much interanimation at the seams where they were joined, the worldviews characteristic of these three orientations differed so widely that relations among their partisans were, in the best of times, fractious. It is not surprising that the publication of *On the Origin of Species* greatly agitated these relations or that the British reception of Darwin's book fell out largely along these preformed lines. The young men who filled the ranks of the third group had long been prone to embrace evolution, not least because its challenge to a fixed universe offered a compelling image of the upward mobility they sought for themselves. They championed Darwin, even if they sometimes mistook their own views for his. Both Tory clerics and Whig gentleman, however, continued for a time to resist evolutionary ideas. If Darwin's stomachaches worsened during the decade in which battles raged around his name, that was in part because, while his intention was to pull evolutionary theory up into the serene legitimacy of uniformitarian Whig science, the relative neutrality of his Whig peers, especially Lyell, made it much more likely that Darwin's most passionate defenders would pull his theory, and him with it, down into the radical evolutionary currents he had so long hoped to skirt. In this chapter, we will follow some of these struggles, and Darwin's entanglement in them, from initial denunciation to belated triumph, when Darwin's body was interred in Westminster Abbey near that of Isaac Newton, Charles Lyell, and John Herschel.

Darwin's friends and mentors at Cambridge did not waste any time rejecting *On the Origin of Species*. Sedgwick wrote that he had read parts of the book—Darwin had sent him a copy—"with absolute sorrow, because I think them utterly false and greviously mischievous. You have deserted . . . the true method of induction, and started in machinery as wild . . . as the locomotive that was to sail us to the moon" (Desmond and Moore 1991, 487–88). Since Sedgwick's view of the "true method of induction" was that of Whewell, which stipulated that citing purposes alone could count as answers to questions about functional phenomena, Darwin's replacement of teleology with the mechanism of natural selection was for him surely a scandalous nonstarter. Other members of Darwin's old circle also used the ploy that Darwin had produced a wild hypothesis to dismiss his work. Henslow wrote to his brother-in-law Jenyns that *On the Origin of Species* "pushes hypothesis (for it is not real theory) too far" (Desmond and Moore 1991, 487–88).

If Darwin's own friends among the Oxbridge clerisy quickly distanced themselves from him when *On the Origin of Species* appeared, it was only to be expected that his enemies among their ranks would have been less polite. Owen, for example, immediately weighed in with a negative article in the *Edinburgh Review*. Taking advantage of Darwin's oblique attempt to soothe creationist feathers by conceding in a few well-placed passages that evolution had proceeded from a single primitive form into which "life was breathed," Owen, still promoting his idealist theory of ongoing creation, asked why, if that were so, Darwin should not admit that the breathing of life into new forms continues to this very day (Owen 1860 in Hull 1973, 175–213)? In a sidewise glance at evolution's connection with disreputable politics, Owen suggested that Darwin's neglect of his alternative view could only be explained by "an abuse of science [like that] to which a neighboring nation, some seventy years since, owed it temporary degradation" (Desmond and Moore 1991, 487–91).

A letter appeared in the *Times*, said to be by Thomas Carlyle (and certainly in his bluff style), that summed up the conservative case against Darwin: "A good sort of man is this Darwin, and well meaning, but with very little intellect. Ah, it's a sad, a terrible thing to see nigh a whole generation of men and women, professing to be cultivated, looking around in a purblind fashion, and finding no God in this universe" (quoted by Himmelfarb 1959, 248). Things got worse when the duke of Argyll, an influential Whig grandee, announced that he would not protect Darwin from the Tories, arguing that Darwin's alternative required a bloody and wasteful slaughter ill befitting a benign Creator.

These reactions illuminate a famous episode that occurred in June 1860. The annual meetings of the British Association for the Advancement of Science (BAAS) for that year chanced to be held at Oxford, home turf of the Tory clerisy. The BAAS had been founded in 1831 by Charles Babbage, John Herschel, and David Brewster, Whiggish gentlemen of science all. They modeled their association on the powerful scientific societies of Germany. In order to infuse a spirit of scientific progress into the provinces and to gain support for professional scientists, meetings were to be held in different cities each year. The annual gatherings would be broken up into different sections, each charged with monitoring and debating new developments in each scientific field. BAAS was an expression of the same spirit that had resulted in passage of the First Reform Bill. Its very existence was a blunt declaration that the clubby days of the aristocratic Royal Society were over for good. From the outset, therefore, the BAAS was the scene of intense confrontations between eminent conservatives like Sedgwick, Whewell, and Owen (all of whom served as its president), Whig moderates, and passionately anticlerical and anticorporatist young men like Huxley, who assumed that a truly professional science could only arise in a resolutely secular milieu. The BAAS, beginning with the meetings at Oxford in the spring of 1860, provides an excellent vantage point from which to view the nerve-racking, and in Darwin's case stomach-churning, controversies that enveloped Darwinism in the decade after its birth.

On the third day of the Oxford meetings, Samuel "Soapy Sam" Wilberforce, the unctious, silver-tongued bishop of Oxford, responded to an inconsequential paper by an American anthropologist on the implications of Darwin's hypothesis for "civilization." Having been coached the night before by Owen, Wilberforce attacked Darwin for treating humans, whom God had made a little lower than the angels, as no higher than the apes. "If the question is put to me would I rather have a miserable ape for a grandfather," replied Huxley, who had shown up to defend Darwin, "or a man highly endowed by nature and possessed of great means and influence, and yet who employs these faculties and that influence for the mere purpose of introducing ridicule into a grave scientific discussion, I unhesitatingly affirm my preference to the ape" (Huxley to Dyster, September 9, 1860, quoted by Desmond and Moore 1991, 497).[1]

Almost seven hundred people were in attendance at this debate. It was a hot and humid day, and the room was stuffy. Here is a highly colored account of what ensued:

The excitement was tremendous. One lady fainted and had to be carried out, while undergraduates leaped from the seats and shouted. Other speakers followed, adding to the confusion and uproar. An Oxford don disputed the theory of development by pointing out that Homer, the greatest of poets, had lived three thousand years ago, and his like had not been seen since. Sir John Lubbock . . . told of a specimen of wheat that had been sent to him as having come from an Egyptian mummy, ostensibly demonstrating that wheat had not changed since the time of the Pharaohs. Admiral Fitzroy got up to describe how he had often expostulated with his old comrade on the *Beagle*. Lifting an immense Bible over his head he solemnly implored the audience to believe God rather than man, to reject with abhorrence the attempt to substitute human conjecture and human institutions for the explicit revelation with which the Almighty had himself made in the book of the great events which took place when it pleased him to create the world and all that it contained. (Himmelfarb 1968, 239–40)

What really happened differs considerably from this and other mythical accounts. The lady fainted because it was hot, not because the topic was scandalous. Huxley's reply, probably meant as a polite joke, went unheard by most of the audience. He was much less effective as "Darwin's bulldog," it appears, than he was soon to become. It was the normally shy and taciturn Hooker who effectively rose to defend the honor of his friend by questioning the credentials of those who had spoken ill of him. In spite of this admirable display of loyalty, many of those present still thought that Wilberforce had got the better of the argument (Gould 1991).

The literal truth of the myth is less important than the reasons for its invention and dissemination. The people who became Darwin's most prominent supporters and protectors were intent on using his book as a weapon in an anticlerical war aimed at taking religion out of the business of science. In the mythic Oxford debate, therefore, professional scientists, the only appropriate defenders of science, are pitted against clerical obscurantists (Wilberforce), aristocratic conservatives (Fitzroy), and humanist ignoramuses (the hapless Homeric don). If Darwin's book had offended the clerisy, Huxley implied, that only showed that their scientific pretenses were but empty and dangerous claims to authority. Huxley meant every word of what he said. He was out to cleanse the Augean stables of science, to displace clerical, and even gentlemanly, domination of its professional institutions. Evolution was an excellent test case, for opposition to it was clearly based on religious, social, and political prejudice. It mattered less to Huxley, therefore, that Darwin's theory differed from Lamarck's, Geoffroy's, Grant's, and Chambers's in nicely Newtonian ways that might commend it to Whig gentlemen than that it was an evolutionary theory put forward by a respectable scientist who offered plenty of empirical support for it. It mattered even less to Huxley whether Wallace's semisocialist view of selection or Darwin's liberal version, or something altogether different from both, was nature's preferred evolutionary mechanism. What mattered was evolutionary naturalism, which in banishing the supernatural banished piety, and in Huxley's view hypocrisy, from science.

In 1864, a semisecret club of self-proclaimed defenders of Darwin, calling itself the "X Society," formed around Huxley. Spencer and the faithful Hooker were among its eight founding members. Surprisingly, Darwin himself seems to have had few scruples about entrusting his interests to their care. Indeed, as the battle was joined, he seems increasingly to have cheered from the sidelines for Huxley's agnosticism and his crusade for professionalism. "I have made up my mind to be well abused," he wrote to Asa Gray. "But I think it of importance that my notions should be read by intelligent men, accustomed to scientific argument, though not [necessarily] naturalists" (Darwin to Gray, December 21, 1859, in Darwin

1887, 2:29). Perhaps Huxley represented a long-repressed side of Darwin himself, the free spirit of his grandfather, for example, confined no longer by the massive inhibitions that two generations of Victorian respectability had imposed on a man who had, in his youth, at least, been something of a free spirit himself. Perhaps, on the other hand, he was just picking up allies where he could find them.

The sea change in Darwin's allegiances after he went public can be seen especially well in his shifting relationship to the liberal clerics and pious scientists who sprang to his defense. Not all of the clerical intelligentsia thought that Darwin had violated the great cultural compromise between mechanism and teleology. In a letter published in the literary journal *Athenaeum* in November 1859, the Christian-socialist preacher Charles Kingsley, in a burst of enthusiasm provoked by reading *On the Origin of Species*, had proclaimed that it was "just as noble a conception of the Deity to believe that he created primal forms capable of self-development [*sic*] as to believe he required a fresh act of intervention to supply the lacunas which he himself had made" (quoted in Desmond and Moore 1991, 477). A few months later, a group of six other liberal clergymen joined Kingsley in publishing what turned out to be a bestseller, in which they claimed that there was no particular reason why natural selection could not be treated as just another divinely sanctioned "secondary cause," something that yet one more Bridgewater treatise might be written about. In claiming Darwin for this view, religious liberals were trying to maintain their slippery footing between Tory conservatism and scientistic anticlericalism. They were much more interested in social ethics, and in maintaining the relevance of religion in the political struggles of the day, than in the Thirty-Nine Articles of the Anglican faith. They were doomed, therefore, if either religious orthodoxy or scientific reductionism triumphed. For their pains, however, these liberal clergymen were tried for heresy. Two of them were convicted, putting them out of work. Biblical literalism, never much of a factor before, was thus born in the orthodox backlash to a religiously liberal version of Darwinism.

The difficulty these liberals faced in denying that Darwin's extension of the Newtonian paradigm was irreligious was that they were forced to think that if evolution by natural selection is to be for the good, then the variation on which it works must be biased in favor of a desired outcome. Like all "secondary causes" that point, through orderly natural laws, to a divine plan, it had to have an overall rationale. Otherwise natural selection, as the duke of Argyll had said, looks like useless, chaotic, and cruel slaughter, suggesting that if God could not find some mechanism less gory and wasteful, he must not be all powerful, or, if he could have found some other means but did not bother to he must not be providentially good.

Darwin's American friend and correspondent Asa Gray took a similar line. Gray was a biologist working in an American intellectual climate that was much more religious, even if it was less ecclesiastical, than England's. He too played up the creative ordering of natural selection, working on directionally biased variations, as a Newtonian mechanism for bringing about higher and higher forms of life, culminating in humans. Gray thus viewed Darwin in a distinctly Paleyesque light. "Let us recognize Darwin's great service to natural science in bringing it back to teleology," he later proclaimed. "Instead of morpology versus teleology we shall have morphology wedded to teleology" (Gray 1874, in Gray 1889, 288). Natural selection, that is, is a mechanism for realizing perfectly adapted Paleyean organisms, and over time ever more perfect kinds, an idea that the American philosopher John Dewey later characterized neatly as "design on the installment plan." In giving a Bridgewater twist to Darwin's argument, Gray was in effect refusing to accept that natural selection meant that it was all over for at least a weak argument from design. He was explicit about this in his review of the *On the Origin of Species:*

It is not surprising that the doctrine of the book should be denounced as atheistical. . . . The theory is perfectly compatible with an atheistic view of the universe. That is true. But it is equally true of physical theories generally. *Indeed, it is more true of the theory of gravitation and of the nebular hypothesis than of the hypothesis in question.* Nor is the theory particularly exposed to the charge of atheism of fortuity, since it undertakes to assign real causes [*verae causae*] for harmonious and systematic results. . . . What is to hinder Mr. Darwin from giving Paley's argument a further a fortiori extension to the supposed case of a watch which sometimes produces better watches, and contrivances adapted to successive conditions, and so at length turns out a chronometer, a town clock, or a series of organisms of the same type? (Gray 1860, 53–184, italics added)

At first, Darwin welcomed support like this. In the hastily prepared second edition of *On the Origin of Species,* he inserted what Kingsley had written in the *Athenaeum* and arranged with his publisher to sew Gray's review into the front of the book. Eventually, however, Darwin recognized the disingenuity of this and had the review taken out. He did not in fact think that variation was biased or that organisms are perfectly adapted, and he became increasingly convinced that if natural selection was to do its best, it could not work with a stacked deck. The implications for the argument from design, and more generally for religious belief and values, that Kingsley and Gray were trying so hard to fend off seemed increasingly correct to Darwin. Thus, as he grew older, Darwin's sense of God's retreat from a world of violence and accident, in which the race was to the strong, grew more pronounced. An unmistakably stoical note can be heard most intensely in those parts of the *Autobiography* in which Darwin recalls his former Romantic self. Especially after the death of his favorite daughter, Annie, in 1850, he seems to have picked up something of the chill that poets like Tennyson had already felt and that his own work helped induce in Matthew Arnold, who represented the "sea of faith" as an outgoing tide that gives forth

a melancholy, long, withdrawing roar,
retreating to the breath of the night wind,
down the vast edges drear
and naked shingles of the world.
(published 1867)

This retreat from teleology was probably less a recoil from the conservative outcry against him, however, than Darwin's honest response to the liberal theodicy of his would-be champions. It is true that as late as 1874 Darwin replied positively to Gray's idea that natural selection explains rather than refutes teleology. "What you say about teleology," he wrote," pleases me especially, and I do not think anyone else has ever noticed the point. I have always said you were the man to hit the nail on the head" (Darwin to Gray, June 5, 1874, in Darwin 1887, 2:367). However, Darwin, who always answered his mail politely, may have vaguely meant something less than Gray did. Contemporary philosophers of biology have shown that the bare explanatory schema "X is there in order to do Y," or "for the sake of Y," can be filled in with design or selection arguments (L. Wright 1973; Brandon 1978). "Selection by consequences" is a culling and inheriting of some variants from a much larger array, with the result that over time the properties that preserve variants build up traits that are really there "for the sake of" something. This is not the result of aim or design, however, but of a purely natural, indeed blind and imperfect, process. Thus, if there is any teleology in Darwin's world, it is, ironically, only because there is also a great deal of chance and accident in it. The more constraint on variation there is, in fact, the less effective the adaptive mechanism is. In that case, however, the very mechanism that causally explains Darwin's teleology blocks the inference Gray wanted to draw from it. Evolution does not, and if natural selection is taken seriously, cannot tend in any predetermined direction. It cannot be a secondary cause in service of a final end.

We take the term *naturalism* to mean not only that supernatural and immaterial entities cannot explain events and processes but that the purely natural processes and laws that do explain them do not point to anything beyond themselves. Darwin's increasing, or at least newly acknowledged, naturalism in this sense led him to identify ever more fully with his materialistic and scientistic defenders and to retreat not only from clerical views of science but from the gentlemanly vision of science of his Whig patrons, Lyell and Herschel. This tendency was doubtless intensified by Lyell's indifferent response to the substance of his book. It was especially stimulated by Herschel's categorical rejection of his theory.

Like Kingsley and Gray, Herschel, and Lyell recognized that unless the direction of natural selection is biased, Darwin's version of evolution would have even worse implications than the old French theories for the inference from uniform natural laws to divine providence. At least Lamarck and Geoffroy had held out a vision of progress. In spite of Gray's and Kingsley's wishful thinking, the doyens of respectable Whig science knew perfectly well that Darwin did not mean that variation is benignly biased and indeed that he could not consistently think that it might be. This was too much chaos and contingency for them. While Lyell hemmed and hawed, however, Herschel straightforwardly denied that Darwin had found an appropriately uniform and causally effective Newtonian law of the kind that Herschel approved of in his famous *Preliminary Discourse,* and so denied that Darwin had met the very criteria he had set for himself. It was perhaps the most severe blow Darwin received.

Darwin had been very anxious to hear Herschel's opinion of *On the Origin of Species.* In Victorian England, the self-invented role of the philosopher of science or "methodologist" had become culturally important. By setting up the criteria according to which the rapidly developing and multiplying sciences, human as well as natural, were to be judged, philosophers of science stood guard over the long-standing cultural consensus that had formed after the Revolution of 1688. They did so in a time when that consensus was under continual pressure. These men did not always presume to decide which hypotheses were true but which could stand as candidates for being true. They were to make a preliminary sort for conceptual and methodogical purity, so that the remaining choices among competing theories could be, or appear to be, a purely empirical one. That was an important aspect of Victorian empiricism itself. Empiricism required that the range of conceptual variation among theories was so narrowed as to make the choice between them rest on matters of fact rather than conceptual leaps ("hypotheses," in the sense in which Newton had said, "*Hypotheses non fingo* [I don't touch hypotheses]," about the nature of gravity). In these matters, Herschel was Darwin's guide. Darwin had written *On the Origin of Species* tailored to Herschel's prescriptions. He had presented natural selection as a close analogue of the artificial selection practiced so effectively by breeders and had argued that, in a world chock full of Malthusian forces, the claim of artificial selection to be a true or actual cause (*vera causa*) would be transferred to natural selection. Darwin had gone on to demonstrate the explanatory fecundity of natural selection considered as a general uniform law of the organic world. Nor, in doing so, had he rigged the game by failing to mention, or trying to address, potential counterexamples and "difficulties on the theory." Herschel should have been impressed (Herschel 1830; Hodge 1983, 1987). Accordingly, it was very painful for Darwin to hear rumors that Herschel had flatly rejected his theory.

He had already intimations that this might happen. He had confided to Lyell in a letter at the end of November 1859 that "Sir J. Herschel, to whom I sent a copy, is going to read my book. He says he leans to the side opposed to me. If you should meet him after he has read me, pray find out what he thinks, for, of course, he will not write; and I should excessively like to hear whether I produce any effect on such a mind" (Darwin to Lyell, November 23, 1859, in Darwin 1887, 2:26). A month later Lyell, whom Darwin was at this early stage still trying to butter up, received the following update from him: "I have heard, by a round about channel, that Herschel says my book 'is the law of higgledy-piddledy.' What this exactly means I do not know, but it is evidently very contemptuous. If true this is a great blow and discouragement" (Darwin to Lyell, December 12, 1859, in Darwin 1887, 2:37).

In order to grasp the significance of this criticism, we must pause to see precisely where the nub of the difference between Darwin and Herschel lies. Darwin's aim was to show that biology could be made to conform to the law-governed, force-filled Newtonian explanatory model, the importance of which as the canonical form of scientific theory in nineteenth-century Britain, or at least Darwin's part of it, cannot be overestimated. Herschel does not reproach Darwin's aim, or his method, or his presentation. The heartbreak in Herschel's remark is his assertion that Darwin had allowed chance to figure in his theory in ways that were not canonical within the ontology of the Newtonian tradition. Herschel is saying that Darwin's theory is not Newtonian enough to count as a candidate. Indeed, Herschel was soon to write: "We can no more accept the principle of arbitrary and casual variations of natural selection as a sufficient condition, *per se*, of the past and present organic world than we can recieve the Laputan method of composing books as a sufficient account of Shakespeare and the *Principia* [*Mathematica* of Newton]" (Herschel 1861, 12).

In spite of his allusion to the low probability that an infinite number of monkeys pounding on an infinite number of typewriters would ever come up with anything intelligible at all, let alone *Hamlet*, Herschel's objection is not the obvious, and mistaken, one that Darwin has allowed chance alone to explain the origin of adaptations and of species. Herschel knew well enough that natural selection, working on chance variation, is the purported causal agent. His point is that natural selection, so construed, cannot count as a law of nature or as a true cause.

In the *Preliminary Discourse*, Herschel had made it clear that true causes (*verae causae*), when generalized to universal laws, cover the cases they causally explain homogeneously. In an orthodox Newtonian law, macroscopic effects are the outcome of adding and subtracting commensurable units of physical quantities, such as mass, force, and momentum. The macroscopic events at the top can straightforwardly be analyzed into the microscopic events at the bottom. Conversely, what is on the bottom can be added up until the top is reached. Universal natural laws, such as the inverse square law of gravity, cover their instances in this way. That is why science can predict deterministically. From a law, and the state of all the variables at one moment, their state at the next moment is completely determined. The same procedure can be followed at the next moment. Adding these results up, any subsequent or prior state of the system can be found. Herschel is saying that Darwin's appeal to the accidental, even if probable, coincidence between variation and adaptive utility fails to meet this criterion, and hence fails to qualify as a natural law from which explanations and consequences follow. It cannot, accordingly, be a true cause. The loose change in the universe cannot account for its order or structure.

Herschel's sensitivity to this seemingly arcane issue derives in part from the fact that he read *On the Origin of Species* at a time when he and many other intellectuals had worked themselves up into a great lather about the work of the Belgian Adolphe Quetelet and his self-proclaimed English disciple, Thomas Henry Buckle, on regularities in social life. We shall return to Quetelet again when considering the probability revolution more systematically and historically in chapter 8. For the present, it suffices to say that, like Herschel, he was an astronomer, trained by Laplace. In the course of his work, Quetelet had had occasion, as Herschel had, to use the fact that large data sets settle around a mean, and fall off regularly on both sides, as a method for detecting observational errors and arriving at a true value. Of this "law of errors" Herschel had written in the *Preliminary Discourse:*

Whatever error we may commit in a single determination, it is highly improbable that we should always err the same way, so that, when we come to take an average of a great number of determinations . . . we cannot fail, at length, to obtain a very near approximation to the truth, and . . . to come much nearer to it than can fairly be expected from any single observation. . . . The useful and valuable property of the average of a great many observations, that it brings us nearer to the truth than any single observation can be relied on as doing, renders it the most constant resource in all physical enquiries where accuracy is desired. (Herschel 1830, sec. 227–28)

Quetelet had used the "law of errors" not only to measure subjective mistakes about natural events, however, but as an accurate way of depicting facts about social life, which show surprising constancy around a mean and regular falloff on either side: the average number of suicides per year, the average number of marriages, the average waist size of soldiers in a Scottish regiment, the average number of dead letters that the post office could expect to accumulate in a given year. Quetelet even postulated the existence of an "average man" to which such regularities apply.

Statistical predictability of this sort seemed from the start to many to imply that what every individual does, including commit crimes, is determined by the sum total of prior influences acting on him or her, which average out to yield the statistical constants found in social record keeping. Although Quetelet himself sidestepped the nagging questions about free will that his views suggested by distinguishing between what people do considered as members or parts of society, and what they do purely as individuals, he did think that an average number of crimes was in any event bound to be committed and that just who was to commit them was simply a question of who happened to be influenced in the right ways. "Society," he scandalously remarked, "prepares the crime" (Porter 1986; Hacking 1990). When Buckle's *History of Civilization in England* (1857) made Quetelet's ideas more widely known in Britain, he did not make the fine distinctions Quetelet did. In fact, Buckle bluntly announced that "social physics" is inconsistent with free will, and no worse off for that. Buckle's version of Quetelet opened up festering wounds about social determinism and moral fatalism just as Victorian society seemed at last to be settling down comfortably into liberalism (Krüger 1987, Hacking 1990). Porter remarks with some justice that throughout the 1860s and 1870s, Buckle's book was discussed at least as much as Darwin's (Porter 1986, 164).

From the very beginning, in the 1830s, statistics, which lay at the root of this disagreeable discussion, had been an uninvited, or at least potentially obnoxious, guest in the respectable, largely Whiggish halls of the BAAS (Hilts 1978). Statistics smelled of radical politics. By portraying the condition of England through a statistical survey, a practice already well advanced in Germany and France, attention would be called to its glaring social inequalities and to the potential inability of gradual reform to handle social change. Engels impressed Marx by using such resources in his *Condition of the Working Class in England*. Marx himself would use statistical reports about factory conditions to great effect in *Capital*. Sedgwick had opposed having a statistical section at all for these reasons among others. He warned statisticans that "if they went into provinces not belonging to them, and opened a door of communication with the dreary world of politics, that instant would the foul demon of discord find its way into the Eden of philosophy" (Sedgwick 1833 in Hilts 1978, 34). It was only with difficulty that Charles Babbage, a statistical enthusiast, prevailed upon the association to form a statistical section, promising that its scientific character would extend purely to its mathematical and logical aspects and that its applications to society, like the applications of theoretical physics by the petty engineer, would play no role in the concerns of the association.

Herschel was generally on Babbage's side. For a long time he had been impressed by Quetelet's efforts to demonstrate just how many phenomena, natural and social, correspond to the "error law." In 1850, he wrote a generally positive article about Quetelet, albeit anonymously, in the *Edinburgh Review* (Herschel 1850). Perhaps because he was so impressed by Quetelet's technical work, however, Herschel took great pains to distance himself from what Buckle at least took to be its implication. Statistical regularities do indeed yield empirical predictions, but prediction is not the same as explanation. In society, as in nature, explanation depends on true causes (*verae causae*), which in the case of nature are uniform laws and in the case of human activity are free choices. To call statistical correlations "laws," accordingly, was to substitute mere prediction for causal explanation as an ideal of science and to short-circuit the search for *verae causae*. It was to give in to the phenomenalist temptation that had spread from "positivist" France to England in the work of John Stuart Mill. The very fact that the exception-ridden regularities of the social statisticians cover their cases heterogeneously was for Herschel evidence, in fact, that social scientists had not found either real laws or true causes. Nor was it likely that any such laws of human nature, analogous or reducible to those governing physical nature, would ever be found, or that the fatalistic consequences for morality that they entail would ensue. Our actions are the result of discriminating intelligence and free will, and statistical generalizations, such as they are, simply reflect intelligent and free responses to similar situations, with deviations from the mean a reflection of the obvious fact that there are some irrational people in the world.

In sum, Herschel brought forward Buckle's work as precisely the sort of abuse of statistical analysis that the founders of the BAAS had feared. With this background fresh in his mind, Herschel read *On the Origin of Species*. Accordingly, what Herschel meant by "higgledy-piggledy" may be glossed as follows. Rather than finding a suitably Newtonian law for biological adaptation and transmutation—a project only slightly less dubious than trying to find social laws—what Darwin had done was push Quetelet's social arithmetic down into the biological world and then claim that he had found in natural selection a law of nature. He had thus compounded Buckle's error about society by reading it into nature. Not surprisingly, Darwin's "laws" do not cover their instances with anything like the uniformity and homogeneity of classical Newtonian laws. Each event seems to lie at the intersection between many separate causal lines or to be a matter of pure happenstance and accident. Natural selection is used as a general idea to cover what survives sorting through this heterogeneity. But natural selection, so construed, cannot be a law, in the canonical Newtonian sense, because it is not connected with its instances in an appropriately lawlike way. Hence, the notion that there are natural laws, as opposed to purposes, governing biology is still just as false as the related idea that such laws govern society or the purposive acts of individuals.

Is this a fair accusation? It is certainly not the case that there is a perfect fit between natural selection and the law of gravitation. For one thing, natural selection is not really a single, commensurably quantifiable force like gravity. The very heterogeneity in the way it covers its instances shows that. "Natural selection" covers many different kinds of causal episodes. Indeed, the process works only when selection is free to "scrutinize" a large array of minutely differing variant traits, whose "components of fitness" differ vastly from one another (Darwin 1859, 83). Nor, for this very reason, is there a really snug fit between Darwin's theory and the assumptions we listed above as governing Newtonian systems. Darwin's biological world is not truly isolated or closed to energy and material flows. Nor is it reversible or deterministic. Nor, finally, in spite of Darwin's bow toward "gemmules," is it atomistic in a sense that permits homogeneous calculations of large effects by using a law to sum over myriad invisible interactions.

These disanalogies have suggested to several contemporary historians and philosophers how Darwin may belatedly be defended against Herschel's attack even within his own framework. Perhaps uniform laws, argues Hodge, are less important than the fact that Darwin looked for, and found, a pretty good candidate for a *vera causa*, lawlike or no (Hodge 1987). Even without going through a law of nature like gravity, what will happen in a natural world conceived as subject to Malthusian pressures in Lyellian environments is sufficiently like artificial selection to justify the extrapolation of a true cause. To this defense of what Hodge calls the *"vera causa* ideal," we may add that the homogeneity Herschel demands is indispensable only if we are to see uniform natural laws as secondary causes that testify remotely to the divine simplicity and intelligence and so proclaim a "religion within the bounds of reason." Where there are many degrees of freedom, accident, and chance, this inference is intellectually, aesthetically, and existentially blocked, as Darwin himself realized. When the tail of natural theology no longer wags the dog of Newtonian paradigm, however, there is no reason not to see Darwin's theory as a legitimate extension of the Whig search for a *vera causa*.

Silvan Schweber's way of defending Darwin is in some ways the counterpart of Hodge's. Rather than causes without laws, Schweber argues that Darwin, as Herschel suspected, was thinking about natural selection as a process governed by statistical laws and in the process insightfully anticipating the direction science was about to take (Schweber 1979). Schweber points out that Darwin owned Quetelet's book, *A Treatise on Man* (1835, translated 1842), had read Herschel's article on Quetelet, and may well have been influenced by both (Schweber 1977). What Herschel takes to be a vice, however, Schweber regards as a virtue. A few years after the appearance of *On the Origin of Species*, Maxwell, the greatest British physicist since Newton, would explain the phenomenal laws governing the temperature, pressure, and volume of gases as the result of the random, and averaging, movement of molecules and atoms, thinking that he had thereby extended the power of classical mechanics. Soon thereafter Boltzmann would use the same techniques to explain the phenomenological laws of thermodynamics. Understandably, Schweber compares Darwin's statement that "natural selection is daily and hourly scrutinizing, throughout the world, every variation, even the slightest, rejecting that which is bad, preserving and adding up all that is good," to the work of Maxwell's Demon, who scrutinizes the energy levels of vast numbers of molecules and sorts them out (Darwin 1859, 83–84; Schweber 1979).

We think this is a misleading defense of Darwin. Schweber's argument is that just as Maxwell wants to use statistical mechanics to reconcile the theory of gases with Newtonian mechanics, so Darwin wants to use natural selection to become the Newton of biology. This is misleading in the first place because Maxwell's point is that such a Demon is utterly improbable (Hodge 1987, 245). Schweber is trading, however, on Maxwell's own reluctance to abandon classical Newtonian determinism. Maxwell's Demon, viewed in this light, is a device employed to make it possible, even if it is improbable, for determination to hold ontologically and for statistical averaging to be an epistemological aid. If Darwin were arguing by analogy to statistical mechanics, it could only be in this way. A Darwinian Demon would be matching variation to utility in the same way that Maxwell's Demon separates fast and slow molecules. There is, however, a crucial disanalogy at this very point. Maxwell's Demon has to scrutinize a much less diverse, and much more homogeneous, number of relevant properties than Darwin's putative Demon would. Indeed, so diverse are the components of fitness with which a Darwinian Demon would have to work that statistical averaging could not be of any epistemological utility at all. A Maxwellian Darwin would have to choose between statistical averaging and biological adaptedness. For us it is clear that Darwin took the second alternative. Indeed, it is far from clear to us that he was even aware of the alternative.

Maxwell, and not Darwin, was already operating in a framework, soon to be made even more explicit by Boltzmann, in which inertial mass is replaced by the most probable behavior of statistical arrays, force is replaced by energy, and equilibrium is construed not as a balance of forces but as the point in an energy gradient where the ability to do useful work is exhausted. If this framework was already bursting through the boundaries of Newtonian thinking, much to the annoyance of its creators, that is because in it time is irreversible and chance events have ordering properties. There can certainly be no doubt that Darwin's theory would eventually find a more congenial home in a framework whose paradigm cases are statistical mechanics and irreversible thermodynamics. We shall attempt to show in the second part of this book, in fact, that Darwinism's continued vitality after Darwin depended on it slowly making its way to that new home. It does not follow, however, that Darwin was already working within this framework. It is true that Boltzmann said that the nineteenth century was "Darwin's Century," pointing out that Darwin had acknowledged the reality of time and time's way of creating order out of chance.[2] If anyone deserves the credit for discovering these things, however, and for seeing Darwin in these terms, it was Boltzmann himself, and perhaps even more justly the American philosopher Charles Sanders Peirce, rather than Darwin.[3] Moreover, in viewing Darwin in these terms, Boltzmann and Peirce tend to screen off Darwin's most important explanatory notion, natural selection, which Darwin considers an analogue of gravitational force, and to stress instead the self-ordering properties of chance setups. In view of the tradition out of which his own thinking came and the criteria he tried to meet, it might be more accurate to say that Darwin was among the last of the great eighteenth century scientists, not among the first of the great nineteenth-century ones. This is an assessment well expressed by Darwin's burial in Westminster Cathedral amid the heroes of the uniformitarian research tradition.

In adjusting Darwin's actual accomplishment to the paradigms within which he thought of himself as working, we believe that the explanatory models Darwin took from the political economists are significant. Although natural selection, like economic competition, is not a single force but a single name for a vast number of different causal transactions sharing an analogous structure, Darwin was able to show that natural selection is a fairly unified and recognizable process, and hence something like a *vera causa*. The representational, explanatory, and ontological resources of the discourse of political economics allowed him to do this in a way that embodied the Newtonian way of thinking, out of which both classical economics and Darwinian evolutionary theory arose. So many different facts and processes, most of which are hidden entirely from view, could not have been collected together and understood unless Darwin had seen his natural analogue of the breeder's art as driven by Malthusian population pressure working on prolific, unconstrained, and directionless variation. Like a free market, which adjusts production, exchange, and consumption, selection creates a moving equilibrium between adapted organisms and environments. Finally, just as a free-market economy expands by diversifying and diversifies by using new technologies to exploit new resource bases, so natural selection builds diverging lineages. The "force" of natural selection, moreover, while it is not governed by a quantifiable inverse square law, certainly falls off with distance from the scene at which organisms and environments interact in complex ecological webs. Indeed, Darwin's objection to Lyell's appeal to God's prevision of preadapted species is that it violates precisely this

Newtonian causal condition (Hodge 1985). The whole process, finally, is as uniform and gradual as Lyell or any Victorian economist could wish for. Those who think that metaphors are not explanatory or that metaphorical explanation is too projective to pick out and describe real processes in nature will remain unimpressed by this achievement. It is, nonetheless, a real achievement.

John Stuart Mill, the third of Victorian Britain's most famous scientific methodologists, and the one most closely associated with the Malthusian liberalism that entered into Darwin's thinking, was no more impressed by *On the Origin of Species* than Herschel. The fact that Mill's difficulties were almost the exact opposite of Herschel's shows, however, how different their ideas about scientific method actually were. Mill had nothing against statistical reasoning in either the social or the natural sciences. Indeed, his *System of Logic* (1843) was one of the first books to set forth criteria for successfully employing statistics in more than an auxiliary role. What Mill objected to was Darwin's attempt to satisfy Herschel by using analogical reasoning to portray natural selection as a *vera causa.* This, for Mill, meant that Darwin's theory could never be any more than a suggestive "hypothesis"—that word again—at least until empirical research and experimentation of a much more piecemeal nature showed natural selection to be a real phenomenon, and not just a clever analogy (Mill 1843). In this way Mill found himself in agreement for once with Whewell. It must have been galling to Darwin to learn that he failed to convince any of Britain's three leading philosophers of science.

This does not mean that Darwin's book was not greeted warmly by Malthusian liberals of Mill's circle on other, largely ideological grounds. The fact that economics played such a large role in its argument made it virtually certain that *On the Origin of Species* would be applauded by the *Westminster Review* crowd. Mill's epistemological scruples notwithstanding, however, Darwin's book was welcomed more for its political correctness than for its scientific validity. "What a book it is!" said Martineau, "overthrowing (if true) revealed Religion on the one hand and Natural (as far as Final Causes and Design are concerned) on the other" (quoted in Desmond and Moore 1991, 486). In addition, *On the Origin of Species* was also welcomed because it seemed to lend support to a theory of evolution that had already been born within the ranks of the Malthusian liberals and that suited their tastes to a tee. Darwin was not the only person in Britain in the 1850s to link political economy and biology in support of evolution, or even the first. That credit goes to Herbert Spencer, who since 1851 had been working out an evolutionary theory made of many of the same conceptual materials as Darwin's but differently assembled.

Spencer was no natural historian. His primary interest, unlike Darwin's, had always been society. He had been an early advocate of the kind of free-market liberalism that in midcentury Britain was perceived to be a force for political reform rather than class oppression. Spencer saw in the free market a way to achieve the social harmony desired by socialists, and in socialism, to which he had been attracted was a young man, a good way not to achieve it. To protect society from the stern tutelege of Malthusianism and from the contingencies and opportunities of the market, was, from his perspective, to threaten the deepest conditions of social progress. A society that softens the blows of population pressures, Spencer proclaimed, will eventually pay a steep price. If, on the other hand, Malthusian competition is allowed free rein, Spencer promised that the social system would develop to a point where its burden is light. People under this kind of pressure will develop moral inhibitions that will enable them to foresee the results of their actions, adopt birth control, change jobs, emigrate, and, in short, become rational economic agents. Rational economic agents are, in Spencer's world, made rather than born. If things are managed well, "In the end pressure of population and its accompanying evils will entirely disappear; and will leave a state of things which require from each individual no more than a normal and pleasurable activity" (Spencer 1852, quoted in Richards 1987, 273). In sum, Spencer arrived at his utopia by way of the magic of the market. His was a social theodicy.

Since societies can evade these truths in the short run by political meddling and socialist schemes, Spencer felt that his points needed to be grounded in universal laws of nature. It was in this spirit that Spencer devised an evolutionary theory. Like Darwin, he had read Lamarck, as well as Lyell's refutation of him, and had sided with Lamarck. As an ardent Malthusian, he was as prepared as Darwin to recognize the beneficial effects of competition in nature as well as society. Finally, like Darwin, Spencer sided with von Baer and Milne-Edwards against the strong recapitulationism of the French evolutionary tradition, and he sought to put an evolutionary spin on von Baer's laws of development. Spencer's theory of evolution is in fact simply von Baer's developmental principle, according to which embryos move from homogeneity to heterogeneity, writ large, used as a language to redescribe every sort of natural and social system, and treated as a universal law of nature. If Newton's laws needed supplementing, Spencer seemed to think, it would not be in the direction of looser statistical laws but in the direction of developmental laws that apply strictly to all systems. The natural state of any system—and it was in terms of whole systems that Spencer usually talked—is toward developmental differentiation. Using this universalized version of von Baer's laws as a premise, Spencer then argued that biological evolution will occur when Malthusian boundary conditions put enough force behind innovation and the transmission of acquired characteristics to result in evolutionary diversification and progress.

Armed with these deep truths, Spencer returned to social policy. He appealed to the inherent developmental dynamics of systems to explain why society is bound to improve when each of its parts is put into competitive relationships. In showing this, Spencer does not begin with the old Hobbesian assumption that humans are inherently individualistic and competitive. That explains nothing. On the contrary, a mature social system is needed to produce genuinely self-interested individuals, whose rational and autonomous actions will preserve the health of the social system. The only way to achieve such a society under the conditions of increasing abundance that such behavior tends to bring about, however, is to maintain competitive pressures in society, for it is these that produce the most effective forms of cooperation through the premium they put on the division of labor. On these terms, progress is bound to happen. In this way, Spencer played a prominent role in the legitimation of late Victorian capitalism under the rubric social Darwinism. Indeed, perhaps one of the reasons for the decline of his reputation as a deep thinker in the twentieth century is the very fact that Spencer's views have entered deeply enough into the ideological constructs of liberal societies to count as common sense, and therefore no longer to stand in need of the elaborate philosophical and cosmological defense he gave them.

When Darwin's book appeared in 1859, Spencer championed it. Spencer was doubtlessly genuine when he referred both to his own and Darwin's theories as "the developmental hypothesis." Darwin and he were, after all, making the same main point. Both were defending, against a very widespread consensus, an evolutionary interpretation of von Baer's developmentalism. If Darwin could admit use inheritance, Spencer could accept a bit of natural selection. Natural selection was, after all, a comprehensible enough process within his framework. All that really divided them was intellectual style; Darwin's inductive search for empirical support contrasted with Spencer's tendency to deduce large consequences from philosophical principles. In spite of considerable overlap, however, the subtle differences between Spencer and Darwin, of which Darwin was more aware than Spencer, make a difference. In his *Autobiography,* Darwin wrote: "After reading any of his books, I generally feel enthusiastic admiration for his transcendent talents. . . . Nonetheless, I am not conscious of having profited in my own work from Spencer's writings. His deductive manner of treating every subject is wholly opposed to my own frame of mind" (Darwin 1958, 108–9; cf. Darwin 1887, 2:84, 152, 239, 301, 371).

Darwin never thought in terms of natural laws that would apply to whole systems, and he certainly did not think of evolutionary laws as inherently developmental. In accord with his uniformitarian background assumptions, the maintenance of equilibrium was for him always an affair between two points whose inertial tendencies are affected by an externally related field of forces. The state of a system is simply the product of the relationships between its component parts. There are no inherent or inertial tendencies in systems as such. Thus, a von Baerian parallel between ontogeny and phylogeny is an effect of causal interactions in a force-filled environment, rather than anything that might be derived in a "deductive manner" from what later became known as "systems theory." What Spencer regards as a first principle, Darwin regards as a contingent effect. From this point of view, it was no accident that natural selection was primary and use inheritance secondary in Darwin's theory. It could not have been otherwise. It was natural selection that keep the accent on external forces. Spencer sometimes comes close to the opposite stress.

Nonetheless, Darwin's failure to dissociate himself more strenuously from what Spencer and the Spencerians were making of his theory meant that what became known as Darwinism in Britain and America was mostly Spencerism. In this way, "Darwinism" incorporated more of the progressive evolutionism of the radical twenties, shorn of its earlier Jacobin resonances and made respectably liberal, than Darwin had in mind. George Eliot, for example, one of the brightest people of her time, thought she was commenting on Darwinism when, in *The Mill on the Floss* and *Middlemarch,* she wrote about how intense, often competitive, social interactions and pressures create and use variations between individuals to produce both more distinctive individuals and more ramified forms of social cooperation as society progresses. She was actually commenting on Spencer, with whom at one point she was romantically entangled, more than on Darwin.

In this chapter, we have seen that Darwin was at first opposed by the Tory intelligentsia, abandoned by his Whig allies, and championed by Malthusian liberals, who turned his theory into a version of evolutionary progressivism. This being so, it is often asked how by the time of his death in 1882 Darwin had managed to become such an icon of respectable British science that he was buried in Westminster Abbey near Newton, Lyell, and Herschel (Desmond and Moore 1991). In general, the answer is that the Whig establishment eventually reconciled itself with evolution, as long as they were assured that it was going somewhere, was going there steadily, and was running on their principles. Former radical liberals were, it seems, quite willing to give those assurances.

In 1866, the growing acceptance of an increasingly fuzzy but generally progressivist Darwinism made a decisive breakthrough in the Whiggish halls of the BAAS. W. R. Grove, a lawyer, physicist, and chemist, was president (he invented the fuel cell and founded the Chemical Society). In his address to the annual meeting Grove remarked on how well Darwinism fit with what was generally acknowledged, in a self-congratulatory way, to be the source of the happiness of England's institutions. "In contrast to the so-called 'natural rights of man,'" Grove intoned, "these were the product of slow adaptations, resulting from continuous struggles." "Happily in this country," he went on, "practical experience has taught us to improve rather than remodel; we follow the law of nature and avoid cataclysms" (quoted in Desmond and Moore 1991, 536). To the extent, then, that Darwin succeeded in meeting criteria for explanatory success laid down by Herschel or Whewell, criteria originally calculated to make a Newtonian biology as difficult as possible, it was because of the gradualism he had inherited from Lyell, to which he had clung to even when his supporters thought it was his weakest point.

In the end, though, it mattered little whether Darwin had or had not met these criteria, for the gradualism that Grove praised now embodied a tacit theory of more or less autonomatic evolutionary progress, an idea once regarded as subversive. The crisis of Darwinism passed quickly enough because the issues that provoked it, and what "Darwinism" was taken to be, changed. Darwin's extension of Newtonian natural science to living things, which had posed such difficulties for Paley's natural theology, had given way to a new kind of theodicy that allowed evolution in nature to symbolize social evolution along Whig constitutional and Liberal economic lines. Two years after Grove's address, Hooker would become president of the BAAS. His presidential address would be a straightforward account of the growing explanatory power of the Darwinian research program. It would be well received. The nasty confrontation at Oxford and the heresy trial of the liberal theologians now seemed like distant echoes of a bygone world. Although conservatives were still around, they had stopped trying to control the production of knowledge, and, with Newman and the Oxford Movement, had taken their stand on faith alone. England had taken the plunge into modernity.

We will see in the next chapter that the trend toward developmentalist theories of evolution increased as the nineteenth century entered its last decades. A religious spin was often put on them. If the evolutionary theories that flourished in this period were called Darwinism, it was in part because Darwinism had been received and reconstructed through developmentalist eyes, in part because Spencerism overlapped both with Darwinism and developmentalism. It would only be later that the statistical alternative anachronistically ascribed by Schweber to Darwin himself would come into its own. That process began with the work of Darwin's cousin Francis Galton (chapter 8). It would not come to fruition, however, until the maturation of the probability revolution, and the severing of Darwinism's old links with developmentalist theories of inheritance, made a genetic theory of natural selection possible (Hodge 1987; Bowler 1988; Gayon 1992).

Reading Guide to Part I

The stress we have placed on the continuity between Aristotle and "neo-classical" biologists like Cuvier and von Baer is intended as a corrective to the uncritical notion that Aristotle was a "typological essentialist." Hull's title "Aristotle's Effect on Taxonomy: Two Thousand Years of Stasis," suggests how easy a target typological essentialism makes. Scholars have largely succeeded by now, however, in disentangling Aristotle from the Platonized Aristotelianism of the medievals, which is indeed typological. To see what Aristotle was up to, read Allan Gotthelf and James Lennox, eds., *Philosophical Issues in Aristotle's Biology;* Montgomery Furth, *Matter, Form and Psyche;* Pierre Pellegrin, *Classification of Animals in Aristotle;* and the essays (mostly in English) in Daniel Devereux and Pierre Pellegrin, eds., *Biologie, logique et métaphysique chez Aristote.*

Understanding Aristotle makes it possible to see how formidable were the forms of essentialism that Darwin and Lamarck actually confronted in Cuvier, von Baer, and Owen and how different their "neoclassical biology," as we have called it, is from the strict recapitulationism that Geoffroy adopted from German idealists. Owen's Hunterian Lectures of 1837, recently edited by Philip Sloan, are a culmination of the Aristotelian tradition's accomplishment in Darwin's era. These lectures can be contrasted with Romantic *Naturphilosophie* in the spirit of Goethe, Humboldt, Schelling, and Oken (as Sloan does in his editorial material). For the intellectual ethos of *Naturphilosophie,* see the essays collected in Andrew Cunningham and Nicholas Jardine, eds., *Romanticism and the Sciences.* On subsequent developments in Germany biology, William Coleman's *Biology in the Nineteenth Century* is a sober guide, and Timothy Lenoir's *The Strategy of Life* is brilliantly illuminating. The early chapters of Robert Richards, *The Meaning of Evolution,* are concise analyses of the relationship between recapitulationism and evolution. On the history of the recapitulationist evolutionary paradigm, also read Stephen J. Gould, *Ontogeny and Phylogeny.*

On the debate between Geoffroy and Cuvier, read Toby A. Appel, *The Cuvier-Geoffroy Debate.* On Lamarck's role in the dispute between Geoffroy and Cuvier, see Pietro Corsi, *The Age of Lamarck,* rev. ed. (1988), as well as Lamarck's *Zoological Philosophy*, which has been made available by Chicago University Press. Robert Richards has interesting and sympathetic things to say about Lamarck, and the Lamarckian influence on Darwin, in *Darwin and the Emergence of Evolutionary Theories of Mind and Behavior*.

Adrian Desmond, *The Politics of Evolution*, has changed the way in which the introduction of the Lamarckian-Geoffroyian tradition into Scotland and England is viewed. Since not everyone is as confident as we are of Desmond's account, especially of the social and political currents swirling around these events, it will be worthwhile to consult Philip Sloan's review of this book in *History of Science* 28 (1990): 419–28. See as well Sloan's treatment of Darwin's relations with Robert Grant in "Darwin's Invertebrate Program." Sloan's introduction to his edition of Owen's *Hunterian Lectures in Comparative Anatomy* is also essential reading.

Until recently there has been no satisfactory biography of Darwin. Now we have John Bowlby, *Charles Darwin*; Peter J. Bowler, *Charles Darwin*; and especially Adrian Desmond and James Moore, *Darwin*. Although the last, like Desmond's *The Politics of Evolution*, has been criticized as social constructionist, we have relied on it, and especially on its treatment of Darwin's Malthusianism, without fully endorsing its interpretive paradigm. Interesting interpretations of special aspects of Darwin's life, milieu, and influence are Dov Ospovat, *The Development of Darwin's Theory*, which is excellent on the Paleyesque aspects of Darwin's thought; Gillian Beer, *Darwin's Plots*; Edward Manier, *The Young Darwin and His Cultural Circle*; and Howard Gruber, *Darwin on Man*. On Darwin's way of life down in Downe and its eerie resemblance to the clerical life he rejected, see the fascinating report by James R. Moore, "Darwin of Down: The Evolutionist as Squarson-Naturalist." On Darwin's illness, and a whole raft of psychobiographical issues we have largely set aside, see Ralph Colp, *To Be an Invalid*.

A complete edition of Darwin's letters is slowly appearing in *Correspondence of Charles Darwin*. It is currently complete only to 1860. For the important letters after that, we must still rely on Charles Darwin, *Life and Letters of Charles Darwin*, edited by Francis Darwin, and on an additional two-volume collection of letters that appeared under the title *More Letters of Charles Darwin*. (We refer to the American versions of these works in the chapter.) Volume 1 of *Life and Letters* also contains an imperfect text of Darwin's *Autobiography*. The definitive text is *The Autobiography of Charles Darwin*, edited by Nora Barlow. We refer to Barlow's edition. A more readily accessible paperback edition, which also contains T. H. Huxley's *Autobiography*, is available from Oxford University Press.

A serious encounter with Darwin's theories might begin with what Darwin himself read. Paley's *Natural Theology* is not readily available in a popular edition, but various editions of what was for a long time a big best-seller can be found in academic libraries. *The Bridgewater Treatises* are worth reading as evidence of the seriousness with which the argument from design was taken and how strenuous were the efforts to integrate it with Newton-inspired science. The treatises can be found under the names of their authors: Thomas Chalmers, Peter Mark Roget, William Buckland, William Kirnby, William Whewell, Sir Charles Bell, William Prout, and Charles Babbage. Perhaps the most important prelude to reading *On the Origin of Species*, however, is volume 2 of Lyell's *Principles of Geology*. A reprint of the very three-volume edition of the *Principles* that Darwin read on the *Beagle* has been reprinted by Chicago University Press. Lyell's troubled notebooks on the species question, which he kept between 1855 and 1861, are also revealing. They have been edited as Leonard G. Wilson, *Charles Lyell's Scientific Journals on the Species Question*. Read Lyell in conjunction with Stephen Jay Gould's *Time's Arrow, Time's Cycle*. On Darwin and Lyell, consult M. J. S. Rudwick, "The Strategy of Lyell's *Principles of Geology*"; Hodge, "Darwin and the Laws of the Animate Part of the Terrestrial System (1835–1837): On the Lyellian Origins of His Zoonomical Explanatory Program"; J. A. Secord, "Discovery of a Vocation: Darwin's Early Geology"; and S. Herbert, "Charles Darwin as a Prospective Geological Author."

For several decades a self-described Darwin industry has industriously been working over Darwin's early notebooks, in the process giving birth to a revised image of Darwin's Darwinism. *Charles Darwin's Notebooks, 1836–1844,* have been published by Cambridge University Press, edited by P. H. Barrett, P. J. Gautreg, S. Herbert, D. Kohn, and S. Smith. The M and N Notebook, together with extracts from Notebooks B, C, D, and E, can more readily be found in Howard Gruber, ed. *Metaphysics, Materialism and the Evolution of Mind.* Varying interpretations of what is in these notebooks and how they affect, or should affect, the interpretation of Darwin's later work have been appearing for some time. Important papers by M. J. S. Hodge, especially "Darwin as a Lifelong Generation Theorist," have shown incontrovertibly that Darwin was never as neutral about reproductive dynamics as later Darwinians made him out to be. The question is how Darwin's "lifelong generation theorizing" is related to Darwin's evolutionary theory. Opinions on this issue still differ too widely for anyone to be absolutely certain. The most recent challenge to the consensus is Robert Richard's startling claim, in *The Meaning of Evolution*, that Darwin's version of the parallel between ontogeny and phylogeny made him not only a progressive evolutionist but a fairly strong recapitulationist. Although we agree that twentieth-century Darwinians have overly downplayed both Darwin's progressivism and his developmentalism, we still think that he was more von Baerian than Geoffroyian. This view is defended by Bowler, *The Non-Darwinian Revolution;* Gould, *Ontogeny and Phylogeny;* and Mayr, *The Growth of Biological*

Thought. For a good overview of Darwin's theory, or rather theories, see Mayr, *One Long Argument.*

One reason we support the consensus view on Darwin's recapitulationism is our belief that Richards neglects or underestimates the conceptual role played by Newtonian thinking in Darwin's work. On Darwin's Newtonian framework, on which so much of our argument rests, read Silvan Schweber, "The Wider British Context in Darwin's Theorizing" and "The Origin of the *Origin* Revisited." For us, as for Schweber, this influence is inseparable from Darwin's relationship to the discourse of political economy. One can find arguments against the influence of the political economists on Darwin, mostly based on the fact that Darwin seems not to have read much economics and confesses to have found what he did read dull and confusing. We think that Schweber's argument in "Darwin and the Political Economists" survives these objections. The matter is vividly summed up in S. J. Gould, "Darwin and Paley Meet the Invisible Hand." For detailed accounts of how Darwin's encounter with Malthus affected the direction of his theorizing in the early notebooks, read M. J. S. Hodge and David Kohn, "The Immediate Origins of Natural Selection" and M. J. S. Hodge, "The Development of Darwin's General Biological Theorizing," and "Darwin as a Lifelong Generation Theorist." These reconstructions may be compared with Mayr, "Darwin and Natural Selection" and *One Long Argument,* as well as with Richard's accounts in *Darwin and the Emergence of Evolutionary Theories of Mind and Behavior* and *The Meaning of Evolution.* For the influence, via Milne-Edwards, of Smith's principle of division of labor on Darwin's mature doctrine of evolutionary diversification, see David Kohn, "Darwin's Principle of Divergence as Internal Dialogue." The fact that economic diversification à la Adam Smith, on which Darwin relies for his picture of phylogeny, supports the notion that Darwin's version of the ontogeny-phylogeny parallel is von Baerian rather than strict recapitulationist. Strict recapitulationism does not look at all like the increasing specialization of an economy (We are indebted to Ronald Giere, *Understanding Science,* for inspiring the abstract model of a Newtonian system that we have constructed.)

The paper Wallace sent to Darwin in 1858 can be found in Darwin and Wallace, *Evolution by Natural Selection,* introduction by Sir Gavin de Beer. Wallace's later essays can be found in Wallace, *Contributions to the Theory of Natural Selection.* Wallace later defended the concept of natural selection in his *Darwinism* against a rising tide of criticism. On Darwin and Wallace, read Malcolm Jay Kottler, "Charles Darwin and Alfred Russell Wallace: Two Decades of Debate over Natural Selection." The manuscript of *Charles Darwin's Natural Selection,* part 2 of the "big species book" on which Darwin worked during the 1850s and from which he mined the material for the *Origin,* was published by Cambridge University Press in 1975, edited by R. C. Stauffer. A facsimile of the first edition of *On the Origin of Species* has been published by Harvard University Press, edited by Ernst Mayr. This is the edition to which we refer in the text except where otherwise noted.

For the argumentative structure of *On the Origin of Species* and its relationship to Herschel's philosophy of science, as well as Lyell's *Principles,* see especially Hodge, "The Structure and Strategy of Darwin's 'Long Argument.'" Work on Darwin's relationship to the methodologists of his time and circle is a helpful corrective to earlier tendencies to insist that Darwin's virtues depended on his uncanny ability to conform *avant le lettre* to the norms of contemporary philosophers of science. Michael Ruse's *The Darwinian Revolution* and Michael Ghiselin's *The Triumph of the Darwinian Method* are better guides to twentieth-century logical empiricism and Poppernian falsificationism than to Darwin, just as Doren Recker, "Causal Efficacy: The Structure of Darwin's Argumentative Strategy in the *Origin of Species*"; Elizabeth Lloyd, "The Nature of Darwin's Support for the Theory of Natural Selection"; and Philip Kitcher, "Darwin's Achievement," opt for postpositivist and post-Popperian perspectives. Hodge, "Darwin's Argument in the *Origin,*" is a neo-Herschelian rejoinder. On Herschel read Schweber, "John Herschel and Charles Darwin: Parallel Lives." On Whewell, see *William Whewell* edited by Fisch and Schaffer. For Mill's philosophy of science and his surprisingly cool reaction to Darwin's theory, read David Hull's introduction to *Darwin and His Critics,* a valuable collection of early learned reactions to *On the Origin of Species.* On Asa Gray's semitheological version of Darwinism, see his collection of essays, *Darwiniana.* More generally, read James R. Moore, *The Post-Darwinian Controversies,* and the relevant chapters in Desmond and Morris, *Darwin.* For a dim, and not altogether trustworthy, view of Darwinism in a Victorian context, see Gertrude Himmelfarb, *Darwin and the Darwinian Revolution.* Gould "Knight Takes Bishop," provides a summary of recent research about what really happened at the famous Oxford debate. On Herbert Spencer's life and theory of evolution, see Richards, *Darwin and the Emergence of Evolutionary Theories of Mind.*

II Genetic Darwinism and the Probability Revolution

7 Ontogeny and Phylogeny: The Ascendancy of Developmentalism in Late Nineteenth-Century Evolutionary Theory

In the last quarter of the nineteenth century, many learned people in Europe and North America, and even some pious folk, were converted to an evolutionary perspective that no more than a generation earlier people of their sort would have resisted strenuously. This has long been called the Darwinian revolution (Himmelfarb 1959). Recently, however, Peter Bowler, who has studied evolutionary theory at the end of the nineteenth century, has argued that, contrary to well-established myth, there never was a "Darwinian revolution" in Britain or anywhere else. Bowler writes:

The *Origin* certainly played a role in converting the English, and to a lesser extent the German-speaking world to evolutionism, often along lines very different from those proposed by Darwin. . . . Darwin's theory should be seen not as the central theme of nineteenth century evolutionism, but as a catalyst that helped bring about the transition to an evolutionary viewpoint within an essentially non-Darwinian conceptual framework. This was the "Non-Darwinian Revolution." It was a revolution because it required the rejection of certain key aspects of creationism, but it was non-Darwinian because it succeeded in preserving and modernizing the old teleological view of things. (Bowler 1988, 4–5)

Whether there was or was not a Darwinian revolution depends, of course, on what is meant by Darwinism. *On the Origin of Species* makes a case, as Mayr has shown, for five distinct hypotheses (Mayr 1982, 1988):

1. Naturalism (or materialism): Questions about the origins of species are explained by natural processes and the laws governing them.

2. Transmutation: Species (and other taxa) are not fixed types.

3. (Monophyletic) descent with modification: All phylogenetic branching goes back to a common ancestor.

4. Natural selection: Transmutation and descent with modification reflect differences in reproductive rates caused by differences in relative adaptedness of chance variants to a given environment.

5. Causal pluralism: Natural selection is the dominant, but not the only, cause of evolution. Subordinate causes include the inheritance of acquired characteristics and sexual selection.

By Darwinism Bowler generally means natural selection. His point in debunking the notion of a Darwinian revolution is to call attention to the fact that widepread acceptance of evolution, in the sense of transmutation of species and descent from an original ancestor, was accompanied by the no-less-widespread opinion that natural selection is *not* its chief motor. Indeed, evolutionists at the turn of the twentieth century generally assumed that Darwinism, in the sense of natural selection, was on its death bed (Dennert 1904). Thus in 1907 Vernon Kellogg, in his still interesting book *Darwinism Today*, wrote:

Darwinism is not synonymous with organic evolution, nor with the theory of descent. . . . Therefore when one reads of the "death-bed of Darwinism," it is not the death bed of organic evolution or of the theory of descent that one is reading. While many reputable biologists today strongly doubt the commonly reputed effectiveness of the Darwinian selection factors to explain descent . . . practically no naturalists of position and recognized attainment doubt the theory of descent. Darwinism might indeed be on its death-bed without shaking in any considerable degree the confidence of biologists and natural philosophers in the theory of descent. (Kellogg 1907, 3)

Bowler's iconoclastic reading of the situation suggests that Darwin's name should not be as deeply associated with transmutation and common descent as it commonly has been. A deep well of evolutionary thinking, more properly associated with Lamarck and Geoffroy than with Darwin, had been around long before *On the Origin of Species* appeared in 1859. Post-Darwinian evolutionary theory resonates better with this old inheritance than with Darwin's own theory of descent. That is because the brands of evolutionary thinking that took hold in the decades after Darwin inclined toward forms of strong recapitulationism and the inheritance of acquired characteristics that run against the grain of Darwin's own thinking.

This is true even of self-proclaimed and highly influential Darwinians like Ernst Haeckel, a central figure in this chapter. In championing Darwinism, Haeckel was championing common descent. Haeckel, however, was a strong recapitulationist, whose famous "biogenic law" held that ontogeny sequentially runs through the *adult* forms of mostly extinct ancestors. Haeckel also assigned a more prominent role to the inheritance of acquired characteristics than Darwin did. Haeckel's success in convincing the world that Darwinism is a good name for common descent conceived along these lines lies at the root of the myth of the Darwinian revolution. This does not, of course, mean that every evolutionist after Darwin was a strong recapitulationist or a so-called Lamarckian. What it means is that in the later nineteenth century, the line between strong recapitulation and the weak recapitulation that Darwin inherited from von Baer, Owen, and Milne-Edwards was blurred in favor of the former, that natural selection was subordinated in various ways to the inheritance of acquired characteristics, and that in the age of progress, new versions of the old Geoffroyian and Lamarckian evolutionary inheritance achieved, sometimes under the name Darwinism, a respectability they had never enjoyed during and after the age of revolution.

In what way, then, did Darwin's book catalyze this shift, as Bowler puts it? This is still an open question among scholars. What we have said in earlier chapters implies, however, an answer. During the 1840s and 1850s, the embryological arguments of von Baer in Germany, Milne-Edwards in France, and Owen in England, according to which vital energy is so exhausted at the terminal point of development that the very possibility of transmutation is precluded, had put evolution on the defensive. Darwin broke down this barrier by giving an evolutionary interpretation of von Baer's, Milne-Edwards's, and Owen's weak recapitulationism. He hoped that by deploying models of evolutionary dynamics based on external forces, the old French forms of evolutionary thinking would remain as peripheral as they had been in the heyday of von Baer and Owen, and therefore that standing objections to evolutionary thought, especially Lyell's geological evidence against linear progress, would be obviated. Instead, what happened is that, having broken through the barrier that divided evolutionists from antievolutionists, Darwin's book had the catalytic effect of unleashing the dammed-up tradition of Geoffroyian and Lamarckian thought that in nearly every European country had for so long virtually defined the idea of evolution. Thus, in France, where evolutionary debates had long had a dynamic of their own, Darwin's work, where it was noted at all, was seen as straightforwardly testifying in favor of the intellectual heirs of Lamarck and Geoffroy and against those of Cuvier (Conry 1974). In Haeckel's Germany, once Darwin had given an evolutionary interpretation of von Baer's laws of development, those laws came to be viewed as timid first approximations to the strong recapitulation that Haeckel took to be equivalent to the nature of evolution itself (Gould 1977; Rasmussen 1991). The catalytic effect of Darwin's book, accordingly, was to allow the older tradition of evolutionary thought from which Darwin had tried to keep his distance to rise again. What Darwin feared might happen did happen. What he did not anticipate was that the old evolutionism, once so redolent with revolutionary overtones, would become so respectable or that it would travel under his name.

Could it have been otherwise? Darwin's acceptance of von Baer's claim that embryos go through only immature stages of ancestral forms was intended to support his picture of phylogeny as a ramified bush rather than a record of directional evolutionary ascent toward humans. Darwin agreed with Milne-Edwards, who bluntly proclaimed that "if [lower animals] were in some way permanent embryos of the [higher animals] it would be necessary to admit . . . a progressive and linear series extending from the monad to man" (Milne-Edwards 1844, 70). In order to avoid just that inference, Darwin relied on his Newtonian philosophy of science to keep the explanatory accent on external rather than internal causes, to ensure that the inheritance of acquired characterististics would remain subordinate to natural selection, and to draw a fairly clear line between strong and weak recapitulation. In France, Germany, and Russia, however, the subtle Whiggish ways of thinking that enforced these distinctions never played any role. They could scarcely be counted on, therefore, to do the work that Darwin rather naively, and ethnocentrically, assigned to them. Indeed, this aspect of Darwin's argument would scarcely have been recognized by most foreign readers of *On the Origin of Species*, especially in translations (Conry 1974). Even in Britain, the Whiggish cult of Newton that had been so important in the delicate period of reform was rapidly being displaced in favor of explicitly developmentalist and progressivist laws of nature, such as those championed by Spencer, and of the statistical laws that so appalled Herschel. For these reasons, Darwin's effort to resist directional and anthropocentric conceptions of evolution by resisting strong recapitulationism and the primacy of inherited acquired characteristics in causing adaptations, fell on deaf ears. Indeed, Darwin himself increasingly stressed the role of acquired characteristics in later years. As a result, the main issue now became whether directional evolution should be envisioned in materialistic, humanistic, and anticlerical terms, as Haeckel and Darwin's French translator, for example, hoped; or as a spiritual process in which, in an evolutionary reinterpretation of the old Romantic idealism, the divine comes to consciousness of itself in and through evolutionary progress, as some liberal Protestants, especially in America, came to believe.

The decline of strong recapitulationism and progressive evolutionism of both materialistic and vitalistic sorts, and the eventual restoration in this century of natural selection to primacy of place in evolutionary theory, is all the more significant, then, because, contrary to common opinion, it was not the outcome of a confident and continuously advancing tradition stemming directly from Darwin. It was instead the product of a protracted crisis in which the long-lived evolutionary tradition that had its roots in Geoffroy and Lamarck went down to defeat. The precipitating cause of this crisis was August Weismann's apparent experimental demonstration in about 1890 that acquired traits cannot be inherited because the "germ line" of reproductive information is sealed off, or "sequestered" like a jury, from outside influences too early in its development for that to happen. The consequences of "hard inheritance," as it came to be called, for most versions of later nineteenth-century evolutionary thinking were disastrous. They were no less profound, however, for Darwinians, who, if they had not already done so, were forced to look for a theory of natural selection that was consistent with hard inheritance alone, and therefore that abandoned Darwin's own causal pluralism and the pangenetic and blending views about inheritance that undergirded it. Genetics eventually proved to be the key. But a genetic theory of natural selection could not triumph until genetics itself was taken from those who thought that it offered an account of speciation through hybridization and macromutation, and therefore a suitably post-Weismannian alternative to both the Darwinian and Lamarckian research traditions. It took fifty years, massive doses of statistical reasoning, and a reconception of Darwinism in terms of probabilistic thinking to achieve that. Accordingly, there is a good deal of truth in Bowler's claim that "there is a sense in which the emergence of the modern synthetic theory can be seen as the first real triumph of Darwinism. If there was a Darwinian revolution, it was not completed until the 1930s" (Bowler 1988, 105).

The slow emergence and maturation of genetic Darwinism is the subject of part II of this book. It is fitting that we should begin with an account of the heyday of evolutionary recapitulationism, for the kind of developmentalism implied by the strong ontogeny-phylogeny parallel formed the background against which the neo-Darwinism of the twentieth century eventually triumphed. In particular, we will retrace the course of evolutionary theory in the two countries that took the lead in grand evolutionary theorizing during this period and that were most immediately affected by the Weismannian challenge that we shall recount at the end of the chapter. These countries are Germany and the United States.

Ernst Haeckel was professor of zoology and comparative anatomy at the same University of Jena in which, long before his time, idealist *Naturphilosophie*, and hence nonevolutionary recapitulationism, had first flourished under the influence of Goethe, Schelling, and Oken. Haeckel showed up on Darwin's door in 1866, proclaiming eternal fealty. Darwin could hardly understand either what he was excitedly saying, or what he had written or, for that matter, what, as Darwin's international champion, he would write in the future. Nonetheless, Darwin liked Haeckel's enthusiasm and frankness, appreciated his undying support, and gave him his blessing (Desmond and Moore 1991, 538–40). From then on, whenever Darwin got close to discovering that Haeckel's *Darwinismus*, which became very popular in Germany and elsewhere, was not Darwinism, he placed the blame on himself. "Perhaps I have misunderstood him," he wrote to George Romanes, another disciple. "His views make nothing clearer to me, but this may be my fault" (quoted in Gould 1977, 79). Even more than in the case of Huxley and Spencer, Darwin's willingness to entrust his theory to Haeckel suggests his desire for support, his willingness to allow advocates of radical scientistic materialism and anticlericalism to become his chief advocates, and perhaps his utter naiveté about what would happen to his theory when it was assimilated to the cultural conditions of countries in which his own genteel Whiggishness was totally absent.

What interested Haeckel most about Darwin's theory was the resources it offered for defending philosophical naturalism in its most radically materialisic and reductionistic form. Transmutation and especially common descent were significant to Haeckel because they backed up reductionistic materialism by forbidding any external agents from entering into the history of life. In fact, Haeckel went further than Darwin's agnosticism about the origins of life. For Haeckel, life is "nothing but"—the reductionist's war cry—chemical synthesis. Natural selection was much less important to Haeckel. At best it provided a way in which new traits, which for him had their origin in the heritability of acquired characteristics, could be amplified and fixed in the developmental sequence.

The motives undergirding Haeckel's reworking of Darwin's theory can be appreciated only after we have recalled something about the development of German intellectual life and science in the nineteenth century. Germany was not a unified country until 1870. Nor did its professional and mercantile classes ever acquire power nearly as successfully as their counterparts in Britain and France. German science, seen from the first as an expression of a distinctive German culture (*Bildung*) that was to form the basis of a distinctive German national identity, had, since the time of Goethe and Humboldt, been framed in Romantic, vitalistic, idealistic, and socially conservative terms. Newtonian mechanism and, relatedly, free-market economics were invidiously contrasted with an organic vision of society that was to be maintained, as Hegel taught, even within modernity. Young Germans from the generation of Heine, Feuerbach, Bauer, and Marx to that of Buchner, Moleschott, and Haeckel accepted the Romantic critique of Newtonian physics and market economics. They chafed, however, under the social and political conservatism that was enforced by idealist metaphysics, which was taught as a state philosophy in high schools and universities. Perhaps they can be forgiven, therefore, for assuming that there is an inherent and mutually reinforcing link between materialism and democracy, even while failing to recognize that there is a more obvious link between democracy and market economies. The point is that every step toward reductionistic materialism was seen by successive generations of young German intellectuals, including Haeckel's, as a step toward human liberation.

Early opponents of Romantic *Naturphilosophie* were understandably attracted to the more pro-Enlightenment philosophy of science that derived from the critical, rather than absolute, idealism of Kant. Indeed, the route by which German biology slid down the slippery slope from the idealistic *Naturphilosophie* of the 1820s to the reductionistic materialism of the 1860s that was at work in Haeckel had a great deal to do with successive changes in prescriptions for biological studies that had been laid down by Kant and his disciples. These methodological changes were intimately connected, moreover, with very concrete and successful research programs in embryology and cytology (the study of cells). The result was the steady replacement of idealistic with materialistic frameworks for scientific research (as well as for social reform).

For Kant, one cannot even begin to ask biological questions unless one has first identified a living thing, and distinguished it from an inanimate object, by seeing it as a functional, purposive whole (see chapter 4). Within an organic context thus presupposed, inquirers can, however, profitably begin to track down the material pathways that sustain the mysterious life of organisms. In this way, we make progress in connecting biology to chemistry and physics, and at the same time come to appreciate, through our utter inability to unravel processes that are both "cause and effect of themselves," the ultimate irreducibility of life to nonlife. Methodologically self-conscious biologists influenced by Kantian thinking, but less intent than Kant on limiting reason to make room for faith, soon modified his guiding idea into what Timothy Lenoir has called the "teleomechanist" research tradition (Lenoir 1982). They hypothesized that while life cannot be reduced to the laws of physics and chemistry, it might nonetheless be governed by natural laws of its own. Since these laws must direct teleological or end-directed processes, teleomechanists sought distinctively biological laws by studying the orderly and apparently purposive differentiation of embryos as they develop toward maturity, the original context in which Aristotle himself had found teleological phenomena.

Von Baer's laws, according to which an embryo develops from an undifferentiated, homogeneous state to a highly articulated and differentiated one, which have already played an important role in the story we have been telling, were conceived within this research program. Von Baer saw development as driven by a distinct vital force (*Lebenskraft*), physically emergent from a physicochemical base and then reacting on it. Operating from the center to the periphery of the embryo, this force guides development toward maturity. The "life force" is supposed to be a force like other Newtonian forces, spreading out from a center after the fashion of an inverse square law, maintaining equilibrium as it goes, but irreducible to gravity, electricity, magnetism, or other canonical Newtonian forces. This is, properly speaking, "vitalism." The enemies of vitalism (and they are legion) have tended to assimilate it to idealistic *Naturphilosophie*, largely as a way of discrediting it even more thoroughly. That is too bad, not only because it makes it harder now to see that vitalistic teleomechanism provided the context in which some of the greatest advances of nineteenth-century biology were achieved, but because many of these discoveries were intended as refutations of *Naturphilosophie*. Von Baer's laws, for example, were aimed at discrediting the strict recapitulationism of the idealist *Naturphilosophen* (Lenoir 1982).

By the 1850s, as Lenoir tells the story, the teleomechanist research tradition had taken a new turn, moving a step closer toward its reductionist materialist antithesis. The new version eschewed vitalism. Life is not a separate force at all, emergent or otherwise, but is simply the effect produced by a specific organization of physical and chemical materials and processes when these occur within a functionally organized context. Nonetheless, life cannot be reduced fully to nonlife, for although constitutuively life is "nothing but" matter and energy organized in certain ways, it relies on already existing organic conditions for its production. Even though there is no life force, there is still an irreducibility to living things because life could not ever have come from nonlife.

The idea that all organisms are composed of cells, and that the history of living things is the history of cell lineages, was developed within this more naturalistic version of teleomechanism. Cell theory was first worked out by Matthias Schleiden and Theodor Schwann in the late 1830s. It was intended as a refutation of von Baer's vitalism by showing that the division and multiplication of cells and their functional aggregation and differentiation explain von Baer's laws by purely chemical means. Schwann and his colleagues were convinced that what a cell does, including replicate itself, is nothing more or less than engage in a series of (then unknown) chemical reactions down a specific set of energetic pathways. Organic chemistry developed in the 1840s within this heuristic frame of reference, especially under the influence of Justus von Liebig. His line of work stands at the fountainhead of modern biochemistry (Holmes 1964). It would be impossible to overestimate the long-term success of this program in actually finding out what the postulated biochemical pathways are. The result would have surprised, and probably dismayed, Kant, for the unraveling of cause-effect chains in cellular mechanisms cast doubt on Kant's religiously inspired hope that there would never be a Newton of such things.

The developers of cell theory had something else in mind too. Their opposition to von Baer's combination of weak recapitulationism with opposition to evolutionism made them sympathetic not only to transmutation and common descent but to strong recapitulationism as well. If a strict parallel between ontogeny and phylogeny was to be carried out, however, it would be necessary to carry the story as far back as the earliest living things that existed before the separation of plants and animals. That point of intersection was the single cell or monad. The first organisms, after all, were single celled, and all later organisms are aggregations of cell lineages, multiplying by dividing, integrating by differentiating. Perhaps the entire recapitulationist story could be plausibly retold and defended from this more basic perspective. Recapitulationism, an idea first hatched up by nonevolutionary Romantic idealists, and subsequently rejected by more sober people like von Baer, now began to recover lost ground under the auspices of cell-centered evolutionary naturalism. In part, this explains why Haeckel, who deeply opposed von

Baer, was a strong recapitulationist and why he wanted badly to read Darwin as one.

With one more step, the teleomechanist tradition was finally transformed into its materialist reductionist antithesis. When one looks at the machinery of the cell, regarding organisms simply as vast assemblages of cells, one begins to suspect that the laws of physics and chemistry, the latter being no more than an application of the former, are fully adequate to explain not only the operation but also the original emergence of all organic functions and beings, for the forces operating within and between cells are none other than known physical processes that synthesize and degrade chemical and biochemical compounds. From the perspective afforded by the energy-centered physics then triumphing in Germany, the life processes of the cell are fully reducible to chemistry, and chemistry to physics.

At the focus of this reductionistic research program was the great German physicist Hermann von Helmholtz, who laid it down that energy is a conserved quantity. If it was impossible to explain blades of grass in terms of force, it seemed distinctly likely that one could explain them in terms of the capture and utilization of energy. Schwann himself had shown that fermentation is a chemical reaction facilitated by a living organism, yeast. Helmholtz now showed that the entire process, more closely considered and measured, is just a case of more general processes of breakdown of organic compounds into inorganic elements. Similarly, Liebig's demonstration that body heat is the oxidation of carbon and hydrogen under conditions of organic respiration was now shown by Helmholtz to be a case where respiration, rather than being a condition of combustion, is "nothing but" an instance of it. Breathing is something like slow burning. Given Helmholtz's law of the conservation of energy, the transformation of living into nonliving, and of nonliving into living, must be energetically equivalent. Perhaps Dr. Frankenstein's fantasy was realizable after all.

This dénouement spelled the end not only of vitalism but of the more functionalist forms of teleomechanism that followed it. Von Baer looked like an old fogey. From the new perspective, vitalism was reidentified as the unsustainable claim that the causes of life violate the law of the conservation of energy and that some additional, nonphysical force is added or subtracted in organic processes. That was, of course, a very prejudicial way of putting the issue, for from this redescribed perspective, people like Schwann appeared no less vitalistic than genuine vitalists like von Baer, whom they had challenged in their day. Even worse, real vitalists were conflated with the idealist Naturphilosophen whom they opposed. It is eloquent testimony to the degree to which reductionism redefined the debate that vitalism is even now a term that vaguely and contemptuously assimilates these importantly different, and in their time highly productive, research programs in their least plausible version.[1]

The ideological lessons implicit in these developments were clear to people like Haeckel. Any social, political, and intellectual power resting on idealistic metaphysics is illegitimate, for idealism and romanticism are illusions from top to bottom. Uncompromising philosophical materialism became in this way a weapon of scientific, secular, republican, democratic, and socialist currents of thought in Germany, and indeed throughout central Europe and Russia as well.[2]

Fatefully, Darwinism entered Germany, mainly through Haeckel's good offices, precisely when materialism of this stamp was most vigorously and obstreperously asserting itself among the educated young. Darwin was immediately hailed by young turk materialists like Ludwig Buchner, Carl Vogt, and Haeckel as the prophet of an uncompromising materialism that extended not only to physics and chemistry but to the study of life and consciousness as well. Ignoring Darwin's studious agnosticism about ultimate origins, Darwin's commitment to common descent was thought by these people to provide a closure condition barring the introduction of any nonnatural causes into the development of life out of physical and chemical forces, and that in fact guaranteed that as evolutionary science matured, life itself would be shown to be "nothing but" chemical interactions utilizing energy gradients in certain ways. The living world is folded into the physical world by laws governing the composition and distribution of matter and energy. Atoms are assembled, by inescapable laws, into molecules, molecules into cells, cells into tissues, tissues into organs, organs into organisms, organisms into colonies, colonies into ecological communities, ecological communities into evolutionary lineages. Everything at a higher level is "nothing but" an aggregation of what lies below it. So construed, *Darwinismus* connoted anticlerical, antimonarchical, antivitalistic, antihierarchical, antispiritual values. Haeckel achieved fame and wealth defending his *Darwinismus* by writing popular science books from its perspective. Young, educated Germans became "Darwinians" in droves. It is not odd, accordingly, that von Baer was very disturbed by these developments and concluded that Darwinism, which he understood in the terms Haeckel described it, had become popular mostly because it attacked religion (von Baer 1873, in Hull 1973, 418).

What one finds in Haeckel, in fact, is little more than the old Geoffroy-Lamarckian developmentalism, complete with revolutionary overtones, cast in terms of mid-nineteenth-century German reductionistic biochemistry and embryology. This set of preoccupations explains why Haeckel wants to say, obscurely enough, that

phylogenesis [the birth of kinds, from Greek *phylum*, "kind," + "*genesis*," birth] is a *physiological* process, which, like all other physiological functions of organisms, is determined with absolute necessity by mechanical causes. . . . Phylogenesis is therefore neither the foreordained, purposeful result of an intelligent creator, nor the product of any sort of unknown mystical force of nature, but rather the simple and necessary operation of . . . physical and chemical processes. (Haeckel 1866, 365)

It also explains why Haeckel wants no less obscurely to say that "phylogeny is the mechanical cause of ontogeny." "The rapid and brief ontogeny [of each organism]," Haeckel writes, "is a condensed synopsis of the long and slow history of the stem" (Haeckel 1905, 415). An embryo develops, according to Haeckel's "biogenetic law," because it runs through this compressed record of the adult stages of its ancestors, each stage in ontogeny completely determining the next. As a result, evolution produces a tree of life based on common descent (figure 7.1).

A moment's inspection of Haeckel's diagrams reveals a tree with a very different look from Darwin's. For Darwin, humans, like any other contemporary species, will be a twig coming off a side branch. For Haeckel, the entire process of phylogeny heads straight up a wide and tall trunk, perched at the top of which sits that most complex and perfect of beings, *Homo sapiens.* Indeed, trees ranking the evolutionary progress, or lack of it, of various human races begin to be extensively drawn in the wake of Haeckel's anthropological work. If then a little of the old Geoffroyian picture of increasing structural complexity lies behind these images, so does a good deal of Lamarckian inheritance. If phylogeny is to be the mechanical cause of ontogeny, and if ontogeny is to be a record of evolutionary complexification, the novel traits that drive lineages up Haeckel's tree of life will have to have been added to the end of the existing developmental sequence. Only in this case can ontogeny be a complete, sequential, speeded-up, compressed record of phylogeny. For this reason, evolutionary novelties cannot arise early in ontogeny. Accordingly, while Haeckel abides by Darwin's judgment that modifications will show up in descendants in the same temporal and spatial position in which they first appear in ancestors ("homochrony" and "homotopy," as Haeckel called these notions), he ignores Darwin's judgment that significant evolutionary changes can arise at any time in ontogeny. Indeed, he contradicts Darwin's judgment that only trivial adaptive modifications to local circumstances are facilitated by traits acquired by adults and passed directly to descendants, for Haeckel relies on the inheritance of acquired characteristics to originate new traits and vaguely assigns to natural selection the role of ensuring that only the most highly adaptive, and hence progressive, of these traits will find their way into the compressed phylogenetic record. He assigns to natural selection the negative role of eliminating novelties that fail to ensure progressive evolution.

Haeckel can hardly be blamed for translating Darwinism into the developmentalist framework in which the most important biological questions in his own culture were being debated. But Haeckel translated Darwinism into these terms for an audience that transcended his own country. He was "the main architect of the late nineteenth-century order-

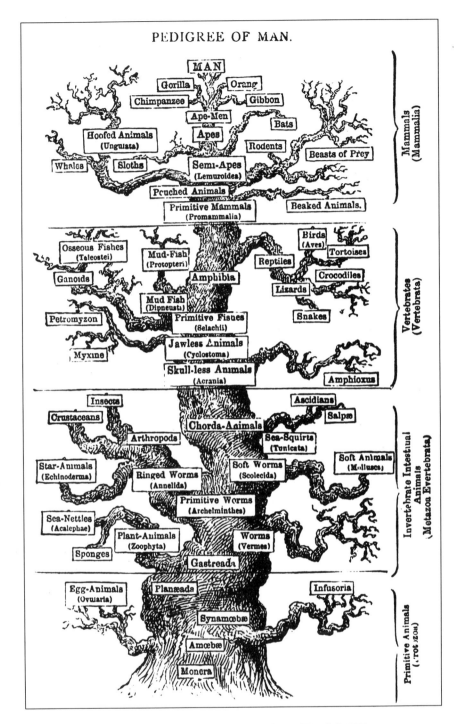

Figure 7.1 "Pedigree of Man" from *The Evolution of Man* (Haeckel 1879).

ing of the life-sciences" in Europe and North America, under whose hand "comparative anatomy, embryology, and paleontology had been unified . . . into a single-minded evolutionary . . . obsession with discovering ancestry" (Rasmussen 1991, 71–72). The result was a research program with a very different ontology and epistemology from Darwin's. For Darwin, the mechanics of generation, important as they are, lie far in the background of a theory whose causal accent falls on the contingent, but patterned, events that occur at the interface of organisms, populations, and environments (Mayr 1982). For Haeckel, and most later nineteenth-century evolutionists, what is background for Darwin comes to the foreground. Even if Haeckel's translation was more faithful to Darwin's original, they would still have very different theories.

Haeckel's coupling of evolutionary recapitulationism with the role of traits acquired by adults and passed to offspring was also pronounced in later nineteenth-century American biology. Professional evolutionary biology was at that time dominated by a school of evolutionary theorists who were pleased to call themselves neo-Lamarckians. The American neo-Lamarckians, Edward Drinker Cope, Alphaeus Hyatt, and Alphaeus Packard, were paleontologists who were interested in finding fossil links that would substantiate an exact parallel between ontogeny and phylogeny. They were happy to accept Haeckel's claim that progressive evolution is brought about by the addition of new traits, acquired and passed on by adults, to the end of the existing developmental program. Indeed, they devoted considerable theoretical effort to answering the puzzles that Haeckel's research program had posed. Precisely how does ontogeny condense and rapidly run through the slow work of phylogeny? How are new traits carried back far enough and fast enough into the developmental program to become permanent additions? Is the rate of recapitulation constant, or does it speed up and slow down? What might be the evolutionary significance of rate changes? What is to be made of the fact that in some clearly higher lineages, earlier traits are deleted rather than recapitulated? Why does development in what are clearly higher species seem to go backward rather than forward (paedomorphosis, neoteny) (Gould 1977)?

In America and Britain, unlike Germany, the term *Darwinism* was associated not so much with evolution as such but with evolution by natural selection. Unlike Haeckel, who finessed or co-opted Darwin's theory of natural selection in order to appropriate his uncompromising affirmation of transformism, common descent, and materialism, the American neo-Lamarckians rejected the name *Darwinian* because they flatly denied that natural selection is a source of evolutionary creativity. They did so in large part, moreover, because they rejected Darwin's materialism, in which they believed natural selection was rooted. This difference in the names evolutionists chose to go by suggests that the ideological context of Darwinism's reception in America differed from its reception in Germany. More closely considered, it also suggests how differently Darwinism was received in America than in Britain. Religious issues had played a role in Darwinism's reception at home only for a time. They remained powerfully at work, however, in America, and indeed remain at work to this day. From the start, the fate of Darwinism in America has depended to a considerable extent on how evolution is viewed as affecting the fundamentally religious view of the world that commands wide loyalty in American democratic culture.

From the start, Darwinian naturalism had its champions in America. Chauncey Wright, an atypically scientifically oriented philosopher, passed his enthusiasm for Darwin to the "Metaphysical Club" that had gathered around him at Harvard. This famous but ill-defined club, whose founding members included Charles Sanders Peirce and William James, became the seedbed of American pragmatism, which in its original incarnation was based on reconstructing philosophy, in the light of the Darwinian principle that the mind is a biological adaptation to an environment.[3] Francis Bowen, a colleague of Wright, was quite hostile to this kind of thinking. He grasped immediately that Darwin's argument had the effect of destroying Paley's argument from design, "the principal argument for the being of a God." Asa Gray, Darwin's friend and American advocate, was thus placed in the position of trying to defend Darwin by arguing that *On the Origin of Species* is not as hostile to the argument from design as it looked. This argument failed in America as badly as it did in England. Many of those who still wanted their religion and evolution too—and they were legion—soon turned to Spencer. Fiske's cosmic philosophy, now all but forgotten but well known in its day, put a spiritualistic spin on Spencer's ideas. That too soon failed, however, as the Malthusian element in Spencer's theory of progress was worked up by William Graham Sumner and others into social Darwinism, a quite materialistic defense of unconstrained capitalism. American intellectuals and clerics of the genteel tradition could still be evolutionists, however, by regarding strong recapitulationism as a sign that evolution is moving toward ever higher degrees of morphological and psychological perfection by an endogenous inner drive. It was among these people that American neo-Lamarckism found its audience. Their hostility to Darwinism and Spencerism prevailed in an American high culture that was traditionally religious until the halfway house of liberal Protestantism was crushed in the early twentieth century between the two rocks of biblical literalism, the country's dominant popular culture, and the brutal, and explicitly secularist competitive capitalism that, sometimes in the name of Darwinism, had transformed it by the end of the nineteenth century.

The work of the American neo-Lamarckians begins with the pervasive influence on professional American biologists, and the learned American public, of the eminent paleontologist and morphologist Louis Agassiz. In 1846, Agassiz, who had studied under Cuvier and later became professor at Neuchâtel in his native Switzerland, came to America to look at the fossils. As befits a biologist and geologist who had grown up in the shadow of the Alps, Agassiz knew a great deal about mountain lakes and glaciers and about the organisms living in and around them. The focus of his work was paleontological. He had written *the* book on fossil fish. It was natural that he would want to visit America, for buried in the vastness of America's glaciated landscapes lay evidence of worlds past. Agassiz stayed on permanently in the United States to become the father of American academic biology. If Grant and Owen contended to be the British Cuvier, Agassiz was unquestionably the American Cuvier. He founded the Harvard Museum of Comparative Zoology, the site to this day of many of the most creative and contentious developments in American evolutionary theory. From Agassiz's hand came the first great scientific product of biological science in the United States, his multivolumed *Contributions to the Natural History of the United States* (1857).

Agassiz was much admired in his new land. To this day there are communities in the United States that preserve a memory of the day Professor Agassiz came to tell them what the rocks found in their neighborhood meant. Agassiz liked America because its culture fitted his philosophical and religious temperament. He brought with him to his new homeland an intensely mystical view of nature, in which God's presence was palpable in the world through the signs he had left there. Like his mentor Cuvier, he was an ardent creationist, deeply convinced that the internal integrity of taxa made them resistant to change, blocking the notion that they could have a purely natural genesis. Unlike Cuvier, however, Agassiz believed in Geoffroy's unity of form and was a strong recapitulationist in the nonevolutionary idealist mold. He was as convinced as any German Romantic that the facts of the biological world arrange themselves into a set of elegant transformations whose systematicity and elegance declare the glory of God. The parallel between development and taxonomic rank was an inspiring witness to what Paley had called "God's constant presence to his creation." But paleontology was to Agassiz an even better window than embryology through which to behold the divine plan. "Before Agassiz," writes Gould, "recapitulation had been defined as a correspondence between two series: embryonic stages and adults of living species. Agassiz introduced a third series: the geological records of fossils. An embryo repeats both a graded series of living, lower forms and the history of its type as recorded by fossils" (Gould 1977, 65–66). This principle would allow the paleontologist to classify extinct forms and to predict what intermediates ("missing links," as they were called) would look like. Paleontology was to the book of natural history what the popular American practice of searching through the Bible for hidden messages was to historical and personal meaning.

From Agassiz's time to that of Gould, paleontology has been a focal discipline in American historical biology, and the site of much of the evidence produced for and against evolution, as comparative anatomy was in France, embryology in Germany, ecology in Russia, and natural history in England. Even today when Americans think of evolution, they first think of fossils (especially dinosaur fossils).

Agassiz brought this sensibility to Boston, a city whose leading figures, descended from God-intoxicated Puritans, saw in history a system of signs to be read even more than a set of Newtonian natural laws to be obeyed. Indeed, if one looks a bit more deeply behind the Enlightenment ideology of the men who wrote the Constitution and represented their infant country in the corridors of power abroad, one finds a society in which, until the early twentieth century, religious congregations controlled virtually all educational institutions, tendencies toward biblical literalism and its applicability to political affairs were strong, and morality was almost universally assumed to rest on religious premises. Accordingly, whereas religious culture in Europe was likely to decline as democratic ideas spread, tied as it was to the albatross of oppressive feudal institutions and to established churches, the opposite is more nearly true in America. A society whose life and discourse were permeated by a disestablished but nonetheless intensely participatory Protestant sensibility was created precisely in proportion as the populist democratic culture that came to power in the Jacksonian era displaced the Enlightened oligarchy. The "Righteous Empire" that Cromwell's armies had failed to bring about in England's green and pleasant land was created in the American states. People who had been on the losing side of the English Civil War impressed their evangelical worldview into virtually every popular institution. Freed from the shackles of an old, corrupt world, set amid a vast array of natural wonders that demanded to be read as a religious or Romantic text, old-stock Americans thought that the sublime display around them had been contrived by a great, good, powerful God to show that he could be counted on to keep his promises to a new nation that had entered into a new covenant in a new world to become a "city on a hill" and a "beacon to all mankind." Views of nature were affected by this ideology. American natural historians and amateur biologists tended toward a transcendentalist or Romantic sensibility. One has only to think of Thoreau, or of the sublime Romantic paintings of the Hudson Valley school and the luminists, or of Audubon's birds. What Professor Agassiz did was tell Americans, from a professional point of view, that they were right about all this.

One can easily imagine how disconcerting *On the Origin of Species* would have been in such an environment. In 1860, accordingly, we find Agassiz sitting around the Boston Society of Natural History and the American Academy of Arts and Sciences with other luminaries taking potshots at Darwin. The minutes of the academy's meetings are full of remarks such as the one for March 27, 1860, the very eve of the Civil War. "Mr. James A. Lowell, Professor Bowen, and Professor Agassiz discussed adversely the hypothesis of the origin of species through natural selection." "The aim of the *Origin of Species*," Agassiz later wrote,

> was to show that neither vegetable nor animal forms are so distinct from one another or so independent in their origin and structural relations as naturalists believed. This idea was not new. Under different aspects it had been urged repeatedly for more than century by E. Maillet, by Lamarck, by E. Geoffroy de St. Hilaire, and others. (Agassiz 1874, in Hull 1973, 434)

This elicits a mighty display of professional trumping:

> Its doctrines contradict what the animal forms buried in the rocky strata of our earth tell us of their own introduction and succession upon the surface of the globe. . . . It has even been said that I have myself furnished the strongest evidence of the transmutation theory [by his admission of transitional types in the fossil record]. This might perhaps be so did these types follow, instead of preceding, the lower fishes. But the whole history of geological succession shows us that the lowest in structure is by no means necessarily the earlier in time. (Agassiz 1874, in Hull 1973, 442–44)

Agassiz goes on to argue that Darwin's version of the evolutionary hypothesis, with its stress on the external contingencies of the environment, flies in the face of facts about ontogeny established by von Baer and Owen:

> Under the recent and novel application of the terms *evolution* and "evolutionist" we are in danger of forgetting the only process of the kind in the growth of animals that has actually been demonstrated [is] the law controlling development and keeping types within appointed cycles of growth which revolve forever upon themselves, returning at appointed intervals to the same starting point and repeating through a succession of phases the same sources. These cycles have never been known to oscillate or to pass into each other. (Agassiz 1874, in Hull 1973, 440)

For all these reasons, Agassiz concludes that the doctrine of transmutation is less plausible than creationism:

> The most advanced Darwinians seem reluctant to acknowledge the intervention of an intellectual power in the diversity which obtains in nature, under the plea that such an admission implies distinct creative acts for every species. What of it, if it were true? . . . The more I look at the great complex of the animal world the more sure I feel I have not yet reached its *hidden meaning,* and the more do I regret that the young and ardent spirits of our day give themselves to speculation rather than to close and accurate investigation. (Agassiz 1874, in Hull 1973, 444–45, italics added)

After the Civil War, American paleontology remained firmly in the hands of Agassiz's students, the American neo-Lamarckians. By then, however, Agassiz's students had put an evolutionary spin on his teachings. "The so-called Haeckelian 'law of biogenesis,'" wrote Alphaeus Hyatt, "is really Agassiz' law of embryological recapitulation restated in terms of evolution" (Hyatt 1897, cited by Gould 1977, 91). Gone are Agassiz's Lyellian scruples about complex fossils' predating simple ones. At the same time, the American neo-Lamarckians' appropriation of evolutionary theory still resonates with Agassiz's conviction that natural history testifies to the working out of God's plan, for the whole point of their theory of inner-driven evolutionary complexification was to defang the idea that the bloody, cruel, chancy, and purposeless process called natural selection is evolution's driver. "The doctrines of 'selection' and 'survival,'" wrote Cope, "plainly do not reach the kernal of evolution" (Cope 1880, quoted in Gould 1977, 85). Natural selection might eliminate unfit variants, but it could not create the intelligible series of new forms revealed by ontogenetic recapitulation as revealed in the fossil record. In this spirit, "Cope's rule," a roughly true empirical correlation in use even today, predicts increasing size over evolutionary time as an index of increasing complexity.

American neo-Lamarckism is Lamarckian much less because of its appeal to use inheritance than because it developed Lamarck's conviction that there is an inherent tendency in living things to move to higher levels of being. These students of the creationist Agassiz simply turned the great chain of being on its side and let it unwind through time toward some inner-driven destination. They put an evolutionary interpretation on the Romantic idealism that Owen had used to argue against evolution. The assumption was that if the old Paleyesque natural theology was no longer effective in a postmechanistic and processive world, a developmentalist natural theology would be. That theology, as the outcome of the theological debate over Darwinism in American and England had shown, could have nothing to do with natural selection, and thus with Darwinism. Hence, unlike both Lamarck and Darwin, the American neo-Lamarckians were loath to think of evolution in adaptationist terms at all, whether Lamarckian or Darwinian. They went so far as to suggest that adaptation is not always a good thing even for the species that is adapted. A species whose developmental program is locked into place will become overadapted, and overadaptation is a prelude to extinction. The poor Irish elk, for example, simply dropped under the weight of his "hypermorphic" antlers (Gould, 1977, 342–43; 1977b, 79–90).

Like pre-Copernican astronomers, who, convinced a priori that there is an order even where it does not appear, explained the apparent retrogression of the planets within an assumed framework of uniform circular motion by claiming that planets ride on circles, which ride on other circles ("epicycles"), the American neo-Lamarckians found very clever ways to resolve what looked like anomalies to evolutionary progressivism. For example, humans seem to be born more helpless than other primates and to retain juvenile characteristics, such as relative hairlessness, in adult-h_ud. If this is true, humans, who must be presumed to be developmentally higher than any other species, would appear to have less complete developmental sequences than apes. Accordingly, the neo-Lamarckians put a good deal of effort, accordingly, into denying that what look like the "neotenous" retention of juvenile characters in humans actually is so. Hyatt argued, for example, that species and lineages have fixed life spans, reaching an acme and then slowly descending. Thus, what looks like the retention of juvenile characteristics is actually the sign of old age, a "second childhood" as it were (Gould 1977, 91–96). In a corollary clearly relevant to the great issue in Hyatt's America, the northern white races were taken to be more apparently neotenous, and thus actually more mature, than the allegedly inferior southern black races (Gould 1977b).

From a Darwinian point of view, American neo-Lamarckism long distorted American paleontology and American ideas about what evolution is and how it works. As late as the 1920s, paleontologists kept looking for nonadaptive but progressive trends, and most systematicists continued to think that species are demarcated by nonadaptive characters.[4] The eventual introduction of genetic Darwinism into the United States meant taking on these views. A sustained effort to do so can be seen in the work of the paleontologist George Gaylord Simpson. It is no accident that Simpson's arguments, as we will discover in chapter 12, are rhetorically surrounded by a great deal of hostility to religion and that this opposition was not directed solely at populist fundamentalism. In the wake of the neo-Lamarckians, theistic evolutionism based on "orthogenetic" or "aristogenetic" views of evolution were taught in semisecularized American philosophy and biology departments well into the twentieth century. Liberal churches became comfortable with evolution of this sort even as they decried Darwinism. As the turn of the century approached, similar ideas gained ascendancy on the Continent, and even in Britain. They have been a permanent feature of evolutionary discourse ever since, springing to new life whenever a novel philosophical theory, such Henri Bergson's "intuitionism" or Alfred North Whitehead's "process philosophy," seems, in the absence of more concrete evidence, to provide wholesale a metaphysical motor and guarantee of their truth.[5]

August Weismann, a radical materialist, secularist, and democrat, was a German Darwinian in Haeckel's mold. He was deeply opposed to the increasing tendency of self-proclaimed Lamarckians to oppose Darwinism in the name of progressive, spiritualistic, or vitalistic versions of evolutionism. In order to stop this tendency dead in its tracks, Weismann defended Darwin's materialism, transmutation, common descent, and natural selection by abrogating the causal pluralism that had allowed Darwinians hitherto to remain tolerant of the inheritance of acquired characteristics. Henceforth, natural selection was to be "all-sufficient" (*Allmacht*). It alone would be the "mechanical cause" of ontogeny and phylogeny. In response, George Romanes, an English Darwinian whose interest in the adaptive value of psychological traits had led him to assign considerable weight to what became known as Lamarckian mechanisms, dubbed Weismann's view "neo-Darwinism." This was the first use of a term subsequently assigned to a widely different succession of theories, at first dismissively, later positively. Writers like Samuel Butler then reinforced the link between the inheritance of acquired traits and directional evolutionary trends by calling themselves Lamarckians and by using "Darwinism," meaning Weismann's defense of evolutionary naturalism by means of "hard inheritance," as a term of abuse (Butler 1879). In an international context dominated by antimaterialist, directional views about evolution, Weismann's intervention not only ended the era of causal pluralism but helped create the turn-of-the-century perception that "Darwinism" was dead.

Weismann's work was taken seriously enough at home and abroad to trigger these watershed changes in opinion in part because his argument rested on what looked like an incontrovertible empirical experiment. In 1888, he demonstrated that rats whose tails had been cut off do not breed tailless rats, and hence that traits acquired by adults cannot be passed to offspring. That is not a novel discovery. Weismann carried it out, however, with impeccable experimental protocols. There was, moreover, a conceptual side to the matter. Due to rapid developments in German cytology in the period leading up to his experiment, Weismann was able to interpret his results as demonstrating that the "germ line"—egg and sperm—is "sequestered" in early ontogeny and so cannot receive environmental influences. (What does or does not happen is always most impressive when it seems to follow from what can or cannot happen.)

The idea that organisms are composed of cells was no older than the 1830s. The discovery that cells come only from other cells was made by Rudolph Virchow in the 1850s. It was in the 1870s that enhanced microscopic power revealed the cell nucleus. In the 1880s the dance of the chromosomes in the cell nucleus was shown to be the birthplace of new cells. In 1883, it was discovered that sex cells, such as sperm and eggs, divide differently from the cells that make up the rest of the body. They each get only half of the chromosomes of the parent cell, making up the difference when they unite and exchange information. Weismann interpreted this difference between "germ cells" and "somatic" or body cells (from Greek *soma*, "body") in accord with long-standing assumptions about the physiological division of labor, such as those of Milne-Edwards. Body cells, he thought, would have only the information to form appropriate tissues. Germ cells, by contrast, are restricted to one function: the making of new organisms by intermingling or blending inherited information from the parents. Weismann did not know that all cells actually contain the total complement of inherited information. Nonetheless, his ignorance of this fact was not crucial, for he was still able to argue that the separation of the germ line from the somatic line occurs so early in the sequence of cellular divisions that constitute development that what goes on in the somatic cells cannot have any effect on this process. That was why, on his view, rats deprived of tails bred rats with tails.[6]

Weismann's rejection of the inheritance of acquired traits destroyed Darwin's own reliance, in his pangenetic speculations, on newly acquired somatic information to affect what is inherited and to explain how adaptations are created. He did this, however, to save natural selection from the excesses of developmentalism, and especially the cryptospiritualism, in which Darwin and most self-proclaimed Darwinians were tacitly complicit to the extent that they too had theories of inheritance that relied on acquired characteristics.[7] Here Weismann seems to anticipate twentieth-century Darwinism. Natural selection works because germinal cells make somatic cells, which in turn make organisms. These are subject to selection pressure in competitive environments. Which germ cells will make it into the next generation, therefore, is a function of how well the bodies carrying those cells do in the struggle for existence. Adaptedness is the result of changing proportions of these cells in populations. Real evolutionary novelty, however, must await internal change in the germ cells.

It is easy to think of Weismann as a prophet of twentieth-century Darwinism. The solitary splendor of Weismann's germ line presaged the Mendelian revival, or at least transformation, of the old idea of "preformationism," according to which offspring (here separate components of offspring) are already hidden in the reproductive material of parents and grandparents like so many Russian dolls. Weismann himself referred to the "immortality of the germ line." The implied severing of reproduction from developmental articulation was so severe that it broke Darwin's own link with the old epigenetic tradition, which, since Aristotle, had treated reproduction and growth as phases of a single developmental process. It is now generally acknowledged that separating Darwin too cleanly from the developmentalist tradition, and prematurely assimilating him to the post-Weismannian, and later the Mendelian, tradition that separates reproduction from development, is the chief culprit in the many misreadings that Darwin, until recently, has suffered (Hodge 1985, 1989a; Bowler, 1988). In retrospect, accordingly, it is Weismann who appears to have virtually refounded the Darwinian tradition by making the break with developmentalism that in the past has anachronistically been ascribed to Darwin. This break is what eventually allowed Darwinism to be united, after a long struggle, with Mendelian genetics (Mayr 1982; Bowler 1988; Gayon 1992). More remotely, Weismann's distinction between germ and soma lines seems to anticipate the distinction between DNA and protein, even though Weismann did not realize that the nucleus of somatic cells contains a complete array of reproductive information as well germinal cells. Indeed, his rejection of the inheritance of acquired characteristics anticipates what later became known as the central dogma of molecular biology, according to which information flows from DNA to protein and not the other way around. Accordingly, all mutations, and hence all evolutionary novelty, must arise from changes in DNA sequences. Seen in this light, Weismann's work seems presciently to foreshadow aspects of the modern genetic theory of natural selection.

Such anachronistic descriptions have their uses. Nonetheless, they obscure the context within which Weismann arrived at his view and many of the quirkier things that he in fact held. For one thing, Weismann did not break entirely with the developmentalist tradition himself (Bowler 1988). He did not break with one of its leading ideas, the strict parallel between ontogeny and phylogeny. Like Haeckel, Weismann was an ardent recapitulationist (Gould 1977). That grand idea, in both strong and weak forms and every shade between, guaranteed the very "order" of the biological world in the nineteenth century. It was just too hard to give up (Rasmussen 1991). Thus, as contemporary scholars have suggested, it hung on for a very long time, was relatively immune to empirical challenge, and can simply be said to have waned (Gould 1977; Rasmussen 1991). No doubt it would have waned more quickly if Weismann and other late nineteenth-century evolutionists had looked at information transmission, selection, and mutation in a populational light (Mayr 1982). Embryology, however, was still the focal point of their science. It was not until statistical methods transformed Darwinian natural history that recapitulationism, a hallmark of nineteenth-century biology from the beginning, at last disappeared.

In Weismann's case, the result of combining strong selectionism with recapitulationism, and hence taking a developmental perspective on what are in actuality populational phenomena, was the degree of conceptual confusion that seriously mitigates our retrospective appreciation of his prescience. Although Weismann himself saw that on his view evolutionary novelty does not have to be the result of terminal addition of new traits, that Darwin was right to think that changes to the germinal material can occur at any point in development, and that new traits will be reflected at the same developmental point in offspring, he still tended to ascribe recapitulation to terminal additions in the orthodox way. That is a testimony to Weismann's fidelity to recapitulationism. Since, on his own theory, terminal additions could not be functions of inherited acquired characteristics, Weismann concluded that they must be due to either natural selection or some other mechanism. Because he did not think that recapitulated traits are necessarily either advantageous or damaging to the possessors, he tended to rule out the former. Somehow, then, the germinal material just manufactures recapitulated traits on its own, including those used to mark off species. These need not be useful to an organism. The all-sufficiency of natural selection does not quite mean, accordingly, that Weismann was a panadaptationist. Indeed, in his later years, Weismann postulated what he called "germinal selection" in order to explain cases of traits that, like the allegedly outsided antlers of Hyatt's Irish elk, can presumably harm the fitness of the organism that possesses them.[8] The idea of germinal selection attempts to address this problem within a selectionist framework by transferring the scene of competition downward to the cellular level, where germinal cells must fight it out with each other for nourishment in an intracellular Malthusian world. What is good for the egg or sperm, however, might not be good for the organism. Thus, Weismann's germinal selection raises the possibility of conflict between "levels of selection." Once again Weismann seems prescient, for contemporary genic selectionism, associated popularly with the notion of "selfish genes," is a descendant of the same approach, while ideas about trade-offs and multilevel selection resemble Weismann's solution to the problem (Dawkins 1976; Buss 1987; cf. Chapter 14). A moment's reflection will show, however, that Weismann lived in a different world. He was a nineteenth-century developmentalist too.

8 Statistics, Biometry, and Eugenics: Francis Galton and the New Darwinism

Eight years after the appearance of *On the Origin of Species,* Darwin was forced to confront a serious objection to his theory put forward by Fleeming Jenkin, a physicist and engineer at the University of Glasgow. According to Jenkin, statistical mathematics renders it highly improbable that variation, selection, and transmission of new traits could ever overcome the conservative effect of blending inheritance. It is more probable that offspring will approximate to a mean distribution of a trait than that they will be like their parents, all the more so over longer periods of time and in large interbreeding populations. New traits, that is, will be swamped by regression to the mean. Put otherwise, the role Darwin had assigned to sexual blending in ensuring populational and species integrity would work only too well. It would drown innovation in a literal sea of mediocrity. "Fleeming Jenkin has given me much trouble," wrote Darwin to Hooker, "but has been of more real use to me than any other essay or review" (Darwin to Hooker, January 1869, in Darwin 1887, 2:379).

If Herschel thought that Darwin had appealed too freely to statistical reasoning, Jenkins was now claiming that he had attended too little to it. In either case, *On the Origin of Species* fairly cries out for a rigorous mathematical assessment of its statistical and probabilistic assumptions (Porter 1986, 134). Indeed, whatever the fate of Darwinism might be amid the vague, speculative, and largely premature macroevolutionary debates that were beginning to take place abroad, at home Darwin's more sober microevolutionary theory would prevail only if the mathematical intuitions that permeate *On the Origin of Species* could be spelled out and adequately defended. Although he was something of what Mayr calls a "population thinker," however, who should perforce have been good at statistical reasoning, Darwin was not a mathematician. He contented himself with claiming only that, given a wide array of variation, an equally wide range of adaptive traits, an enormous amount of time, and few internal constraints, it is highly unlikely that variations that happen to be useful in a particular environment will not occur with sufficient regularity to produce evolutionary change (Darwin 1859). The mathematics of *On the Origin of Species* is restricted to the simplest of calculations, some of which are botched (Parshall 1982).

This chapter is an account of how, and why, Darwin's immediate successors in England met the mathematical challenge that he did not. The central figure is Darwin's cousin Francis Galton. Several themes of surpassing importance in the subsequent history of the Darwinian research tradition make their first appearance in Galton's reworking of Darwin's theory. First, in contrast to the clashes between statistics and Darwinism visible in Herschel's and Jenkin's objections, Galton's work represents the first positive intersection between Darwinism and the probability revolution that by the middle decades of the nineteenth century was transforming the social and the natural sciences and would continue to do so (Hacking 1990). Second, although most British Darwinians subscribed to Darwin's causal pluralism, Galton's hereditarian stress on nature over nurture made it possible for him, and for the research program he created, to greet Weismann's refutation of so-called Lamarckian inheritance with more equanimity than other Darwinians. Third, Galton's hereditarian research program was an integral part of his attempt to turn Darwinism into an applied science of human breeding, "eugenics" (from Greek *eu* + *genos*, "well born," "good birth"). It is an inescapable part of Darwinism's history that statistical Darwinism, in both its pregenetic and genetic forms, was for a very long time an instrument in the service of the eugenics movement.

Fleeming Jenkin held no brief for Platonic or typological essentialism, or even for what we have called Aristotelian or constitutive essentialism. He was prepared to concede to Darwin that in principle, "There shall be no limit to the possible differences between descendants and their progenitors" (Jenkin 1867, in Hull 1973, 305). His objection to Darwin was that in a freely interbreeding population, the distribution of variation will conform to the "law of errors" that we encountered in chapter 6, according to which variation around a mean will follow the bell-shaped, normal, or Gaussian curve (after the mathematician Christian Gauss) that the French statistician Quetelet had used to give an orderly portrait of many apparently random distributions. In saying this, however, Jenkin puts forward a sort of essentialism after all. What we will call "statistical essentialism" makes it a mathematical necessity (and mathematical necessity is no mean sort of necessity) that large classes of apparently random distributions will cluster around a mean and will fall off regularly toward both edges. The result of applying the mathematics of statistical "errors," as they were still called, to randomly interbreeding or freely crossing biological populations would be, according to Jenkin, that "any individual *may* produce descendants varying in any direction, but is *more likely* to produce descendants varying toward the center of the sphere" (Jenkin 1867, in Hull 1973, 308, italics added). Regression to the mean became in this way a stable background expectation, according to which, for example, "a set of race [horses] of equal merit indiscriminately breeding will produce colts and foals of inferior rather than superior speed" (Jenkin 1867, in Hull, 1973, 308). It would be nearly impossible, therefore, for useful variants to spread through a population over time. They would be swamped ever more fully in each generation. Jenkin proposed the following highly revealing thought experiment to illustrate this result:

Suppose a white man to have been wrecked on an island inhabited by negroes, and to have established himself in friendly relations with a powerful tribe whose customs he has learnt. Suppose him to possess the physical strength, energy, and ability of a dominant white race, and let the food and climate of the island suit his constitution; grant him every advantage which we can conceive a white to possess over the native. . . . Yet from all these admissions there does not follow the conclusion that after a limited or unlimited number of generations, the inhabitants of the island will be white. Our shipwrecked hero would probably be king; he would kill a great many blacks in the struggle for existence; he would have a great many wives and children. . . . Yet he would not suffice in any number of generations to turn his subjects' descendants white. (Jenkin 1867, in Hull 1973, 315–16)

This startling passage gains its rhetorical force from the opportunities for racist, sexist, and imperialist fantasies it offered its author and his audience. It betrays a great deal about the historical circumstances of late Victorian Britain. The plunge into free-market capitalism was now sustaining itself by living off a growing colonial empire, developing in the process a pervasively racist ideology to facilitate that project. Setting aside the perverse example, however, Jenkin's technical point is that blending inheritance will always predominate over selection pressure when the latter is exerted at one point in the array. Thus, Darwin's spasm of variation might well take place. But it remained to be shown how selection could push the mean permanently to a different, more adaptive peak in the face of a law of diminishing returns no less iron-clad in biology, it appeared, than in economics. The outlook did not seem promising.

The difficulty did not lie in the long-prevalent misreading of this incident, according to which Darwin was unaware until then of the swamping effect of blending inheritance. As early as 1842, Darwin himself had written that "if in any country or district all animals of one species be allowed freely to cross, any small tendency in them to vary will be constantly counteracted." We have seen, in fact, that Darwin actually relied on swamping to provide populational integrity in the face of the tendency of individuals and populations to vary and depart from type (Hodge 1987; cf. chapter 5). What was annoying about Jenkin's objection was its use of a statistical argument to suggest just how difficult it would be for single novelties, or "sports," as breeders called them, to fight off the gravitational force of blending.

Darwin's struggle to respond to Jenkin on this point led him to distinguish more clearly than he had between distinctive "single variations," or "sports," and what he called "individual differences," that is, a wide spectrum of very small phenotypic differences continuously distributed through a population. Jenkin's argument against Darwin, as the racist fantasy of the white man shows, was based entirely on what would happen to sports. Darwin had given Jenkins his chance to register his complaint by treating natural selection as applying indifferently to both sorts of variation and by presenting most of the examples in *On the Origin of Species* in terms of single variations. He now repented. In a letter to Wallace, Darwin wrote, "I was blind and thought that single variations might be preserved much oftener than I now see is possible or probable" (Darwin to Wallace, February 2, 1869, in Darwin 1887, 2:288). He now conceded, therefore, that "if a bird of some kind could procure its food more easily by having its beak curved, and if one were born with its beak strongly curved, and which consequently flourished, nevertheless there would be a very poor chance of this one individual perpetuating its kind to the exclusion of the common kind" (Darwin to Wallace, February 2, 1869, in Darwin 1887, 2:288; cf. Darwin 1859, fifth edition, 1869, 176–77).

Darwin's slowness to see this point probably reflected his residual attachment to the idea that variation will be comparatively scarce because, on his theory of reproductive and developmental mechanics, it is triggered only when developmental stress occurs. Darwin, still complicit with the developmentalist tradition, never changed his theoretical mind about that. Instead, he resolved the issue with Jenkin simply by enhancing (in the spirit of Lyell) the array of continuous variation, the raw material of selection, by sheer fiat. If you increase the amount of variation, provide a continuum of variants, and track changes over a number of generations, you should get evolution by natural selection. Darwin had no mathematical or biological evidence, however, to offer in support of this hypothesis. All he did was systematically amend the fifth and subsequent editions of the *On the Origins of Species,* so that the singulars referring to individuals were changed to plurals. Thus, what appears in the first edition as "any variation" turns in the fifth into "variations." "An accidental deviation in the size and form of the body" in "an individual" becomes "individual differences . . . too slight to be appreciated by us" (Darwin 1859, 145, 177, fifth edition, 145, 183).[1]

It is not surprising, then, that after about 1870, regression to the mean, and the consequences Jenkin and others drew from it, formed an issue to which committed Darwinians felt they had to make a persuasive response. That fact provided the context in which many of the second generation of British Darwinians became statisticians. The most important of these was Francis Galton, whose mother was Erasmus Darwin's daughter and thus Charles Darwin's great aunt. Galton had been born in 1822. That made him thirteen years younger than Darwin. Galton's reverence for his cousin knew few bounds. He suffered from an "anxiety of influence" strong enough to fire him with desire to solve all the outstanding problems of Darwinism, even if he had to modify Darwin's views in order to do so. The fact that Galton, unlike Darwin, was a trained mathematican, with a degree from Cambridge, helped him to achieve his goal.

In his young manhood, Galton had done some geographic explorations in Africa in the fashion of Charles Darwin's travels, carrying with him not only Humboldt but the *Voyage of the Beagle* as well. On his African adventures, he had become convinced of the inherent and unamendable inferiority of the nonwhite races. Unlike Darwin, he did not content himself with long-term visions of human progress or insist that native populations and cultures must remain free enough to develop intellectually and morally. On the contrary, he worried that the higher birthrates of the inferior races and the long-term effects of miscegenation would dissipate the superiority of the white Europeans with whom they were now in contact. These were the worries of an imperialist. They contrast vividly with Darwin's conviction that, even though the races are unequal, sexual selection (in this case, the suspiciously Victorian tendency of females to select reliable and virtuous mates) and the cultural transmission of traits that favor the development of some moral sensitivities, would make it possible for all to progress slowly upward (Darwin 1871). Darwin was a child of the great age of democratic revolutions. Galton was a son of the age of imperialism.

A widening generational gap had in fact opened up between people like Darwin and people like Galton. As the passage from Jenkin suggests, the world of the latter began to approximate that of Rudyard Kipling, Edgar Rice Burroughs, and Colonel Blimp. The attitudinal shift between Darwin and his cousin is on display in a letter the aged Darwin wrote to him about the inheritance of intelligence. "You have made a convert of an opponent in one sense," Darwin confesses, "for I have always maintained that excepting fools, men did not differ much in intellect, only in zeal and hard work; and I still think [this] is an eminently important difference" (Darwin to Galton, 1869, in Darwin and Seward 1903, 2:41). Darwin's mind clearly did not readily accommodate itself to the new view. Nonetheless, he managed to convince himself that Galton was right.

This bifurcation was also reflected in a widening split among the ranks of British Darwinians about Darwin's causal pluralism. All subscribed officially to that doctrine. But while Spencer, Romanes, and others headed off in a "Lamarckian" direction, Galton thought that behavioral traits like intelligence, artistic genius, and political skill are as hereditary as morphological ones. If they were to be amended, therefore, natural selection, working on hard inheritance, would have to do more work than the inheritance of acquired characteristics.[2] Galton would have fit Romanes's dismissive epithet "neo-Darwinian" almost as well as Weismann, its intended target. Long before Weismann had erected his famous barrier, in fact, Galton had attempted experimentally to disconfirm Darwin's theory of pangenesis, on which his commitment to soft inheritance had been built. On the assumption that Darwin's "gemmules" were transmitted from somatic to sex cells through the blood, Galton attempted to determine whether massive blood transfusions would affect the appearance of new traits. That they did not meant to Galton that heredity was hard. (When Galton claimed to have confirmed his hypothesis, Darwin replied weakly that perhaps the gemmules were not transmitted through the blood after all.)

Galton's hereditarian inclinations made Jenkin's argument even more galling to him than to Darwin, but for a different reason. If means of perpetuating the extraordinary qualities concentrated at the good end of the error curve could not be found, not only would the superior races be swamped by the higher reproductive rates of inferior peoples but the inherited excellences of good families would dissipate into the mediocrity that had been politically ordained by the rise of democracy. Democratic reforms, which by the time of the Third Reform Bill had extended the vote to most male heads of households, and free markets had given inferior families and classes the means to propagate more successfully and had dismantled informal caste barriers to intermarriage. The problem was made worse by the fact that most talented families and classes restricted their birthrate in the interest of protecting inherited wealth. Galton became a devotee of statistical analysis so that he could analyze these problems and do something about them.

Galton's greatest mathematical triumph was his so-called law of ancestral regression, which quantifies precisely how much a trait will be lost in each generation if nothing is done to stop the leakage. If you start with the normal curve as a baseline and assume blending inheritance (which Galton did not question), it mathematically follows (as a crude first approximation) that in each generation, "each unit of peculiarity in each ancestor taken singly is reduced in transmission according to the following average scale—a parent transmits 1/4, and a grandparent 1/16th" (Galton 1889, 138). That is a pretty big swamping effect, making it even clearer than Jenkin had supposed that "the ordinary genealogical course of a race consists in a constant outgrowth from its center, a constant dying away at its margins, and a tendency of the scanty remnants of all exceptional stock to revert to that mediocrity, whence the majority of their ancestors originally sprang" (Galton 1875–1876, 298).

The melancholy rhetoric of this passage suggests how badly Galton wanted to do something about the problem it posed. This would require, in the first instance, a way of measuring the normal distributions of all sorts of traits and deviations from the mean, a way of keeping tabs on potentially superior lineages by figuring out the "hereditiary genius" of families, and methods of calculating "correlation coefficients" and "regression cofficients" across many generations. To this end, Galton founded a biometrical laboratory at the University of London, which eventually became the first department of statistics in the world. Thus, was born the science of biometry, the measurement of biological and psychological variables. Meanwhile, Galton started to keep records in which "hereditary genius" in families was tracked. He wrote a book with that title. In addition to offering empirical support for Galton's mathematical predictions, the book and the records would presumably encourage the right people to marry the right people, and perhaps suggest laws that would prevent the wrong people from marrying the wrong ones. In 1907 Galton founded the Eugenics Education Society of London, thereby becoming the father of the eugenics movement.

Galton and his disciples, right down to the enthusiasts who even today still solicit Nobel Prize winners to deposit their sperm in banks, never had much doubt about which characteristics were heritable and most fit or about who had them: They were the traits admired and cultivated by their own class. Not for a moment did they consider that in a purely Darwinian sense, a class that cannot reproductively outcompete another class might not, by definition, be superior. Nor did they reflect that traits assumed to be adaptive were simply the preferences of their own subculture, or that these traits might not be heritable, or that so-called superior people occupied positions of influence largely because they were born with contingent advantages. It is nature, they generally assumed, rather than convention, that arranges things in societies, at least in societies that, through free, competitive, and meritocratic institutions, have not tried to hold natural laws back by false conceptions of the common good, whether classically conservative or socialist.

Until recently, the widespread influence of the eugenics movement in Britain, North America, and Europe has been systematically underestimated in view of the tragic consequences to which one branch of the movement eventually led in Nazi Germany. Until then, however, the movement was spreading rapidly among the educated classes on both sides of the Atlantic (Kevles 1985; Adams 1990). Charles Lindbergh's sympathy with the Nazis, for example, was born largely out of his eugenic preoccupations. It spread, moreover, without much respect for the political and ideological lines that usually divided conservatives from liberals and liberals from leftists. They all developed one form of eugenics or another. What differences there were centered for the most part on the two different sorts of possible remedies that eugenicists could propose, which corresponded to what was happening at the two tail ends of the error curve (Paul 1984).

One alternative was to prevent people presumably at the bad end of the distribution from breeding. This was "negative eugenics." It was under this description that the eugenics movement had its greatest effect in America, where, until the rise of the Nazis to power in Germany, legal sterilization and immigration restrictions were more widespread than in any other country. Alternatively, you could encourage the talented tenth, assuming you knew who they were, to marry only within their own kind and to breed extensively, whether directly or through artificial means. This became known as "positive eugenics." British eugenics after the fashion of Galton and his followers was largely of the positive sort. What Galton and his disciples wanted was a voluntary effort on the part of the naturally superior—they knew who they were, or else could look it up in *Hereditary Genius*—to reverse the tendency of ruling classes to marry inappropriately and limit their issue. British eugenicists advocated a "eugenic life-style," a sort of secular religion whose devotees hoped to skew the melancholy regression to the mean by exerting self-initiated selection pressure at the good end of the curve (Kevles 1985).

Given the way he posed the issue, however, Galton believed that neither positive nor negative eugenics was likely to accomplish much if his cousin had been right about continuous variation. Continuous variations, each of which has only a small effect, were the most likely to be swamped, and swamped quickly. The mean point would never change. Thus, Galton thought that Darwin should not have conceded Jenkin's point about sports. Admittedly, sports do not generally fare well. Most monsters are, in Richard Goldschmidt's colorful phrase, "hopeless." But that is just their beauty. Why couldn't natural selection work quickly and to great effect on the discontinuous variations that can be found at the tail ends of the normal distribution, some of which may have marked adaptive advantage? That presumably is how superior families became superior families. If marriages are consciously fostered among such families, accordingly, their edge can be maintained in spite of their smaller numbers. In this way, Galton took up again the "saltationist" theme that Huxley had broached at the birth of Darwinism, when he warned Darwin on the eve of the publication of *On the Origin of Species* that "you have loaded yourself with an unnecessary difficulty in assuming that *natura non facit saltum* [nature does not make leaps]" (Huxley to Darwin, November 23, 1859, in Darwin 1887, 2:27).

The fact that twentieth-century Darwinism was nurtured in its formative years within the framework of the eugenics movement invites sober meditation. That is not only because many Darwinians were eugenics enthusiasts but more significantly because the theoretical development of the Darwinian research tradition owes something to eugenics, mostly of the positive brand.[3] It might even be said that eugenic problems provided the stimuli that pushed the tradition forward. Galton stands at the fountainhead of this sequence of research programs because his strong hereditarianism, intimately bound up with his eugenic preoccupations, led directly to the crucial debates about continuous variation versus sports, hard versus soft inheritance, and particulate versus blending hereditary units, whose successful resolution, although not without difficulties and ruptures, propelled the Darwinian tradition into the twentieth century. Moreover, it was because, within an officially pluralist view of evolutionary forces, Galton stressed natural selection and hard inheritance that at least one wing of British Darwinism was not stopped dead in its tracks by Weismann's neo-Darwinism, even if the other wing wandered more and more into the orbit of inner-directed, neo-Lamarckian theories. In this way, Galton and his successors preserved Darwin's crucial stress on external forces in a post-Newtonian conceptual environment.

Galton contributed less to the continuity of the Darwinian tradition by his substantive views, however, than his conceptual and methodological ones. Galton's eugenic concerns and mathematical training led him to take statistical distributions of traits very seriously. As a result, *he was the first to reconceive Darwinism as a theory in which statistical arguments, rather than posing problems for Darwinism, acquire explanatory power and fecundity in their own right* (Hacking 1990, 186). Whereas classical and neoclassical biologists assumed that offspring would closely resemble parents unless something interfered, Galton argued that it was intrinsically more probable that they would vary in ways that slowly approximate to the normal distribution. He did this by treating the normal distribution not as a deviation from a developmental norm but as itself a norm from which deviations are to be measured. In this way, Galton turned statistical generalizations into the premises of explanatory arguments. (If the "law of errors" had been known by the Greeks, he proclaimed, they would have deified it [Galton 1888, 86].) The normal distribution and the expected regression constitute a new sort of inertial baseline from which systems might deviate under certain conditions. It is the deviations from what is expected against this background that will tell you what needs to be explained. Moreover, what does the explaining are not the general laws that structure the expected background but the particular set of circumstances or events that modify the expected distribution. Natural selection might be among these causes. This conceptual setup made it possible for Galton to argue that it might be possible for natural selection, working on exceptional traits lying at either tail of the distribution, to fight successfully against the tide of regression and to shift the mean itself.

Galton's shift to a statistical framework gave new life to the hypothesis of natural selection by deepening the theory of natural selection beyond what Darwin could achieve (Gayon 1992). Galton could defend something close to Darwin's original stress on the potential effectiveness of discontinuous variations or sports, for example, whereas Darwin himself was forced to retreat on the issue because of his different conceptual and methodological background. If Herschel had objected to Darwin's "law of higgledy-piggledy," Darwin, deeply embedded in the same set of background assumptions, could not reply effectively. Galton, however, was no longer looking at natural selection through Darwin's Herschelian lenses. He was looking at it through the lens of his account of statistical norms and local deviations. What he saw through this lens was not a smooth sphere over which different "races," incipient species, and species are distributed with blurring and in many ways factitious boundaries, as Darwin did, but facets of a many-faceted polyhedron (Galton 1869, 1889). There is a great deal of continuous variation and, in virtue of the "correlation of parts" on which Darwin, as well as neoclassical biologists, insisted, a large number of compromises that preserve the overall fitness of organisms without maximizing any single trait. It would not take much to tilt such systems by shifting the mean to another face of the polyhedron (Gayon 1992, 176–80). Discontinuous variations would do the job. For this reason, Galton's hopes were high for eugenic success by matching up exceptional individuals with strongly inherited good-making traits.

Galton passed his statistical way of looking at things, even if not always his particular interpretations and conclusions, to his disciples, notably Pearson and Weldon. Later, but still within a context of positive eugenics, their ideas were put in the service of genetic Darwinism by Fisher (chapter 10). None of this would have happened, however, unless, beginning with Galton, the statistical and probability revolution was at the same time being imported ever more deeply into evolutionary biology. We pause, accordingly, to reflect on the development of statistics and probability to Galton's time.

Historians of science increasingly concur that roughly between 1830 and 1920 there took place a significant shift in the conceptual and methodological underpinnings of many sciences. It has been called the probability revolution. Ian Hacking suggests something crucial about this shift by calling it "the taming of chance" (Hacking 1990). What since antiquity had been irrational surds in an otherwise rational universe now became not only intelligible and explicable but considerations that could be used to explain other things. This change was perhaps too pervasive, too slow, and too diffuse to count as a scientific revolution or a paradigm shift in the sense of the term on which Kuhn finally settled (1970). That should not deter us, however, from calling "the probability revolution" a scientific revolution (Hacking 1987, in Krüger et al. 1987). Indeed, if our paradigm case of a scientific revolution is the shift from Aristotelian and

Ptolemaic assumptions to those of Galileo and Newton, we have every reason to think of the probability revolution in the same way. What happened between 1830 and 1920 took place on a similar scale and had similar consequences to what happened in the period leading up to Newton and thereafter. Both revolutions were based on the spread of models from field to field, and both changed the general orientation of culture. Moreover, just as the first scientific revolution was the mother of the first industrial revolution, in which the science of forces was used to convert water power and iron into the steam engine, so the probability revolution ushered in a second industrial revolution, in which fossil (and radioactive) fuels were married to electricity though a science of energy (Smith and Wise 1989).

The probability revolution began with a statistical revolution, that is, with a mania for collecting and analyzing quantifiable data about people and their doings that got underway at the turn of the seventeeth century but entered into full flood in the early nineteenth (Hacking 1975, 1990; Porter 1986). The notion of analyzing records in order to make generalizations, predictions, and bets about people goes back to early seventeenth-century England. There, records of births and deaths had been kept in parishes ever since the Domesday Book. The first person to make something of all these data was John Graunt, who analyzed weekly births (christenings) and deaths (funerals) in London in order to ascertain the effects of a particularly nasty plague that had swept through London in 1603. Graunt summed up his results in *Natural and Political Observations* (1662). In the same year William Petty's *Political Arithmetic,* which was suspiciously indebted to Graunt's work, bestowed the first canonical name on this sort of inquiry. Petty went on to make a statistical survey of Ireland in order to help Oliver Cromwell divide up the spoils of that humiliated land among the English and Scots colonists he so tragically planted there. Graunt and Petty recommended in the 1680s that the Crown construct a permanent statistical office. It was not long before all European states had them.

It was only after the rise of the intrusive state after the French Revolution, however, that what Hacking has calls an "avalanche of numbers" about people began to rise exponentially. That avalance, and the concomitant bureaucratic rationalization of life that the sociologist Max Weber has referred to as the "iron cage" of modernity, was at least as pronounced a feature of states trying to avoid revolutionary infection, by keeping tabs on everybody, as it was of revolutionary states intent, like France, on harnessing their populations to perform heroic collective feats. Either way, as the French social historian and critic Michel Foucault has suggested, the very notion of an "individuated" person, with a unique "personality profile," was to some extent a product of all this categorization and record keeping (Foucault 1971). We are individual persons in something of the same way that our cars can be "customized" and our haircuts and clothes "personalized." Accordingly, as classical conservative thinkers like Edmund Burke and Alexis de Tocqueville thought at the time, the kind of freedom we associate with unique personalities was also attended by a corresponding loss of other sorts of freedom. It is not irrelevant to these considerations that the term *statistics* was coined in the early eighteenth century by Prussian bureaucrats. The term meant a numerical portrait of the state of a state. Nor is it surprising that it was German professors at Göttingen who first treated *Statistik* as a science, the mission of which was to find the proper categories, data bases, and methods of number crunching with which to prepare an accurate analysis of the realm.

If the statistical revolution was one component of the probability revolution, probability theory was the other. Probability theory too has been around for a long time, at least in the form of advice about how to act rationally (or morally) under conditions of limited knowledge. In antiquity, probabilism was a version of, and a response to, philosophical skepticism. How should one act if one knows nothing? Presumably in the most probable, that is testably effective, way (from Latin *probare*, "to test" "try out"). In the sixteenth century, *probabilism* was a word applied to the casuistic resolution of moral problems whose answers are hard to deduce from principles. (The word *casuistry* itself comes from Latin *casus*, "what accidently or contingently befalls or happens.") It was associated by the religiously tortured philosopher and mathematicical genius Blaise Pascal with the Jesuits, who were reputed to take advantage of moral uncertainties to go easy on influential sinners, thereby acquiring influence over their policies, and it has come to have negative connotations by this route.) In the Enlightenment, probabilities became both more computational, and in that sense rational, and more practical. The probabilist would try to compute what a prudently rational man, such as an insurer of cargoes, should do under x conditions of risk and y conditions of possible gain. In general, the answer was that the bigger the payoff, the greater the risk worth taking. "Pascal's Wager" is a clever and ironic example of this kind of reasoning. There may be a small chance that God, and therefore heaven or hellfire, exist. But an infinite payoff justifies the risk of believing, and so suggests to the libertine that it is more reasonable to be a believer than to cultivate a devil-may-care life-style. That presumes, of course, that you are a rational person. But in the Enlightenment, all bets were off for the irrational man in any case.

In this context, mathematicians like Jakob Bernoulli and Thomas Bayes attempted to quantify how past success could predict future knowledge. Bernoulli's most famous theorem was a formula showing how the true proportions of a population of future alternatives—black balls or red balls, heads or tails—would be revealed as the number of attempts went up. What is known today as the central limit theorem says that frequencies will stabilize around the true proportions as the number of tries increases. A good track record on events of a certain kind therefore gives you a good sense of the true, underlying "chances," at least to a certain calculable level of confidence. Bayes was interested in a far more intractable problem. From a given number of samplings or tries, what can we infer about how existing knowledge affects the probability of an unconfirmed hypothesis? The probability revolution in the sciences depends on the availability of quantifiable solutions of these questions. In particular, the problem worried over by Bayes mirrors the situation of the scientist whose observations and experiments "disturb the universe" only around the edges and who must know, therefore, when he or she is entitled to infer from a sample to a general claim about the remaining population, or from the existing body of scientific knowledge whether science is nearing its cognitive goals.

It was only when classical probability theory of this sort hit the cascade of numbers flowing from the statistical explosion of the early nineteenth century that probability and statistics were integrated into a more powerful science, and the probability revolution properly so-called got underway. Statistical claims can, of course, nearly always be translated into probability claims. To use a simple example, the claim that "34% of adult American women are smokers can easily be transformed into 'The probability of an adult American woman being a smoker is 34%'" (Giere 1979b). For this reason probabilities and statistics have had an affinity for one another from the beginning. Accordingly, although a statistic is merely a piece of data ("I don't want you to become a statistic," says the mother issuing warnings to her teenager), the term *statistics* has come to mean not the data themselves but what probability reasoning makes of them and, by extension, the set of techniques for doing this. This is what is meant when we refer to a "statistics text" or a "class in statistics." The content of these texts and classes is probability reasoning applied to statistics in the sense of arrays of data.

Graunt himself was already applying probability reasoning to his mortality data and doing it for a very concrete, even low, motive. He drew up a table that told him what proportion of the population (of London) would (probably) die in any given decade of their lives. The chance of someone surviving to age ten, for example, was 64 percent. The chance of making it to age seventy-six was 1 percent. That is extremely useful information if you are, for example, an insurance company, for it is possible to raise money from folks if they have a well-founded expectation that they will get a large cash annuity if they beat the odds on their own death and survive past the predicted time. The fact that many of them predictably will not means that it is the insurer who will pocket most of the money—if he has calculated correctly. If you disregard, or build in, the effect of inflation, two factors determine the fair rate for an annuity: the mortality curve for the age cohorts in the population that buy annuities on themselves and the rate of interest on long-term loans. That is the meat and potatoes of insurance schemes to this day, providing much of the motive for the development of actuarial tables into a very fine art indeed.

As one might expect, these practices, and mathematical reflection on them, were fairly advanced in seventeenth-century Holland, the first genuinely bourgeois society, run by and largely for solid middle-class citizens. Ingenious Dutchmen in high places, like John Hudde, mayor of Amsterdam, who had studied mathematics with Descartes, and Christian Huygens were interested in actuarial schemes, however, not because they owned insurance companies, but because the early modern state, and later capitalist firms, used this method to raise cash by issuing bonds. What you need to know if you are a state or firm is how much to sell the bond for, how big to make the payoff over what increments of time, and what the expected price of money will be. What you need to know if you buy one of these instruments is the financial affairs of the state or firm, the current and expected rate of inflation, the chances of war—in other words, many of the things we now call "the news," most of which arose out of this very cognitive interest.

Philosophical interest in probability theory has accompanied it since its birth. In general, however, there has been a shift over time in the philosophical interpretation of just what probabilities are. Roughly put, what were classically regarded as subjective beliefs about the unknown future became expected frequencies of events. These ideas still have their advocates. But in the twentieth century, the idea arose that probabilities are based on objective propensities of real things. Each step in this process embedded probability further into the structure of the world.

Until at least the middle of the nineteenth century, probabilities were presumed to reflect ignorance, and therefore to be merely subjective estimates, if only because classical probability theory was concerned with how a rational person might deal with his or her ignorance about future events. It was obvious that if you knew the underlying distribution of objects, past, present, and future, like God, you would not need probabilistic reasoning. This interpretation was intensified and semisecularized by the Enlightenment. If the underlying order of things was controlled by deterministic Newtonian laws, the future, as well as the past, could be calculated precisely. When people talked about chance and chances, therefore, that signified only that they lacked precisely what Laplace's demon had: the ability to compute, backward and forward, the entire world order from his data base. Darwin and nearly everyone else until the middle of the nineteenth century assumed such an "ignorance interpretation" of chance.

Given this assumption, it is not difficult to see why the discovery that such things as divorce rates, suicide rates, crime rates, and the steady rates at which undeliverable letters collect in post offices are constant might have caused an epistemological crisis. Ever since antiquity, such matters had been assumed to be a jumble of contingencies, not amenable to the regularity of natural laws precisely because they are subject to inconstant passions and individual free will. By the end of the eighteenth century, however, the ever-alert Kant, for one, had already noted the existence of social regularities on the basis of the work of academic German *Statistik* (Porter 1986). The first paragraph of Kant's *Idea for a Universal History from a Cosmopolitan Point of View* (1784) contains the following sentences:

Since the free will of man has obvious influence upon marriages, births, and deaths, they seem to be subject to no rule by which the number of them could be reckoned in advance. Yet the annual tables of them in the major countries prove that they occur according to laws as stable as those of the unstable weather, which we likewise cannot determine in advance, but which, in the large, maintain the growth of plants, the flow of rivers, and other natural events in a unbroken, uniform course.

Kant did not have much trouble reassuring himself that these regularities, causally considered, are effects of free will. That was made all the easier because, as this passage suggests, he did not appreciate just how predictable such regularities actually are. It was only when Quetelet revealed in the 1840s that social regularities take the form of the "error curve" that astronomers and physicists like himself had been using to compensate for observational and experimental error, and showed just how many phenomena, both natural and social, fall along Gauss's bell-shaped curve, that the regularity of things previously thought inconstant became an object of real puzzlement, wonder, inquiry—and scandal. Quetelet introduced mathematical predictability into the issue. This raised the possibility that probabilities are not merely measures of subjective error after all, but objective frequencies with which classes of events do, and will, occur. It also raised the possibility that they must occur, for in a world where Newtonian (=Laplacean) causal determinism was assumed to hold, the mathematical predictability of marriage, divorce, murder, and robbery suggested that human affairs are largely determined, that free will is an illusion, and that predictability implies fatalism about human actions.

We have alluded to the intense discussion of these issues that Buckle's *History of Civilization in England* (1857) triggered in Britain in chapter 6. These issues were just as intensely discussed all over Europe, however, in the context of the novel idea that there could be a science of society, a "sociology," that would reveal as much order in the human world as the scientific revolution had revealed in nature. This project, which Buckle and other positivists were attempting to introduce into England, was born in France. It was the product of a number of thinkers, including Quetelet, who wanted to legitimate the rule of the middle-class merchants, professionals, and intellectuals who had come tenuously to power with Louis Philippe, the "bourgeois monarch," in the Revolution of 1830. The idea of such a social science, conceived as an extension of law-governed natural science, was latent all along in Enlightenment thought. The task now was to strip these ideas of their former radical, revolutionary, materialist hue and to reconstrue them as instruments, like the constitutional monarchy itself, of respectable, moderate, gradual progress. (The project parallels, in the different intellectual idiom of French thought, that of the reformed Malthusians in England.)

An idea for a "sociology" along these lines was first put forward explicitly by the French philosopher Auguste Comte, who thought of himself as taming the socialism of his mentor, Saint-Simon. Comte thought that civilization progresses in three stages. It begins in a primitive, theological worldview dominated by religious authorities, moves through an evanescent metaphysical stage in which pseudorationalist philosophers inherit the priestly role, and culminates in a "positive" stage, in which scientists, having given up as illusory the search for metaphysical truths by armchair methods, extend the inductive and experimental methods that have proved so successful in natural science to human affairs. Vast amounts of data about society were to be collected. "Facts, facts, facts," says Mr. Gradgrind in Dickens's *Hard Times*, an attack on this kind of thinking. These data are to be analyzed by finding predictive correlations between different classes of facts. Then the problem is to ameliorate the situation by changing the social conditions in which "human subjects" predictably behave in this way or that. For this purpose, a class of "social engineers" will be empowered to take the place of ignorant priests and fatuous moral philosophers. The idea lives today in the pages of B. F. Skinner's *Walden Two* and *Beyond Freedom and Dignity*.

It was easy for Comte to swallow the idea that social life, down to its most intimate details, is governed by social laws because, as a positivist, he did not think that the prediction of one set of facts from another set has any metaphysical implications at all. He was not worried, therefore, about determinism, fatalism, and free will. Determinism, as in physics, simply means that any event is calculable from a prior state of affairs and the relevant laws of motion. From Comte's phenomenalist perspective, there is nothing beyond or below that to worry about, at least anything that could be known. These problems were artifacts of the very "metaphysical stage" from which positive science frees us. By giving up such fruitless concepts, philosophers, scientists, and policy makers would be able to recognize that the important issue is not abstract freedom but finding ways to bolster the solidarity of society as it passes through the fiery trials of modernization, which destabilizes and weakens traditional social bonds. Sociology has been preoccupied with the problem ever since.[4]

Comte was not impressed by statistics. He was much more inclined toward a developmentalist perspective, as his grand historical laws themselves suggest. It was Quetelet's achievement, and after him Emile Durkheim's, to suggest that the science Comte was after would take a statistical form. Quetelet's Laplacean background, together with worries about modern society very much like Comte's, inspired in him a desire for a "social mechanics" having the epistemic power of "celestial mechanics" or astrophysics. In order to achieve that end, however, Quetelet had to do something that was without precedent. Titles such as *Letters on the Theory of Probabilities* (1846), "On the Influence of Man's Free Will on Social Facts" (1847), and "The Social System and the Laws That Govern It" (1848) show that Quetelet was trying to demonstrate that in the nature of the case, societal facts, in Comte's sense, are distributed along the bell-shaped curve and are therefore under the control of "social forces." The law of errors gives a good description of social reality.

Quetelet's view of why this is true reveals that he was still thinking of social deviation as an error of some sort, like the errors in an astronomer's observations, which cancel each other out to give an accurate account of a phenomenon that can in principle be measured precisely. The fact that behaviors average out means for Quetelet that a very large number of people are similarly affected by circumstances and respond to them in similar ways. Indeed, he thought of society as composed of a large cluster of "average men" (*l'hommes moyens*) surrounded by a decreasing number of outlying "deviants." Whereas average men have a balanced and harmonious set of inclinations, which both reflect and preserve social order, deviants have lopsided passions, drives, cognitions, and habits, which perturb it. The domination of society by average man is, then, a very good thing. In fact, the average man was for Quetelet a kind of social ideal (Porter 1986). A firm believer in the cult of moderation that characterized the "bourgeois monarchy" and as concerned as Comte about the destabilizing effects of modernization, Quetelet thought that the point of social policy must be "to compress more and more the limits within which the different elements relative to man oscillate. The more Enlightenment is propagated, the more will deviation from the mean diminish" (Quetelet 1835, 2:342, quoted by Porter 1986, 104). The existence of a smaller number of people with a "propensity [*penchant*] for crime," meanwhile, although guaranteed by the error law itself, is nothing to worry about. For the very fact that such people are abnormal implies that society is presumptively stable, will not be readily jarred out of its moderate ways, and hence is fit to be governed by a moderate regime. Quetelet's theory was meant to be comforting.

It would have been immoderate in itself for Quetelet to deny free will. Thus, Quetelet distinguished between what people do as individuals and how they behave as fragments of the social whole. Sociology sees only the latter. Free will might well exist for the former. In making this distinction, Quetelet was trading on a typically French way of thinking about how society and individuals are related. Rousseau, for example, had thought of the social contract as a way in which the individual becomes a public person, a citizen, by giving up or at least privatizing his natural, egoistic self. The Jacobin idea of a social contract plays no role in Quetelet's thought. He thinks, however, that the socialization that Rousseau derives from the fictional social contract is brought about in fact by the statistical properties of the error law.

In Britain, where individualistic and nominalistic habits of thinking had always screened off reified talk about social reality and where social contracts were presumed to facilitate the passions and interests of private persons, no such bifurcation between individual freedom and social determinism was likely to take hold. That is one reason why Buckle could not, or in any case did not, take advantage of Quetelet's attempt to preserve free will but instead happily drove home the deterministic side of his sociology. It was Buckle's interpretation of social regularities that ignited resistance to statistical approaches to social phenomena, even among people like Herschel, who had earlier been quite in tune with Quetelet's work (Porter 1986). It was precisely when debate about these important issues had reached white heat that Darwin's *On the Origin of Species* had the misfortune of coming off the press, provoking Herschel's harsh rejection.

It was not just in England, however, that debate about what the Germans called *Queteletismus* took place. Sparks flew all over Europe. The Russian reaction to *Queteletismus*, for example, was particularly intense and characteristic of the way in which Russian intellectuals of the 1860s tended to take Western debates to extremes. Tolstoy used Quetelet's ideas to turn the tables on Western contempt for traditional Russian passivity and fatalism. These traits, quintessentially manifest in the supposedly superstitious and ignorant Russian General Kutuzov in *War and Peace*, now revealed themselves to be true wisdom. They showed why Kutuzov, an embodiment of all the forces running through and determining the aggregate, unintended actions of the Russian people, had defeated that fool Napoleon, whose illusory belief in his own free will was indistinguishable from his fatuous self-importance. Dostoyevsky, for his part, agreed with Tolstoy that the Western cult of social progress is an illusion. He revolted, however, at Tolstoy's concessions to determinism and moral fatalism. In *Notes from the Underground*, Dostoyevsky wrote that the individual is free to reject the material happiness that modern (Western) science promises and will consciously choose to be unhappy if that is the only way to protect his arbitrary free choice. Dostoyevsky's bitter and unhappy hero is explicitly contemptuous of Quetelet's *homme moyen* and of the happy-talk to which science-oriented cults of progress are addicted. Even his own rationality, he feels, is a constraint on his precious freedom.

The fact that social scientists and policy makers still talk without any sense of metaphor about "social pressures," "socialization," "social deviance," and "social conformity" shows that sociology has affected our way of conceiving the human condition. It was within this discourse that the phenomenon of modernization, the great topic of sociology, began to be constituted. Political change, classically conceived as the subject of narratives and practical rules of action on the part of the powerful, now became a dependent variable determined by forces governing other parts of the "social system." Indeed, the very idea of society as a system composed of subsystems is no older than these ways of talking. In discussing modernization, however, professional sociologists soon began to express less confidence than Quetelet in the average man. Durkheim, for example, turning his attention to populations that deviate not simply from average behavior but from an expected distribution, showed that an abnormal suicide rate is found in subpopulations undergoing rapid transformations from agrarian to industrial life. What was happening at the extremes was perhaps more worrisome than Quetelet wanted to think. By the end of the century, Max Weber's worries had deepened further. The question was what the causes of this departure were, and whether they were ameliorable. Talk of alienation and anomie filled the increasingly disenchanted air.

Galton's eugenic theory can be viewed as statistically based social science like Durkheim's. In both cases, attention shifts from the mean to the outliers, and particular causes are sought for deviation from an expected distribution, like Durkheim's suicides. Whereas Durkheim has social pressures that impinge on people, Galton has the effects of hereditary traits. Determinism, rather than being a problem for Galton's brand of "sociobiology," is only to be expected at this deep level. In consequence, the kind of social engineering recommended by sociologists and sociobiologists differs. For the latter, it will not be centered on changing environmental conditions so much as on figuring out who should and should not have progeny. Our story thus moves to Galton's disciple, Karl Pearson, who advanced this program in significant ways, and to those he in turn influenced, Walter Frank Weldon and, before he became a Mendelian, William Bateson.

Pearson was a mathematician, philosopher of science, and social theorist. As a mathematician, he developed statistical analysis enough to make "biometry" academically respectable and to take Galton's eugenicist program out of the hands of amateur clubs, giving it a secure academic home at what became the Galton Laboratory at University College of the University of London, where Pearson became the first Galton Professor of Eugenics. Pearson's fondness for statistical analysis went hand in hand with his admiration for the methodological views of positivist philosophers, especially the ideas of the Austrian physicist Ernst Mach. (Pearson had a liking for things Teutonic, including the letter *k*, as in Karl, for his given Carl, and *Biometrika*, for the name of the important journal he founded and edited.) Pearson's *The Grammar of*

Science (1892) is a positivist tract. In this work empiricism, the ancient and respectable view that all of our knowledge is built up out of our sensory experiences, and is justified or falsified solely by reference to sensory experience through experiment and controlled observation, turns into a more radical phenomenalism. When scientists talk about atoms, genes, forces, or any other such things, they should realize that they are not entitled to infer some reality behind appearances. The "theoretical entities" that ostensibly name the values to which their variables refer are really just convenient fictions that make it easier to collect, measure, predict, and control sensory phenomena. (A more radical phenomenalism and antirealism still is instrumentalism, according to which the very meaning of theoretical terms is "nothing but" the measurements that justify their use. Instrumentalism was popular in the 1920s.) Empiricism, properly articulated in a positivist framework, is not, then, merely the claim that we begin, as Aristotle recommends, with sense data but that we end there as well. For Pearson, accordingly, measurement was not a means to or method of science. Nor was it merely a clue to something deeper, like Herschel's causes. It was the whole thing. That was the vision that informed biometry.

By the time of Mach and Pearson, it was becoming increasingly clear that statistical regularities, based on ever more fine-grained analysis of correlated phenomena, could generate predictions and even guide intervention. In this connection, it becomes relevant to say that, like Comte and Quetelet, Pearson was more than a scientific methodologist. He was a social theorist, whose version of rule by scientific expertise was a primitive form of National Socialism, according to which an elite would manage the affairs of the state in such a way that its population benefits from the scientific control of society, including the economy, and a well-managed state triumphs in the Hobbesian war of all against all that prevails, by a fierce law of nature, among nations, cultures, and races. Eugenic management of the population, both to prevent its stock from deteriorating and to create a caste of intelligent rulers, was a major plank in Pearson's platform. He combined the socialism of his youth, which he had learned in Germany, with ardent support for imperialism abroad, which would provide the wherewithal for increasing welfare at home by extracting a surplus from inferior races. Amid the fevered atmosphere of prewar Europe, social Darwinism, hitherto identified with semibiological arguments for laissez-faire economics, was slowly mutating by way of eugenics into National Socialism.

It is worth mentioning that in the same year that Pearson published his *Grammar of Science*, the aged Huxley, a scientistic firebrand in his own time, drew back from this dimension of Darwinism. In a famous lecture, "Evolution and Ethics," in 1892, Huxley reflected long and hard on the "administrators," as he calls them, who would be charged with "pruning" the human garden by this new version of artificial selection. He thinks theirs would be a pyrrhic victory, for

I do not see how such selection could be practiced without a serious weakening, it may be the destruction, of the bonds which held society together. It strikes me that men who are accustomed to contemplate the active or passive extripation of the weak, the unfortunate and the superfluous; who justify that conduct on the ground that it has the sanction of the cosmic process, and is the only way of ensuring the progress of the race; . . . whose whole lives are, therefore an education in the noble art of suppressing natural affection and sympathy, are not likely to have any large stock of these commodities left. (Huxley 1894 [1989, 36])

Huxley implies that if Darwin's hopes of naturalizing morality without undermining it are not sustainable in the light of what it would take to achieve a "fit" population, then Darwin's hopes are best abandoned. The whole point of culture, and the institutions of morality, philosophy, and religion, is to hold down the whisperings of nature and the Malthusian competition that prevails there. To extend such protections to the chosen of one's own nation and to enforce the law of the jungle elsewhere is to undermine the very point of doing so. The fact that such matters worried Huxley suggests how radically the terms of social discourse had changed. The very air in the decades before the great suicide of Europe broke out in 1914 was fevered. The dream of reason was about to bring forth monsters.

Pearson's positivist disdain for theoretical entities, processes, and dogmas allowed him to wear Darwin's pluralistic view of the causes of adaptation lightly. Whether adaptation is caused by inherited traits or natural selection was a matter for experiment in particular cases rather than speculation about nature as a whole. Whatever the causes of adaptation are, however, Pearson was convinced that the variation on which it works was continuous rather than a matter of individual differences or sports, as Galton thought. That was because Pearson had developed statistical analysis far enough to discover continuous variation in many phenomena. To validate that biological change was fueled by continuous variation was, in fact, the research program that Pearson gladly left to his biologically trained colleague, Weldon.

Weldon was a highly trained comparative anatomist who reacted early and strongly against the developmentalist tradition in nineteenth-century biology in which he had been schooled. Under Pearson's positivist tutelege, Weldon came to see how full of "life forces," "entelechies," "germ plasms," "gemmules," and other metaphysical monstrosities biological science was. Wisdom lay in forgetting about assigning reality to these "theoretical entities," and largely in forgetting about such terms themselves. One should spend one's time instead measuring the distribution of variation instead. The rapidity and enthusiasm with which

Weldon abandoned his morphological training is a good example of Kuhn's claim that change from one scientific paradigm to another is often the work of young scientists who suddenly and passionately revolt against the "puzzle-solving" tradition in which they were nurtured, not by answering its old questions in a new way but simply by forgetting the old questions, formulating new ones, and answering them instead. "It cannot be too strongly urged," Weldon wrote, "that the problem of animal evolution is essentially a statistical problem" (Weldon 1893, 329, in Provine 1971, 31).

Weldon set out to measure everything in sight. With the gusto of a biologizing Quetelet, he showed that the bell-shaped or Gaussian distribution was to be expected of vast arrays of traits found in natural populations. Weldon then noted where this expectation was defeated and tried to find arguments supporting or falsifying the hypothesis that the deviation was due to the work of natural selection. What subsequently became canonical method among later Darwinians is already present in full bloom in Weldon's work: expected distribution, statistically significant deviation, followed by adaptationist explanation of the deviation based on field observation bolstered by experiment. It can and has been argued that Weldon was the first person to demonstrate natural selection in nature (Gayon 1992, 249).

Of particular significance in this research program was a paper Weldon published in 1894, which showed that a large number of traits in a particular crab found both in Plymouth (England) Bay and the Bay of Naples followed the normal distribution in both places, but that one trait, the width of the crab's front edge, differed in statistically significant ways between the two populations. The distributed width of the crabs in the Bay of Naples fell into two humps of different means with a saddle in between. Was this evidence that selection pressure was pushing these populations apart? To verify that it was, a concrete source of the selection pressure would have to be found. If such a source could be located, it would be a highly significant discovery for another reason as well. It would suggest that natural selection can occur on a curve of very small, continuously differing variations after all, as Darwin had divined, and not (only) on the basis of Galton's discontinuous sports.

The next year Weldon published another paper claiming that the average width of the crab's front edge in Plymouth Bay was actually decreasing; this time he proposed a plausible selectionist interpretation of this fact. The bay was rapidly silting up due to its increasing role as a dump for human sewage and industrial waste. Weldon conducted an experiment to show that exposure to silt of this sort kills off crabs with larger front edges faster than those with smaller edges (cf. Kellogg 1907, 157). There was the source of selection pressure. Accordingly, Pearson and Weldon were soon suggesting that Darwin needn't have worried about swamping in the first place and that Galton did not have to retreat to discontinuous variation to get around objections to it. Natural selection could be fully effective, they claimed, where subpopulations exhibiting a continuous range of variation are interbred for several successive generations with each other. It was good news for Darwinians, and for eugenicists.

By now the wheels had been spinning on the issue of continuous and discontinuous variation for a long time. Darwin had come to believe, in response to Fleeming Jenkin, that variation was continuous. Galton held that variation was discontinuous. Pearson and Weldon had returned to Darwin's view armed with enhanced statistical analysis. Soon William Bateson, a friend of Weldon since boyhood, and like him a youthful convert from morphological to statistical biology, would take up, for reasons that we will review in the next chapter, Galton's view again. What was to make Bateson's return to discontinuous variation more than just one more spin of the same old wheel, however, was the connection he was to draw between his theory of saltationist speciation and a theory of hard inheritance first proposed by an Austrian monk named Gregor Mendel, according to which units of inheritance come in atomlike units that are combined and dissociated in various mathematically regular ways. None of the English giants of evolution had held anything like that. Upon learning about Mendel's ideas, as reconstructed (or perhaps reinvented) by several contemporary continental scientists, Bateson immediately jumped to the conclusion that speciation does not occur through natural selection at all but through sudden, changes in the units of heredity that he was the first to call "genes." This was too much for Weldon and Pearson. One might favor continuous variation or sports, and still be a Darwinian. In fixing this variation, one might even favor the inheritance of acquired characteristics over natural selection and still be a Darwinian. But to say that speciation occurs without any of the processes that kept the Darwinian tradition within the gradualist and adaptationist bounds laid down by the founder was to put oneself beyond the pale. Just why Bateson was happy to put himself beyond the pale is a topic for the next chapter. The immediate result, however, was that Bateson and the biometrical establishment went to war. The war lasted, as ideologically driven ones often seem to, for thirty years.

9 Mendel, Mendelism, and the Mendelian Revolution: Natural Selection versus Genetics

In 1865, Father Gregor Mendel, an Augustinian monk living in Brno (Brunn), which was then part of the Austro-Hungarian Empire and until recently was part of Czeckoslovakia, published a paper in *Proceedings of the Brunn Natural Sciences Society.* It was entitled "Experiments on Plant Hybridization." In it Mendel described the results of experiments that he had begun on garden peas almost a decade earlier (Mendel 1865). When this hitherto obscure paper was rediscovered by a number of biologists and botanists in 1900, a scientific revolution ensued (Brannigan 1981). The key concept of the Mendelian revolution is that heredity comes in discrete units that are combinable and dissociable in mathematically predictable ways.[1] This chapter is an account of what its later admirers thought they found in Mendel's paper and how the debate about Mendelism affected issues about hard versus soft heredity, blending versus particulate inheritance, continuous versus discontinuous variation, and regression to the mean versus constancy of inherited traits.

Peas have a number of distinct and variable traits. There are peas that are wrinkled and peas that are smooth; peas that are yellow and peas that are green; pea plants that are tall and pea plants that are short. In all, Mendel claims to have followed seven sets of alternating traits like these. Suppose, he hypothesized, each parent contributes one or the other of these alternative "factors" from each set to its offspring. Let it be equally probable that he could get either alternative from either parent. Suppose further that each set of traits works independently of the others. Let the fact that a pea is wrinkled, for example, have nothing to do with whether it is short or tall. Finally, suppose that not every factor that one inherits shows itself in the phenotype. Let factors go "underground" for a while and perhaps reappear. These three assumptions yield what would eventually be called Mendel's laws (or rules):

Law I: Independent Segregation. Alternative "factors" (as Mendel called them) of a trait are obtained by offspring as the result of inheriting one factor from each parent. Which of the two parental factors offspring receive from each parent is a matter of chance.

Law II: Independent Assortment. Each pair of factors is not affected by what is happening to other pairs of factors.

Law III: Dominance and Recession. A hybrid individual shows the trait associated with only one of the two factors at any locus, this factor being by definition dominant.

If these laws are true, then at least one known fact—that when you cross, for example, true-breeding lineages of tall and short plants you get all tall plants in the first (F_1) generation, but when you in turn cross these F_1 plants with each other, you get a mix of tall and short—suddenly looks intelligible with a little help from combinatorial algebra. Suppose you call the factor for tall A and the factor for short a. Because of Law III, $A + a$ in F_1 will always yield an A that "shows" in the offspring, for A is, by definition, dominant. But if you cross the next generation, F_2, with another member of that generation, $(A + a) \times (A + a)$, then it follows from the binomial theorem that you can have offspring of four sorts: $A^2 + 2Aa + a^2 = AA, Aa, aA$ and aa. Now put these two inferences together. Since A is dominant, Law III dictates that A will show, on average, three times out of the four. On average, every fourth time, however, an aa will show up, and thus the recessive trait will reappear. This hypothesis would explain, if it were empirically verified, why regression to grandparental traits occurs, something about which even Aristotle had been puzzled. The recessive traits were there all along but hiding.

Suppose now that you follow two traits simultaneously. If round is dominant over wrinkled, for example, and yellow over green, for the seeds of the plants, how many times will you get round and yellow seeds, round and green seeds, wrinkled and yellow seeds, and, finally, wrinkled and green seeds? If Law II holds and each pair of factors does indeed work independently of other pairs, the expected ratios for two traits, each of which has two factors, will be 9:3:3:1. This can easily be seen by consulting a Punnett square, a device for representing and calculating Mendelian ratios invented, to the applause of generations of beginning genetics students, by Reginald W. Punnett, a pioneer British geneticist (figure 9.1).

Mendel's paper reports on his attempts to test these hypotheses on his peas. The technique is fertilization by hand, with the experimenter taking the place of the bees and insects that usually do the job. Mendel had a carefully thought out experimental design. He says he used thirty-four varieties of peas from the genus *Pisum*, with twenty-two varieties serving as control groups. For each of the traits studied, he employed hundreds of individual plants. When he tested his hypothesis on smooth versus wrinked seeds, Mendel reported astonishingly good agreement between his experimental values and his theoretical assumptions. He observed the following distribution in his F_2 generation: 5,474 round seeds and 1,850 wrinkled. That is a 2.96:1 ratio, well within even intuitive levels of

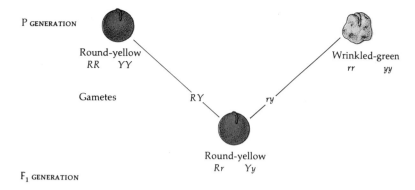

Round-yellow
RR   YY

Wrinkled-green
rr   yy

Gametes

RY   *ry*

Round-yellow
Rr   Yy

F₁ GENERATION

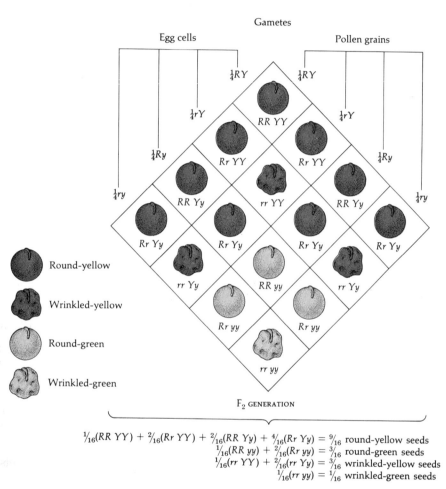

Gametes

Egg cells   Pollen grains

¼*RY*   ¼*RY*

¼*rY*   ¼*rY*

RR YY

¼*Ry*   ¼*Ry*

Rr YY   Rr YY

¼*ry*   ¼*ry*

RR Yy   rr YY   RR Yy

Rr Yy   Rr Yy   Rr Yy   Rr Yy

Round-yellow

rr Yy   RR yy   rr Yy

Wrinkled-yellow

Rr yy   Rr yy

Round-green

rr yy

Wrinkled-green

F₂ GENERATION

$$\tfrac{1}{16}(RR\ YY) + \tfrac{2}{16}(Rr\ YY) + \tfrac{2}{16}(RR\ Yy) + \tfrac{4}{16}(Rr\ Yy) = \tfrac{9}{16}\ \text{round-yellow seeds}$$
$$\tfrac{1}{16}(RR\ yy) + \tfrac{2}{16}(Rr\ yy) = \tfrac{3}{16}\ \text{round-green seeds}$$
$$\tfrac{1}{16}(rr\ YY) + \tfrac{2}{16}(rr\ Yy) = \tfrac{3}{16}\ \text{wrinkled-yellow seeds}$$
$$\tfrac{1}{16}(rr\ yy) = \tfrac{1}{16}\ \text{wrinkled-green seeds}$$

Figure 9.1 A Punnett square diagram illustrating independent assortment. In the experiments represented here, plants with round-yellow (RRYY) seeds are crossed with plants with wrinkled-green (rryy) seeds. The F₁ generation of hybrids are all round-yellow phenotypically, but their genotypes are (RrYy). When these hybrids are crossed to produce the F₂ generation, sixteen possibilities are possible, as shown. From F. J. Ayala and J. A. Kiger *Modern Genetics*, Menlo Park CA: Benjamin/Cummings, p. 41. Reprinted with permission.

confidence for statistically significant results. In another experiment, this time with yellow versus green seeds, he counted 6,022 yellow seeds and 2,001 green seeds. That is a 3.01:1 ratio. For two traits at a time, Mendel obtained 315 round yellow, 108 round green, 101 wrinkled yellow, and 32 wrinkled green, approximating a 9:3:3:1 ratio. These results seemed to Mendel, and Mendelians, to prove that inheritance is particulate. Like the atoms and quanta of physics, it comes in discrete packets. Moreover, like the underlying atoms postulated by physicists to explain the visible world, these packets, and their constant conjunction and disjunction, are hidden from view.

For our purposes, it matters less how much of this was actually in Mendel's paper than with what Mendel's intentions were and how consonant they were with the purposes of those who belatedly lionized him. The textbook image of Gregor Mendel hybridizing peas in his monastery garden is one of monkish industry and piety cloaking isolated and heterodox scientific genius. As an official hero of true science, Mendel is often portrayed in what Stephen Jay Gould calls "textbook cardboard" as desperately trying to signal to Darwin that he had *just* the theory of inheritance that would make natural selection, and evolution by natural selection, finally work.[2] It flies in the face of reality, however, to imagine in the first place that a man who would eventually be elected abbot of his monastery would ever have been a crypto-Darwinian. In German-speaking countries Darwinism *meant* Haeckel's reductionistic and materialistic *Darwinismus*. This does not mean that Mendel was not interested in the origin of species. At the end of his paper, he wrote that his work would provide a "solution to a question whose significance for the evolutionary history of organic forms must not be underestimated" (Mendel 1865, in Stein and Sherwood 1966, 2). Mendel, however, had a different, indeed a rival, theory of transmutation. His motive for undertaking his experiments with peas was not to find a universally true set of laws about inheritance that might support Darwin but to see whether the recombination of factors could provide a mechanism for production of new species by hybridization. That Mendel did not articulate this clearly in his famous paper is due to the fact that his results did not support his hypothesis. Inheritance came in discrete packets, but new combinations of these factors did not stick together well enough to make a new species.

The bare facts of Mendel's life are these. He was born Johann Mendel in 1822, the son of peasants. (Gregor was his monastic name.) He entered the monastery largely because it was the only way he could acquire an education and some leisure in which to pursue the study of science. That did not mean that Mendel was not religious or that he was an isolated genius, a scientific sport in a world of ignorant, pious rednecks. Brno was in fact a center of Moravian culture and intellectual life in that part of the Austro-Hungarian Empire. Many of the monks, including Mendel, taught in the local high school. Some even left the monastery to hold professorships and lectureships at major universities and colleges, as Mendel aspired to do. (In a country where Catholicism was the established religion, it was no more odd for priests to be professors than for Oxford dons to be Anglican parsons.) Between 1851 and 1853, Mendel attended the University of Vienna, where he studied physics with some of the best scientists of his day, including Andreas von Ettinghausen and Christian Doppler, discoverer of the famous Doppler effect. He did well in these studies, even working for a while as a demonstrator at the Physical Institute in Vienna. Mendel also studied, however, with Franz Unger, a professor of plant physiology, from whom he learned botany. From Ungar, Mendel seems to have inherited the conviction of many botanists, in a tradition to goes back to Linnaeus, that if species come from other species it is by means of hybridization. Unger, it turns out, had been trying to refute several botanists who had denied that hybrids are ever stable enough over generations to become fixed species. That was to be Mendel's research program as well (Brannigan 1981). Mendel was taken ill, however, just before he was to take his exams. He returned to his monastery in Brno without an advanced degree, taught in the local monastery school, and undertook experiments on hybridization on his own. Mendel's efforts to explore these issues beyond what he said about inheritance in his 1865 paper were, however, cut off by his election as abbot in 1868. (Most of an abbot's time is taken up with administrative problems.) He died in 1884.

Mendel's experimental methods show his training by methodologically self-conscious physicists like von Ettinghausen and Doppler. The mental habits of a physicist are also evident in Mendel's way of disregarding interfering factors when he is looking for mathematical patterns of inheritance. Mendel used idealization as a way of setting aside annoying factors like incomplete segregation, much as Galileo set aside such things as friction and air pressure. Finally, Mendel's willingness to countenance the idea that there are discrete units of heredity also betrays his training in physics. The idea that there are atoms of inheritance owes something to the Democritean imagination of modern physics. Mendel's attraction to an atomistic conception of inheritance does not necessarily mean, on the other hand, that he harbored a materialistic interpretation of these units. The notion that alternative factors are physical particles locatable in the chromosome never crossed Mendel's mind for the simple reason that it had not yet crossed anyone else's either. That would only be shown by Thomas Hunt Morgan in the early decades of the twentieth century. It seems increasingly likely, in fact, that Mendel interpreted the factors of inheritance in terms of the Catholic neo-Aristotelian philosophy in which he had been trained. Form, in the sense of abstract information, is the organizing and vivifying principle of matter. Recessive traits are potentialities. Dominant traits are their actualizations (Kalmus 1983).

What Mendel tried to do was cast this traditional scholastic ontology in the more quantitative terms of contemporary physics.

That is what he did to Unger's ideas about hybridization as well. But whereas Unger was a full-fledged evolutionist, whose belief in common descent and the materialist penumbra that surrounded this idea almost got him fired from the University of Vienna, turning him into a hero of radical students, it would be perilous to ascribe this position to Mendel. The contention that hybridized varieties can be fixed enough to count as species does not, by itself, make one into a full-fledged evolutionist. There is no reason to think that Mendel believed that transmutation by hybridization produces higher taxa. We may presume that when it came to evolution on the grand scale, Mendel was a creationist of some sort.

The research tradition according to which new species arise through hybridization had been passed down from Linnaeus to successive generations of botanists. What selective breeding is to zoologists, hybridizing is to botanists. Even today many botanists make reluctant Darwinians. What Mendel uniquely brought to his tradition, and to Unger's attempt to prove that hybrids can become species, was the physics-inspired hypothesis that unit characters might make new species in something like the way stable compounds arise from combining elements in fixed proportions in chemistry. Seen in these terms the results that Mendel reported in his famous article probably suggested to Mendel himself almost the opposite of what they have been taken to have meant (Callendar 1988). The experiments with peas proved that unit characters are transmitted through inheritance and that they combine and dissociate in organisms. But the clean segregation and indepedendent assortment that Mendel found in his peas suggested that these traits at least are too easily reversible to form permanent true-breeding species. They were more like chemical mixtures than true compounds. When Mendel went to study another, more complex organism, then, it was not to confirm and generalize the laws of inheritance that today we associate with his name but to find forms of hybridization more constant than those he had turned up in the pea (Olby 1985).

Mendel's experimental method reflects the rules for quantitative analysis of data that he had learned from von Ettinghausen, who had written texts on the mathematics of combinatorial analysis, and on methodological norms for scientific research. Von Ettinghausen stressed the importance of well-organized experiments based on a clear model of the phenomenon being studied and the use of the error curve to reduce measurement mistakes. R. A. Fisher, one of the masters of modern statistical analysis, started a still-ongoing controversy, however, by showing that Mendel's results are actually too good to be true (Fisher 1936). In one hundred repetitions of these experiments, there would be only a one-in-twenty chance of getting data as close to the predicted value as Mendel reported! Fisher came close to accusing Mendel of fraud. Debate about this issue even today is hampered by the fact that Mendel never published his raw data, putting scholars in the position of having to reconstruct logically what must have happened.

One possible explanation is that Mendel, following von Ettinghausen's experimental dictates, collected only about two-thirds of the data he could have, stopping at a point when the errors cancelled out sufficiently to give a number near what he was looking for. There was no fudging of data, on this view, but merely a difference of practice—admittedly an inadequate practice on Mendel's part—about how to apply statistics to data (Olby 1985). Even then, however, Mendel's results defy probability. The problem is made worse by the fact that the likelihood of Mendel's picking seven traits lying on seven different chromosomes, and hence segregating as independently as he reported, or on the same chromosome but sufficiently separated as to not affect each other, is only 1/163 (Di Trocchio 1991).

One version of what happened that seems plausible is the following. None of the hybrids Mendel was following in his pursuit of new species bred true. Disappointed, he searched his data for patterns and discovered the 3:1 ratio in some of them. Then he looked for others. Out of the twenty to thirty characters he was following, Mendel found seven conforming to the 3:1 pattern. Neglecting linkage as an annoyance, he did the calculations and reported the results (Di Trocchio 1991). Was this fraud? If Mendel had been interested in finding a set of universal natural laws governing inheritance, they might well have been. It was because Fisher read him that way that he could raise the issue as a serious one. When it is recognized, however, that Mendel was reporting what was left of a failed experiment about speciation by hybridization, and that his results, even when fudged a bit, did support particulate inheritance, a different light is thrown on the issue. Debate continues about it, however, and is likely to continue in the future.

Three European biologists—Hugo de Vries in Holland, Carl Correns in Germany, and Erick von Tschermak in Austria—claimed in 1900 to have independently and almost simultaneously rediscovered Mendel's ratios, only to find afterward that they had been anticipated in Mendel's neglected paper. It now appears that the three codiscoverers were not entirely independent of each other and that Mendel's paper may have guided rather than followed some of their thinking. The myth of simultaneous independent rediscovery probably emerged to head off a nasty priority dispute among them and to give their shared or overlapping research programs the patina of surprising empirical confirmation.[3] We need not believe the orthodox version of this spectacular case of scientific convergence, however, to appreciate that there was a deeper connection between Mendel and the Mendelians.

De Vries and Correns, the two important of the rediscoverers, were both botanists. Like Mendel, they worked in the botanical evolutionary tradition, which favored speciation by hybridization. Mendel and his rediscoverers were, accordingly, members of the same research tradition, which was then stirring to new life. What is of special interest is just why this tradition was so active at the turn of the century. One answer is that the collapse of the heritability of acquired characteristics at the hands of Weismann had deprived anti-Darwinians of a plausible alternative research tradition. Former self-described Lamarckians, convinced by Weismann's experiments, would probably jump sooner to speciation by hybridization. At the same time the Mendelians were now working in a post-Weismannian environment in which the revitalization of their tradition depended on finding and confirming a general theory of hard inheritance (Brannigan 1981). For them, Mendel's ratios appeared to support the idea that inheritance is universally fixed in particulate units in the germ line. It is not strange, then, that the three codiscoverers anachronistically assumed that Mendel himself, like them, was looking for a general theory of hard inheritance on the basis of particulate factors. Such a theory would provide a clear alternative to the soft inheritance of the Lamarckians by construing speciation by hybridization as novel combinations of hereditary units, or as the effect of the implantation of genetic sports—what de Vreis called "mutations"—into reproductive lineages. Whatever the details turned out to be, Mendelism would provide a more plausible mechanism for speciation than Darwinian gradual selection, clearing the way for the ultimate triumph of the much-maligned botany-based tradition.

A central figure in the battle of the Mendelians with the biometrical Darwinians was William Bateson, the brilliant son of the master of a Cambridge college, who became an early convert to the Mendelian cause and a traitor to his former biometrical self. The young Bateson had been a participant in the turn-of-the-century "revolt against morphology." Like Weldon, his close friend until Mendelism drove them apart, Bateson had been trained in embryology, comparative anatomy, and phylogeny at Cambridge. Even when he was publishing papers in the old tradition, however, Bateson was hedging his bets. "Of late the attempt to arrange genealogical trees involving hypothetical groups," he wrote in 1886, "has come to be the subject of some ridicule, perhaps deserved" (Bateson 1928, 1:1, in Provine 1971, 37). These doubts had first been stirred by the American embryologist W. R. Brooks, a former student of Agassiz with whom Bateson worked at the Chesapeake Bay Biological Laboratory for several summers, who had begun to turn toward variation in natural populations. A few years later, like Weldon, Bateson simply left the old ontogeny-phylogeny research program to work in the statistical analysis of variation. He went off to Russia to measure variations among organisms in isolated lakes, being among the first to appreciate that the closed and isolated conditions of such lakes create natural experimental laboratories. On his travels, Bateson found little to convince him of natural selection's role in producing change within a population of the same species, at least when it was working on an array of measurable continuous variation. If natural selection was going to work at all, he came to believe, it must be on the basis of discontinuous variations. This was even more true when speciation was at issue. The biometricians seemed to have put aside that thorny issue, assuming that speciation would take care of itself over the long run if natural selection could be shown to be at work in the short run. Bateson, however, was preoccupied by the old question. He could not see how one could be an authentic disciple of Charles Darwin and at the same time merely wave one's hands at the problem of descent. Bateson thus drifted back toward Huxley's saltationism and to Galton's stress on discontinuous variation. In 1891, Bateson wrote, "It is difficult to suppose both that the process of variation has been a continuous one, and also that natural selection has been the chief agent in building up the mechanisms of things" (Bateson 1928, 1:128, in Provine 1971, 41). Relations grew strained with his old chum Weldon, and with Pearson as well, who had worked hard to bring the Darwinian tradition back to Darwin's stress on continuous variation.

These relations were to grow positively vitriolic when Bateson took another step. Sensing that blending inheritance was the Achilles' heel of the biometricians, Bateson began casting about for a theory of particulate inheritance as a way to defend the emphasis he was now placing on discontinuous variation. If inheritance comes in discrete units, discontinuous variants would be less likely to disappear in the blending process. If the units were big enough, in fact, speciation might be achieved almost instantly through hybridization. Bateson's burning interest in Mendelism, once he heard about it, arose from his burning desire to prove to his former colleagues Weldon and Pearson that the units of hard heredity are particulate and that the agents of speciation are discontinuous variations or sports.

The way Bateson heard about Mendel and Mendelism has been the subject of another of the dubious legends that seem to track the early history of the genetic revolution. In 1899, Bateson attended the International Congress on Hybridization, where he was much taken by de Vries's claim that speciation occurs when mutations are spread through a population by hybridization. On May 8 of the following year, while he on his way by train to report some his own ideas about hybrids to the Royal Horticultural Society, Bateson read a paper de Vries had published the previous month reporting the 3:1 ratio. He immediately changed the text of his speech to say that the 3:1 meant that Galton's regression law would have to be amended. He did not read Mendel's paper itself on the train, as Beatrice Bateson reported in her 1928 biography of her husband, and he certainly did not march into the Royal Horticultural Society and throw Mendel in the face of his former biometrical friends and colleagues (Olby 1987). It was only afterward that Bateson read Mendel, to whom de Vries had made reference, and became an even more ardent Mendelian than the three codiscoverers, claiming that Mendelian inheritance, in refuting Darwinism's dependence on blending, showed that speciation is a function of sudden large mutations rather than of the gradual effects of natural selection working on continuous variations. It was then that a "thirty years' war" broke out between Bateson's band of Mendelians and the Darwinian biometricians (Provine 1971).

From a strictly logical point of view, this war need never have occurred. Almost immediately, the Cambridge biologist G. Udny Yule, for one, recognized that particulate inheritance is consistent with continuous variation if it takes many slightly differing units of inheritance to determine a single trait—and if the traits themselves are genuinely single traits rather than the mélange on which our received "folk biology" tends to light. Yule's insight eventually turned out to be the key to reconciling Darwinism and Mendelism, once the distinction was drawn between "phenotypes," the often-continuous traits that show and are measured by biometricians, and underlying "genotypes," which are combinatory units of inheritance. However, the conceptual, and especially the statistical, machinery to fill that position out would not be available for some time. In any case, neither Bateson nor his opponents would have been willing to make the required concessions. Bateson's larger aim in using Mendel against his former friends was to affirm de Vries's mutational account of speciation. To do so he felt he had to deny that you could have both particulate inheritance and selection, whether on continuous variation or individual differences. Affirm the first and you falsify the latter. This meant both to Bateson and his former colleagues that he had put himself beyond the Darwinian pale. You might stress continuous variation or individual differences and still be a Darwinian. You might stress natural selection or use inheritance as means of fixing variation and still be a Darwinian. But if the cost of hard inheritance was particulate inheritance and if the cost of particulate inheritance, as Bateson insisted, was speciation by mutation, then, as the Darwinian E. B. Poulton overtly claimed, the gradualism that, for these people at least, lay at the indispensable core of the Darwinian tradition would have to be protected by abandoning hard inheritance. Pearson's staunchly positivistic agnosticism about theoretical entities and processes would help if it came to that. Darwinism would do without any theory of inheritance. But even without going so far, Bateson's biometrical enemies dug in their heels against particulate inheritance in order to maintain the primacy of selection on continuous variation as the explanation of all evolutionary phenomena. Moreover, the way Bateson had chosen to frame the issue meant that vindication of his views would result in the instant collapse of the entire discipline of biometry. Under such conditions, those under attack usually do not give an inch for fear that their opponents will take a mile, particularly when friendships have been ruptured—and when ideological issues like eugenics, which Bateson disliked, were at stake.

Bateson set out to prove his point by replicating Mendel's experiments and by generalizing them from plants to animals. In his new capacity as professor of zoology at Cambridge, he worked on sweet peas and chickens with the help of a number of young research assistants. One of these pioneers was Punnett (of Punnett square fame) who remembers the excitement of forming a new research community, a research program, indeed a new discipline, as follows:

The set up was primitive, for money was scarce. The poultry occupied a small paddock split up into about two dozen little pens. . . . Every afternoon one of us went out collecting and marking the eggs from the various pens. . . . Though we reared some hundreds of chicks each year the great majority of the eggs incubated were never allowed to hatch, for some of the characters on which we were working were sufficiently developed for determination at about the eighteenth day. So we had periodical "openings" which were recorded in a separate notebook known as the "Book of the Dead". . . . On the day of an "opening," we adjourned to the outhouse in which the incubators were kept, having previously collected Mrs. Bateson to clerk for us, a function which she performed with the greatest efficiency and devotion. . . . Bateson took off his coat and produced his knife with the big blunt blade, while I stood by with a pair of scissors. He then took up an egg, read off the numbers of the pen, the hen, the date of hatching and . . . proceeded to stab and peel off the shell and call out the peculiarities of that particular embryo, such as . . . "light down, no colored ticks seen, rose comb, no extra toes, feathering on leg." . . . Sweet peas were another main line of inquiry. . . . Having made Mrs. Bateson comfortable we proceeded to . . . pull up the plant and sing out its characters, all duly logged by Mrs. Bateson. (Punnett 1950, 5–6)

While attending to this tedious work, Bateson was writing an address in which he was to report success in finding Mendelian ratios to the BAAS meetings for 1904. He had already had trouble getting his papers and his rejoinders to attacks on Mendelism published, in large part because of Pearson's opposition. In a scene that cannot help but recall an earlier and no less dramatic debate in the BAAS, Bateson finally had it out with Weldon. Punnett recalls:

Even the window sills were requisitioned. For word had gone round that there was going to be a fight. . . . Weldon spoke with voluminous and impassioned eloquence, beads of sweat dripping from his face. . . . Bateson replied . . . .Toward the end Pearson got up and the gist of his remarks was to propose a truce to controversy for three years, after which the protagonists might meet again for further discussion. On Pearson resuming his seat, the Chairman, the Rev. T. R. Stebbing, a mild and benevolent looking little figure . . . rose to conclude the discussion. In a preamble he deplored the feelings that had been aroused, and assured us that as a man of peace such controversy was little to his taste. We all began fidgeting. . . . But we need not have been anxious. . . ."You have all heard," said [Stebbing], "what Professor Pearson has suggested. . . . But what I say is: Let them fight it out." On that note the meeting ended. Bateson's generalship had won all along the line and thenceforth there was no danger of Mendelism being squelched out through apathy and ignorance. (Punnett 1950, 7–8)

During these years, Bateson began to bestow on the new science some of its canonical nomenclature, since much depends in an emerging research program on standardizing terminology. The term *genetics* was his invention. He also called Mendel's alternative factors *allelomorphs,* eventually shortened to *alleles* (from Latin *alius,* "other" or "alternative"), and contributed *homozygote* and *heterozygote* to name, respectively, cases where offspring receive the same and different alleles of a gene from parents.

It was William Johannsen, however, a Danish botanist whose work was closely related to that of de Vries, who coined the term *gene*—after de Vries's *pangenes* and more remotely Darwin's *pangenesis*—and who, by distinguishing between the phenotype and the genotype, formulated the crucial distinction on which the autonomy of genetics as a distinct science eventually came to depend. The way Johannsen conceived the phenotype-genotype distinction, however, handed genetics its first conceptual crisis. Johannsen meant by *gene* whatever unit of heredity underlies a single perceived trait. That meant that the reference of gene was fixed by the phenotype. It also meant that, by definition, there had to be one gene for one trait (Carlson 1966; Kitcher 1982b; Burian 1985). This one-to-one mapping caused no end of trouble, since in fact the normal relationship is many-to-many. Nonetheless, Johannson's definition served his own theoretical purposes well enough. His idea was that intensive crossbreeding, aimed at turning up all hidden variation, would ultimately reveal "pure lines," in which there is a one-for-one correspondence between genes and traits, and in which, accordingly, Galton's regressions would find their limit. Since each individual in a pure line has the same genotype, the pure changeless stream of transgenerational information would be there for all to see. Diversity in the living world could then be accounted for by combinations of the genotypic-phenotypic units that make up the pure line and by modifications to their structure (mutations). Inheritance being held constant in this way, one could go on to test how phenotypes are correlated with environmental conditions. If, on the other hand, environments are held constant and change still occurs in what one knows to be a pure line, then one may suspect that something disruptive had occurred in the germ line. Pure lines, on this view, form the stable background against which evolutionary change is to be measured.

It is just here that resonances between Mendelism and the old preformationist tradition make themselves felt most strongly. In both cases, genetic information remains presumptively intact generation after generation, and novelty comes from disruption, presumptively by way of de Vries's mutations. This gives some substance to Mayr's belief that there was more than a whiff of essentialism, even of Platonism, in much early Mendelism, which contributed not a little to delays in reconciling genetics and Darwinism (Mayr 1980, 1982). Perhaps that was the price Johannsen had to pay, however, for refusing to treat regression to the mean as the computational baseline for evolutionary studies, as it had been and continued to be for Galton and his disciples, and for proposing that pure lines should henceforth serve that role. That was a first crucial step in dislodging the notion that selection, which modifies the expected regression curve, is the primary agent of evolutionary change.

Nevertheless, even if we grant that every new paradigm is born in a "sea of anomalies," which it eventually resolves in and through the articulation of its original inspiration, as Kuhn would have it, pure lines posed so many problems that they were not destined to provide the new computational baseline against which geneticists eventually learned to measure evolutionary change (Kuhn 1962, 1970). In the first place, pure lines are hard to come by. Many organisms lose viability as they approach this condition. (This is the counterpart of hybrid vigor.) Again, it is hard to know when and whether one has actually isolated a pure line. Much wind was taken out of the sails of this research program when de Vries, confident that he had found a pure line of evening primroses (*Oenothera lamarckiana*) that had suddenly speciated by mutation, proved to be wrong. Primroses are clever organisms, hiding variation in unsuspected places.

But the deepest problem with pure line was that it is only very special organisms, or a limited number of traits, that seem to obey Mendel's third law, as the discouraged Mendel himself had had to admit. In many cases, there is a continuum between dominant and recessive traits, making blending inheritance the obvious hypothesis. Early Mendelians tried to resolve this difficulty by modifying Johannsen's conceptual framework in either of two ways. William Castle, an early geneticist at Harvard's Bussey Institute, insisted that inheritance is particulate and combinatorial but denied that genes have stable unit characters. Variability goes all the way down. On this view, incomplete dominance is a function of contamination by variable units. Bateson, on the contrary, waved his hand at the problem by challenging the assumption that there is one gene to one trait. Blends can be accounted for by hypothesizing more genetic determinants and by further separating phenotypic characters into multiple traits. The second tactic eventually proved so robust that the very notion of pure lines receded into an irrelevant background. At first, however, it was no less of a hand-waving gesture than Castle's.

It was not only Mendel's third law that seemed to be obeyed more in the breach than in the observance, however, but the second as well. Many genes do not segregate cleanly. They come in linked groups that fail to follow Mendel's rules, a phenomenon now called *gene linkage*. In the course of resolving why this is so, Thomas Hunt Morgan and his students, working at Columbia University in the 1910s, achieved some of the most beautiful experimental results in the history of science. In the process, they not only explained linkage but demonstrated conclusively that genes are physical units lying at particular places along chromosomes. The result was a vindication of the science Bateson had been calling genetics. No longer could biometricians say that genes were just theoretical entities.

Morgan was a son of the Old South. Born into a wealthy family in Lexington, Kentucky, his uncle was John Hunt Morgan, leader of Morgan's Raiders, a famous Confederate guerrilla unit. Morgan attended Kentucky public schools and then studied at Johns Hopkins. Like Weldon and Bateson, Morgan looks at first like another instance of the turn-of-the-century revolt against comparative morphology. In Morgan's case, however, interest in genetics did not lead to disinterest in embryology, as it did in Weldon and Bateson. On the contrary, Morgan was a lifelong developmental biologist and close friend of its eminent contemporary practitioner, Hans Dreisch. Morgan also possessed useful personal virtues—an open and attractive personality combined with organizational ability—that enabled him to run the most productive and cooperative laboratory in the history of genetics and to empower his students, especially Robert Bridges, Alfred Sturtevant, and Hermann Muller, to make important discoveries in their own right and to pass the torch to another generation of American geneticists, most of whom were trained at Columbia.

Morgan's sensitivity to developments in embryology and the rapid advances in cytology guided his genetic work. Four cytological discoveries in particular were relevant to his research program. First, it had been shown that the chromosomes inside the nucleus of the sex cells (sperm and eggs, for example) divide in such a way that each parent contributes one of two paired or homologous chromosomes to its offspring. (This is called meiosis, meiotic division or reduction division, in contrast to the mitosis that occurs in somatic cells, which lose nothing in the transmission.) Second, microscopes had been used to observe bands of dark and light along the chromosomes. This was especially true in the humble fruit fly, *Drosophila melanogaster*, which has anomalously large chromosomes, allowing Morgan and his students to turn this now most studied of organisms into an instructive object and tool of genetic research. Third, a number of biologists, notably Walter Sutton, Theodore Boveri, and Edmund B. Wilson, had discovered that one of the chromosomes—the so called X chromosome—is linked to the sex of its bearer. Finally, F. A. Janssens, a Belgian cytologist, showed that in the crossing over that reunites the sex cells, bits and pieces of chromosomes are "reshuffled" as cards are reshuffled through a deck. This fact gave a concrete explanation of why Mendel could find two factors for each of his traits, and why traits segregate independently. Taken together, this suite of discoveries made it possible for Morgan to show that genes are located on chromosomes, to explain gene linkage, and to set his students to work making the first genetic maps.

Surprisingly, Morgan was at first no more of a physicalist than Mendel, Bateson, Johannsen, and other early Mendelians when it came to speculating about the nature and ontological status of genes. Genes may be particulate in the sense of being unit characters but might not be made of particles. Bateson and Johannsen, in fact, thought vaguely of genes as harmonic resonances or Pythagorean ratios governing stable energy levels in the organism, since they doubted that a mere material entity could direct pattern formation. So, at first, did the developmentally sensitive Morgan. Noting, however, that sex-specific chromosomes carry information about other traits as well, such as eye color, body color, and wing deformation in the case of fruit flies, Morgan hypothesized that if you kept track of these traits in crosses, correlated them with sex and with changes in the pattern of dark and light banding in the chromosome, you would be able to show sufficient correlation between trait and band to suggest that loci on the chromosome are the site of the genetic information. This proved to be correct.

Morgan was then able to use this result to verify a hypothesis that even more strongly suggested a cytological conception of the gene. Perhaps incomplete segregation, or gene linkage, was correlated with Janssens's discovery of crossing over. Crossing over, and hence segregation, might occur anywhere along the chromosome. Mendel was right to think that there is no inherent connection between one trait and other, but Mendel and the early Mendelians were wrong to presume that segregation would occur with equal probability between any two genes. Given the phenomenon of crossing over, the probability that genes lying next to each other would segregate was low. It was at its highest for genes lying at opposite ends of a chromosome. When this hypothesis was verified, linkage between groups of genes became an expected fact rather than an embarrassing anomaly for the Mendelian tradition, or something that had to be explained away by positing ever receding pure lines.

An air of triumph and energy pervaded the cramped, but now legendary "fly room" at Columbia, where this work was done. It was in this atmosphere that Sturtevant used Morgan's hypothesis to make the first genetic map. In what college students call an "all-nighter," Sturtevant in 1913 used the proportions of crossovers as a guide to the distance between genes on sex-linked chromosomes, enabling him to compute where genes lay relative to each other. Genetic mapping of the chromosomes of fruit flies, work on which culminated in about 1915, won Morgan a Nobel Prize in 1933, giving America a preeminence in genetic research it never lost. In the process of making genetic maps, it became clear, moreover, that Yule and others had been right all along. A single phenotypic trait can be controlled by a large number of genes. Conversely, a single gene can affect several phenotypic traits (pleiotropy). Mappings between Johannsen's phenotypes and genotypes, accordingly, are many-to-one, one-to-many, and even many-to-many. Genes and traits were no longer bound by the one-to-one idea that they had when they were purely theoretical entities. The very conception of what a gene is was changing.[4]

Morgan's research was not the only source of pressure on early Mendelism. Habits formed by thinking about Galton's revered law of regression, and the desire to defend it, stimulated British biologists to pose hard questions to Mendelians about the long-run fate of recessives. Some of these questions were based on conceptual misunderstandings of concepts like dominance. If, for example, brown eyes are dominant over blue, Punnett was asked after a talk to the Royal Society of Medicine in 1908 why blue eyes were not getting rarer still. The answer Punnett gave was that these alleles are now in stable equilibrium in the relevant populations as a whole. But how were these proportions established and calculated?

For his part, Yule volunteered that the proportion of dominant and recessive alleles strewn through a freely interbreeding population in which two alleles are in equilibrium would always be the old Mendelian 3:1 ratio writ large. Punnett sensed that this was wrong, for this would happen only when the proportion of dominant to recessive genes was fixed in a population at 50 percent apiece: Mendel's 3:1 ratio worked only on the assumption of his first law, according to which the chance that offspring have either parent's version of a particular trait is .5. The challenge, however, was to explain why, in the absence of internal change like mutation or external change like selection pressure, even small proportions of genes stay stable. Punnett's instincts were right. His problem was that he did not know how to represent mathematically the general rule that holds when there are different, and indeed varying, proportions of alleles floating around in a population. In another incident now legendary in the annals of the genetic revolution, Punnett brought up the problem to his old friend the Cambridge mathematician G. H. Hardy, with whom he had played intermural cricket. Hardy almost instantly gave him the solution. If you assume Mendel's laws, the resulting combinations will expand into the binominal distribution. Hence the distribution of the genotypes will be the Hardy-Weinberg equilibrium formula

$$p^2(AA) + 2pq(Aa) + q^2(aa),$$

where p is the initial frequency of the dominant A in a population, q the initial frequency of the recessive a, and (since there are only two alleles at the locus) $p + q = 1$.

In the F_1 generation, the frequency of A in the population is given by $p^2 + 1/2(2pq)$. That equals $p(p + q)$, which in turn equals $p(p + \{1 - p\})$, which equals $p^2 + p - p^2$, which equals—p. That is, the frequency of A among the progeny is exactly the same as it was among the parental generation. It is in equilibrium. Thus, it does not matter whether a gene is dominant or recessive, or in what proportion it is either. *Its frequency in the population will remain the same unless it is perturbed by some other factor:* random fluctuation ("sampling error"), mutation, recombination of bits and pieces of the chromosome during meiotic division, migration into the population of individuals of the same species with a different gene frequency, or the pressure of natural selection. This formula was also derived by the German physicist Wilhelm Weinberg. Hence, it is called the Hardy-Weinberg equilibrium formula, or principle, or, more grandiose still, law. The theorem is perfectly general and can be modified to keep track of the distribution of alleles in genetic systems that are not diploid (that is, do not have their genes lying on two homologous chromosomes). It can be shown in fact that any population obeying Mendel's laws will, to the extent that it does so, also obey the Hardy-Weinberg principle to precisely the same degree (Beatty 1981).

Punnett quipped that "whether the battle of Waterloo was won on the playing fields of Eton is still, I gather, a matter of conjecture. Certain it is, however, that 'Hardy's Law' owed its genesis to a mutual interest in cricket" (Punnett 1950, 9–10). In saying this Punnett was commenting less on the roles of chance, friendship, or sports in scientific progress than on the importance of the Hardy-Weinberg equilibrium formula for the formulation of what has long been called population genetics (as distinguished from the concentration of the Morgan school on transmission genetics, the mechanics of heredity). People like de Vries or Johannsen, who fetishized pure lines, had, it seems, been looking in the wrong place for a new baseline against which to measure evolutionary change. Their roots, like Mendel's, were in plant hybridization. They crossed the results of successive generations of hybrids with each other and took a look at what came up. What they failed to ask was how Mendelian factors or Batesonian alleles would be distributed in a freely interbreeding population. Pursuing the question Punnett asked Hardy, on the other hand, proved to be the key to Mayr's "population thinking" in the most fertile sense of the term (Mayr 1980, 1982).

Shifting the terms of debate to the level of populations made possible, in the first place, a resolution of the quarrel between biometricians and Mendelians. The terms of reconciliation were these: Mendelians must give up their fascination with pure lineages and their arbitrary restriction of the sources of evolutionary change to forces other than natural selection. Mendelians, who had hitherto used statistics simply in the traditional role as estimates of error, would also be asked to apply to the distribution of genotypes in a population the techniques of statistical analysis and probability pioneered by the biometricians in their quest to measure continuous phenotypic variation. For their part, biometricians would have to give up their exclusive concentration on what is happening to the phenotype and, yielding their resistance to theoretical entities, would have to bring their developed use of statistical techniques to the analysis of populations of genotypes as well. Indeed, they would have to see, for the first time, what they had been doing in terms of the reformed phenotype-genotype distinction, in which many-to-many mappings between genes and observed traits were expected.

Within the broad confines of this understanding, the Hardy-Weinberg equilibrium formula now began to play the role Johannsen had unsuccessfully proposed for pure lines. It was to serve as a new inertial background of expected stability against which evolutionary change was to be measured. In this connection, we hazard a generalization that is relevant to the major themes of this book: *Changes in computational background are correlated with successive and competing research programs within the wider Darwinian tradition.* The Hardy-Weinberg formula collects into a certain unity a succession of research programs that collectively have come to be called the modern evolutionary synthesis. Forces acting to perturb the Hardy-Weinberg equilibrium, in which gene frequencies in a population are otherwise presumed to remain unchanged or in equilibrium, now take the place of forces disturbing the Malthusian population parameter, as Darwin had it, or the forces skewing the Gaussian curve along which phenotypic variation is distributed, as Galton and the biometrical school had it, as the zero-state or computational baseline of evolutionary hypotheses (Sober 1984a).[5] The project of evolutionary population genetics, then, is to determine the variety, the nature, and, in particular cases, the values of the various evolutionary forces or processes that disturb the expected Hardy-Weinberg frequencies of genotypes in a population. Athough we will see that more were to come, three such processes drew attention from the start: mutation pressure, migration, and selection. For the makers of the modern synthesis, the task was clear enough: "Plug in non-zero values for the variables associated with natural selection, mutation, or migration, and you can calculate the evolutionary consequences. Thus, for instance, one can calculate the evolutionary consequences of moderate mutation with heavy migration, or heavy mutation together with heavy selection and moderate migration, etc." (Beatty 1986, 127).

To say that the Morgan school focused on transmission genetics is not to say that Morgan and his students were indifferent to questions about population genetics and the forces that operated to move gene frequencies out of Hardy-Weinberg equilibrium. Indeed, Morgan and his students were the first to combine transmission genetics and an early sort of population genetics into a revised theory of evolution, known since as the classical view. This work begins with Morgan's redefinition of *muta-*

tion in *The Physical Basis of Heredity* (1919). It now came to mean the appearance of a new allele within a population, not the single large change of great effect that de Vries had designated by this term. One reason for this change was that large mutations were nearly always fatal or immediately maladaptive, while small mutations enter into the gene pool as alternative form of traits, where they can be amplified or eliminated by natural selection or other forces. Either way, however, mutations did not tend to set up interbreeding barriers, and hence were not as closely connected with speciation or other forms of significant evolutionary change as earlier Mendelians had thought. A more proximate source of evolutionary change was in any case implicit in the phenomenon of crossing over and Morgan's theory of linkage. Crossing over and linkage made new genetic combinations constantly and normally possible. Natural selection would submit these to the test and fix the ones that proved superior in the population.

Morgan's brilliant but difficult student Muller then took Morgan's views about mutation a step further. He proved that bombarding chromosomes with X rays could induce large mutations, a feat for which he won the Nobel Prize. Not surprisingly, such mutations were almost always fatal or damaging. Muller came to the conclusion that most mutations have their origins in some sort of insult and therefore nearly always have maladaptive effects. The variant alleles that have collected in natural populations are therefore already well adapted. All the abnormal ones would have been quickly eliminated. For Muller, the "wild type" is "normal" and fairly homogeneous. From this perspective, the job of natural selection is to remove abnormal genes from a presumptively adapted population. This view has been known ever since as "purifying selection," the idea that lies at the foundation of the classical view of evolution by natural selection. The quickest summary of subsequent developments in genetic Darwinism is that they have consisted in an increasingly radical series of ways of proving the classical view wrong.

J. B. S. Haldane, one of the pioneers of theoretical population biology, was instrumental in providing experimental results that set up the conditions under which the classical view was challenged, especially by successive generations of talented English genetic Darwinians. Haldane, who became professor of genetics at the University of London, was an interesting character. His father was a chemist, indeed a biochemist, whose chief effort was to discover the role played by carbon dioxide in the bloodstream. Haldane literally grew up underfoot the lab bench. He did not have to learn experimental science when he was at school because he had already learned it at his father's knees. His father's way of teaching was, moreover, characteristically experimental. The elder Haldane served as a mine safety officer for the British government. He often took young Haldane into dangerous mines and taught him applied science in ways the boy was not likely to forget. For example, to get across the useful fact that in a mine disaster methane rises while breathable oxygen will still be available closer to the ground, the father had Haldane stand up, under appropriate experimental conditions at the bottom of a mine shaft, and recite Mark Antony's speech from *Julius Caesar*. Even before he got to "Lend me your ears," the boy had collapsed in a heap on the floor of the mine, where he quickly revived by inhaling the oxygen available there (Clark 1968). Experiences like this, as well as his capacity to endure the insufferable bullying of British public (that is, private) schools and the generally tough-minded inheritance of a family of warlike, Scottish border lords, led Haldane to discover in the trenches of World War I a truth about himself that he was not embarrassed to acknowledge: He was not only physically courageous but really liked killing enemy soldiers. As the leader of an irregular squadron of sapppers, he devised sometimes fiendishly clever means of blowing up Germans. Haldane's men, needless to say, admired him.

Haldane's reductionistic materialism freed him from scruples about this. He was a firm believer in natural selection, of a rather hard variety, and thought many unpalatable human practices, such as war, were deeply rooted, hence more or less natural, consequences of early adaptations. At Oxford, Haldane had shown himself to be not only a first-class mathematician, but a good classicist, who found the blood and guts ethos of the Homeric epics much to his taste. He hated cant and did not suffer fools gladly. Haldane never ceased pointing out how inconsistent with evolution's ways were those of the Christian God. He is famous for his answer to a Paleyesque question about what he had learned, over a lifetime of inquiry, about the Creator from his creation. Noting the spectacular variety of these insects, Haldane famously replied, "An inordinate fondness for beetles." He also said this: "Blake expressed some doubt as to whether God had made the tiger. But the tiger is in many ways an admirable animal. We now have to ask whether God made the tapeworm. It is questionable whether an affirmative answer fits either with what we know about evolution or what many of us believe about the moral perfection of God" (Haldane 1932, 159).

It may seem odd that Haldane sympathized with socialism and communism, whose roots lie in secularized versions of the compassionate and inclusivist ethics of the higher religions that he disdained. Their shared atheism no doubt had something to do with this. Attraction to communism, and to the struggling Soviet Union, was common among Haldane's generation of British upper-class intellectuals, however, mostly because their sort of socialism was based not on any belief in the good sense and nobility of the working class, or any deep sympathy for the underdog, but in their aristocratic contempt for the vulgar rule of the bourgeoisie and their elitist attraction to the idea of a scientifically planned society, a program that would include reproductive policies.

Toward the end of his life, Haldane emigrated to India, where he thought that a less self-delusional view of the human condition than that entertained by the European ruling classes still held sway. There he helped build up the scientific prowess of Indian universities.

Haldane's first paper, sent to press as he went off to World War I, was a study of gene linkage in guinea pigs, work done not long after Bateson's and Morgan's and in full awareness of what was at stake. The data for this paper were acquired when, at the tender age of sixteen, Haldane had persuaded his uncomprehending sister to help him fill up an entire tennis court at the family estate with promiscuous guinea pigs of several strains, so that he might discover genetic links between albinism and other traits. Haldane's results documented gene linkage in animals. After the war, Haldane went to work on the mathematical analysis of population genetic problems in a serious way. In a series of papers published between 1924 and 1934 and in his book *The Causes of Evolution* (1932), Haldane applied statistical methods to the theory of population genetics in a way that focused on the frequency with which specific genes would be propelled through populations under various scenarios. His results (together with those of Fisher, whose work we will consider in the following chapter) suggested that the Morgan school had underestimated the extent and role of variation in natural populations, had underutilized statistical methods of analysis, and, in the notion of adapted wild types and crippled mutants, still harbored remnants of the semiessentialism that had plagued Mendelism from the beginning.

Among Haldane's and Fisher's most important conclusions were these. It does not take much mutation for natural selection to work, so mutation is not the primary agent in evolutionary change. Indeed, very small mutation rates can have significant effects on the distribution of gene frequencies, even where selection pressures are quite weak, because useful variation exists in populations, stored, for example, on the recessive side of heterozygotes. It is, in fact, very difficult to extinguish a recessive gene from a population fully. Moreover, the random or chance extinction of new genes in a finite population is rare (Haldane 1932). Since vast quantities of variation are available, therefore, and new gene combinations are constantly arising through sexual recombination, adaptation is not a matter of occasional purification but a constant and creative process, in which variation serves as fuel for natural selection. This was a vision much closer to Darwin's than Muller's and Morgan's. This, unlike Muller's halfway house—so reminiscent of Wallace, another socialist—is genetic *Darwinism*.

Haldane's analysis of gene frequencies in *Biston betularia* (Haldane 1924) gave genetic Darwinism its most famous paradigm case. It been known for some time that the darker or melanic (from Greek *melanon*, "black") form of the pepper moth (*Biston betularia*) was rapidly displacing lighter moths. Biometrical analysis had pinpointed the cause. The trees on which the moths live had quickly been covered with soot by the "great Satanic mills" of rural England's industrialization. The birds that ate the moths were less likely to find dark moths than light moths. Hence, the proportion of white moths in the population was decreasing rapidly. Like Weldon's crabs in Plymouth Bay, "industrial melanism" was another case in which the industrial revolution had turned the environment into a great laboratory through providing intense selection pressures. Haldane, a superb mathematician, now added an analysis in terms of gene frequency changes to these biometrical results. He proved that the black pepper moth then proliferating was being pumped up by selection pressure in ways that matched data to model, and hence gave empirical support to treating natural selection in terms of changing gene frequencies. From this perspective, Darwin's two-stage theory, in which selection would give direction to originally undirected spasms of variation, was converted into a two-level theory, in which mutation of genotypes would be amplified or diminished by the action of natural selection working on phenotypes. The process could be observed in populations in a finite number of generations and could be analyzed by way of statistics and probability theory.[6] If Weldon was the first Darwinian to prove the action of natural selection, Haldane deepened the proof by moving the mathematics down to the gene-frequency level (Gayon 1992).

Mendelism was originally associated with arguments against, or at best indifferent to, eugenic policies. Bateson's opposition to Galton's and Pearson's biometry included deep distrust of eugenics. At the same time, Pearson and Weldon were loathe to link their eugenic ideas with genetics. Pearson's positivism fueled his opposition to Bateson's "gene talk." The theoretical entities called genes struck Pearson as metaphysical, and hence as scientifically retrograde, all the more so because in the days before Morgan's work, genes really were purely theoretical entities. Pearson worried that holding eugenics hostage to a metaphysically suspect theory would ultimately prove disastrous for the eugenics movement. Genetics might reinforce nature over nurture, but it might also make genetic eugenicists appear as dogmatic and antiempirical crackpots, whose abstract theories about abstract entities would soon be held to be both dangerous and silly.

It did not take long, however, for eugenicists to take advantage of the enhanced rhetorical power offered by positing a genetic basis for inherited traits, or for Pearson's fears that genetic eugenics would attract crackpots to prove well grounded. This can be seen most readily by reviewing the tangled interweaving of early genetics and eugenics in the United States. An instructive figure to follow is Charles Davenport, an American biologist who was originally trained in the post-Agassiz embryological mode but who soon joined the biometrical and eugenicist revolution. Davenport taught at Harvard's Museum of Comparative Zoology, the institution Agassiz had founded, and then created the Laboratory for Experimental Evolution (later Experimental Genetics) at Cold Spring Harbor on Long Island. To British eyes, Davenport might appear as an early, rather frightening example of the American entrepreneur-scientist. He raised money from the Carnegie Institute, the Harriman family, and John D. Rockefeller, Jr., to establish his laboratory. Davenport's talent for raising money for his project was closely connected to the fact that a Eugenics Record Office was to be established cheek by jowl with the laboratory, and hence to the fact that the ascendant families in America, riven by palpable fears that unrestricted immigration was ruining the country's vigor (and morality), were eager to support his work. The iconoclastic lawyer Clarence Darrow later put the point with elegant irony, although he vaguely held some such view himself: "The good old Mayflower stock is suffering the same unhappy fate as the good old pre-Prohibition liquor. It is being mixed with all sorts of alien and debilitating substances" (Darrow 1926, quoted in Richards 1987, 516). The WASP ascendancy was being threatened by its own commitment to free-market and free-trade capitalism. Free-market capitalism required a steady supply of cheap labor, not only to work for low wages but to counter the rising power of the labor movement. Indeed, during the early twentieth century, the United States was the site of a protracted war between labor and capital. Since there was no question either of killing the goose of open immigration that was laying the golden egg or, on the other hand, of keeping the old Northern European Protestant culture nonpluralistic, the problem might be resolved by preventing degenerated natives as well as inferior immigrants from breeding in order to protect the healthy components of both populations. That was a route to pluralism without regression. For this reason, it was, in general, middle-class Progressives and intellectuals who fueled the fires of negative eugenics in America because they welcomed the opportunity of Solomonically managing the economic and cultural contradictions of American society through the cult of scientific expertise. Davenport and his cohorts shared none of Huxley's scruples about the beneficent role of these administrators.

Although he had secured the blessings of Galton, Weldon, and Pearson for his eugenical mission, Davenport converted easily to Mendelism. His deepest commitment was to eugenics, not to one side or the other in the war between biometry and genetics. If Mendelism supported eugenics better than biometry by reinforcing the domination of nature over nurture and inheritance over environment, then Davenport would become a Mendelian. De Vries and Bateson were, for this reason, welcomed at Cold Spring Harbor. Even if Mendelian elements could not be changed by selection pressures, selection as a source of evolutionary novelties would be missed mostly by British eugenicists, whose preoccupations centered on improving the happy few rather than eliminating the unfit.

In America, however, and especially in the mind of Charles Davenport, the eugenics movement was primarily about how to stop the defective from breeding in a polyglot and panmictic mass society. Davenport did not need natural selection as a creative force. He needed eugenics as a purifying force. Thus, in a sort of reversed mirror image of Galton's search for hereditary genius, Davenport set out to find hereditary defects, so that those possessing them could be prevented from mating—or could be turned back by the Lady who continued to lift her lamp to the wretched refuse of many teeming shores. Under Davenport's organizational auspices, a research program devoted to identifying hereditary traits as functions of specific genes was initiated, for humans and for other organisms.

Morgan's work at Columbia, that of Raymond Pearl and Herbert Jennings at Johns Hopkins, of Castle and Edward East at Harvard, and of plant geneticists at Cornell acquired an increasing supply of public and private funds for genetic research because, in this Progressive era, genetics promised to improve agricultural stocks. Most of these laboratories, and many of their patrons, were also sympathetic to eugenics. In raising such funds no one else was as prominent as Davenport, nor was anyone else as prone as he to making the extravagant claims that turned eugenics into an American cult devoted to passing eugenic laws, especially laws about sterilization of the "weakminded." Kevles writes of Davenport:

He combined Mendelian theory with incautious speculation. . . . Davenport thought in terms of single Mendelian characters, grossly oversimplified matters, and ignored the force of the environment. Sometimes he was just ludicrous, particularly in the various post-1911 studies on the inheritance of "nomadism," "shiftlessness," and "thalassophilia"—the love of the sea [from Greek *thallasos*, "sea," + *philia*, "love"] he discerned in naval officers and concluded must be a sex-linked recessive trait because, like color blindless, it was almost always expressed in males. (Kevles 1985, 48–49)

On the basis of this kind of propaganda, sterilization laws began to pass through state legislatures. This was made easier by Davenport's interpretation of the family history of the Jukes. The history of this poor northern Appalachian family had been traced first in a book by Robert Dugdale in 1877. This book was reissued by Davenport's Eugenics Record Office and was later updated by Arthur Estabrook (Estabrook 1916). In 1916, a similar study was made on a family named Kallikak by Herbert Goddard, who used IQ tests to prove inherited mental weakness. Of the Jukes, Davenport reported:

Their descendants show a proponderance of harlotry. . . . Two of the three sons were licentious and criminalistic in tendency and the third, while capable, drank and received out-of-door relief. All of the three daughters were harlots or prostitutes and two married criminals. . . . The difference in the germ plasm determines the difference in the prevailing trait. But however varied the forms of non-social behavior of the progeny of the mother of the Jukes girls the result was calculated to cost the state of New York over a million and a quarter dollars in seventy five years—up to 1877, and their protoplasm has been multiplied and dispersed during the subsequent years and is still marching on. (Davenport 1911, 233–34, in Richards 1987, 513)

Muller's view of "purifying selection" might be seen as supporting Davenport's sort of negative eugenicists. It too could be used to justify ridding the population of defectives. Muller's eugenical theory was so different from Davenport's, however, that he could denounce Davenport's eugenic cult to its face at the International Eugenics Congress in 1932, even while preparing to issue a "geneticists' manifesto" in which he would enlist the support of the seventy other geneticists for the eugenic views he had set forth in a best-selling book, *Out of the Night* (1935). As the "geneticists' manifesto" makes clear, even in its allusion to Marx's famous title, Muller was a left-wing or socialist eugenicist, who saw in both Davenport's and Fisher's brands nothing more than capitalist distortions of a very good idea indeed, and whose vision was concentrated on producing excellent humans, and as quickly as technological intervention made possible, rather than on becoming complicit in yet one more ideologically motivated assault on poor, working-class people.

Leftist eugenicists like Muller were not against the inheritance of good-making traits. They were against the fallacious and fatuous identification of good-making traits with "family values" and capitalist economics (Paul 1984). Recall Davenport's way of talking about the deadbeat and promiscuous Jukes family. Leftist eugenicists, including the famous playwright George Bernard Shaw as well as Haldane and Muller, argued that only in a classless society could real good-making traits and differences in individual ability be recognized and fostered. These differences were screened off and ground down in class societies, aristocratic or bourgeois. Only in a society where the daily Malthusian grind has been repealed and where religious myths have been exploded can children be raised in state nurseries to become sexually liberated adults, free from religious, economic, and family bugaboos. For this reason, only a planned and engineered socialist society, like the one presumably emerging in the Soviet Union, can take full advantage of the technological revolution to create a "new humanity" by artificial insemination. That was Muller's pet idea. It depended on the fact that males produce billions of usable sperm, whereas women produce only a few eggs. Why not inseminate many females with the sperm of superior males? "It is easy to show," Muller wrote, "that in the course of a paltry century or two . . . it would be possible for the majority of the population to become of the innate quality of such men as Lenin, Newton, Leonardo, Pasteur, Beethoven, Omar Khayyam, Pushkin, Sun Yat Sen, Marx (I purposely mention men of different fields and races) or even to possess their varied faculties combined . . . [instead of] a population composed of a maximum number of Billy Sundays, Valentinos, Jack Dempseys, Babe Ruths, even Al Capones" (Muller 1935, 113–114).

Muller went to the Soviet Union and dunned Stalin with letters outlining his plan. He heard nothing. That was probably not, however, because Stalin had not read his letters. In fact, Stalin had had *Out of the Night* privately translated. There were probably many things in this book that Stalin did not like, such as the vision of a sexually liberated society. The main problem, however, was that T. D. Lysenko had persuaded the dictator that no kind of genetics could work fast enough to create the new man, or new food sources, fast enough to protect the struggling Soviet Union from its enemies. Research should focus instead on Lysenko's own agricultural experiments, which, with their Lamarckian stress on the self-determination of organisms and nurture over nature, were supposedly quicker as well as conveniently more consistent with Marxian thought, which derived, after all, from French "active materialism" like that of Lamarck. Thereafter genetics itself, including Muller's, became nothing but "bourgeois ideology," and Russian geneticists—among the most accomplished and numerous in the world, as we will see—became an endangered species. That was because Muller's dream of a sexually liberated, eugenically superior socialist "new man" was tragically concurrent with the beginning of Stalin's terror. Muller beat a retreat, finding his way first to Spain to fight against the Fascists, and then back to the United States, where he became a greater object of public suspicion than before he left.

Haldane's views about eugenics were not entirely different from Muller's. The student of late nineteenth- and early twentieth-century intellectual history soon learns, in fact, that amid a vast, hypertrophic explosion of ideas and ideologies let loose by the collapse of traditional worldviews and the old social order and the apparent failure of the great bourgeois revolutions of the nineteenth century to take hold, there was hardly any theory so wild that some otherwise sensible person did not commit to it body and soul. This is a point worth bearing in mind as we turn in the next chapter to an account of the life and work of Fisher, whose enthusiasm for Galton's sort of positivist eugenics, and unshakable belief that genetics must support the Galtonian program even if Galton's biometrical disciples did not think so, led to the most thoroughgoing application yet of the probability revolution to evolutionary biology.

10 The Boltzmann of a Blade of Grass: R. A. Fisher's Thermodynamic Model of Genetic Natural Selection

By the beginning of the 1920s, population genetics was becoming a respectable science. Support was also growing for the idea that natural selection could significantly modify frequency distributions of genetic factors in populations, which will otherwise remain in Hardy-Weinberg equilibrium. In America, Morgan and Muller were advancing versions of genetic evolution according to which the primary role of natural selection was to prune maladaptive genotypes from an otherwise fit population. In Britain, on the other hand, support was increasing for brands of genetic Darwinism in which natural selection was allowed to play a more creative role. Punnett had said in 1915 that "natural selection . . . must . . . operate with extraordinary swiftness where it is given established variations with which to work" (Punnett 1915, 96, and quoted in Provine, 1971, 139). In that case, an array of small, continuous variants could be used by natural selection as fuel to drive the slow but steady creation of phenotypes that have to keep working to improve or stay fit. This tendency reflects the fidelity of the founders of British population genetics to the creative gradualism of the Darwinian tradition. By these lights, the Morgan school, with its stress on the difference between already adapted populations and destructive mutations, did not seem to give natural selection enough variation on which to work. It was in this context that Haldane set out to show that natural selection had the power to affect the frequencies of genes in a population, even when mutation rates are low, so long as variation is abundant enough to drive it. Similar concerns motivated Ronald A. Fisher. Fisher's approach differed from Haldane's, however, in a number of important respects.

Haldane was interested primarily in the conditions under which a single gene, originating in a mutation, could be amplified through a population. As the son of a chemist, he approached genetic evolution in the spirit of a biochemist. He wanted to find the biochemical and physiological mechanisms by which individual genes work. The ultimate payoff of his work in physiological genetics was the molecular revolution in genetics that in large measure took place in Cambridge after World War II. Fisher, by contrast, sought to understand the dynamics of enormous arrays of genes in a population, rather than the causal pathways of single genes. He did this by importing into evolutionary biology models taken from statistical mechanics and thermodynamics. That is because Fisher was interested in turning the genetic view of natural selection into a highly general, law-governed theory of evolution that could complete and supplant Darwin's account of natural selection in the new age of post-Newtonian physics. Fisher, that is, tracked the trajectories of genes in the same probabilistic spirit in which Maxwell, Boltzmann, and Gibbs tracked arrays of gas molecules. Just as there is a difference between the microstates in which molecules and their energy levels can be arranged and the overall observable effects, or macrostates, of these various underlying arrangements, so, for Fisher, arrays of genotypes are related to arrays of phenotypes.

In Fisher's own mind, however, he was not simply borrowing models from physics. He was trying to find, or at least pointing the way toward, a theory of natural selection that would unify evolutionary biology and physics by simultaneously expanding physics to accommodate evolutionary biology and reducing evolutionary biology to a comprehensive vision of the cosmic dance (Hodge 1992a). On this basis, Fisher attempted to articulate a view of the human condition, including a large role for positive eugenics, that he explicitly took to be Nietzschean in character, but which, curiously enough, he thought of as Christian as well (Turner 1985). Fisher is the central figure in this chapter, accordingly, not because his theory succeeded but because it is illustrative, on a grand scale, of how the probability revolution, after it had been extended to physics by Maxwell and Boltzmann, and to the Darwinian tradition by Galton, Pearson and Weldon, intersected at last with the Mendelian revolution.

Fisher was an earnest and distinctly uncharming middle-class boy, quite unlike the flamboyant, self-confident, aristocratic Haldane. Fisher recalls that when he came to Cambridge as a mathematics major in 1909, "Darwin's birth and the jubilee [fiftieth anniversary] of the publication of the *Origin of Species* were being celebrated. The new school of genetics using Mendel's laws of inheritance was full of activity and confidence, and the shops were full of books good and bad from which one could see how completely many writers of this movement believed that Darwin's position had been discredited" (Box 1978, 23).

In their famous quarrel, Fisher immediately sided with Pearson against Bateson on a crucial point. As a convinced Darwinian by the time he had reached his teens, Fisher was all for continuous variation if that is what it took to drive natural selection. As a nascent eugenicist and hence a strong hereditarian, however, Fisher was certain a priori that there could be no contradiction between genetics and natural selection. In 1918, accordingly, Fisher wrote a paper entitled, "The Correlation between Relatives on the Supposition of Mendelian Inheritance," in which he showed that Pearson was wrong in thinking that correlations between relatives for various traits, such as height, necessarily contradict Mendel's theory of inheritance. The details of the argument are unimportant. What matters is that it is all a question of how many genes it takes to specify a trait. In general, thought Fisher, it takes a lot, the differences between which are minute and continuous. This was certainly adverse to the stress Bateson and other Mendelians placed on large mutations. It was no less critical, however, of Pearson, since Fisher was proposing that one could have continuous variation, natural selection, and Mendelian genetics all at the same time. Indeed, Fisher mathematically derived Galton's regression curves from deeper Mendelian principles by making continuationist assumptions about genes. This was the beginning of a very rocky relationship between Fisher and Pearson.

What was most novel about Fisher's paper, however, was its author's use of statistical analysis to divide variations due to nonheritable factors, such as environmentally induced modifications to development, from heritable factors, and his further partition of heritable variation into nonadditive and additive components. The nonadditive portions would include the effects of dominance, linkage, epistasis, pleiotropy, and other constraints.[1] What was left after all that was set aside must, when added up, represent the variation in a trait due to the effects of single alleles on which natural selection could act directly, proportionately, additively, and deterministically.[2] Those effects are, admittedly, very small, for each allele that is preserved in a population has only a minute role in determining phenotypic traits. But for Fisher it is precisely under these conditions that natural selection would be most effective in fixing traits.

Fisher had devised his now-standard method of analysis of variance through his work at an agricultural research station, where he was expected to figure out the effects of different fertilizing and watering regimes on plant growth. That involved setting certain factors aside and summing over what was left. Fisher was soon to become one of the great statistical mathematicians of the twentieth century. Much of what is found in statistics textbooks today was first presented in his classic little book, *Statistical Methods for Research Workers* (1925). From the first, however, Fisher's eye was on the evolutionary problem. Fisher was already convinced that everything other than natural selection working on additive inherited variation could be set aside as trivial. It was on this point that Fisher proposed to assume Darwin's gradualist mantle by marrying Darwin's gradualist vision of natural selection to genetics. Not since Haeckel, in fact, had anyone believed so thoroughly in the "all-sufficiency of natural selection."

Fisher's use and development of statistical methods, his willingness to think in terms of idealizations, and his insouciance about complicating factors had all been fostered not only by his eugenics-driven knowledge of Galton, Pearson, and Weldon's biometrical research programs but by a postdoctoral year he had spent at Cambridge in 1912–1913 under the tutelege of the physicist James Jeans. James imbued Fisher with the spirit of Maxwell's statistical theory of gases, Boltzmann's statistical thermodynamics, and the new quantum mechanics, much as von Ettinghausen had imbued Mendel with the spirit of combinatorial algebra. In statistical mechanics, you do not follow the trajectory of each molecule, like some overworked Laplacean Demon, but assume that macroscopic effects like temperature and pressure are the result of averaging over the energy levels of millions and millions of separate atoms and molecules. From this perspective, it may not be impossible that all the molecules of oxygen in this room will bunch up in a corner, leaving one gasping for air. Nor is it impossible that if I put a pan of water to boil on the stove it will freeze instead. For in each case, the movement of each atom or molecule is, in principle, reversible. In both cases, however, the collective reversal of a molecular motion is immensely improbable. Probability theory had in this way licensed fundamental physical conclusions, and the power of physics was extended to classes of systems composed of large aggregates that had hitherto remained beyond its analytical competence. Fisher believed deeply that something like this had to happen for evolutionary theory, that the biometrician's application of the probability revolution to mere phenotypes was insufficient to do the job, and that he was destined to do what was required by working at the level of genotypes.

This spirit of idealization and of looking below the phenomenal surface is evident in a paper Fisher published in 1922 on dominance. There he made explicit his conviction that to do evolutionary theory right, you must treat populations as arrays of genes rather than as groups of visible organisms and should, in turn, treat arrays of genes in the way Maxwell and Boltzmann treated large arrays of gas molecules or, more precisely, the average velocities and energy levels of such molecules. "It is often convenient," Fisher wrote, "to consider a natural population not so much as an aggregate of living individuals but as an aggregate of gene ratios" (Fisher 1922, 340). Accordingly, Hodge talks rightly about Fisher's "two hero history," in which Boltzmann and Darwin "are the authors of the two great probabilistic insights that have . . . set the decisive precedents for all subsequent thinking about the inanimate and animate creation, including man himself" (Hodge 1992a, 242). It was a connection that was probably suggested to Fisher by Boltzmann's own remark about "Darwin's century," amended by Fisher's clear recognition that the twentieth century belonged to Boltzmann.[3] However that may be, Fisher's model of how various factors affect the Hardy-Weinberg equilibrium in his 1922 paper is taken directly from statistical mechanics:

Fisher considers such causes of change as selection and random sampling error insofar as they influence the statistical distribution of gene frequencies. A population is treated as a collection of genes, with each gene having a certain frequency because it is present in a certain proportion of individuals; and it is inquired what the statistical distribution of those gene frequencies is. Thus if that distribution is a Normal distribution, so called, as represented by the familiar bell curve, then many genes will be present in about half the individuals, while only a few will be present in either a great majority or a small minority. . . . Evolution on such a representation can be analyzed as change in the distribution of gene frequencies. For under Mendelian assumptions, the distribution is stable in a large population with random mating, and no mutation, selection or migration. In this paper, Fisher did what no one had done before. He asked how such factors as dominance relations, mutation, selection and random extinction of genes in finite population would affect the distribution; and he devised expressions for the effects of various mutation rates or selection intensities and so on. He hinted at a conviction he would never give up, namely that adaptive evolution is most effectively produced in a large randomly breeding population subject to sustained natural selection of very small heritable differences. (Hodge 1992a, 235)

A chief advantage of this way of looking at the matter is that the very method of representation allowed Fisher to set aside constraints on the power of natural selection working on arrays of separate alleles. Fisher is explicit about this:

Special causes, such as epistacy, may produce departures, which may in general be expected to be very small, from the general simplicity of the results; the whole investigation may be compared to the analytic treatment of the Theory of Gases, in which it is possible to make the most varied assumptions as to the accidental circumstances, and even the essential nature of the individual molecules, and yet to develop the general laws as to the behavior of gases. (Fisher 1922, 321–22)

While Fisher is correct in saying that additive variance can propel adaptive natural selection in such a way that the more alleles there are of a given genotype in a population the more of the corresponding trait there will also be, and indeed that linkage and other constraints on Fisher's model are less directly connected to the rate of selection, Fisher clearly underestimated the degree to which you can have too much of a good thing, or that two good things can make a bad thing. That is because, in his eagerness to brush aside the organism-environment level, Fisher underestimated, among other things, the relativity of fitness to different environments, and cases in which natural selection can operate without adaptive benefit (Lewontin 1978; Sober 1985). The point is that Fisher's physics envy is inseparable from his rather high-handed, and ultimately fatal, desire to minimize context dependency.

Fisher developed this line of thought until, in 1930, he summed up his results in one of the Darwinian tradition's seminal books, *The Genetical Theory of Natural Selection* (Fisher 1930). In consequence, in 1933 Fisher became Pearson's successor in the Galton Chair of Eugenics at London University. Fisher's ascendancy did not come about without a good deal of struggle, most of it with Pearson. Pearson, unalterably opposed to Mendelism in any form, in large part because of his ideological animus against theoretical entities, had refused to publish Fisher's important paper of 1922 paper in *Biometrika*. When it appeared elsewhere, Pearson added insult to injury by taking time to drop a note to Fisher proclaiming, "I am afraid I am not a believer in cumulative Mendelian factors as being the solution of the heredity puzzle" (Box 1978, 82). This has the sound of a man worried by footsteps at his back. Nonetheless, Fisher's technical work in statistics was so brilliant that a year later Pearson offered him a job in the Galton laboratory, on condition that he forget about this Mendelism business and work on Pearson's own biometrical projects. Fisher refused. Soon he sent Pearson yet another paper, provoking a rejection letter full of even more editorial arrogance. "I am regretfully compelled to exclude," Pearson wrote, "all that I think is erroneous on my own judgment" (Box 1978, 83). In the end, however, Fisher's sheer competence eventually got him Pearson's job.

His own competence aside, however, Fisher got the job because the genetic revolution was no longer capable of being evaded by the now aged biometrical school and because Fisher was as devoted to eugenics, the driving force of biometry, as anyone else alive—and had allies who appreciated that fact. Indeed, Fisher proposed his theory of natural selection within the framework of a sort of eugenics that had been familiar in England since Galton, a right-leaning positive eugenics that was more steeply meritocratic than it was either vulgar capitalist, like Davenport's, or utopian socialist, like Muller's. In Fisher's case even more than in theirs, "eugenics was the dog that wagged the tail of population genetics and evolutionary theory, and not the other way around" (Norton 1983, in Grene 1983, 21). Indeed, Fisher's principal supporter and patron throughout his life was Major Leonard Darwin, Charles Darwin's son, who was president of Galton's London Eugenics Education Society. Fisher had met him in 1912 when the younger Darwin came to Cambridge to address the Cambridge Eugenics Society, of which Fisher (as well as Punnett, John Maynard Keynes, and Horace Darwin, another of Charles's sons) was a cofounder. Leonard Darwin saw in Fisher not only a competent mathematician but a man wholly devoted to the Darwinian and Galtonian inheritance and to the eugenic cause. Fisher became a sort of intellectual son to Leonard Darwin. He was pleased to see himself in this light. His connection with Leonard (and Horace) Darwin made him, in some sense, Darwin's grandson.

By temperament, though, Fisher was a very different man from Darwin, and a good deal less easy to like. He was one of those people whose intellectual idealism is made all the more intense by the fact that they lack the emotional capacities that are the normal carriers of a person's values and that allow them to live with others in sympathy and a shared commonsense view of the world. "He grew up without developing a sensitivity to the ordinary humanity of his fellows," writes Fisher's daughter and biographer. "He was unaware of the effects of his own behavor, and often expressed his love ineptly" (Box 1978, 10). Yet Joan Fisher Box also has this to say:

He gave much to friendship. . . . His loyalty was absolute. . . . Having formed a largely intuitive opinion of any man, usually but not always sound, he was fully committed. A similar loyalty bound him to his country, his church and his profession. He was a patriot, a political Conservative, a member of the Church of England . . . loyal to the ideals he perceived in the various establishments. The peculiarity of his blindness to emotional tones was to set him apart as in some sense a difficult person to know, to some natures baffling, to some intolerable, to some "beyond good and evil." (Box 1978, 11)

Fisher had an intense vision of human destiny in which Mendelian genetics, Darwinian evolution, positive eugenics, Boltzmannian physics, Anglican Christianity, England's imperial interests, and, as the last phrase quoted shows, Nietzsche's talk about "supermen" were blended into an extraordinary, and perilously incoherent, compound. The bottom line, however, was eugenics. The entire second half of *The Genetical Theory of Natural Selection* is devoted to problems about the means by which a righteous and disciplined eugenic nation might win in international and intercultural struggles in serene confidence that such things were achievable. Accordingly, Fisher's stress on additive fitness was not only dictated by the dynamical models he borrowed from physics but by the desire of a positive eugenicist to identify and favor particular good-making traits by "doubling the dose" through facilitating good marriages. If context dependency or internal genetic constraints were much of a factor in evolution, the basic idea might begin to seem not only far fetched, but dangerous. (We will see in the following chapter that Theodosius Dobzhansky's appreciation of context dependency is inseparable from his rejection of eugenics.)

In spite of these ominous concerns, Fisher was less of an authoritarian than it might appear. He was not for intensifying imperialism by oppressing subject peoples, as Fascists and Nazis were soon to be, but for strengthening the virtues and talents that would be needed under conditions of rapid modernization, when power was passing from the old, degenerate aristocracy to a new, meritocratic, middle-class establishment. Fisher's most cherished recommendation for public policy, accordingly, was to find ways to fire the professional and commercial classes with voluntary enthusiasm for the healthy, natural, fecund, "eugenic life" that Fisher himself, as well as Galton and the Darwin sons, treated as a virtual religion. Fisher was as firmly set as any Spencerian against providing subsidies for the unfit poor, nor did he mind a little legal sterilization here and there. But what he really wanted was for the state to help support the solid, self-reliant middle class, out of which he himself came, so that they could have more children, from whose work-ethical upbringing and eagerness for competition would quickly spring a new meritocratic aristocracy that would take over the functions of the declining landed aristocracy, whose "hereditary genius" was already pretty well dissipated. Fired with enthusiasm for "the eugenic life," the young Fisher told the members of the Eugenics Education Society, to which Leonard Darwin invited him to speak on numerous occasions, that

eugenicists are the agents of a new phase of evolution. Eugenicists will on the whole marry better than other people, have higher ability, richer health, greater beauty. They will, on the whole, have more children than other people. Their biological type [is] characterized by their solicitude for human betterment, their scientific insight, above all their intense appreciation of human excellence. . . . Absorbing more and more the best qualities of our race, [they] will become fitted to spread abroad, not by precept only . . . the doctrine of a new, natural nobility of worth and birth. (Box 1978, 32)

The problem, Fisher wrote in 1914, was that

the qualities of all kinds, physical, mental and moral which go to make up what may be called "resultant sterility" tend, other things being equal, to rise steadily in the social scale; so that in such a society, the highest social strata, containing the finest representatives of ability, beauty and taste which the nation can provide, will have, apart from individual inducements, the smallest proportion of descendants; and this dysfunctional effect of social selection will extend throughout every class in which any degree of resultant sterility provides a social advantage. (quoted in Box 1978, 31)

Fisher hoped to offset the effect of "resultant sterility" by financially subsidizing commercial and professional families so that they could have more children, an idea that would prove an utterly naive idea in a postwar environment in which the Labour party and the working class was to rise to power.[4] In any case, Fisher tried to live up to his own ideal. "He was the only man I know," wrote one of his colleagues, "to actually *practice* eugenics" (quoted in Norton 1983, in Grene 1983, 19). To this end, Fisher took his young wife to a farm near the Rothamsted Experimental Station, where, after nine years of part-time teaching and other menial work (he had been rejected by the army in World War I on grounds of poor eyesight), he had finally secured his first good job as a statistician. There she was expected to bear children and to support them on the bounty that the well-managed land was supposed to supply. Eileen Fisher had came from the Guinness family of beer and records fame. She was only seventeen in 1917, when she married Fisher. Eventually they had eight children. His wife cooperated willingly in this life-style until much later, noticing that she had missed something, she sued for divorce.

Fisher's claims about the dynamics of natural selection are summed up in the opening chapters of *The Genetical Theory of Natural Selection*. In the first chapter, Fisher puts an end to an old idea by mathematically demonstrating that blending inheritance would so quickly deplete the store of variability that selection, on the basis of Darwin's theory of blending inheritance, could occur only in the presence of improbably high, and mostly damaging, rates of mutation. Mendelian inheritance, on the other hand, with its capacity to retain large arrays of genes in populations, has tremendous amounts of variation to work with even with low, and benign, mutation rates. Fisher interprets the capacity of Mendelian systems to conserve variation as a principle formally analogous to the conservation laws of classical physics. What he wants, however, and in the second chapter proceeds to give himself, is a rate-changing counterpart of his conservation principle, a principle of acceleration formally analogous to Newton's second law ($F = ma$). Fisher wants the rate of natural selection to be equal or proportioned to some other quantity so that differential equations governing the instant-by-instant (generation-by-generation) change of gene frequencies in populations could be written and solved just as handily as in any other mature, highly quantified science. Fisher is working in the tradition of grand physical theory. He will have nothing but predictive laws governing rate changes in measurable quantities. He is a genetic dynamicist. Fisher's fundamental theorem of natural selection proposes that:

The rate of increase in fitness of any organism at any time is equal to its additive genetic variance in fitness at that time.

In the most general terms, Fisher's theorem implies that the rate of selection in a population of alleles is a function of the amount of available independent, cumulative, and therefore additive, genetic variation. Fisher is painting a picture in which natural selection speeds up as usable variation is fed to it. Moreover, he means to say that as natural selection acts on variation, it necessarily does so in such a way that it increases the fitness of a population from what it was at the instant before the integration of the action of selection on the genetic array. (By "fitness" Fisher means comparative reproductive rate, which is presumptively equivalent to relative adaptedness because the rate of reproduction is assumed to be a function of adaptive advantage.) The system moves naturally toward a state of maximal fitness, even if it never quite arrives because as it approaches maximal fitness, it runs, by definition, out of fuel. Thereupon it settles down into Hardy-Weinberg equilibrium until new variation is fed into the system (Price 1972; Hodge 1992a).

There is a good deal of the same sort of thinking in this picture that we have seen in Darwin's theory. As in classical economics, Fisher thinks that maximization leads to constantly renewed equilibrium.[5] Change is gradual over an array of continuous variation. Just as his relationship with Leonard Darwin secures the personal continuity of the Darwinian tradition, so Fisher's commitment to the primacy of selection, to gradualism, and to equilibrium ensures its intellectual continuity. This continuity had been threatened, indeed disrupted, by untenable ideas, such as blending inheritance, and by false dichotomies, such as that between Mendelism and selection. That had happened in part because Darwin's Darwinism had proved incapable of converting the hypothesis of natural selection into a fully adequate theory of natural selection (Gayon 1992). Some of this unproductivity had been underwritten by Newtonian analogies, which now constituted a conceptual drag on the tradition. Darwin, for example, thought of blending inheritance as an analogue of gravity. Galton's program had also failed at crucial points. The biometricians' opposition to the Mendelians' stress on particulate inheritance was driven by a false view that blending inheritance, and hence regression, was at the core of Darwinism. What was needed was a new conceptual framework in which the core of the Darwinian tradition, the theory of natural selection, could be defended and used to solve evolution's main problems, while accumulated liabilities were tossed aside. It is Fisher's change from Newtonian and Smithean to Maxwellian and Boltzmannian models, we believe, that more than anything else allowed this to happen. At the same time, it soon became clear enough that Fisher's proposals carried too much baggage of their own. In particular, Fisher's desire to resist the saltationism of most early genetics, and his desire to get back to the authentic Darwinism, provoked him to treat gradualism as part of Darwinism's core, or at least to conceive of gradual natural selection in a way that was too dependent on additive variation.

It was around the idea of equilibrium that Fisher's crucial emendations were centered. Like Darwin, Fisher puts a good deal of stress on this idea. His conception of equilibrium, however, differs from the Newtonian ideal of an earlier age because it is worked out in terms of a different model, a model derived from statistical mechanics and thermodynamics. Millions of alleles, differing slightly from one another and capable of determining slightly differing macroscopic traits by entering into various combinations, are portrayed as moving in an excited field, in which small differences between variant genetic combinations are exploited as the system slides down toward a state in which no more work will be done, that is, in which no more selection will take place. That makes Fisher's law of natural selection look less like Newton's (second) law of acceleration, which governs what happens when force is applied to a mass point, than like Boltzmann's statistical version of the second law of thermodynamics, which governs what happens when energy gradients are turned into work in an array of molecules. Equilibrium is imaged as a balance of forces against a background of inertial motion. It is conceived as a point to which a system composed of a large array of entities spontaneously moves when it can no longer do any work, when there are no more gradients to exploit, when all potential energy has been converted to kinetic energy. Fisher's equilibrium is equilibrium conceived in terms of a physics of energy rather than a physics of forces.

But there is more to it than that. Equilibrium is not only the point in the history of a system when no more energy can be converted into work but the point when something called entropy has reached a maximum. According to classical thermodynamics, the second law of thermodynamics dictates that as energy is utilized to do work, some other quantity, entropy (from Greek *en* + *tropos*: "turned inward," rather than outward, toward real work) is necessarily maximized. Phenomenologically considered, the second law implies that no energy-utilizing machine is perfectly efficient. Even though total energy is conserved, some energy is dissipated as heat in the very process of doing work, and so does no work itself. More generally, entropy is a measure of the disorder created when the energy in molecules has been used as equilibrium is approached. It is supposed to be constantly increasing in the universe, "striving toward a maximum," as the physicist Rudolf Clausius put it. Boltzmann provided a profound explanation of entropy and of the deep reality of distinctly non-Newtonian irreversible processes in nature, by using probability theory.

It was on this model that Fisher predicated the more basic parts of his theory. In Fisher's law, "fitness" is supposed to be maximized like entropy. Immediately after stating the law, in fact, he goes on to say:

It will be noticed that the Fundamental Theorem . . . bears some remarkable resemblances to the Second Law of Thermodynamics. Both are properties of populations, or aggregates, true irrespective of the nature of the units which compose them; both are statistical laws; each requires the constant increase of a measurable quantity, in the one case the entropy of a physical system and in the other the fitness . . . of a biological population. . . . Professor Eddington has recently remarked that "The Second Law of Thermodynamics . . . holds . . . the supreme position among the laws of nature." It is not a little instructive that so similar a law should hold the supreme position among the biological sciences. . . . Entropy changes lead to a progressive disorganization of the physical world, at least from the human standpoint of the utilization of energy, while evolutionary changes are generally recognized as producing progressively higher organization in the organic world. (Fisher 1930, 36–37)

We have little doubt that Fisher's appeal to this model lets him solve some problems that older forms of Darwinism could not, or at least to remove some of the accumulated clutter that had stalled the problem-solving prowess of the Darwinian tradition. Whether Fisher's Maxwellian and Boltzmannian reconceptualization of natural selection helped advance his vision of the human condition, however, is far more doubtful. Any assessment of what Fisher accomplished or failed to accomplish by recasting the Darwinian theory of natural selection in these terms depends, however, on first having in hand a bit more knowledge of the conceptual framework within which he was working. This means resuming our story of the probability revolution at the point at which physics got in the act. We will then be prepared to sum up Fisher's successes and failures.

By the end of the nineteenth century, Newton's luminous explanation of the system of the world had been honorifically retired as an exemplar of great physics. Two new, but closely related, paradigm cases had taken its place: Maxwell's reduction of the phenomenological gas laws, relating temperature, pressure, and volume to statistically calculable collisions between millions of molecules, and, hard on its heels, Boltzmann's reduction of thermodynamics to more or less probable arrays of molecular motion. These were achievements of the 1860s and 1870s. The culmination of this new research tradition was the quantum mechanics of the early twentieth century, in which the statistical properties of ensembles utilized by Maxwell and Boltzmann were pushed even further down into the structure of the universe. At this point, it appeared that the continuity that Maxwell and Boltzmann had sought with the earlier Newtonian tradition, a continuity that was to be achieved by changing to a new ontology, had actually resulted in the collapse of Newtonianism and the reorientation of statistical mechanics around the deeper indeterminism of quantum mechanics. Even in Boltzmann's time, physics was beginning to address the fact that the world contains irreversible processes that cannot be rolled backward and forward like a movie in the fashion Laplace assumed. By the time of Werner Heisenberg and Niels Bohr, however, it appeared that the reason for that was that at rock bottom the world is governed in significant measure by laws of chance. There was no longer any question that God played dice. The only question was the name of the game. Howls of pain were heard from classicists like Einstein, who, like Herschel before him, could not believe that chance had such a good hand.

Fisher, under Jeans's influence, regarded the probability revolution's crowning achievement, the indeterminacy of quantum mechanics, as confirming and deepening what Maxwell almost knew: that irreversible systems are in fact irreversible, that statistical systems are inherently statistical, and that probabilities are objective features of the world rather than subjective states of mind (Hodge 1992a). This does not mean that Fisher was anachronistically pushing quantum mechanics back into the nineteenth century, or certainly that his own theory of evolution was stochastic rather than deterministic. It is simply that, in the light of modern physics, Maxwell and Boltzmann had at last been delivered from their own residual hankerings to posit hidden laws that would make irreversible systems reversible or statistical systems subject to the calculative imperialism of Laplace's Demon. What was new was that Fisher and his contemporaries were now in a position to appreciate what had happened, to stress Boltzmann's and Maxwell's breaks with classical physics rather than their continuities, and so to deliver evolutionary theory from a limited Newtonian matrix into a more liberating and theoretically powerful one. It was in these terms that Fisher looked to Maxwell and Boltzmann for guidance in bringing the probability revolution in physics to bear on evolutionary biology:

Perhaps the most dramatic development was when Boltzmann restated the second laws of thermodynamics, the central physical principle with which so many of the laws of physics are interlocked, in the form that physical changes take place only from the less probable to the more probable conditions, a *form of statement which seemed to transmute probability from a subjective concept derivable from human ignorance to one of the central concepts of physical reality.* More concretely, perhaps, we may say that the reliability of physical material was found to flow, not necessarily from the reliability of its ultimate components, but simply from the fact that these components are *very numerous and largely independent.* (Fisher 1932, quoted in Hodge 1992a, 253, italics added)

The roots of the revolution in physics led by Maxwell and Boltzmann lie at the intersection between the probability revolution in the human sciences and the concomitant elevation of the concept of energy to primacy in physics. By the middle of the nineteenth century, energy was becoming the organizing concept in that discipline. It was an old pattern, which we have had occasion to observe before in the development of physics. What was once treated as an inviolable constant suddenly becomes a variable. It is not force *simpliciter* that is conserved, as with Newton, but energy. Helmholtz thought that life is simply chemical reactions and that chemical reactions are themselves "nothing but" instances of physical laws because he, as well as James Joule and several other physicists in the 1840s, had laid it down that energy, which drives biochemical systems, is a conserved quantity in all physical and chemical interactions. In 1854, William Thompson (Lord Kelvin) told the BAAS that Joule's work on the equivalence of work and heat (1847) "leads to the greatest reform that physical science has experienced since the days of Newton" (quoted in Harmon 1982, 58). For Lord Kelvin, "Energy had become . . . the primary concept upon which physics was to be based, and his generalized use of the concept to apply to all phenomena of physics expressed this primacy. The fundamental status of energy derived from its immutability and its convertibility, and from its unifying role in linking all physical phenomena within a web of energy transformations" (Harmon 1982, 58).

Just as the new social sciences of the nineteenth century did not really get underway, however, until, at the hands of Quetelet, Buckle, and Durkheim, the probability revolution met an "avalanche of numbers" flowing from the statistical bookkeeping and head-counting practices of modern states, neither did the energetics revolution in physics achieve its full promise until it intersected with the methods of statistical analysis pioneered by the social scientists (Porter 1986; Hacking 1990). In a culture in which it is often taken for granted that science progresses by extending its analytic prowess from the "hard" natural to the "soft" social sciences, it is sobering and instructive to bear in mind that statistical analysis was the gift of the emergent social sciences to what were regarded as already mature natural sciences. Admittedly, this was a gift handsomely repaid, for not only did natural scientists, including Fisher, significantly advance the tools of statistical analysis, which were then put to good use by social scientists, but the very fact that the hardest of the hard sciences was soon rendered even more mature by employing statistical laws made it possible for the social sciences, stuck with probable generalizations and cases, to acquire a legitimacy they might otherwise not have enjoyed and to undergo the explosive development they have had in this century. That is what was set afoot by the work of Maxwell and Boltzmann.

The site at which the crucial developments took place was the mid-nineteenth-century energetics revolution. That revolution began with an intensification of the interest of physicists in heat flows, and with what Sadi Carnot called "the motive force of fire." This interest was driven by a compelling desire to improve the efficiency of steam engines. The industrial revolution was in full swing. Since heat was at that time assumed to be a quantity ("caloric") that spontaneously flows from one body to another in order to maintain equilibrium, heat flows were initially thought to behave like, or even to be an instance of, fluid motion. Under these assumptions, Carnot, a French engineer, provided a mathematical description of the behavior of heat in a steam engine as it goes through its various phases. The engine works entirely because of temperature differences in four phases. It can never be totally efficient because heat is lost in this cycle. Carnot still assumed that heat was "caloric," a conserved quantity that flows like a fluid. Early in the new century, however, Count Rumford demonstrated that heat is not like fluid flow at all. Instead, it is matter in motion, the excitation of a mass of particles moving chaotically, and not coherently like a river. In 1847, Joule helped to integrate this understanding of heat into classical physics by casting the entire subject in terms of the concept of energy. It was this notion that excited Lord Kelvin (Smith and Wise 1989). Energy is divisible into potential and kinetic energy. Water at the top of a dam is potential energy; water falling over the spillway is kinetic energy. Work, then, is a process in which one of these forms of energy is converted into the other. The total amount of energy in such conversions, however, remains constant, since at each instant the amount of potential energy given up is picked up by the energy of motion. Thus, the principle of conservation on which physics depends was transferred from momentum to energy, yielding a much larger conceptual framework.

The culminating moment in the energetics revolution was Helmholtz's generalization of the postulate of interconvertibility to include all kinds of energetic transformations, and not just heat. Helmholtz argued that although energy is interconvertible, it cannot be created or destroyed.

Thus, the total energy of an isolated system, one that cannot exchange matter or energy with its surrounding environment, will remain constant, however that energy is apportioned within the system, whether into heat or work. This principle has become known as the first law of thermodynamics:

The total energy of a closed and isolated system is conserved; the energy of the universe, closed and isolated as it is, is constant.

From the emerging perspective of energetics, Joule and Clausius were able to show that Rumford's conception of heat as matter in motion meant that heat *is* kinetic energy, billions of atoms moving around and bumping into each other like the Keystone Kops—and being unable to do useful work precisely because of this fact. Clausius went on to claim that the kinetic way of looking at things made similar sense of the macroscopic laws governing pressure, temperature, and volumes of gases. The pressure of a gas, he said, is due to the random collisions of gas molecules against the walls of the containing vessel. An increase in both pressure and temperature reflects the intensification of this bombardment.

It is at this point that Maxwell, and the statistical or probability revolution with him, enters the picture. Maxwell's problem was to reconcile quantitatively the "phenomenological" laws governing the volume, pressure, and temperature of ideal gases, which went back to Boyle, with the fact that heat is (nothing but) random molecular motion. The problem is that a given quantity of gas consists of an array of separate molecules and motions so vast and heterogeneous as to defeat the skill even of Laplace's Demon in keeping track of them. Just as the social scientist can average over peoples' independent, chaotic actions, however, thought Maxwell, who was fresh from an intense bout with the vexed issues surrounding Quetelet and Buckle, so the physicist might average over the motions of myriad molecules of a gas as they move and collide with one another with no cost to predictability or the reliability of deterministic equations. In particular, the physicist could identify by probabilistic reasoning the average or "mean kinetic energy" of the molecules, and from this average derive the phenomenological gas laws. Maxwell wrote: "The modern atomists have adopted a method which is, I believe, new in the department of mathematical physics, though it has long been in use in the section of statistics. The data of the statistical method as applied to molecular science are sums of large numbers of molecular quantities. *In studying the relations between quantities of this kind we meet with a new kind of regularity, the regularity of averages*" (Maxwell 1890, 2:373–74, quoted in Gigerenzer et al. 1989, 62, italics added).

Maxwell did his work in steps. In papers published in 1859 and 1860 the Scottish physicist, who in 1871 became the first professor of experimental physics at Cambridge, freed himself from Clausius's assumption that all molecules in a container of gas would have the same velocity and would simply be traveling in random directions. He proposed a statistical formula for the distribution of the velocities of the molecules in a gas at uniform pressure at a given temperature T. That meant a redefinition of the energy levels of these particles. For kinetic energy, the energy of a system in action, is (and since the mid-eighteenth century had been) defined as $1/2mv^2$, that is, one-half of the mass multiplied by the square of the velocity. For Maxwell, temperature was, by this means, reconceived as the average or mean kinetic energy of the constituent molecules. Although the distribution of kinetic energies that Maxwell obtained was not identical to a Gaussian curve (the energy curve is skewed), he was sufficiently emboldened to keep working along these lines. By 1867 he had extended his derivation to include collisions between the molecules of the gas themselves, and not just between the molecules and the walls of the container. He thereby provided a causal basis for the changes in energy distributions in the statistical ensemble of the gas. When two molecules collide, they exchange energy. One may leave the scene of the crash more energetic or charged up, while the other may have lost energy. These energy exchanges average out.

Much of this creative thinking originated in contexts we have already encountered. Maxwell had read and written his way through the crisis of *Quetelismus*, as intensely as anyone else and a good deal more insightfully than most. Indeed, the same 1850 *Westminster Review* article by Herschel on Quetelet that Darwin read seems to have inspired statistical mechanics (Porter 1986, 118). On one crucial point, however, Maxwell took issue with Herschel. Herschel praised Quetelet for recognizing the extent to which both natural and cultural phenomena conform to Gaussian distributions of data. At the same time, Herschel's enthusiasm was constrained by his unwillingness to admit any interpretation of these facts that was inconsistent with his subjective interpretation of probabilities. Maxwell now recognized, however, that the very patterns that social scientists like Quetelet and Buckle were discerning in social reality also yielded beautiful models of the microstructure that lies beneath well-known macroscopic physical laws! Averaging that is this pervasive, and at levels of nature as deep as basic mechanics, suggested to Maxwell that the old "ignorance interpretation" of probabilities, which was demanded by the assumptions of Laplacean and Herschelian Newtonianism, was now so useless as to be wrong. For certain classes of phenomena, statistical averaging is what we will call a required form of representation. If you do not represent the phenomena this way, you do not see them at all.

This was, to put it mildly, an important moment in the history of thought and certainly one of its great intellectual pirouettes. The writ of the "law of errors," rather than being restricted to subjective measurements of error, actually runs in the opposite direction, even if not to the opposite extreme. This extension of the explanatory range of physics by using statistical analysis led Maxwell to stand on its head the old demand that would-be quantitative social sciences must conform themselves to the Galilean, Cartesian, and Newtonian models of the hard natural sciences. On the contrary, the natural sciences must now inscribe the representational resources of the quantitative social sciences inside their own domain. "Doubtless it would be too brave," writes Porter, "to argue that statistical gas theory only became possible after social statistics had accustomed scientific thinkers to the possibility of stable laws of mass phenomena with no dependence on predictability of individual events. Still, the actual history of the kinetic gas theory is fully consistent with such a claim" (Porter 1986, 114).

By the same token, Maxwell thought that if Herschel had not been objectivistic enough about statistics, Quetelet and Buckle had been no less shortsighted in implicitly importing the outdated metaphysical determinism of the old physics into the social sciences, thereby instigating all sorts of unnecessary and confused debates about statistics and free human actions. Maxwell, who thought he was extending the range of classical physics rather than replacing it, certainly held that at some very deep, purely theoretical level it was possible, perhaps even required, for a demon who operated in his statistical world rather than in Laplace's world—a demon who has ever since been called Maxwell's Demon—to interfere with the expected distribution of molecules in gases, to reverse the irreversible, in short, to save classical mechanics by frustrating statistical mechanics. "The statistical method," Maxwell conceded, "involves an *abandonment of strict dynamical principles*" (Maxwell 1890, 2:253, quoted in Porter 1986, 201, italics added). Nonetheless, statistical averages conform to laws and equations deterministic and predictive enough in practice to render more basic forms of determinism unnecessary in theory. Indeed, Laplacean determinism, and the metaphysical determinism with which it intersects, now threatened to stand in the path of inquiry, both natural and social, by casting too cold an eye on the explanatory power of statistical mechanics. To save classical mechanics and to extend its scope did not mean, then, that we—and perhaps even a Laplacean or a Maxwellian Demon himself (Sober 1984a)—can freely dispense with statistical methods. They are necessary if we are even to see phenomena and processes that cannot be seen in any other way, and therefore to explain them. "In the present state of our knowledge," Maxwell wrote, statistical averaging is "the *only available method* of studying the properties of real bodies" (Maxwell 1890, 2:253, quoted in Porter 1986, 202, italics added). What must theoretically and ontologically be presumed to be reversible, accordingly, is in no way practically or epistemically so.

This meant to Maxwell that the entire debate about human freedom that had been unleashed by Quetelet, or at least by Buckle's interpretation of Quetelet, was laboring under a false assumption and that human affairs need not remain under siege from the reductionism ("x is nothing but a y") and eliminationism ("since x is true, y does not even exist") of an outdated natural science. The more he thought about it, in fact, the more Maxwell inclined to the view that the metaphysical excesses of the materialistic and reductionistic matrix within which the scientific revolution was born were no longer required for its further development. In fact, "If the scientific doctrines most familiar to us had been those which must be expressed in this way [statistically], it is possible that we might have considered the existence of a certain kind of contingency a self-evident truth, and treated the doctrine of philosophical necessity as a mere sophism" (Maxwell 1890, 2:253, quoted in Porter 1986, 202).

Maxwell was not arguing, to be sure, that there is room in the world for the radical causal indeterminacy of Augustine's, Descartes's, and Kant's spiritual *liberum arbitrium.* That desperate idea was as incoherent and unnecessary as the causal determinism that had provoked it. Maxwell was arguing, along with other Victorian advocates of what has come to be called soft determinism, such as George Eliot and John Stuart Mill, that our intuitions about human freedom are not so inconsistent with basic physics as to be illusory and, more positively, that our sense of freedom and responsibility can be reconstructed in terms of the field of gentle irregular influences, including those of our own prior actions, that surround us and impinge on us. (Maxwell's brief speculations on these matters might well be called a "field theory of freedom.")

Boltzmann took up statistical mechanics where Maxwell left off. His work is at least as revolutionary as Maxwell's because, in using statistical methods to explain the second law of thermodynamics, Boltzmann showed that nature is full of processes that are inherently (if only probabilistically) irreversible. Indeed, by giving a powerful statistical explanation of the second law, Boltzmann deprived himself of the comfort that Maxwell still took in the thought that reversibility is in principle possible and that irreversible phenomena are in some sense subjective illusions (Porter 1986, 208).

The problem of irreversible thermodynamics is this: Carnot's analysis of steam engines had shown that it was impossible to construct a device that would do nothing but cool a body at one temperature and heat another at a higher temperature. When you put a piece of ice in a glass of warm water, you never cool down the ice more and warm the water up. When you put a pan of cold water on a stove and ignite the gas, the water never starts to freeze. In sum, you never see heat flowing from a cold body to a hotter one but only from a hotter body to a colder one, even though there is nothing in Newtonian mechanics that prohibits it. Yet this is a very deep prohibition. The question was why.

In 1852, Kelvin put forward the idea that for closed systems—those that permit exchanges of energy but not of matter with the outside world—a portion of the total energy is dissipated as heat, and thus is unavailable to perform work. The analysis of heat as molecular motion made it easier to see why this is so. The particles are not, as it were, lined up in a sufficiently coherent way to do anything useful. There is always a dissipative loss. This is the reason that there is no such thing as a perfectly efficient machine or a machine that produces more energy than it uses. In 1865, Clausius gave this dissipated form of energy a name. He called it "entropy," using S as a symbol to denote it. Any system that shows no change in entropy ($S = 0$) is taken by definition to be in "thermodynamic equilibrium." For any real process in which work is done, on the other hand, there will be an increase of entropy ($\Delta S > 0$), and so a decrease in its further capacity to do work. Thus Clausius arrived at the second law of thermodynamics, which states: The entropy of an isolated system never decreases; the entropy of the universe strives to a maximum. Change in entropy, ΔS, was defined as Q/T, where Q measures the heat absorbed by a system at temperature T. Later, Q was taken as a measure of the amount of energy degraded into a less accessible form. As a system approaches equilibrium, where no further changes in T will occur, the entropy of the system increases to the maximum available under existing boundary conditions. There can be no further dissipation within the system. Taking the universe as a whole as an example of such a closed and isolated system, Kelvin concluded that the ultimate fate of the universe would be a "heat death" in which everything comes to equilibrium at some constant temperature. There would be no further heat flows and no energy gradients by means of which work could be done.[6]

Boltzmann used Maxwell's statistical approach to explain why the second law carries the deep prohibitions Clausius and Kelvin said it did. From Maxwell, "Boltzmann knew the average velocity of the molecules. . . . But many of the molecules were, of course, moving much more slowly than the average, others much faster. Boltzmann wanted to know what proportion of them were moving at, say, 1/2 the average velocity, what proportion at 4/3 the average, and so on" (Kuhn 1987, 16).

To answer his question, Boltzmann represented all the possible energy states as little cells. (He was inspired by Abraham De Moivre, who had worked on probabilities in card games.) A cell defining a very high energy state must constrain the distribution of the others if the same overall average kinetic energy is to be maintained. Thus, it is highly unlikely that most of the molecules will be bunched up into a few high-energy cells and that all the rest will happily cooperate by concentrating themselves into low-energy cells. It is much more probable that the energy states will explore the entire range of possibilities. (This is called an *ergodic* system.) Thus there are many different ways in which the same energetic macrostate, or overall energy state, can be achieved. Some of these, however, are intrinsically more probable than others.

Consider a simplistic but helpful example of a system made of three particles, each of which can be in either one or the other of two compartments or cells. There are eight possible ways in which the particles can be distributed. Two of these are macrostates in which all three particles can be in one compartment or the other. The other six possibilities are ways in which one particle can be in one compartment and two in another. Thus, the probability of a microstate that puts all the particles in one compartment is $2/8 = 1/4 = .25$. Similarly, the probability of a microstate in which two particles are in one compartment and one in another is $6/8 = 3/4 = .75$. It is thus three times more likely that the particles will be spread out into both compartments. From the point of view of the macrostate, however, it makes no difference which of these two or six possibilities is instantiated. As the number of possible microstates corresponding to a given macrostate increases, the macrostate becomes increasingly degenerate, in the sense in which a code or a language is degenerate when it contains multiple, and thus ambiguous, ways of encoding the same information. Boltzmann called this measure of degeneracy W. He then defined entropy, considered as the spread-outness of energy states of the molecules, as $S = k \ln W$, where $\ln$ is the natural logarithm ($= 2.303 \log_{10}$), k is a constant named in Boltzmann's honor by Max Planck, and W is the probability of a macrostate. Thermodynamic equilibrium was now reconceived, even redefined, in these terms as that configuration of molecules that has the highest probability at temperature T. That is precisely the state of the system in which no more work can be done without opening the system up to the outside world. At equilibrium, the entities in the system will have accessed the highest number of possible arrays. On this view, the tendency of a closed system to move toward equilibrium with the passage of time, which is what Clausius's version of the second law implied, is simply *the inherent and spontanous tendency for its energy states to assume the most probable distribution.*

Boltzmann's explanation accords well with our everyday experience. Air molecules remain dispersed in a room. They do not suddenly rush up to one corner or form a crystal. Although the trajectories of the gas molecules are indeed reversible, each molecule having a fifty-fifty chance of being positioned in the left or the right half of a containing vessel, the population of molecules as a whole will be distributed close to expectation over both chambers and will not spontaneously move only to the left or the right. The power of Boltzmann's explanation will be missed, however, unless we forsake our highly idealized little example of three molecules and two configurations and consider instead just how many molecules in how many different configurations there are in about a gallon of some gas. We are not talking here about following a few molecules or collisions. Rather, we have a very large number, such as 10^{23} (1 followed by twenty-three zeros) molecules going to a much more probable state. To comprehend the magnitude of 10^{23} two comparisons might help. The U. S. national debt was about 10^{13} dollars in 1989. The estimated lifetime of the universe since the big bang is 10^{12} years (10 to 20 billion years) or 10^{17} seconds. If all the molecules of air suddenly rushed into one corner, that would be like tossing a fair coin and getting heads 10^{23} times in a row. It might happen, but you would not want to bet on it.

A large and speculative implication of this analysis is that it is possible to say that time is an arrow pointing in the direction of increased entropy. This idea begins to explain the objective reality and irreversibility of time, a phenomenon so baffling to Newtonian philosophy of science that its apparent unidirectionality is often treated in classical physics as a subjective illusion. It is even possible that in some way, time itself is a function or expression or effect of the second law (Prigogine and Stengers 1984; Denbigh and Denbigh 1985; Coveney and Highfield 1991).

Although Boltzmann recognized and hence "saved" irreversible phenomena, his own emphasis was on how he had saved classical mechanics from contradiction with basic facts about the world. Boltzmann did not challenge the classical postulate that each collision is in principle reversible. Indeed, he was much more of a classical mechanist than Maxwell. He was not tempted, as Maxwell was, to use the new physics to back off from Laplacean dreams and to leave room for traditional views about human actions and values. About such matters the chips would have to lay where they fell. Accordingly, Boltzmann argued that the probability of systemwide reversals in thermodynamic phenomena is so low that it amounts to a physical impossibility, rather than the mere practical impossibility that Maxwell took it to be. Therefore, although he was living in a world where phenomenalistic, positivistic, and instrumentalistic interpretations of physics were so prevalent that almost everyone took atoms to be purely theoretical constructs, Boltzmann became a convinced scientific realist. If irreversible physics rested on the postulate of atoms, the inescapable nature of irreversible phenomena was almost sufficient in itself to entail that atoms must be real entities rather than theoretical constructs.

To get some notion of the difficulties Boltzmann faced in taking a position that to us seems commonsensical, we should recognize that positivistic attitudes, and in particular phenomenalism, often become fashionable in the philosophy of science at times when there seems to exist a clash between what science tells us and our other embedded beliefs about the world, when the best theories in different sciences seem to contradict each other, and when equally indispensable, but apparently inconsistent, assumptions, such as the wave-particle duality in the theory of light, seem required within a single science. Copernicus's heliocentrism, for example, with its apparent conflict with both the Bible and common sense, survived its first half-century under the understanding that it was just an elegant device for calculating orbits—until Galileo upset the apple cart by giving it a realistic interpretation. In the later nineteenth century, similar appeals to phenomenalism blunted conflict between classical physics, with its assumptions of reversibility, determinism, and decomposibility, and our deep intuitions that time and many other processes are irreversible, as well as our intuitive conviction that free choice and responsibility, however they are interpreted, are real. Positivists and their cousins the pragmatists correlate progress in science with a willingness to push aside the metaphysical conundra that come from taking a realistic rather than an instrumentalistic view of these conflicts. For them, metaphysical dogmas about the ultimate nature of reality, and the permanent framework within which science must be interpreted, elevate what are merely temporary conflicts into permanent contradictions, creating unnecessary hang-ups and putting roadblocks in the way of scientific discovery. One may have a variety of background beliefs and be attracted to this view. Among its most eminent advocates, for example, was a succession of French Catholic physicists from Ampère to Duhem, who adopted phenomenalistic and positivistic ideas about science not to advance Comte's secularizing ideological program, but, on the contrary, to preserve their Roman Catholic beliefs about the nature of reality. Appearances, accessible by scientific method, are one thing. Ultimate reality, accessible by religious faith, is another.

It was in such an environment that Boltzmann had to defend his realistic interpretation of atomism and the molecular approach to gases, statistical mechanics, and statistical thermodynamics. Resistance to his statistical explanation of entropy became entangled with Boltzmann's failure to get his colleagues, enmeshed in such matters as the wave-particle duality, to take a philosophically realistic view of science generally. Other people, such as the ardent positivist Ernst Mach, were much better than he at defending their views. Eventually Boltzmann came to feel that his life's work had been futile. Always prone to manic-depressive cycles, he sank into a deep depression and committed suicide in 1906. Ironically, the year before, Einstein had developed a convincing statistical atomistic theory to explain the observable fact of "Brownian motion," the constant random motion of small particles in fluids. Several years later, when Einstein's interpretation was confirmed, skepticism about the reality of atoms dissipated. At about the same time, Planck and other creators of the quantum revolution began to use the little cells into which Boltzmann partitioned energy levels for purely analytical purposes to analyze "quanta" or packets of energy—and to show that their distribution is governed by objective laws of chance rather than subjective estimates and that light could be, in some sense, both particulate and undulatory (Kuhn 1987).

The epitaph on Boltzmann's grave in the Central Cemetery in Vienna is still there: $S = k \log W$.

By extending the probability revolution down into the depths of the natural world, albeit in different ways, Maxwell and Boltzmann were trying to protect classical physics by incorporating a range of phenomena hitherto resistant to its basic principles. Whether they thought in these terms or not, Maxwell and Boltzmann did this by changing the received ontology of physics into an explicitly statistical one. What we have been saying about the Darwinian tradition, therefore, is visible in physics as well. To protect the integrity of a scientific tradition, ontological revision is sometimes called for; it is, however, risky, for ontology can just as easily facilitate research that magnifies the very conflicts it was intended to resolve. It is here that the trajectory of the Darwinian tradition departs from that of classical physics. For while statistical analysis led ultimately to the demise of classical physics, it greatly helped revive the fortunes of Darwinism in the twentieth century.

In large part, statistical mechanics and thermodynamics helped heal the tears in the Darwinian fabric because they helped genetic Darwinians think of equilibrium in a way that could apply to genetic arrays. Boltzmann and Maxwell proposed a new conception of equilibrium by providing physics with a new ontology. In the old Newtonian model, a system is typically composed of two bodies, where "body" stands for any distinct and invariant mass point on which exogenous forces act: a Newtonian planet and the sun, a Smithean buyer and a seller, a Darwinian finch and its cousin who compete for the same bits of seed but whose beaks may be shaped just a bit differently. The paradigm of equilibrium is a balance among the forces that define the position and momentum of the two bodies. Equilibrium follows the path of "least action" (mass × velocity × distance). It is threatened and restored at every instant. In Maxwell's and Boltzmann's models, on the other hand, the entities countenanced are vast, heterogeneous populations, nonadditive and nonproportionate effects of which are presumed to wash out by way of the law of averages. It takes statistical representation even to recognize the systems in question as systems and to explain their dynamics. Equilibrium in such a system can still be represented as a condition of resolved forces, even though the position of each particular entity in the ensemble is determined by chance, in the (limited) sense that that entity might just as well (equiprobably) be in one position or cell as in another. But equilibrium can more interestingly be thought of in a way that reflects the greater sense of temporality and process in a Maxwellian or Boltzmannian world, namely, as the end-point of a process of change from which all potential for further change has been drained. (This is certainly how the American chemist Josiah Willard Gibbs saw chemical equilibrium.) In the Boltzmannian model, equilibrium is a state toward which the system will spontaneously tend as it rids itself of asymmetries in energy distribution by flattening out gradients.

Darwin's Darwinism, as we have already seen, was shoehorned into the Newtonian model with the greatest of difficulties. This fact partly explains the subsequent attraction of endogenous developmentalism in evolutionary thought, for this tradition preserved the phenomenal irreversibility and directionality of life, although at the cost of ignoring or even flatly contradicting what classical physics tells us about the world. The introduction of irreversibility into the physical world in the new physics helped dissipate this tension. Although Darwin's thought is sometimes allusive to the work done by Carnot's heat machine, he did not have in hand Boltzmann's analysis of the irreversible physics of heat machines. Appeal to the new statistical way of thinking would have put this analogy in a different light. It would have stressed the inevitability of entropic dissipations, and so would have set free Darwin's intuitive commitment to the irreversibility of lineages from the Lyellian cycles into which it was continually pulled back. The new physics would have allowed Darwinism to find a more congenial conceptual home than the one it was born in and in which it was forced to seek scientific legitimacy.[7]

Maxwell himself had no interest in using his models to update Darwinian evolutionary biology in this way. Indeed, Darwinism of any kind seems to have bothered him. Pearson tells the following anecdote: "The conversation turned on Darwinian evolution. . . . I spoke disrespectfully of Noah's flood. Clerk Maxwell was instantly aroused to the highest pitch of anger, reproving me for lack of faith in the Bible. I had no idea at the time that he had retained the rigid faith of his childhood, and was, if possible, a firmer believer than Gladstone in the accuracy of *Genesis*" (quoted in Porter 1986, 200). Such scruples not standing in his own way, Fisher took it upon himself to rework Darwinism by using Maxwell's and Boltzmann's models and methods over arrays not of molecules of gases but of genotypes in populations. In a rough and ready way, Fisher's fundamental theorem of natural selection is true enough. In nature, all other things beings equal, the rate of selection does roughly go up as additive variation is available.[8] That may be enough to sustain Fisher's analogy between the laws of physics and the laws of evolutionary biology. In Fisher's mind, however, the fundamental theorem is not satisfied by rough qualitative correlations and striking analogies. His is, for one thing, a strictly quantitative law, each term of which has a technical meaning that needs to be explicated. By "gene" Fisher means an allele. A gene is fitter than another if and only if it is gaining in proportional representation within a population in comparison to one or more alternatives. An allele at a given locus may have a .5 representation in the population at t_1. At t_2 it may have .6 or .7, or may fall to .4. The phenomenal or phenotypic explanation of increased representation is presumably that individual organisms carrying this gene will, all other things being equal, have more viable offspring because they are flourishing in this environment.[9] Hence, the gene itself will have greater representation in the next generation and will therefore be fitter. Fisher, however, prefers to represent the whole affair in terms of how arrays of genes are behaving over time. He likes to work in genotype space. We are now invited to assume that there is a more or less steep gradient of variation in fitness corresponding to the range of conditions that make one allele do better than another. As long as there is any variation in fitness, Fisher is saying, selection will take place. Indeed, Fisher's definitions are reversible. What he means by natural selection is what happens when there are fitness gradients. The rate at which this process will take place, moreover, is measured by the height of the fitness gradient. Fisher says this can be measured.

Variation in fitness, that is, reproductive success, is itself variable. It shows what is technically called variance, that is, degrees of variability or spread. It is, for instance, intuitively clear that while 5 is the mean between both 10 and 1 and 6 and 4, there is more variation between 1 and 10 than between 6 and 4. Nine integers (intervals) stretch between 1 and 10, while there are only two between 6 and 4. There is, then, greater variance between 1 and 10 than between 4 and 6. Fisher proposes to measure genetic variance in fitness by using this yardstick. When he says he is looking at the "additive genetic variance in fitness," he means that he is setting aside the part of overall variance that is due to nongenetic factors, and indeed looking only at the portion of genetic variance that is contributed by each gene separately and whose effects are, like the efforts of rowers in a boat race, cumulative. Add up the additive genetic variance of natural selection working at all loci, divide to get the average contribution of each gene, and distribute this across the entire interbreeding population. The result will be a measure of the rate of natural selection for that entire population. For Fisher this is equivalent to the rate for all the populations that constitute a species, which, in a peculiarly unfortunate use of terms, he refers to in his law as "any organism" (i.e., the total population of any sort of organism).[10]

The power of a theory of this scope and ambition depends on its ability to generate fertile consequences, what Whewell called a consilience of inductions and we have termed explanatory fecundity. Newton was able to deduce Kepler's laws from his own, deeper laws. In addition, Newton's laws gave accounts of phenomena that had hitherto resisted elegant explanation. Newton's theory was in these respects highly fecund. Fisher, holding criteria like these before himself, went on to deduce from his fundamental theorem a proposition that the new Darwinism badly needed. Natural selection, he showed, works best on an array of small, continuous Mendelian variants. *Thus, Darwin's mature insistence on continuous variation is rendered consistent with particulate inheritance through the ability of the probabilistic concept of natural selection to yoke the two ideas together.*

Fisher's proof is a mathematical one. Its results can be seen using visual imagery developed by Sewall Wright in his meditations on Fisher's theory. Wright imagined fitness as using variation to climb up a hill in an "adaptive landscape" where there are lots of hills. (Richard Lewontin has parodied Fisher's theory by calling it "The Ascent of Mount Fitness.") At the top of this hill is complete adaptedness or fitness—and zero variance in fitness. Suppose that on its trip up the hill the population develops a very large genetic variant, a macromutation. If it happens in these conditions to be fitter than its competitors, that is, is highly correlated to the production of more descendants, the population will land much further up the adaptive hill, but the chances of this happening diminish with the size of the variation. Just as a literal hill contains an increasingly small area on which to land safely if you jump, since the hill slopes away ever more steeply beneath you, so most genetic changes, which, as Darwin had postulated and as Fisher now sees more clearly with Maxwellian gases in mind, can occur in any direction, will be deleterious. Thus, the bigger the variant is, the more deleterious it *probably* will be. Conversely, though, if the population is presented with an array of small steps, it can successfully inch its way up the hill with decreased risk. Thus, Fisher uses probability reasoning to argue that the width of genetic variance, and hence the rate of selection, is proportional to the smallness of the genetic differences on which selection works. Fisher's Boltzmannian model is being used to put some flesh and bones on Darwin's nonmathematical intuition that "it would be a most extraordinary fact if no variation had ever occurred useful to each being's own welfare" and on his mature, but conceptually undefended, assumption that natural selection works on what he called "continuous variation."

Fisher was also eager to show that his theory had the capacity to deduce and explain propositions that had been discovered and accepted by working geneticists but had previously lacked theoretical grounding. In retrospect, Darwin's attempt to turn his hypothesis of natural selection into a theory in which natural selection could explain and unify a large number of biological phenomena, in the manner of Whewell's "consilience," had failed because it did not rest on an adequate theory of inheritance and because it was mathematically impoverished (Gayon 1992). These two stumbling blocks having been removed by genetics on the one hand and statistics on the other, Fisher now tried again what Darwin had tried before him.

First, Fisher used his theory to give explanations of basic phenomena that any adequate evolutionary theory had to account for by that time. He used it, for example, to explain in purely probabilistic terms why variation will be maintained in populations of heterozygotes and why, accordingly, heterozygotes are often fitter; why it is so difficult fully to flush recessive genes out of a natural population, even by intensive and selective cross-breeding in a laboratory; how dominance could have evolved bit by bit, like everything else, out of heterozygosity; why sexes would normally come in a 1:1 ratio; and how mimicry arises.[11] In addition, Fisher argued that his picture of selectionist dynamics alleviated a number of conceptual or philosophical impediments that had too long plagued the Darwinian tradition and held it hostage to its enemies. His claim is most compelling when we note how Fisher's theory affects the Malthusian element in Darwinism.

In Darwin's Newtonian scheme, it is assumed that populations are normally at or near their Malthusian limits and that the force of natural selection operates by securing differential deaths under actual conditions of scarcity. The Malthusian population parameter is better conceived, however, as a limit concept. Within its limits, competition is not normally a matter of differential deaths but of comparative fecundity, based on comparative efficiency in the use and partition of resources. Factors such as better ability to find mates, to survive longer, to have more offspring, and so forth become components of fitness, an idea that cannot and need not be reduced to a single measure or factor. That idea was always implicit in Darwin's ideas about diversification through resource partitioning. However, these conceptions were hard to deal with so long as Darwinism was conceived in a way in which differential deaths provided both a propelling force and a paradigmatic scenario. Fisher clearly recognized that Darwin's background assumptions were an unwarranted drag on the power of his selectionist explanations. He writes: "The historical fact that Darwin and Wallace were led through reading Malthus' essay on population to appreciate the efficacy of selection, though extremely instructive as to the philosophy of their age, should no longer constrain us to confuse the consequences of that principle with its foundations" (Fisher 1930, 44).

The new model allows Fisher to see that differential fecundity will occur whenever there is a marginal fitness gradient. Thus, selection can and will take place even under conditions in which population numbers are not at, or even near, their Malthusian limits. It is differences that are important and relative fitness that is central. The grim scenario that was paradigmatic for Darwin now suddenly becomes a marginal case of a more powerful general theory. That is what Fisher means when he says that the classical Malthusian scenario is a consequence of the principle of natural selection, and not its foundation. As Sober puts the point, "Excess reproduction with a finite carrying capacity is a special case" (Sober 1984a, 195). Increased generality is what Fisher is claiming for his genetic, probabalistic theory of natural selection. As a result, natural communities do not normally look like Manchester, and Malthusianism, in its most powerful form, is merely the recognition that environments have "carrying capacities."

Fisher's reasons for dwelling on this issue are closely connected with his eugenic concerns. Darwin's interpretation of Malthusian constraints in nature had posed a difficulty for extending Darwinism to human societies, particularly modern ones. Under modern conditions of production, distribution, and political intervention into economic and social issues, the principle of scarcity is suspended to such an extent that the presumably less fit are free to outreproduce the presumably most fit, and indeed are free to intermarry with them. This fact, or even hope, could be appealed not only to falsify Darwinism as a general theory of evolution but to make it possible to look for similar scarcity-suspending mechanisms elsewhere in nature, such as internal regulation of population numbers. As we will see in the following chapter, Russian biologists calling themselves Darwinians had in fact predicated research programs aimed as looking for such mechanisms on rejecting Malthusianism as a mere ideological drag on the idea of natural selection. There have been, moreover, British evolutionists eager and willing to say the same thing.[12]

Galton and Pearson were so concerned with this conundrum that they distinguished among natural selection, sexual selection, and what Pearson called "reproductive selection" (Pearson 1896; Gayon 1992). If eugenics is necessarily concerned more with relative fecundity than with differential deaths, this is particularly true of positive eugenics. By working at the level of gene frequencies, rather than at the messy level of phenotypes, Fisher ended this halfway house between two conceptions of natural selection by proposing to redefine natural selection in terms of relative reproduction. His concern, moreover, was with how fitness is improved in an entire species. Thus, the unquestioned reproductive success of *Homo sapiens* might well be due to those who created modern conditions of production and distribution in the first place and who might well be able to do something about its unintended consequences if they are able to acquire political power.

This sense of liberation from what was intellectually limited about early Darwinism unleashed a certain prophetic strain in Fisher. He wanted to claim even more for his fundamental theorem than that it is suitably quantitative and explanatorily fecund to serve a good heuristic for a general theory of evolution. He expected nature to obey his theorem, considered as a law of nature, as assiduously as it does Newton's or Boltzmann's laws. He wanted to say, in fact, that although the law of entropy and the law of fitness move in opposite directions, the one toward decreasing order and the other toward increasing order, they are connected in a more profound way. According to Hodge, Fisher's "two-hero" history, in which Darwin and Boltzmann play the starring roles, reflects his belief that we live in a "two-tendency" universe (Hodge 1992a; Turner 1985). Natural selection bears upward a cosmos that would otherwise run downhill through the action of entropic dissipation. These are not unrelated or even contradictory processes, however, as they are often made out to be. The idea that there is contradiction between the second law and the emergence of organic form may have been inviting before statistical dynamics was around to show that, however different these processes are, they exhibit the same statistical and probabalistic formalism and utilize the same general mechanisms. Just as the world moves downhill by the exploitation of energetic gradients, so it moves uphill by the exploitation of fitness gradients. Fisher recognizes that "for the present" there are profound differences between thermodynamic and selectionist laws. But their dependence on the same statistical and probabalistic forms emboldens him to hope that "both may ultimately be absorbed by some more general principle" (Fisher 1930, 37). Fisher is clearly a reductionist. He explicitly holds that that the selection of the fittest is "nothing but" a case of the more general process of the selection of the most stable (Hodge 1992a, 255). But that is not because Fisher thinks that evolutionary theory can be reduced to current physics, but because he hopes for an expanded physics that can do justice to biology.

Fisher, when he philosophized, did so in Maxwell's and Boltzmann's shadows. Thus, he may well have known that Maxwell, who was always tinkering with possible mechanisms by which subtle physical processes might work, had spent some time thinking about our individual and collective efficacy in producing new sequences of actions in terms of "switching mechanisms" which, like the swerve that old Lucretius had talked about in his *De Rerum Natura*, would take a system off onto a different trajectory. In particular, he may have known that in a letter to Galton, Maxwell had written:

There are certain cases in which a material system, when it comes to a phase in which the particular path which it is describing coincides with the envelope of all such paths, may either continue in the particular path or take to the envelope. . . . When the bifurcation of path occurs, the system, *ipso facto*, invokes some determining principle *which is extra physical (but not extra natural)* to determine which of the two paths it is to follow. . . . When it is on the enveloping path it may at any instant, of its own sweet will, without exerting any force or expending any energy, go off along that one of the particular paths which happens to coincide with the actual condition of the system at that instant. (Maxwell to Galton, February 26, 1879, quoted from Maxwell's papers at University College, London, in Porter 1986, 206, italics added)

Since the pious Maxwell opposed Darwinism, and evolutionary theory in general, he did not see that natural selection itself may be the most effective of these switching mechanisms in a "two-tendency" universe. That, however, is precisely the role that Fisher now proposed for natural selection (Hodge 1992a). The creative novelty of the world is preserved and amplified by selective retention of what would otherwise dissipate into disorder. In this role, natural selection would also help Maxwell in his project of reducing the distance between the determinism of physics and the freedom of culture. It is very difficult to get directly from basic physics to the free acts of enculturated humans. Natural selection, however, by ranging over chance combinations of genes, is a creative process that fills in much of the space between nature and culture by creating adaptations that allow organisms, and in particular humans, to use intelligence to mimic natural selection, turning what seems accidental into beneficial consequences. The hyperdeterminism of classical physics implies that our actions, insofar as they are subject to the laws of nature, are usually dissipated into ineffective sequences, as (in extremis) in Tolstoy's *War and Peace*. Causal chains once initiated run quickly into the sand. The deeper understanding of the nature of the world permitted by statistical thermodynamics and statistical natural selection, however, permits the management of human affairs that literally add up. Eugenic policy is Fisher's prime exhibit. Fisher's employment of dynamic models from statistical mechanics and thermodynamics should not, therefore, be taken as mere metaphor. Metaphor it is, but in Fisher's mind, at least, it was a good deal more than that.

In reflecting on these matters, it is important to recognize that Fisher was not appealing to Maxwell and Boltzmann to introduce some sort of pure indeterminacy or stochasticity into biology. What Fisher was interested in acquiring from the probability revolution was what Maxwell and Boltzmann themselves were seeking in it: an enhanced idea of natural law that becomes available only when the idea of natural law is taken out of Laplace's and Herschel's hands (Hodge 1992a). Even though it must repeal Darwin's ignorance interpretation of chance, the heterogeneity and chanciness of natural selection need not be any less causalist on Fisher's account than on Darwin's own. The multiplicity of causes that introduce chance into the exploitation of fitness gradients now turns natural selection into a lawlike process that cannot invidiously be compared to the smooth and homogeneous sphere of physical law, as Herschel had done when he called Darwin's principle of natural selection a "law of higgledy-piggledy." Fisher is now able to take Darwin's revenge for him and to turn what had appeared to be a defect into a virtue. By changing ontologies, Fisher seeks to preserve, defend, and enhance the Darwinian tradition by converting natural selection into a universal law of nature on a par with the best and deepest physical laws, which are irreducibly statistical in nature. That is why for Fisher natural selection and entropic dissipation are two sides of a single coin.

Fisher knew that to say these things, and to encourage leaders to think of themselves as switching the entire human race onto a different trajectory by thinking in terms of Fisher's ideas, was to place humans, or at least some humans, in the hubristic position against which Greek tragedians and Christian moralists ceaselessly issued warnings. It was to transcend the bonds of common morality. It was, in fact, to place oneself, one's eugenic progeny, and one's hoped-for leaders "beyond good and evil." Fisher was neither unaware of this consequence nor resistant to it. In his youth, he belonged a a circle who called each other by names taken from Nietzsche's *Thus Spake Zarathustra*. Indeed, in the Cambridge of Fisher's day, aestheticist cults, the main point of which was to dissociate oneself and one's peers from common morality, flourished. In G. E. Moore's *Principia Ethica*, a Cambridge classic, beauty and loyalty to friends are regarded as joint components of the highest good. As the

Bloomsbury circle of artists read it, these values are intertwined because only those who contemplate each others' attempts to make beautiful persons out of the materials of selfhood and beautiful lives to display to one another should be loyal to each other or can be "beautiful people." Thus as his classmates went off to the slaughter of World War I, Fisher was writing in the *Eugenical Review* that although morality and aesthetics are both grounded in sexual selection, those who rightly rule in a society know that beauty is a higher value than morality. It is freer and more responsible because it makes self-cultivation, and the loyalty to one's superior peers, the highest good (Hodge 1992a, 255–56). The flower of European bourgois society, defending work that could no longer be automatically sustained by relying on the laws of the free market and the liberal state, was about to take a decided turn to the right.

It is difficult to imagine how such a vision might be coherently combined with Fisher's proclaimed devotion to Christianity and his steadfast loyalty to the Anglican church. Nietzsche at least paid Christian values a complement by recognizing that they were directly opposed to his own. They stood agonistically on the same level of profundity. Fisher's attempt to have it both ways seems correspondingly tacky. No doubt there is in Fisher's support for the Anglican establishment more than a hint that his new ruling class will recognize and preserve the achievements of the old. Fisher would thereby show the worthiness of the new middle class to inherit the world made by the landed aristocracy, a claim that had long been doubted. It is all a bit arriviste. It is also true, however, that Fisher tried to square the circle by thinking of eugenics as an instrument for continued creation, a sort of eighth day in which humans freely take over the work begun by God (Turner 1985). "I think," Fisher wrote to a correspondent, "that we must regard the human race as now becoming responsible for the guidance of the evolutionary process acting on itself" (quoted in Turner 1985, 192). In any case, Fisher's extraordinary combination of seemingly inconsistent ideas shows that Aristotle was probably right: One who cannot fully participate in human practices, and whose own life does not exhibit the normal range of virtues, probably cannot think about the human condition very coherently.

11 Giving Chance (Half) a Chance: Sewall Wright, Theodosius Dobzhansky, and Genetic Drift

Fisher's invocation of probability theory leaves room in his theory for agents of evolution other than natural selection. In particular, Fisher's theory countenances the possibility that in populations smaller than the large, panmictic ones Fisher himself assumed, genes can become established by chance rather than by natural selection. This can occur by a process called genetic drift. Drift is the genetic equivalent of the fact that a roulette wheel might land on red ten times in a row without violating the law of large numbers, which says that in the long run, red will come up only 50 percent of the time (assuming a fair wheel). The smaller the sample is, the more likely are such departures from the underlying distribution. In the same way, genes can "go to fixation" in a small interbreeding population without benefit of natural selection.

Fisher's assumption of large panmictic populations kept the possibility of drift from marring the beautiful universality of his fundamental theorem. This implied, however, that in treating his theorem as an empirical law, Fisher was tacitly making an empirical assumption about how biological populations live. That genetic drift might be an important process in real biological populations was an idea first advanced by Sewall Wright, an American geneticist who is commonly linked with Fisher and Haldane as one of the founding fathers of mathematical population genetics. Wright did so because he was convinced that many organisms live in relatively small breeding groups, between which there is often little gene flow and within which genes can spread without selection pressure. Because he approached genetic natural selection from the point of view of a biologist rather than that of a chemist, like Haldane, or of a physics-admiring statistician, like Fisher, Wright brought to population genetics a more realistic view of how organisms actually live (Provine 1986). From this perspective, Wright thought he could explain much more persuasively than Fisher how populations make the Ascent of Mt. Fitness. The fact that on this view populations are scattered and of finite size was enough, together with Wright's experimental work on breeding, to allow him to contest Fisher's theory of adaptation by natural selection with his own "shifting-balance" theory of adaptation in natural populations.

Not long after arriving in the United States, the Russian geneticist Theodosius Dobzhansky became enthusiastic about Wright's work. He sought to put Wright's shifting-balance theory to the test by studying wild populations of fruit flies in western mountains and deserts. On the combined basis of fieldwork on wild *Drosophila* populations, experimental analysis in the laboratory, and mathematical analysis (mostly by Wright himself back in Chicago), Dobzhansky at first thought he had verified Wright's shifting-balance theory in nature.[1] Drift, he reasoned, plays a significant role in maintaining variation. It allows the available variation to be spread around in different small populations. Successful variants can then reconquer larger populations by adaptive natural selection. Eventually, however, Dobzhansky found reasons to propose an alternative, selectionist explanation of how fitness is achieved. His model of "balancing selection" (as opposed to Wright's "shifting balance") proposed that maintaining variation in heterozygotes is an adaptation, or at least a direct consequence of an adaptation, which allows populations to ride over sometimes large fluctuations in changing environments.

Dobzhansky and Wright brought to their experimental and theoretical work different sets of assumptions about natural history, ecology, biogeography, and speciation. These influenced the different paths they ultimately took. Wright's views about natural populations were tacitly informed by a group of turn-of-the-century American naturalists, led by David Starr Jordan, who argued strenuously against claims by Mendelians like de Vries that large mutations can produce new species. As Darwin had originally suspected, and as his disciples Romanes and Moritz Wagner had made a career of insisting, Starr maintained that geographic isolation, often by physical barriers, was necessary, and perhaps even sufficient, for producing species. Although Wright thought of his shifting-balance theory as a theory of natural selection, his openness from the very beginning of his career to the role isolation plays in the evolutionary dynamics of natural populations made him more open to processes that at least begin with chance partitioning of environments and genetic variance.

Dobzhansky, for his part, brought with him to America a characteristically Russian set of views about natural populations, according to which nature stores variation in heterozygotes as a mechanism for preserving adaptive flexibility. Accordingly, when Dobzhansky found evidence that populations of flies that, following Wright, he had assumed were separated by drift and geographic isolation also possessed adaptive mechanisms that seemed to ensure isolation and the maintenance of variation, he was pleased to abandon drift for selection, at least of a certain kind (Beatty 1987). Dobzhansky's conviction that natural selection itself could produce variation-maintaining mechanisms came to dominate the American wing of what is called the modern synthesis. In their contests with the more adaptationist British versions of synthesis that derived from Fisher, the American synthesists, starting with Dobzhansky himself, increasingly downplayed their Wrightean inheritance. Much to his irritation, Wright thereupon receded somewhat until his ideas and those of his students reemerged, in radically transformed ways, in the 1970s, when both wings of the modern synthesis entered a period of crisis under the impact of the revolution in molecular genetics.

In both Wright's and Dobzhansky's cases, the probability revolution in Darwinism took a new and deeper turn. Wright's genetic drift exploited the possibility, dismissed by Fisher, that genes can be fixed in populations stochastically. Dobzhansky's balancing selection made fitness relative to constantly changing environments, and hence not nearly as linear, additive, or predictive as Fisher assumed. It is significant, moreover, that in both Wright's and Dobzhansky's cases, a sense of nature's complexity seemed to dampen enthusiasm for talking about universal laws of biology, as Fisher did. In this chapter, we will tell these stories and consider the lessons to be learned from them.

Sewall Wright was as American as Fisher and Haldane were, in their differing ways, British. He came from a long line of high-minded Congregationalist intellectuals, the sort of folk with whom Agassiz might have been comfortable. His ancestors included a judge in the Salem trials, as well as one of its victims, who was hanged as a witch. In later generations, descendants of these New England Puritans would become earnest Unitarians and ardent abolitionists. These were Sewall Wright's people. His father, who was his mother's first cousin, was professor of economics, mathematics, astronomy, and much else at Lombard College, a Unitarian school in Galesburg, Illinois. From him Wright learned mathematics, creatively and a bit idiosyncratically. As the father's interests were centered in the humanities, however, it was Wright's mother, whose maiden name provided his first name, who encouraged his scientific bent. She gave him Darwin's *On the Origin of Species* to read while he was still in high school.

In his senior year at Lombard, Wright had the good fortune to encounter a knowledgeable biology teacher, Wilhelmine Key. Key's husband, a descendant of the author of the "Star Spangled Banner," had died soon after their marriage. She never remarried, becoming instead the very model of the devoted, spinsterlike college teacher. Wilhelmine Key had been trained at the new, Progressive University of Chicago. She eagerly shared with Wright all the controversies in evolutionary biology that had been raging since the turn of the century. In one course, Wright read Wallace, Galton, Vernon Kellogg's insightful *Darwinism Today*, and Punnett's exposition of Mendelism in the *Encyclopedia Britannica*. He also absorbed experimental reports on inheritance and genetics coming out of American laboratories at Columbia, Harvard, and Cold Spring Harbor.

It was in this context that Wright also encountered the work of the research community around Jordan, a student of Agassiz and like him an ichthyologist. Kellogg, with whom Jordan cotaught what may well have been the first philosophy of biology course in the country, was a member of this ecologically minded West Coast research community. Their center of activity was the newly founded Stanford University in northern California, where Jordan served as president, while still managing to conduct research into isolation in the Sierra Nevada and other California mountain ranges. Other members of this research community were Charles Gilbert and Joseph Grinnell. These people were not anti-Mendelians, even though they were Darwinians. Nor does the fact that they were naturalists imply that they were not experimentally minded (Magnus, unpublished).[2] They simply thought the inference that self-proclaimed Mendelians like de Vries and Bateson were making from macromutations to speciation was conceptually and empirically wrong and that a dissident Darwinian tradition that went back to Romanes, Wagner, and John Thomas Gulick, which stressed the role of geographic isolation of populations in speciation, could do the trick instead (Magnus, unpublished).

A forceful debate about the issue had been conducted in the pages of *Science* in 1905, a few years before Wright studied with Key. He learned all about it from her. The California naturalists had provided evidence showing not only that two species do not share the same environment but that the number of species in a genus is correlated with the number of physical barriers in the range of the genus. From this point of view, it was easy to see not only that speciation is related to isolation but that species are, by definition, interbreeding populations. In talking about speciation by macromutation in single organisms, people like de Vries were not only empirically wrong but had a conceptually defective view of what a species is (Mayr and Provine 1980). The role of isolation also suggested that the diagnostic characteristics that distinguish separate species might very well not be adaptive, a view that resonated with the American neo-Lamarckians' notion that adaptation can be harmful to its possessors. Wright grew so comfortable with this view that, much to his later chagrin, he did not seriously question it.

Key managed to arrange a summer fellowship for Wright at Davenport's Cold Spring Harbor after he had graduated from college, so that he could get up to speed before entering graduate school at the University of Illinois, to which he had won a modest fellowship. This was in 1911. In 1912, William Castle visited Illinois and arranged to bring Wright back to Harvard to work with him and East at the Bussey Institute. Wright's experimental work in Castle's lab was on the inheritance of coat color in guinea pigs. Castle's unsuccessful defense of the idea that units of heredity are inherently variable had been waged in defense of the actual biological complexity that he found in his laboratory animals. Perhaps something of Wright's biological realism sprang from his experiences in Castle's lab. Throughout his long life, in any case, Wright kept and worked with his experimental guinea pig population.

In the following summer, Wright returned to Cold Spring Harbor, where he formed a friendship with Sturtevant, the leading light, along with Bridges and Muller, in Morgan's *Drosophila* lab. After acquiring his Ph.D. in 1915, Wright went to work for ten years as a geneticist for the U.S. Department of Agriculture (USDA) in Washington, D.C. The government, already involved in improved breeding and hybridization since the 1890s, was beginning to show interest in the practical implications of genetics. In 1926, Wright was offered a job as an experimental geneticist at the University of Chicago. Upon his somewhat forced retirement in 1955, he moved to the University of Wisconsin, where he assembled his papers and notes into a massive work on genetics and evolution (Wright 1977).

It is worth mentioning that, although Wright was associated with well-known eugenics enthusiasts, took considerable interest in animal breeding, and was in fact a member of the Eugenics Society of America, he was never sanguine about human eugenics. In 1931 he wrote:

Positive eugenics seems to require . . . the setting up of an ideal of society to aim at, and this is just what people do not agree on. In the South before the [Civil] War, an ideal eugenic program would doubtless have been one that tended to develop certain admirable individual qualities in a relatively small white population, eliminated troublesome poor whites and equally troublesome intelligent and aggressive negroes, and developed a large population of docile good natured negroes. . . . It has been easy enough for those of us who have been working with guinea pigs or corn or paramecia to see the shortcomings of many eugenicists in the field of pure genetics, but not so easy to recognize our own shortcomings in applying our findings in lower organisms to the very special case of man. (quoted in Provine 1986, 181)

This passage exhibits Wright's puritan sense of human frailty, the moderation of a good upbringing, the abolitionist sentiments of a loyal Unitarian, and a distinctively American egalitarianism.[3] In the case of Sewall Wright's father, Philip Green Wright, this egalitarianism could take on a mildly socialist hue. It also reflects his sense of biological complexity. Wright's conviction that Fisher had underestimated this complexity led him to enter into a productive correspondence with Fisher and to publish a number of articles taking issue with him in the early 1930s. The issues between them were conceptual and empirical, not mathematical. Once they grasped how to read each other's mathematical and statistical methods, Fisher and Wright understood and respected each other, at least until their relations broke down into stony silence and mutual contempt. Their computations, if not their notation, were usually in agreement. Although Wright was a creative mathematician, and even devised statistical techniques that are now used by social scientists, he was not by a long shot as elegant a mathematician as Fisher. His mathematics was home-made.[4] On the other hand, in spite of his years of working at the Agricultural Research Station, Fisher was far less biologically sophisticated than Wright. His model made several assumptions that to Wright's mind were unrealistic and that would not offer the most favorable conditions for adaptive evolution.

Fisher assumed in the first instance that selection takes place in a large, randomly interbreeding population on a trait-by-trait, gene-by-gene basis. This is called mass selection. Wright even wheedled out of Fisher the admission that the number N of an interbreeding population that is climbing an adaptive peak must usually be the entire population of the planet! There were two problems with this supposition. First, organisms do not live like that. As Jordan and others had shown, they generally live in smallish, clannish, relatively bounded interbreeding populations, which Wright called "colonies," but which later came to be called, following a suggestion by Gilmour and Gregor (1939), "demes" (from Greek *demos*, people linked through intermarriage into villages and neighborhoods who form the population base of demo-cracies, rule of the people).

Second, Fisher's genetic atomism led him to neglect what would happen to selection when the degree of connectivity among genes is very high and when their effects are relativized to different conditions and environments. Wright's work on coat (hair) color in guinea pigs in Castle's laboratory had taught him that mass selection has a tendency to lower fitness in the population by turning up all sorts of unwanted gene combinations, normally hidden in heterozygotes, eventually inducing infertility. For this very reason, breeders had long avoided mass selection in favor of a more sophisticated technique. Skilled breeders take the best specimens that can be produced by inbreeding and then outbreeding them with the best specimens of a separate population. From a genetic point of view, this has two advantages: Inbreeding turns up adaptive gene combinations by uncovering homozygotes, and outbreeding prevents infertility and the loss of the vigor that is associated with excessive homozygosity.

These breeding practices inspired Wright's shifting-balance model of how natural selection would work to best effect in realistic populations. Nature itself seems to do what breeders do. If you break up a large population into many small demes, whose members interbreed mostly with each other, two things will happen. Given enough demes, each of which is small enough, variations will sometimes get an initial toehold and spread through a deme by the process Wright called genetic drift. This idea can best be explained by analogy to gambling. If in playing roulette red comes up five times in a row, that certainly does not reflect the true proportions of red to black. Since the observed proportions accurately reflect the true frequencies and underlying propensities only as the number of tries increases significantly, the law of large numbers will easily tolerate runs of five, or even ten, reds or blacks in a row, for the probability of each toss is independent of the others. (What is called the gambler's fallacy consists in forgetting this fact.) In a similar fashion,

Wright reasoned, genes can be spread around in small populations. The idea of drift is at least logically implicit in the Boltzmannian framework within which Fisher had resituated natural selection. Fisher wanted to exclude it as an annoying constraint on natural selection. By recognizing that statistical arrays are open to stochastic or random processes, however, Wright proclaimed that drift might be part of the solution, not part of the problem. With Wright's appeal to the explanatory role of genetic drift, the probability revolution in evolutionary biology took a new turn.

Now a second aspect of population structure comes into play in Wright's theory of adaptive natural selection. Since most unexpressed variations, whether stored as recessives in heterozygotes or due to new mutations and recombinations, can be assumed to be at least mildly deleterious, many small demes will suffer severe losses when drifting genes, especially recessives turned up by inbreeding, begin to circulate through populations. Extinctions of demes will often occur. Sometimes, however, useful variants will establish themselves in one or another deme. Natural selection, working on individual organisms, can then go to work, spreading a new adaptive gene complex one from population to another. What intentional outbreeding accomplishes in laboratories and breeding pens occurs in the wild through the migration of individuals into new demes.

It is tempting to call this group selection, a process in which group-level traits are selected for (Brandon and Burian 1984; Sober 1984a). For his part, Wright always resisted this temptation. The salient fact in Wright's shifting-balance theory is that natural selection works effectively because, in combination with genetic drift and migration, it is working on individuals, and presumably for their enhanced fitness, within structured demes. Although there is differential survival of groups, therefore, selection is not for group-level traits.[5] Wright was not always able to circumvent this misunderstanding, however, in part because he tended to think of selection within a deme as genic selection, that is, selection *for* separate genes, à la Fisher, and to think of organismic selection as selection occurring in and through outbreeding between demes (Provine 1986). Even though what Wright called organismic selection would not take place effectively unless there were groups, it is not group selection.

Wright was aware that, according to Fisher, species, considered as large panmictic populations, could be pushed up an adaptive peak only step by small step, gene by gene. That conception may have restored Darwinian gradualism to favor against Mendelian macromutationism, but Wright was convinced it was a pyrrhic victory. On the basis of his work with guinea pigs and, in his USDA days, short-horned cattle, Wright appreciated that genes are tied together into adaptive bundles or complexes. They are not atoms, as Fisher's thermodynamic models led him to imagine, or beans in a bag, as Ernst Mayr later contemptuously characterized Fisher's view. Given the internal complexity and connectivity of what Wright called "adaptive gene complexes," populations might easily get locked into ineffective, even harmful, solutions. Fitness, measured by production of offspring, is not nearly as well correlated with additive variability as Fisher thinks. The drive of a gene toward greater and greater representation in a population often carries with it all sorts of other linked effects, not all of them good. The connectivity of genes is not something that can be brushed aside. That is how Wright analyzed the maladaptive effects of sustained inbreeding. (It also explains why he called inbreeding "genic" selection, and outbreeding "organismic or genotype selection." The former corresponds to Fisher's atomistic, gene-by-gene model, the latter to Wright's recognition that a whole organism is determined by a system of genes [Hodge 1992a, 268].) Given Wright's more realistic conception of both genes and populations, Fisher's recipe for fitness will lead to the exact opposite.

Nonetheless, Wright's solution pays a certain homage to Fisher by adapting his convention of thinking of populations as climbing up an adaptive peak. Subtracting the thermodynamics, and hence physics, that lies behind that representational model, Wright's "adaptive landscapes" complexify the notion of genetic gradients. Fisher's idea was that selection is the opposite of entropic dissipation. Just as the latter slides down a gradient or hill, so the former climbs up a hill. Because genes are tied together into bundles in genotype space, however, and organisms are tied together into populations in ecological space, Wright concluded that there must not normally be a single adaptive peak but a variety of them, no one of which is perfect in every dimension. That is what Wright meant by an adaptive landscape. In this way, Wright modified and visually complexified Fisher's energetics-inspired metaphor of genes climbing up a hill to show that Fisher's model of adaptation is bound to end in evolutionary (and, tacitly, eugenic) disasters (figure 11.1). Here is George Gaylord Simpson's description of Wright's idea:

Wright has suggested a figure of speech and a pictorial representation that graphically portray the relationship between selection, [population] structure and adaptation. The field of possible structural variation is pictured as a landscape with hills and valleys, and the extent and direction of variation in a population can be represented by outlining an area and a shape on the field. Each elevation represents some particular adaptive optimum for the characters and groups under consideration, sharper and higher or broader and lower, according as the adaptation is more or less specific. The direction of positive selection is uphill, of negative selection downhill, and its intensity is proportional to the gradient. The surface may be represented in two dimensions by using contour lines in topographic maps. (Simpson 1944, 89)

In this variation of Fisher's Boltzmannian model, the trick of adaptive success is not to trudge wearily up the side of one mountain, one allele at a time, but, when advance in one dimension spells trouble in another,

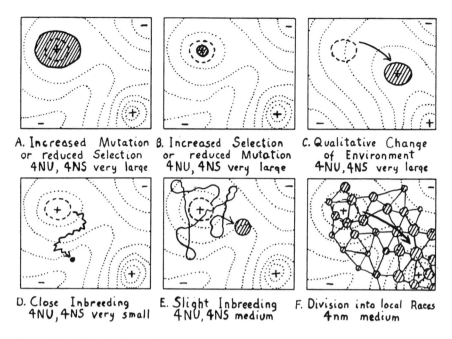

Figure 11.1 Wright's illustration of adaptive landscapes, showing the field of gene combinations occupied by a population within the general field of possible combinations. Type of history under specified conditions is indicated by relation to initial field (heavy broken contour) and arrow. This was figure 4 in S. Wright (1932), page 166. Reprinted with permission.

linked dimension, to shift from one slope to another. The problem is how to get through the nonadaptive valleys between the peaks. It is here that genetic drift plays an explanatory role. In the first place, given relatively small demes, such a thing will, or can, be in actual play. Second, drift is what gets a population through the valleys from one adaptive peak, or local equilibrium point, to another. Sheer chance fixes variants that would otherwise be swamped by the forced march of mass selection in a big population. Migration and subsequent natural selection then spread the adaptive wealth around. Evolutionary success is a matter of exploring possibilities in which various combinations of genes are tested. Wright's model of adaptive evolution enjoyed not only the advantage of biological realism over Fisher's but enhanced problem-solving power as well.

A "much more favorable condition" for adaptive evolution than Fisher's, then, Wright wrote

would be that of a large population, broken up into imperfectly isolated local strains. The rate of evolutionary change depends on the balance between the effective size of the population in the local strain (N) and the amount of interchange of individuals with the species as a whole (m), and is not therefore limited by mutation rates. The consequence would seem to be rapid differentiation of local strains, in itself non-adaptive, but permitting selective increase or decrease of the numbers in different strains, and thus leading to relatively rapid adaptive advance of the species as a whole. (Wright 1930, 354–55)

On his view, Wright proposed an amendment to Fisher's fundamental theorem:

The rate of increase in fitness of any population at any time is equal to its genetic variance in fitness at that time, *except as affected by mutation, migration, change of environment, and effects of random sampling* (Wright to Fisher, February 3, 1931, quoted in Provine 1985, 206)

In spite of its syntax, this is no mere codicil to Fisher's theorem, tricked out with a few qualifications. It is a polite way of annihilating Fisher's whole argument. It is, for one thing, a thoroughgoing attack on Fisher's thermodynamic analogy. The atomicity that makes the analogy to entropy work for Fisher is so severely compromised by Wright's biologically more realistic, structured, lumpier field of interconnected genetic arrays that it disappears into disanalogy, never to be heard from again. It becomes a buried metaphor in subsequent evolutionary theory. Maximization of fitness, on the model of entropy "striving to a maximum," is replaced by a heuristics of exploration, and even a touch of good old American "problem-solving talk," in which selection, and the organisms whose activities the model represent, "try out" various "solutions."

In saying this, Wright is also making a tacit point about the philosophy of biological explanation. He implicitly challenges the notion that a law for evolution by natural selection can be written that is in any way closely analogous to any laws of physics. The alternative "forces" that Wright makes use of, such as genetic drift and gene flow, form the conceptual background against which the "pluralism" about evolutionary forces that was subsequently acknowledged by the makers of the modern synthesis makes sense. These "forces," if that is a good word at all, can play no coherent role unless the background picture is disanalogous to a container full of molecules. As a result, the explanatorily relevant causes of evolutionary change are not the rate of mutation, as for Haldane, or even the rate at which selection eats up genetic variation, as Fisher has it, or any other simple maximizing quantities, but rather, as Wright clearly says, population size (N) and migration rate (m). Ecological parameters move to the explanatory foreground, universal laws to the background. Wright's recognition of biological complexity anticipates a view of scientific explanation in which laws are much less important than models and parameters.

The irony is that Wright's appeal to genetic drift as an evolutionary factor was made conceptually accessible precisely because it was already built into Fisher's model by his Boltzmannian assumptions. In showing that drift could play a role in biological evolution, Wright undermined the empirical applicability of the very model that made stochastic fixation of genes picturable in the first place. Similarly, Wright's notion of an adaptive landscape was a complexification of Fisher's thermodynamic metaphor because Wright wanted to make Fisher's model of adaptive hill climbing actually work. In doing so he negated the very meaning that Fisher attached to his model.

Wright's contestation with Fisher extended to the contrary philosophical interpretations of biology that they built out of similar conceptual materials. Wright was as convinced as any positivist that science is restricted to explaining and predicting phenomena. He too had learned his philosophy of science from Pearson's *Grammar of Science*. But like the French Catholic physicists Ampère, de Broglie, and Duhem, whose views we mentioned in the previous chapter, Wright took the phenomenalist status of science to mean that scientists examine only the outer husk of a deeper, hidden reality that is accessible to faith or, in Wright's case, metaphysical reasoning. There is in this belief a whiff of Agassiz's idealist conviction that the visible world is the external manifestation of deeper, hidden realities. Idealism was, in any case, pervasive in the genteel ambience of Wright's background. In addition, Wright's idealist inclinations had been sustained by the "process philosophies" of Henri Bergson, and especially of Wright's colleague Charles Hartshorne, who taught in the University of Chicago's distinguished (but quirky) philosophy department. Hartshorne persuaded Wright to publish his views about biology and philosophy in the *Monist,* a respected philosophical journal with an idealist genealogy (Hodge 1992a; Wright 1964). When Wright did so, it became clear that he was metaphysically a "pan-psychist," that is, one who thinks that the glimmerings of life, consciousness, and freedom are present in all physical beings.

Wright saw in the probability revolution a way to reduce the gap between phenomenal science and metaphysical reality. In brushing Laplacean determinism aside, a probabilistic universe offered even more openings for creative initiations to Wright than to Maxwell or Fisher. The looser structure of a probabilistic world allows initiations and "occasions" to be distributed by degrees throughout the universe. It is probably this greater sense of the world's openness to novelty that encouraged Wright to take a more generous view than Fisher of the empirical possibility of chance fixation of genes in the small populations in which organisms actually live. Thus, unlike Fisher, Wright did not restrict the role of uncoupled causal events to providing an opportunity for natural selection to drive the world toward higher peaks of adaptedness. Like Fisher, and Maxwell before him, what Wright explicitly called "switch-and-trigger mechanisms" could drive a system into a new, unpredictable trajectory. Unlike Fisher, however, Wright thought this could happen without relying so exclusively on the deterministic intervention of natural selection (Hodge 1992a). Statistical processes themselves did some of the work. In these respects, Wright's philosophy of biology is reminiscent of the evolutionary views of Hartshorne's philosophical hero,

Charles Sanders Peirce, who believed that statistical arrays have ordering properties of their own.

For Wright, as for Fisher, the point of bringing the probability revolution in physics to bear on biology was not to countenance radical indeterminacy. "Even with indiscriminate sampling," writes Hodge, "no ball jumps spontaneously out of the bag into the sampler's hand" (Hodge 1987, 252). In every case there is a causal story that can, in principle at least, be told about why these or those genotypes get fixed by drifting into this or that population, even if it is hard to tell it (Hodge 1987, 253; Horan 1994). Indeed, we can think of no research program in genetic Darwinism that has radical indeterminacy as either its aim or its consequence. Both natural selection and drift are causal processes that bias the fixation of genes. They are distinguished by the fact that in the former case, the property that is selected for is "causally relevant to its own differential reproduction," whereas in the latter case it is not (Hodge 1987, 253).[6] Adaptive value is not among the reasons this or that subset of drifting genes gets amplified in a particular population. This difference has consequences for biological explanation. Plausible etiological or adaptationist stories can be told, if only retrospectively, when natural selection is at work. In the case of genetic drift, such stories are harder to come by. Causal tracks, complicated and highly particularistic, are forever buried. Hence we resort, somewhat misleadingly, to talk about "chance."

From this perspective, we can see that what was at stake in the conflict between Fisher and Wright was how many of the conceptual resources of statistical models are relevant to causal explanations of biological processes rather than causality versus indeterminacy. This issue is still very much alive today. One of these conceptual possibilities is selection, in which fitness gradients are flattened out and fitness is maximized. Another is drift. Both possibilities are built into the method of representing genetic arrays. There are two ways of looking at the resulting situation. In one, selection, a deterministic process in which equations fully govern successive states of a population, operates largely alone. In the other, the process of natural selection operates in a sea of noncorrelated causal events. Fisher thought the possibility of drift could be ignored. Wright did not. Two epistemological considerations will make one sympathetic to Fisher's ambition to rule drift out in practice. First, it is not easy to distinguish drift from selection in a real, finite population (Beatty 1984). Second, if drift is a real possibility, you cannot get directly from natural selection working on genetic variance to evolution (Sober 1984a). If the biological world is as causally complex, as we have every reason to think it is, however, it is more plausible to say that that natural selection is a deterministic process operating in a stochastic world, so long as we remind ourselves that "deterministic" does not mean what Laplace meant and that "stochastic" does not mean "uncaused" (Burian and Richardson 1992). But in this case, "adaptationist programs" of even the most flexible sort must fail because there is no presumptive inference from the presence of selection to the fact of evolution, and considerable heuristic risk in thinking that there is (Dyke and Depew 1988; Richardson and Burian 1992). Even if one restrictively, and stipulatively, defines evolution as "change in gene frequencies in a population," that covers drift as well as selection. Sewall Wright opened up this Pandora's box.

Earlier we caught a brief glimpse of Morgan's difficult but brilliant student Muller emigating to the Soviet Union in 1932, his head full of socialism, sex, eugenics, artificial insemination, and other brave new world ideas. It turns out that Muller had been to Russia before, but with more profitable consequences. In 1922, he had brought some *Drosophila* populations from Morgan's laboratory to the Institute of Experimental Biology in Moscow. This turned out to be an important event in the history of genetics. This institute was in the hands of an effective scientific organizer named Sergei Kol'tsov. After the years of world war, revolution, and civil war, Russian biologists were eager to get back to work and to see what had been happening abroad. As the country was unbelievably poor, there was no possibility of acquiring sophisticated equipment to do "big science." There were, however, Muller's flies. They did not eat much, they bred fast, and lots of questions about them, and their cousins in the wild, could be answered without complicated equipment. What these questions were had been outlined by I. A. Philipchenko in a 1922 report to his colleagues about what had been happening in experimental genetics in Morgan's laboratory and elsewhere. Soon there were three major institutes in Russia working in genetics. Kol'tzov's group in Moscow was led by Sergei Chetverikov. Philipchenko and N. Vavilov directed separate laboratories in Leningrad. The Muscovites shared their flies with these groups. By the end of the decade, Russia contained the lion's share of all the geneticists in the world. Nor in the optimistic and experimentalist culture of the new Soviet Union in the early 1920s had anyone yet suggested that genetics was a bourgeois plot or that dialectical materialism logically entailed the empirical truth of Lysenkoism.

By these routes, a number of talented young scientists went into genetics who might not otherwise have done so. Among them was Theodosius Dobzhansky, an entomologist and avid butterfly collector from Kiev. Dobzhansky was trained in Leningrad by Philipchenko, although he was fully aware of what was happening in the other groups. In particular, he was aware of Chetverikov's research program in Moscow, for Chetverikov was attempting to answer certain questions posed by Philipchenko's review. The most pressing of these was how far the variation that had been uncovered or produced by laboratory genetics existed in natural populations. There prevailed a suspicion, originating in Morgan's and Muller's work at Columbia and in Muller's classical view of population genetics, that variation was much less common in nature than in laboratory populations and that what there was of it was generally well adapted. Laboratory populations, it was felt, inbred to the point of exhaustion and bombarded by Muller with X rays and other insults to induce mutations, were unnatural, maladaptive monsters—laboratory artifacts. Accordingly, Morgan and Muller held that variation must be low in wild populations and that whatever variant alleles were floating around in populations were not, on the whole, deleterious. Otherwise a population would be carrying around what became known as a genetic load. In the days before Haldane, Fisher, and Wright challenged it in the 1930s, the purpose of natural selection was assumed to be to "purify" a population of these maladaptive mutations.

Were these claims about natural populations true or false? The Russians set out to find the answer. Under current social conditions many scientific hands were available to find out by doing the labor-intensive work of finding field populations of flies and bringing them back to the laboratory for analysis of their hidden genetic composition, often by controlled interbreeding with Muller's known commodities. What started to turn up was announced in 1926 by Chetverikov in one of the classic papers of population genetics, and a turning point in the history of the field:

In nature, the process of mutation proceeds in precisely the same way as it does in the laboratory. . . . All these mutations, originating within a "normal" species pass, as a result of crossing, into the heterozygous state, and are thus swallowed up, absorbed by the species, remaining in it in the form of isolated individuals. As a result we arrive at the conclusion that a *species, like a sponge, soaks up heterozygous mutations*, while remaining from first to last externally (phenotypically) homogeneous. (Chetverikov 1926, 178, italics added)

This paragraph states the central issue that was to dominate population genetics—theoretical, experimental, and ecological—for the next fifty years. Genetic load or no, Chetverikov was hypothesizing that natural populations not only carry around enough variation to provide natural selection with fuel, as Fisher was soon to require, but enough to make one suspect, as Dobzhansky eventually did, that natural selection generates a variation-maintaining, as well as a variation-depleting, mechanism. Either way, the days of the old Morgan-Muller "classical" school of those favoring "purifying selection" were numbered.

It is worth noting, or at least speculating, that the eagerness of Russian geneticists to do fieldwork, and their readiness to defy the conventional wisdom by allowing nature to tolerate a large amount of genetic variability, reflects at a distance traditions that had been present among Russian Darwinians from the very beginning. *On the Origin of Species* had come to Russia at the high tide of one of its periodic liberalizing and Westernizing phases. In the feverish 1860s, when the country was casting off the yoke of serfdom, young, recently radicalized intellectuals became self-proclaimed Darwinians in droves. As in Germany, Darwinism meant materialism, and materialism meant modernity and progress. But whereas in Germany evolutionary science was centered in comparative anatomy, embryology, and cytology, in Russia it was weighted from the first toward field studies, the sorts of applied science that students educated abroad, or gifted amateurs, could hope to carry out in the vast, underpopulated regions of the Russian empire (Todes 1989).

It was also distinctive of Russian Darwinians that, with the keen ideological sensitivities of marginalized intellectuals in an undeveloped country, they were convinced from the outset that Darwin's Malthusian picture of nature's economy was simply an ideological reflex of Western commercial and capitalist societies, and so hardly a good picture of nature, or of natural selection, at all (Scudo and Acanfora 1985; Todes 1989; Vucinich 1989). Malthusianism, and more generally British philosophical individualism, had never played well with Russian intellectuals. One could thus despise Malthus from the Right or the Left with impunity. If one approved of the idea that a benevolent authoritarianism can and should keep market society, with its "war of all against all," at bay, one might be a conservative anti-Malthusian. If one sought to find a way to modernize Russia without the discontents of market society, as socialists, nihilists, and populists did, one could find plenty of reasons in Malthus for doing so. Chernesevsky, an influential leftist editor, held that Malthusianism was nothing more than an attempt to "legitimize and justify the unsightly phenomena of modern life." "When purchasing power is in the hands of one man and hunger in the stomach of another," he wrote, "then food for the latter will not be produced, although nature presents no obstacles for its production" (quoted in Todes 1989, 28). English tolerance of, or even complicity in creating, the Malthusian conditions that resulted in the Irish potato famine provided a case in point. What, then, could the great Darwin really have in common with the "pastor thief"? Not much, Chernesevsky replied:

Poor Darwin reads Malthus, or some Malthusian pamphlet, and animated by the brilliant idea of the beneficial results of hunger and illness discovers his America: "Organisms are improved by the struggle for life." But Darwin forgot that privation always harms organisms, killing some and weakening others. . . . The vileness of Malthusianism passed into Darwin's doctrine. . . . The result was the same as if Adam Smith had taken it unto himself to write a course in zoology. (quoted in Todes 1989, 37)

Could, then, one be a Darwinian, and hence a progressive materialist, and even an advocate of natural selection, without being a Malthusian? It is still a good question. The Russians thought the answer was yes—but only if adaptations favoring cooperation could be selected for by a process in which individuals or groups or "races" struggle against the same harsh environmental conditions rather than primarily against one another, an idea that had first been articulated by Wallace (Kottler 1985; Gayon 1992). The Russian Darwinians thought that anyone looking at the cooperative ecological arrangements that nature had made in the steppes would be compelled by empirical evidence to become a non-Malthusian Darwinian along these lines. In the sparsely populated steppes of Central Asia, Russian field ecologists saw precious little population pressure and an array of cooperative arrangements between species in complex food chains, as well as within species through various forms of group cooperation.

The most well-known advocate of this vision was Prince Piotr Kropotkin, a world-famous journalist and advocate of cooperative anarchism. In his younger days, Kropotkin had been a geographer and geologist in the traveling tradition of Humboldt. "In the steppes," he wrote,

> we vainly looked for the keen competion between animals of the same species which the reading of Darwin's work had prepared us to expect. . . . We saw plenty of adaptations for struggling, very often in common, against the adverse circumstances of climate, against various enemies. . . . We witnessed numbers of facts of mutual support, especially during migrations of birds and ruminants; but even in the Amur and Usuri regions, where animal life swarms in abundance, facts of real competition and struggle between animals of the same species came very seldom under my notice, though I eagerly searched for them. (quoted in Todes 1989, 129)

In such regions, the most frequent limiting factor does not appear to be population pressure but usable energy sources, a fact that was not lost on field scientists who were daily measuring energy flows on the materialist assumption of Helmholtz's principle of the conservation of energy. What was crucial in sustaining this dynamic was not that populations will always produce more offspring than can be supported. That was Malthus's model, which had been based on atypical, marginal agricultural cases and on competitive individualist philosophical assumptions. Rather, what drives diversification is the rate at which matter circulates through an ecosystem. This rate increases as a community comes to be organized to exploit it (Scudo and Acanfora 1985, 740). Accordingly, there is selection pressure for adaptations that increase the rate of flow through the system. This means that the struggles of plants and animals must be examined holistically as energy-processing communities and systems held together in food chains.[7] Energetics starts to become a language in which adaptations and fitness can be quantitatively and commensurably measured. It was, in sum, their anti-Malthusianism that led the Russian ecological Darwinians to see, or to think they saw, that natural selection will normally reward populations that find ways to cooperate in energy extraction. This included group selection for traits that allow animals to regulate their own numbers in relation to available resources, an idea that, for better or worse, clearly involved the notion that natural selection can favor not only group-level traits, but among them altruistic or self-sacrificing traits in individuals whose behavior helps the group. Thus was born the distinctively Russian evolutionary idea of "mutual aid," which became Kropotkin's political rallying cry. If Darwin had failed to see this, it was because of his Malthusian ideological blindnesses—and because he had "primarily studied the coastal zones of tropical lands, where overcrowding is more noticeable," rather than "regions . . . where the struggle of species against natural obstacles, early frosts, violent snowstorms, floods, etc., was more obvious" (Todes 1989, 123).

Against a remote background like this, it would not have been hard for a Russian genetic Darwinian like Chetverikov to assert that nature might well harbor a good deal more potentially adaptive variation than other Darwinians, with their Malthusian and eugenic preoccupations, had suspected. None of the other founding fathers of the modern synthesis had in any case suggested anything like this. It is now generally conceded that Chetverikov was one of the founders of the modern synthesis in virtue of having this crucial idea. That might not have happened, however, if the young Dobzhansky had not in 1927 received an invitation to go to America to study in Morgan's laboratory. In reversing Muller's path, Dobzhansky brought Chetverikov's idea with him (Mayr and Provine 1980; Adams, 1968, 1980). It turned out to be a timely move for Dobzhansky personally, as well as for Philipchenko's and Chetverikov's intellectual legacies. Dobzhansky was an Orthodox Christian who detested the regime. In 1929, Chetverikov was arrested, and all work on genetics was shut down as antisocialist, on the ground that anything smacking of genetic determinism must be an ideologically motivated attempt to throw cold water on the role of environmental change in producing the Soviet new man.

Morgan himself was no longer at Columbia. He had been asked to organize the biology division of the newly founded California Institute of Technology in Pasadena. When he arrived in New York, accordingly, Dobzhansky was taken in hand by Sturtevant, who had been Wright's friend ever since their days at Cold Spring Harbor. Dobzhansky was no sooner inducted into Columbia's famous "fly room," however, than Sturtevant was invited to join Morgan at Cal Tech. Dobzhansky went too. Eager to inform the Americans about Chetverikov's research program, Dobzhansky and Sturtevant devised a field research program based on collecting and analyzing populations of flies from all over the West, from Seattle (where one of their students lived) to Mexico, but mostly in the mountains and deserts near southern California. In this way, Dobzhansky successfully transplanted the Russian research program in which he had been schooled to America. Not surprisingly, the same breadth of variation, banked largely in heterozygotes, that had been found in Russia showed up in American fruit flies.

But Dobzhansky soon learned something new in America. In 1932, he became enthralled by Wright's shifting-balance theory when he heard Wright talk at the Sixth International Genetics Congress at Cornell University, in Ithaca, New York. He later confessed effusively to Wright's biographer that he immediately "fell in love" with Wright (Provine 1986, 341–65). The reason is simple: He saw in Wright's theory an explanation of the extensive variation in natural populations that he and Sturtevant were finding, variation that is difficult to explain on Morgan's and Muller's "classical" assumption that very much of it becomes a harmful genetic load, which nature will presumably shuck off as quickly as possible. If nature can fix genes by genetic drift, Dobzhansky thought, it can also store variation by making it invisible to natural selection until changed circumstances turn it into material for adaptation. That was already to put a rather unique spin on genetic drift.

Wright's theory gave Dobzhansky the idea of looking out for population structure in nature as he chased fruit flies up and down the San Bernardino Mountains, for Wright's shifting-balance idea implied that there must be isolating mechanisms between demes if genetic drift is to lead to adaptation on a different peak in the adaptive landscape, and ultimately to reproductive barriers strong enough to drive an incipient species away from its parental stock. Because Dobzhansky's and Sturtevant's program would offer empirical evidence for the shifting-balance theory, which Wright was eager to have in his increasingly difficult correspondence with Fisher, Wright agreed to analyze Dobzhansky's data, which he would send by mail to Wright in Chicago. Dobzhansky was no mathematician but was more than willing to take Wright's word for it.[8] When species containing significant, but seemingly nonadaptive, genetic differences began to be correlated with natural barriers like mountains, Dobzhansky took that as further confirmation not only of Wright's shifting-balance theory of adaptation but of the concomitant suggestion, so deeply rooted among American naturalists, that geographic isolation is the mother of speciation (Provine 1986).

On the basis of the significant monographs that he and Wright were publishing about variation in natural populations, Dobzhansky was invited in 1936 to give a series of lectures at Columbia on the genetics of natural populations. In the following year, these lectures became the first edition of *Genetics and the Origin of Species*, a book considerably different from the many later editions that gradually transformed it into *the* canonical textbook of the modern evolutionary synthesis. What Dobzhansky had to say about species and speciation in this book, and later versions of it, we will consider in the following chapter. Here our concern is with his more general views about adaptive change within populations.

What Dobzhansky stressed about Wright's picture reflected his inheritance from Chetverikov and Philipchenko. He focused on the role of population structure in preserving variation by "scattering the variability" across many small populations through drift and subsequent migration:

In an isolated, self-sufficient population [the scattering of hereditary variability] leads toward genetic uniformity, loss of variance, and consequently to restriction of the adaptive potencies. In a species subdivided into numerous semi-isolated colonies, the same process leads toward a greater differentiation of the species population as a whole, which may mean an increase instead of a decrease of the potentiality for adaptation. The process of migration, which means in this case an exchange of individuals between semi-isolated local colonies, counteracts the differentiating effects of isolation, and prevents the approach toward a genetic uniformity in the separate subgroups. (Dobzhansky 1937, 185–86)

Dobzhansky's originality is most visible in his recognition that changes in gene frequencies take place in a volatile ecological theater. This was a view of nature that his Russian naturalist inheritance and his recent adventures as a naturalist in California encouraged but which is hardly noticeable at all in Wright's experimental work. Because the environment is always changing, in part because of the resource-depleting activities of organisms themselves, natural selection will be required to ensure that species remain attuned to the demands of a changing environment:[9]

The environment does not remain constant, either in terms of geological periods or even from one year to the next. Selection and mutation rates, and hence genetic equilibria, are therefore in a state of perpetual flux. The nature of the genetic mechanism is therefore such that the composition of the species population is probably never static. A species that would remain long quiescent in the evolutionary sense is likely to be doomed to extinction. (Dobzhansky, 1937, 179)

The whole point of natural selection, in fact, is to keep populations tuned to changing environments by allowing them to roll successfully over periodic cycles of various kinds. If natural selection is to reward tolerance of such cycles, which are pervasive in nature, it must be able to presuppose, and if possible preserve, the variability that is necessary to do so. Thus, it is natural to speak of organisms, local populations, and even species as highly dynamic systems that are faced, in highly labile environments, with "challenges" they must "respond" to and, at the cost of perishing, "problems" they must "solve":

The environment is not uniform throughout the distribution areas of most species. Each habitat has a set of environmental coordinates which are, as a rule, more or less constant during short time intervals. The variation of the environment from habitat to habitat constitutes a *challenge* similar in principle to that arising from environmental variations in the course of time. The species *responds* to the challenge of diversified habitats by becoming differentiated into local races. Each local race consists of a group of biotypes having the highest adaptive value in the environment prevailing in the particular class of habitats. (Dobzhansky 1941, 197 italics added)

Dobzhansky's approach implies his unquestioned confidence that gene frequency changes closely reflect events occurring in real environments and that there is a dynamic interaction between genotypes and phenotypes. He believed that each species occupies a certain adaptive peak, a harmonious combination of genes, or suites of more or less equally harmonious combinations, keyed to a specific multidimensional niche.[10] This conviction undergirds the sweeping boldness with which Dobzhansky immediately proceeds to map Wright's abstract idea of adaptive landscapes onto the very real ecological niches to which he refers in the passage just cited. The mountains and valleys of which Wright spoke somewhat metaphorically now become mapped onto the actual peaks and valleys of the San Bernardino Mountains and other geographic locations. Dobzhansky was, in effect, using organisms as analogue computers to link mathematical fitness functions with natural history and to test evolutionary hypotheses about the relative proportions of different evolutionary "forces," like selection, drift, and migration, in the wild. With this further concretization of Wright's imagery of adaptive landscapes, Dobzhansky, whether he knew it or not, was putting yet one more spin on representational devices that had their origins in statistical mechanics and thermodynamics but had subsequently been transformed into Fisher's adaptive peaks and Wright's adaptive landscapes. Fisher's thermodynamic metaphor is buried and at the same time resurrected as a representational device with increased explanatory power. *The statistical properties of genetic arrays are transferred to actual populations in real ecologies.*

In the first edition of *Genetics and the Origin of Species*, Dobzhansky's support for Wright's hypothesis that drift plays a significant role in fixing and maintaining variation in small populations, and in helping adaptive change spread by migration across demes, was firm. Soon, however, difficulties began to crop up. In their correspondence, Wright had always worried that the populations Dobzhansky was finding were too big for variation to have been caused by drift (Provine 1986). A way of testing the issue was found when experimentalists working on *Drosphila* discovered that the chromosomes in the fly's salivary glands are many times larger even than the generally large chromosomes of fruit flies. The better look at the location of genes on chromosomes thus afforded allowed C. C. Tan, working at Cal Tech under Sturtevant's and Dobzhansky's direction, to discover that gene segments of the chromosome in one of the two partially reproductively isolated populations of *Drosophila pseudoobscura* were inverted. Chromosomal rearrangements like inversions are created during the crossing-over phase of meiosis, when alleles from each parent are scrambled and reassembled. At first this seemed to confirm that isolation and preservation of variation are functions of drift. In 1937, Dobzhansky was writing to a fellow geneticist that "unless the data are deceiving we have a proof of the genetic differentiation of the population due to isolation without, and even despite, the influence of natural selection" (Dobzhansky to L. C. Dunn, September 3, 1937, quoted in Provine 1986, 349).

Soon, however, Dobzhansky, working with populations on the side of Mt. San Jacinto, near San Bernardino, found that the frequency of this inversion pattern in *Drosophila pseudoobscura* was correlated to changing seasons. This eventually led him to think that inversion was an adaptive response. Inversions, a remarkably fertile source of variation and a mechanism that might easily be favored by selection pressure, represented adaptations precisely for changing environments. Indeed, for Dobzhansky the most evolutionarily significant adaptations would henceforth be ways a lineage develops to respond to a world chock full of environmental contingencies and irregular cycles. By 1940, Dobzhansky, and in the subsequent editions of *Genetics and the Origin of Species,* began backing off from genetic drift. Even Wright himself began demoting the significance of drift at any other than a purely local, demic level. Both completely stopped talking about species' being formed by drift and marked by nonadaptive characteristics.

By then, Dobzhansky had moved to Columbia, where he trained most of the next generation of American population geneticists, who were expected to be naturalists, experimentalists, and theorists in about equal measures. During these decades, Dobzhansky developed his adaptationist theory of "balancing selection," according to which the superior fitness of heterozygotes, a phenomenon long known to breeders as hybrid vigor, was construed, like chromosomal inversion, as an evolutionary adaptation that provides flexible response to changing, cycling, and complex environments. The case of sickle cell anemia illustrates Dobzhansky's model. Sickle cell anemia is caused by a gene that diminishes oxygen transport and so, over the long run, is correlated with premature death. It is heavily concentrated in populations of African origin. The reason for its persistence in that population is that its heterozygous form affords some protection from the malaria that is endemic in that region without unduly raising mortality. (The reason is that the organism responsible for malaria has a higher oxygen requirement than the host.) A double dose of the gene for sickling, however, far from being modestly helpful, will be lethal. That is precisely what Dobzhansky meant by "balancing." This makes fitness highly sensitive, and relative, to particular environments. Change the environment enough, and what was selected to permit organisms to meet environmental contingencies may be ineffectual, or even harmful. In Africa, it is a good thing, on balance, for the sickling allele to be floating around in populations, maintained by selection in the higher fitness of heterozygotes that carry it. In America, where malaria is no problem, it is not a good thing, contributing to lower life expectancy among Americans of African descent.

Dobzhansky proposed balancing selection as an alternative to Wright's shifting-balance theory. In the process, he proved himself a more thoroughgoing Darwinian selectionist than Wright, even if he had to invent a particularly subtle sort of selection to do so (Beatty 1987). Since that time, populations have been shown to be generally polymorphic, that is, to contain at least several variant alleles for each trait, out of which adapted heterozygotes are fashioned, giving substance to Dobzhansky's hypothesis. As we will see in chapter 14, Dobzhansky's heirs and disciples have continued to develop his perspective, egged on by ever-increasing estimates of the amount of genetic polymorphism harbored in natural populations (Lewontin 1974). The result of this research program has been a much more ecologically sensitive view of the changing and cycling conditions that real populations of real organisms face, as well as a more subtle view than Darwinians had before of the kinds of adaptation that selection can produce to deal with these phenomena.

It should be recognized, however, that variation can in principle be maintained in natural populations in several ways. It might come about by drift, as Dobzhansky and Wright originally thought. It might come about through balancing selection and the superior fitness of heterozygotes. It might even be the result of selection for properties that belong irreducibly to groups. Whatever its causes might be, Dobzhansky's abiding concern, and lasting contribution, was to insist not only that populations are in fact full of diversity but that *diversity is good for them.* "*Drosophila* flies," he wrote, "are doing nicely in their natural habitats, despite the fact that they bear enormous genetic loads"—and so are humans (Dobzhansky 1962, 295–96). From his lofty position as Morgan's successor at Columbia, Dobzhansky pilloried Muller's view of population genetics, which assumed that natural selection keeps presumptively well-adapted populations tuned to the environment by "purifying" them of maladapted outliers—and depriving them of diversity. Dobzhansky's stress on the maintenance of diversity led him to distance himself not only from Muller but from Fisher as well. He acknowledged that Fisher's theory presupposed enough variation for natural selection to get anywhere, and thus opposed Muller's classical view in all the right ways. Fisher too thought that natural selection is a creative force. Fisher's theory still neglected, however, the ways in which nature might try to ensure and enhance variation, as well as use it up. For Dobzhansky that was not good enough.

Dobzhansky's fidelity to this theme, even as he kept changing his mind about the mechanisms of variation-maintenance, had an important moral dimension (Beatty 1987).[11] Dobzhansky wanted the human species in particular to be chock full of variation because that is the raw material for the problem solving that is required if populations are to fix adaptations for changing environments, that is, to roll over cyclical changes and to act in ways that meet changing conditions: "Environmental instability presents challenges to the organism. . . . To maintain itself in harmony with a changing environment, the organism must not only be adapted, but also adaptable. . . . A species should not only possess genetic variety, but must also be able to generate variety. It may then respond to changing environments by genetic changes" (Dobzhansky 1962, 289).

Behind the science is, in fact, an ardent defense of the pluralistic society that Dobzhansky had come to love and a deep desire to refute eugenics not only as a blot on the Darwinian tradition but on American democracy as well. "Equality is necessary if a society wishes to maximize the benefits of genetic diversity among its members," he wrote (Dobzhansky 1973, 44–45, quoted in Beatty 1987, 304). A democratic society is not a passive experimental population living under highly controlled and constraining conditions but a highly diverse community of active, problem-solving organisms—a point that the philosopher John Dewey, who also taught at Columbia, had also been insisting on as the deepest and most benign "influence of Darwinism on philosophy" (as opposed to the superficial and malignant influence of social Darwiniam) (Dewey 1910). From this perspective, the eugenicist preoccupations lying behind Muller's and Fisher's theories bothered Dobzhansky a great deal. Eugenicists worried that pluralist democratic cultures let too much variation accumulate in populations. Real eugenic wisdom, however, was to let the world, and pluralist democracies, take their bumptious but creative course, as nature itself did everywhere.

By the time Dobzhansky was in a position to appeal to population genetics to prove this view, the eugenicist craze in America had passed, brought down politically by an odd coalition of Catholic conservatives, who opposed any tinkering with reproductive practices, and repentant Progressives, whose earlier enthusiasm for social engineering had been blunted by the rise of behaviorism in psychology and of cultural relativism in anthropology. The scales tipped toward the side of nurture over nature in a social and political atmosphere dominated by the ascendancy of Roosevelt's liberal coalition and universal horror at Nazi policies (Degler 1991). The movement to legislate forced sterilization abated, although the practice never entirely stopped, and restrictive immigration policies continued for some time. Respectable geneticists had run for cover long before. Now, however, Dobzhansky was able to seal the fate of eugenics by putting genetics itself on the side of the angels. He threw the weight not only of empirical evidence and theoretical reasoning into the cause but also a moral passion for democracy that only a refugee from totalitarianism could muster. Precisely what Pearson feared so long ago had came to pass. Eugenics had gone down with phony genetics. Proponents of American-style eugenics went underground. They did not dare stick their heads up again until the collapse of the liberal ascendancy and a new conservative tide began to flow some four decades later. Then the fight had to be fought again—by Dobzhansky's students.

12 Species, Speciation, and Systematics in the Modern Synthesis

The phrases *evolutionary synthesis* and *synthetic theory of evolution* are commonly used to designate an alleged consensus around which research into evolutionary problems has been conducted for the last half-century or so. These names are variants on modern synthesis, which appears in the title of Julian Huxley's *Evolution: The Modern Synthesis* (1942). This book is one of a series of works, all appearing within a few years of one another, that announced that the genetic theory of natural selection, which had come of age in the 1930s, was henceforth to be the obligatory framework within which evolutionary problems should be, could be, and would be solved. The classical works of the modern synthesis all proclaimed that the Darwinian tradition was up and running again, and poised to triumph over contending research traditions. Three other books require mention in this connection: Dobzhansky's *Genetics and the Origin of Species* (1937), Ernst Mayr's *Systematics and the Origin of Species* (1942), and George Gaylord Simpson's *Tempo and Mode in Evolution* (1944).[1]

Nonetheless, phrases like *evolutionary synthesis* and *modern synthesis* are problematic and ambiguous. When the synthesis is considered as a theory, there is a tendency to treat these sobriquets as referring to the synthesis between Mendelism and Darwinism achieved in the 1930s by Fisher, Wright, Haldane, and Chetverikov. Although he asserts that there was more to the synthesis than just theory, John Beatty writes in this spirit that "the core of the synthetic *theory* is pretty much just the theory of population genetics developed by Sergei Chetverikov (1926), J. B. S. Haldane (1924–1932), R. A. Fisher (1930), and Sewall Wright (1931), based on the so-called Hardy-Weinberg principle" (Beatty 1986, in Bechtel 1986, 127, italics added).

A slightly different connotation comes from thinking of the synthesis as a synthesis among fields (Darden 1986, 1991; Darden and Maull 1977). In that case the modern synthesis appears as a call for explanatory unification among a variety of disparate disciplines in biology, such as biogeography, paleontology, systematics, and morphology, on the assumption that population genetics in one or another of its variant forms now made that unification possible. This is, in fact, the original usage of the phrase *modern synthesis*. In his book of that title, Huxley, the grandson of Darwin's champion, wrote:

Biology in the last twenty years, after a period in which new disciplines were taken up in turn and worked out in comparative isolation, has become a more *unified science*. It has embarked upon a period of synthesis, until today it no longer presents the spectacle of a number of semi-independent and largely contradictory sub-sciences, but is coming to rival the unity of older sciences like physics, in which advance in any one branch leads almost at once to advance in all fields, and theory and experiment march hand-in-hand. (Huxley 1942, 26, italics added)

A third sense of *modern synthesis* arises when the stress falls on problems. The architects of the modern synthesis were convinced that even the most troublesome evolutionary problems would eventually find a solution within the broad confines of a synthetic theory. For Huxley, Dobzhansky, Mayr, and Simpson, however, the focal point was one cluster of problems in particular. If population genetic theory was to unify the fields of biogeography, systematics, and paleontology and to become a general theory of evolution, it could do so only by solving a set of problems about species: what they are, how they are formed, how they are distributed, how they should be classified, and how their historic relationships are to be exhibited and explained. The Darwinian tradition had gone into eclipse in part because it had lost ground on these crucial issues to its developmentalist and mutationist competitors. The sense of triumph one cannot avoid feeling when reading the classic texts of the modern synthesis derives to no inconsiderable degree from the confidence their authors exude about issues misleadingly lumped together as "the species problem." To the extent that the modern synthesis is about "the species problem," Dobzhansky, Mayr, Simpson, and Huxley, rather than Fisher, Wright, Haldane, and Chetverikov, are, in fact, its fathers. So considered, the phrase *modern synthesis* refers not to the theoretical premises from which the species problem was solved but to a set of conclusions drawn from those premises.

There is a fourth sense of *modern synthesis* that emerges from the focus on problems. The makers of the synthesis assumed, indeed asserted, that progress in solving evolutionary biology's problems would be steady and reliable only under certain constraints. From this perspective, the modern synthesis looks more like a treaty than a theory (Depew and Weber 1985). It is a set of interdisciplinary agreements, some tacit, some explicit, that generated the conceptual space within which a whole raft of specific research programs became possible, while other approaches were ruled out by common consent as a waste of time. The modern synthesis prescribed that the fundamental units or entities with which evolutionary biology was to deal were populations rather than individual organisms or abstract types. This "population thinking," as Mayr calls it, ruled out the macromutationist Mendelism of Bateson and de Vries. The last stand of that tradition, as Mayr and Simpson saw it, was the work of the German geneticist Richard Goldschmidt, whose marginalization by the makers of the modern synthesis was itself one of the foundational acts of the synthesis. Also ruled out was any regression to the old morphological-developmental model of evolution, in which evolutionary process is associated with some grand overall pattern. Casting out heretics, making a canon, and constructing a genealogy were important ways of forging the modern synthesis in this sense of the term.

From this perspective, the integration of population genetics, biogeography, systematics, and paleontology through overlapping solutions to the "species problem" was to be the first fruit of a more ambitious synthesis among biological fields still to come. Developmental biology, ecology, and even social science, which for the time being remained within the orbit of the developmentalist tradition, would be pulled into a new Darwinian synthesis. This grand synthesis has, however, been a long time coming, and results have been more mixed than in the successful case of "the species problem." According to Mayr himself, the most enthusiastic advocate of a hegemonic synthesis, it was not until 1947, at a conference in Princeton, that representatives of most biological and some anthropological disciplines agreed to these terms (Mayr 1980). What they agreed to, moreover, was only a common framework for pursuing research within their various fields, a framework with its fair share of promissory notes and defined at least as much by what it opposed as by what its signatories agreed to. Botany was not brought into the modern synthesis until the efforts of Ledyard Stebbins in the late 1950s and early 1960s. Ecology was not integrated with population genetics until the sustained effort of population ecologists like Robert MacArthur and John Maynard Smith to apply population genetics to relations among species in a community. That occurred in the 1960s. Developmental biology was never really brought in because it awaited, and still partly awaits, progress in developmental genetics. S. I. Washburn, who like Dobzhansky taught at Columbia, was influential in bringing anthropology into the synthesis. "Cultural anthropology," however, whose roots are Lamarckian, remained distinct from neo-Darwinian "physical anthropology." The new synthesis between biology and the social sciences announced in the 1970s by sociobiologists like E. O. Wilson is best interpreted as an attempt to use new developments in population ecology to end this bifurcation by pulling cultural anthropology into the conceptual orbit of neo-Darwinian physical anthropology. The proposal has faced gales of resistance.[2] The reach of the modern synthesis seems to many to have exceeded its grasp.

We see no reason to decide among these four conceptions of the modern synthesis. The idea is clearly a contextually sensitive one, imposing an obligation on writers to be sensitive to context. In this chapter, we will focus on the overlapping but subtly different ways in which Dobzhansky, Mayr, and Simpson approached the linked cluster of problems about species, speciation, and systematics. If there is any single claim that forged a unity among these writers and gave substance to the belief that there already was, or soon would be, a single modern synthesis, it is the now seemingly obvious, but then novel, idea that species are real entities, spatially and temporally bounded populations held together by genetic links in a well-defined ecological niche. The importance of this idea, and the notion that related problems can be solved on its basis, is easy to overlook until one realizes that it is far from clear whether Darwin himself believed it, or at least could express it coherently. Our first claim, then, will be that this insight into the populational nature of species is the living heart, if not the theoretical brain, of the modern synthesis. Our second claim is more directly related to the main theme of this book. Real insight into the nature of species became possible only when species could be treated in statistical and probabilistic terms that Darwin himself did not possess but that population genetics now extended to Darwinian naturalists. Mayr claims that Darwin's greatness lay in the fact that he was a "population thinker" rather than a "typologist" (Mayr 1963; cf. Grene 1990). We do not entirely disagree. We think, however, that one will be a much better population thinker if one is a probability thinker— and Darwin was not yet that.

In following this theme, three additional points will be made. First, in spite of significant overlaps, Dobzhansky, Mayr, and Simpson do not treat species, speciation, and systematics in exactly the same way. Some of these differences reflect the differing cognitive interests of their differing subdisciplines. Other factors, however, some of them personal, were also at work. The second point is historic. The very fact that Mayr, Simpson, and Huxley wrote prolifically and persuasively about Darwin suggests that they were in a position to articulate elements of Darwin's own thinking that he could not clearly express and to resolve problems that he left unresolved. At the same time, in portraying themselves as Darwin's heirs and defenders, these authors all see slightly different things in Darwin, and their understandable enthusiasm for Darwin's genius leads them to project onto him anachronistic anticipations of their own achievements.

A final point is philosophical. Modern Darwinism is a historic science and does not fit well with models of scientific method, or criteria for successful science, devised for physics. This fact is ineluctable when attention turns from the general principles of population genetics to issues about species, speciation, and systematics. Some philosophers of biology have sprung to the defense of the modern synthesis in this matter, claiming that its patent maturity serves to show that philosophers of science have wrongly privileged certain of the roles of universal laws of nature in their physics-biased discussions of scientific method and progress (Beatty 1981; Brandon 1981; Grene 1985; Lloyd 1988; Thompson 1989). These defenses of the conceptual coherence and scientific maturity of modern evolutionary theory have succeeded roughly in proportion as the key concepts of evolutionary theory, such as fitness, have been analyzed in probabilistic terms. By this route, the probability revolution has penetrated even into the conceptual structure of the reconfigured Darwinian tradition.

To grasp the significance of the idea that species are populations, and hence real entities in nature, between which there are "bridgeless gaps," we must review the "species problem" as Dobzhansky, Mayr, Simpson, and Huxley inherited it. The species problem is actually a nest of problems, beginning with the critical question of what a species is anyway. Species, considered as taxa at a certain low level in the classificatory hierarchy, are names of classes. As such they are defined in terms of the necessary and sufficient conditions that must hold if something is to be a member of one of those classes. The tendency to think of species as fixed and unchanging is to a large extent a reflection of the fact that criteria for membership in a class are themselves unchanging; change the criteria and you have a different thing. Essentialism is the metaphysical correlate of this idea. Essentialism in biology is the idea that nature holds organisms to this criterion preeminently at the level we call species but at higher levels as well.

On a strict rendering of the traditional view, Darwin's and Wallace's notion that biogeographic distribution in space and time supports transmutation of species seemed to many of their contemporaries little short of absurd. Whatever it was that was changing, Lyell thought, it could not be species. However, Darwin and Wallace thought of species not only as systematists but as biogeographers. Species, from this perspective, are populations held together by barriers to interbreeding. This was not a new idea. Buffon, for example, held it. Yet Buffon saw no problem in being an essentialist as well. The reason is that the taxonomic and reproductive conceptions of species are consistent with each other so long as Aristotelian, rather than Platonic, essentialism holds—that is, so long as reproductive mechanisms are faithful enough to hold members of species, except for the odd monster, to type.[3] To affirm transmutation and common descent, however, and to do so as Darwin did, by insisting on a wide array of individual differences, was to force a choice between thinking of species as entities held together by reproductive links and entities defined by a fixed set of characteristics. This choice seemed to many in the nineteenth century to be reason enough to resist Darwinism.

The problem was only slightly less puzzling to Darwin himself than to his opponents. Still in thrall to the taxonomic conception of species, Darwin's appreciation of the continuum of differences in space, and the even more marked changes in lineages over time, led him to deny the reality of species. "In vain do naturalists consume their time in describing new species," he wrote in *On the Origin of Species*. "We shall have to treat species as . . . merely artificial combinations made for convenience" (Darwin 1859, 485). This seemed all the more true to Darwin because a taxonomic system is a grid somewhat arbitrarily laid over the biological continuum. Whether something should be called a subspecies, a species, or a genus depends in part on how many ranks there are to be in one's taxonomic hierarchy. To some extent, that is a pragmatic, even an arbitrary, issue. (It is well known that some systematists are so-called lumpers and some are splitters.) Admittedly, the further taxa are apart from each other, the more faithful to reality they are. On this criterion, however, species taxa are, oddly enough, less objective than genera, families, or phyla—a precise reversal of the traditional, Aristotelian view. Darwin's doubts about the reality of species were, moreover, reinforced by his typically British nominalist epistemology, according to which particulars alone are assumed to be real and kinds are treated as arbitrary mental constructs. In making individual differences the motor of evolution, and in regarding them as analogues of Newtonian atoms, Darwin committed himself even more fully to this view. Individual organisms are real. Species are not.

On the other hand, Darwin sometimes talks as if he thinks species are real after all. On these occasions, he is tacitly regarding them as relatively distinct populations, held together and separated from others by barriers to interbreeding, and driven apart from each other by occupying and exploiting different economic niches. On this view, the entities to which Darwin refers as species must be the same things that professional biologists find in nature and then describe in terms of taxonomic nomenclature. From this perspective, Darwin differed from his naturalist colleagues not in what they were talking about but in what they were saying about it (Beatty 1985). If anyone is to give ground, it is systematists. They should listen to their naturalist colleagues. If systematics was to be reformed from this evolutionary perspective, however, it should be by way of what Darwin called natural classification, according to which taxonomic hierarchies map onto degrees of distance from a common ancestor, whether or not this corresponds to descriptive similarity. Under the name "phylogenetic systematics," or "cladism" (from Greek *clade*, "branch"), this idea has now come into favor (Hennig 1966; Wiley 1981). Phylogenetic systematics is open to sometimes radical revisions of traditional classifications because, unlike them, it is based not on overall degree of resemblance—a criterion that may well reflect analogies (similar traits with independent origins) rather than homologies (common ancestrry) and is probably infected with residual essentialism—but solely on evidence about where lineages branched.

There is good reason to think that this way of conceiving systematics is more consistently Darwinian than Darwin's own. Darwin himself, however, did not take this route, for two reasons. First, he assumed that evolutionary relationships would fit pretty well into traditional classifications. Indeed, he took it as a burden of proof that his own theory had to meet that it should be consistent with standard taxonomic practice, adding explanatory depth to what was already presumed to be explanatory adequacy. Second, Darwin did not think that the branching off of one species from another was a process with clear enough beginnings and endings to individuate species. He really did think that only individuals are real, and that populations, races, subspecies, and species refer to temporal and spatial blurs. Caught between his individualist ontology and a linguistically entrenched essentialist definition of species, Darwin never fully resolved his ambiguities about species. He bequeathed the problem to his successors, where it would not be resolved until species, considered as populations, came to be seen as real, discrete units in nature. That was the great triumph of the modern synthesis.

The idea that geographic isolation leads to speciation, which came into favor among American evolutionary naturalists early in the twentieth century, helped resolve this impasse by favoring a populational conception of species over a taxonomic one. Biogeographic boundaries between populations would be a good guide to their status as discrete species. Darwin had come close to this view. He originally thought that variation is scarce, that adaptedness is normal and nearly perfect, and therefore that stress at the biogeographic boundary of a population would be required to trigger variation and to get it fixed by directional selection. Darwin had trouble, however, mapping natural selection onto systematics in these terms. When he finally became convinced that large, widely distributed genera speciate more often than smaller, isolated ones and do so by opening up new resource niches in the same terrritory, he was able to map his theory of origins onto the received systematics. In doing so, however, he changed from what is now called an allopatric (from Greek, "another country") to a sympatric ("from a common homeland") theory of speciation. That made him resist the claim of his fellow Darwinians Romanes and Wagner, later taken up by Gulick, Jordan, and others, that "the constant tendency of individuals to wander from the station of their species is absolutely necessary for the formation of races and species" (Wagner 1973, 4).

The subsequent revival of allopatric speciation by geographic isolation gave renewed hope that populationism would triumph over classificationism and that systematics would be reformed in the direction of what Darwin himself called "natural classification" based on descent. This development was delayed by the conflict between the geneticists and biometricians, even if it was also provoked by that conflict. The dispute between Mendelians and Darwinians was unfortunate for the issue of speciation because both sides were equally blind to the real issue. The biometricians tended to set the problem of speciation aside or, much to

Bateson's annoyance, wave their hands at it. Mendelians, however, took speciation to be a function of sudden macromutations rather than as an ecological or populational phenomenon at all. It was naturalists like Jordan who first pointed out that de Vries and Bateson were using the term *species* incoherently, or at least in such a different way that their claims had no real relevance to "the species problem" (Jordan 1905). To use the term to refer to a single individual differing by some large degree from its fellows was a triumph of the classificatory over the populational conception of the species concept if there ever was one. In its sheer indifference to the populational side of the species question, says Mayr, this was tantamount to restoring classical essentialism (Mayr 1980). It was in this connection that Jordan and his associates challenged the Mendelians by using information about the distribution of species to argue that speciation does not, and probably cannot, take place without geographic isolation of a population.

Their first line of proof was that the number of species of a given genus in a given region is well correlated with the amount of geographic heterogeneity in that region. In and around the California mountains, for example, there are many more kinds of frogs, trout, rodents, and other groups than in the homogeneous environments of the Great Plains, where there is often one species for thousands of uninterrupted miles. Jordan thus concluded that, as a matter of general law, "The nearest related species is not likely to be found in the same region, nor in a remote region, but in a neighboring district separated from the first by a barrier or some sort," and that if the "conditions of life are greatly changed so that a new set of demands are made on the species," the process of speciation will take place rapidly (Jordan 1905, 547).

Wright's stress on demes and migration was informed by the views of these American naturalists. But his theory also developed and strengthened this research program, for isolation of a deme by genetic drift and gene flow from one deme to another might be precisely the mechanism by which geographic isolation leads to speciation. Wright himself probably thought something like this. His own interests, however, focused on providing an alternative to Fisher's theory of adaptedness rather than on problems connected with speciation. Fisher did not even have a theory of speciation that might have provoked Wright to refute him. Accordingly, the resources of Wright's shifting-balance theory for problems about species went underutilized for some time.

This was the state of affairs when Dobzhansky and the other Jesup lecturers took up "the species problem." Although each of them went on to develop theories of his own and to distance himself from Wright, the first editions of the various Jesup lectures have one thing in common: They all tried to link Wright's genetic drift to the problem of speciation. In doing so, they brought the probability thinking that already lay at the heart of population genetics to bear on the nature of species. Species, rather than being the most arbitrary of taxonomic ranks, became, in the form of arrays of biogeographically distributed groups of organisms and genotypes, the real entities on which classification is securely anchored. On this issue, Bolzmannian Darwinism differs from Newtonian Darwinism in ways that should not be minimized.

Dobzhansky made no claim to theoretical originality for the principles of population genetics to which he appeals in considering problems about species in the first edition of *Genetics and the Origin of Species* (1937). Years later, in fact, he remarked:

The reason why the book had whatever success it did was that . . . it was the first general book presenting what is nowadays called . . ."the synthetic theory of evolution." . . . I certainly don't mean to make a preposterous claim that I invented the synthetic or biological theory of evolution. It was, so to speak, in the air. People who contributed to it most I believe were R. A. Fisher, Sewall Wright, and J. B. S. Haldane. Their predecessor was Chetverikov. What that book of mine, however, did was, in a sense, to popularize this theory. (Dobzhansky 1962, 398–99, in Provine 1986, 345)

Dobzhansky speaks here of the synthesis as theory and names the appropriate theorists. In fact, however, Fisher and Haldane play hardly any role at all in *Genetics and the Origin of Species*. Wright, and at a considerable distance Chetverikov, are clearly the theorists whose views most fully inform the first edition of *Genetics and the Origin of Species* (Provine 1986, 345). What is novel about *Genetics and the Origin of Species* is that in "popularizing the theory," as he puts it, Dobzhansky rather boldly applies various components of that "theory" to the central problems of evolution that Darwin himself faced but on which the fathers of population genetics had so far not made a frontal attack. Indeed, the outline of Dobzhansky's book more or less parallels that of *On the Origin of Species* itself. It begins with variation, moves to natural selection, takes on speciation under the rubric of "isolating mechanisms," shifts to macroevolution, and ends with problems of classification.

In tackling these problems, Dobzhansky situates Wright's views within the frame of reference of his own concerns. Dobzhansky's abiding interest was in the mechanisms by which nature maintains variation for adaptation to changing environments. His approach echoes Chetverikov's hypothesis that nature uses heterozygotes to "soak up" and preserve variation for a rainy day. His own "balancing selection" theory of how nature does this is absent from the first edition of his book. It is Wright's shifting-balance theory that does the crucial work. What Dobzhansky stresses about Wright's theory, however, is precisely the phenomenon that his later theory was designed to account for: the role of population structure in preserving variation by "scattering the variability" across many small populations by way of genetic drift and subsequent migration, all occurring in an "adaptive landscape," both mathematical and real, that is presumed to be extremely mutable (Beatty 1987). "The species population is probably never static," Dobzhansky writes. "A species that would remain long quiescent in the evolutionary sense is likely to be doomed to extinction" (Dobzhansky 1937, 202).

In thinking along these lines, Dobzhansky treats adaptive landscapes as models of actual biogeographic distributions and organisms as analogue computers of the various evolutionary forces that will be at work in the volatile ecological theaters thus posited. The most striking consequence of these representational and analytic decisions is that Wright's inadaptive valleys, or at least the most barren of them, come to be identified, willy-nilly, with actual barriers—geographic, ecological, and physiological—that divide species, as well as their component populations, into discreet units. This picture of species distribution is maintained even after Dobzhansky develops an adaptationist mechanism for variation-maintenance. As late as 1950, he writes:

The enormous diversity of organisms may be envisioned as correlated with the immense variety of environments and ecological niches which exist on earth. But the variety of ecological niches is not only immense, it is also discontinuous. . . . The living world is not a formless array of randomly combining genes and traits, but a great array of families of related gene combinations, which are clustered on a large but finite number of adaptive peaks. Each living species may be thought of as occupying one of the available peaks in the field of gene combinations. The adaptive valleys are deserted and empty. (Dobzhansky 1951, 9–10)[4]

Dobzhansky's 1937 picture of how species come into existence is roughly as follows. In small enough populations, genetic drift begins to push races onto different adaptive peaks (Dobzhansky 1937, 134). This is facilitated by external "isolating mechanisms," such as geographic isolation—getting trapped on this or that side of a mountain, for example, or being unable to get across a valley because of, say, a temperature gradient that the organisms in question cannot tolerate. If you map Wright's genetics back onto what naturalists like Jordan had shown about the geographic conditions about speciation, that is precisely what you would expect. Even in 1937, however, speciation does not fully occur until natural selection produces internal or "physiological" isolating mechanisms that erect strong barriers to interbreeding if and when diverging populations are reunited.[5] This is important because genetic isolation seals in the variation that keeps species tuned to a particular adaptive peak by enabling them to ride over the challenges that natural cycles present to them. Accordingly, Dobzhansky's later shift to balancing selection does not require all that much change in his original theory of speciation. It is simply that natural selection, or the subtle form of it that Dobzhansky developed, does more of the work, and does it earlier in the process, than drift.

Dobzhansky's ideas about the dynamics of speciation constituted the strongest effort to date to subordinate a taxonomic to a populational conception of species. That is why hybrid sterility, long recognized as a determinant of species, was for Dobzhansky exactly that. It marks off a species in and through the very act of sealing in variation by physiological isolation. The term *species* thus acquires an ontological more than an epistemological meaning. The term does not refer primarily to a category that helps us know individual entities, or to the mere collection of instances of that category, but to an ensemble of organisms held together by genetic as well as ecological bonds. Species, that is, are distinct spatiotemporal entities with reasonably well-marked beginnings and endings and finite geographic and ecological ranges. This perception led Dobzhansky to try his hand at formulating a populational definition of species that has since become known as the biological species concept (BSC). A species, Dobzhansky asserts, exists when "a once actually or potentially interbreeding array of forms *becomes* segregated into two or more separate *arrays* which are *physiologically* incapable of interbreeding" (Dobzhansky, 1937, 312, italics added).

Three points are worth making about this definition. First, the term *arrays* signals that we are dealing with an entity whose boundaries are defined statistically. Second, notice that *physiological isolation*, which is evident in hybrid infertility, is put into the definition. Geographic isolation is not enough. Second, the verb *becomes* suggests that species are taken to be stages in an ongoing process of differentiation. This implies that the separation to which Dobzhansky's definition refers is achieved by degrees. The definition, therefore, is not fully operational. It does not provide a way to tell definitively one species from another at any given point. To the objection that this constitutes a weakness in his definition, Dobzhansky has a quick reply. He simply downplays the importance of epistemology, which he treats as an issue primarily of concern to professional systematists, whose worries should not dominate evolutionary theorists:

Species is a stage in a process, not a static unit. . . . Our definition cannot pretend to offer a systematist a fixed yardstick with the aid of which he could decide in any given case whether two or more forms have or have not yet reached the species rank. This drawback is unavoidable. . . . The stage when physiological isolating mechanisms develop, and at which genetic discontinuity reaches a state of fixation, undoubtedly occurs in evolution. Therefore, there is no doubt that our definition of species refers to a *real and important phenomenon in nature*. (Dobzhansky 1937, 312–13, italics added).

Dobzhansky's ideas were reinforced and at the same time challenged by Ernst Mayr, whose own series of Jesup lectures appeared in 1942. Mayr, an accomplished biogeographer and systematist, moved from Germany, where he had been trained as an ornithologist at the University of

Berlin, to New York to classify the bird collection at the Museum of Natural History. Later, he went to the Museum of Comparative Zoology (MCZ) at Harvard, where, as director, he helped lay to rest the ghost of its typologically obsessed founder, Louis Agassiz. Mayr's object in *Systematics and the Origin of Species* (1942) was to bring systematics into the synthesis. By systematics, Mayr did not just mean classification. He saw that classification is inseparable from views about what species are, where they are, and how they come to be. About these topics Mayr developed strongly held positions.

Mayr's initial relation to Dobzhansky is somewhat comparable to Dobzhansky's early relation to Wright. In both cases, an initial identification is followed by a reconfiguration of key conceptual elements, reflecting latent differences that were soon to be magnified. Mayr has written:

When Dobzhansky gave the Jesup lectures at Columbia University in 1936, it was an intellectual honeymoon for me. He came came down to the Museum and I was able to demonstrate to him the magnificent geographical variation of South Sea Island birds. . . . I was delighted with the book that came out of his lectures . . . and found that Dobzhansky's interpretation agreed on the whole extremely well with the ideas I had formed independently. (Mayr 1980, 419–20)

It is important to recognize that Mayr is not, like Dobzhansky, an experimental geneticist. On the contrary, as an ornithological biogeographer, he is instead an almost fanatical naturalist, whose chief contribution to the modern synthesis has been to shift attention from genotypic aspects of evolution to phenotypes as they appear in biogeographic space. From this perspective, Mayr has argued incessantly for many decades that the modern synthesis has triumphed over its developmentalist and mutationist opponents because, whereas they retained elements of premodern "essentialism" (a term Mayr uses with enormous definitional and rhetorical plasticity), the new Darwinism vindicates and develops Darwin's status as a natural historian, thereby treating evolutionary biology as sophisticated natural history (Mayr 1980, 1982, 1985, 1988). So strongly does Mayr feel about this that he categorically distinguishes between the ultimate biology of origins, that is, the historical science of evolutionary biology, and the merely proximate functional biology of physiology, including genetics.

Mayr's starting point seems to have been an intense aversion to the unthinking essentialism of traditional systematists, who thought of individual specimens as exemplars of species types (a conception inherent in the very word *specimen*). Mayr sees in the Darwinian tradition, and especially in Darwin himself, an antidote to essentialism (Mayr 1988, 1991). Darwin's biogeographic perspective, we are given to understand, enabled him to discern what Mayr now sees even more clearly: phenotypical variation within populations is the same in kind as phenotypical variation between species. There is, for this reason, no typical member of any species. This insight, which Mayr calls population thinking, is inseparable in his view from Darwin's recognition that the study of evolution is a historical discipline.

Mayr's shift in perspective to the phenotypic level at which naturalists work, and to a historic conception of evolutionary biology, affects every subject he touches. For our purposes, the leitmotiv is that Mayr's naturalist-oriented stress on what is going on at the phenotypic level, where organisms living in populations interact with environments, is that the metaphor of adaptive landscapes, whose adventures we have been following since Fisher lifted the idea of gradients from Boltzmann's thermodynamics, takes yet one more crucial turn. In Mayr's hands, adaptive landscapes become landscapes pure and simple. They become biogeographic distributions of adapted populations of organisms.

This shift in perspective affects, in the first instance, Mayr's version of the BSC. His final version (for there have been a number of them) reads as follows: "Species are groups of actually [or potentially] interbreeding populations, which are reproductively isolated from other such groups" (Mayr 1963, 19).[6] This formulation is important more for what it does not say than for what it does. Taken as a whole, in fact, it constitutes a quarrel with Dobzhansky. Dobzhansky took it to be a question to be settled by experiment whether two populations are physiologially, that is genetically, prevented from interbreeding. By removing *physiologically isolated* from the definition, and substituting the broader *reproductively isolated*, Mayr registers his conviction that even if such experiments could be carried out, they would show little about how reproductive isolation is attained in natural populations. Hybrid sterility can more readily be a laboratory artifact than a natural phenomenon. Observed ecological separation is, accordingly, a better guide to marking off species than anything else.

This does not mean that Mayr thinks that speciation is merely a consequence of geographic isolation followed by adaptive differentiation. On the contrary, Mayr, like Dobzhanksy, thinks that speciation cannot occur without considerable genetic change, most of which takes place quickly. What is happening at the physiological, hence genetic, level is not, however, a matter of definition but of causality. With respect to what a species is, it suffices to say that it is a set of biogeographically distributed populations that do not in fact (or would not if they were brought together) interbreed with other populations. Our attention is thereby drawn to the spatial distribution of real organisms in real populations in real communities in real places. It is also drawn to the art of the systematist, whose skilled judgments and wide experience are implicitly maligned by Dobzhansky's stress on experimental criteria that he himself did not think could be reached in practice. The work of field systematicists like Mayr himself is not to be waved aside.

Mayr's formulation of the BSC also removes the process verb *become* from Dobzhansky's formula and replaces it with the static verb *to be*. This is because as early as 1942 Mayr explicitly disagreed with Dobzhansky's gloss on the ontology of the BSC. Species, he says, are the reproductively isolated results of a process, not stages of one (Mayr 1942, 119). The reasons for this difference are telling. Mayr's species are biogeographic entities. They are thus relativized to space, and hence to other contemporaneous, synchronic groups of populations, more than to the blurring effects of time. Mayr does not want to think of species as stages of anything. To do so would run the risk of mixing the reality of species, as populations, with taxonomic categories, as Dobzhansky does when he talks of species as stages between races and higher taxonomic ranks.

On Mayr's view, early and late, speciation, when viewed through this lens, is presumed to be fairly common. It happens in small populations in places that afford isolation, and especially at the peripheral boundaries of a species' range, where, as the American naturalists had shown, speciation most often occurs. Athough he never denies the theoretical possibility of sympatric speciation, allopatric speciation is Mayr's preferred model. So true is this that Mayr has been quick to dismiss most other scenarios (Mayr 1988, 376–77). Moreover, for Mayr speciation is sudden enough to be spoken of more as an event than as a process.

In 1942, Mayr seems to have assumed that allopatric speciation occurs mostly through genetic drift, for he was then willing to suggest with Wright, and many naturalists before him, that the marks that divide species are often nonadaptive (Mayr 1942, 86). Mayr, along with nearly everyone else, would soon change his mind about that. It is indicative of Mayr's stress on phenotypic rather than genetic descriptions, however, that even in this early stage he talks not about genetic drift but about "the founder principle," an idea that refers to organisms and their spatial distribution under the phenotypic notion of migration more than the genotypic notion of gene flow: "The reduced variability of small populations is not always due to accidental gene loss, but sometimes to the fact that the entire population was started by a single pair or by a single fertilized female. These 'founders' of the population carried with them only a very small proportion of the variability of the parent population" (Mayr 1942, 237).

The contemporary belief among evolutionists that modern humans all descended from a very small population, perhaps from a single "Eve," and that many other populations of hominids did not make it to first base, or at least to second, is an illustration of, and a testimony to the influence of, Mayr's theory.[7] It is almost a corollary of the populational conception of species that they originate in very small populations, within a much wider array of similar populations, only so many of which will, statistically, have a future. The implication is that successful small populations become widespread species by reentering the range of their parent populations and displacing what is left of them. Higher taxa correspond at an even higher level to what Mayr calls "adaptive radiation." Most adaptive radiation, on Mayr's view, occurs through biogeographic dispersal.

By 1954, Mayr had distanced himself even further from Wright than Dobzhansky had (Mayr 1954, 1963). Just as Dobzhansky's views were changed by his discovery of seasonal variation in chromosome inversions, so Mayr's were affected by his discovery that populations living at the most peripheral point of a species' biogeographic range differ most radically from other populations in the group. This sort of change cannot, Mayr reasoned, result from geographic isolation alone. Rather, isolation and stress must trigger off a wholesale reorganization of the genetic structure of the population. Mayr calls this "peripatric" speciation in order to distinguish it from other models of allopatric speciation. He later sums up his claims this way:

1. The founders (in many cases a single fertilized female) carry only a fraction of the total genetic variability of the parental population.

2. The extreme inbreeding of the ensuing generations not only leads to increased homozygosity, but also exposes many, if not most, of the recessive alleles (now made homozygous) to selection.

3. The elimination of many of the previously existing allelic and epistatic balances may result in a considerable loosening up of the cohesion of the genotype.

4. Such genetically unbalanced populations may be ideally suited to shift into new niches.

5. The genetic reorganization might be sufficiently drastic to have weakened genetic homeostasis sufficiently to facilitate the acquisition of morphological innovations.

6. The drastically different physical as well as biotic environment of the founder population will exert greatly increased selection pressures. (Mayr 1988, 446)

Mayr's view puts a genetic spin on something remarkably like Darwin's early ideas about peripheral populations. Accordingly, the Darwin worshipper in Mayr has tried to say that in later editions of *On the Origin of Species* Darwin returned to something closer to his first, allopatric view (Mayr 1988, 180–81, 366). However that may be, Mayr's peripatric model certainly differs from Dobzhansky's account of speciation. It implies that isolation is, or may be, completed before a new species reconnects with its parent population. Dobzhansky, by contrast, held that the process of speciation continues after renewed contact (Mayr 1988, 361). Mayr does not claim to know precisely how the "genetic revolutions" to which he refers are accomplished. For him, genetics is a black box whose mechanisms are less important than what naturalists can observe about populations. He simply infers that it must happen and implies that geneticists, by restricting their views about natural history to known genetic mechanisms, may put unwise constraints on progress in evolutionary theory, which, on the whole, will more profitably be led by naturalists than by geneticists, particularly overtly theoretical ones.[8] (This perception accounts for Mayr's considerable contempt for Fisher, and for his increasing tensions with Wright [Provine 1986, 477–484].)

At the same time, Mayr's belief that gradual adaptation and branching does not create species should not be construed as minimizing the role he assigns to adaptation in evolution. Quite the opposite is true. Mayr is at least as pronounced an adaptationist as any other architect of the modern synthesis (Mayr 1988). While speciation involves (rapid) adaptation, not all adaptation involves speciation. Before a lineage branches into a new species, adaptive change is constantly accumulating. It continues, moreover, to accumulate after speciation. This is called (by Simpson) evolution in the "phyletic" mode. Plenty of evolution goes on between the splitting of branches. Evolutionary time tracks adaptive change in the phyletic mode much more than it tracks speciation events. Speciation events, accordingly, occur as temporal and spatial dots in a veritable ocean of adaptive change. One might even say they punctuate adaptive evolution.

This adaptationist view carries consequences for systematics. In a sort of radicalized version of what Darwin called "natural classification," a consistently Darwinian reform of systematics might abandon preevolutionary practices based on degrees of similarity altogether, replacing them with a strictly genealogical criterion. What Julian Huxley called "clades" (from Greek, "branch" or "twig") would in that case entirely displace "grades" (Huxley 1942). For some contemporary systematicists, called "cladists," they in fact have (Hennig 1966; Wiley 1981; Hull 1988a; Sober 1988). Cladists have good reason for at least wanting to take this path. For grades, based on criteria of similarity, can reflect mere analogies as well as homologies, and "convergent evolution," in which different lineages find their way to the same solution to a design problem, is common. For a cladist, reptiles may not be a real group, for the category is apparently composed of different lineages with similar or convergent features. What may be initially startling is how strongly Mayr resists such a purely genealogical conception of systematics. Although they admittedly run the risk of turning similarities into false genealogies, grades track massive amounts of phyletic evolutionary change to which cladism, Mayr thinks, is blind. For Mayr, adaptive natural selection has indeed created a grade of reptiles. This kind of evolution, protracted across taxa, is called "anagenesis" (as distinguished from "cladogenesis"). Mayr believes that our received categorical system, as Darwin rightly thought, is much more sensitive to this kind of evolutionary change than cladism is. Mayr's non-Darwinian view of species is thus coupled with a very traditional Darwinian view of systematics, and of gradual adaptation. One may suspect, accordingly, that one of Mayr's reasons for so strenuously insisting on the discrete reality of species, which emerge at discrete points in small populations, is to allow traditional systematics comparatively free range to track adaptive evolution by transformation of lineages rather than by splitting.

In insisting that competing criteria of genealogy and similarity must both be honored by systematists, Mayr is reinforcing his view of evolutionary biology as natural history and of systematics as an art (Mayr 1988, 283–84). This may carry a suggestion that systematics is not a science. (In fact, cladists, reacting strongly to Mayr, have prided themselves on having a *science* of systematics that meets Popper's falsificationist criteria.) This charge, however, has merely incited Mayr to a vigorous defense of the scientific status of natural history by arguing against the tacit presumption that the methods of evolutionary science should resemble those of sciences like physics (Mayr 1985, 1988). To the argument of the philosopher J. J. C. Smart, who insisted that immature sciences should use the same methods that led to progress in physics, our most mature science, and who argued that evolutionary biology, lacking laws comparable to those of physics, is nothing but a new version of the old narrative tradition of natural history, Mayr has replied that evolutionary biology is indeed natural history. If it does not rely on universal laws, so much the worse for laws. If it is not strongly predictive, this simply shows that explanation cannot be reduced to prediction (Smart 1963; Mayr 1985). The biological world is more complex, variegated, and full of novelties, contingencies, and surprises than the world of physics. Accordingly, what is needed, Mayr thinks, are not simple laws and reductionist ideals into which all this variety can be "shoehorned," as Gould has it, but a series of concepts that offer help in interpreting and explaining problematic phenomena one by one (Mayr 1985; Depew and Weber 1985). Dobzhansky's vision of organisms as problem-solving creatures, we might say, is trumped by Mayr's problem-solving conception of biological inquiry itself.

Mayr's stress on adaptive evolution, in combination with the primacy he accords to natural history, has stimulated reconstructions of natural selection's two-step process that place the causal accent decidedly on the phenotypical level (Mayr 1978, 1988, 97–100). An organism must succeed in its environment well enough to be in a position to reproduce and to pass on its traits. Thus a first selection process takes place in a single generation. Following the philosopher of biology Robert Brandon, who has given a conceptual reconstruction and defense of Mayr's kind of Darwinism, we will call this "environmental selection" (Brandon 1990). Only after successful reproduction does the heritability of the trait making for success in environmental selection begin to be converted into the transgenerational process of natural selection. Even then, the fact that environmental selection has occurred in step 1 is causally responsible for the transmission of genes from one generation to another in the second step. However true it may be that an organism succeeds in environmental selection because it has a certain array of genes, as well as many other physiological abilities, it is no less true that unless an organism succeeds and breeds, its genes will not be passed down (Brandon 1990). That is a causal, developmental, and historical event that is open to many contingencies, and is not entirely captured by genetic calculations. Accordingly, Mayr's stress on what is happening at the phenotypic level is also a stress on the activities of organisms, and on the particular circumstances of their interaction with their biotic and abiotic environments. From the organismic perspective, in fact, natural selection does not "see" genotypes, and a fortiori individual genes, at all, as Fisher's genotypic perspective seems to suggest. It "sees" whole organisms, or at least their larger features (Mayr 1988, 100–103). "The target of selection," Mayr writes, "does not consist of single genes, but rather of such components of the phenotype as the eye, the legs, the flower, the thermo-regulatory or photosynthetic apparatus, etc." (Mayr 1984, 76; see also Mayr 1988, 423–38).

By shifting attention to the phenotypic level, where environmental selection occurs on whole organisms, Mayr makes an even more strenuous case than Dobzhansky for the relativity of fitness. Fitness is relative not only to Dobzhansky's constantly changing environments but to the fact that the genome itself is an internally connected set of adaptive gene complexes. Having delivered himself from a gene-centered point of view and from physics envy, Mayr is quite hostile to what he calls "mathematical," as opposed to "ecological," population genetics, even to Wright's version of it (Provine 1986, 479–81). The very notion of the context-independent, additive fitness of a single gene is incoherent from Mayr's perspective. "Since the fitness of a gene depends in part on the success of its interaction with its genetic background, it is no longer possible to assign absolute selective value to a gene" (Mayr 1984, 76). This hostility is especially directed at Fisher, whose atomistic conception of genetic natural selection, based as it is on his thermodynamic model, derives from the very physics worship that Mayr thinks of as distorting evolutionary biology. Mayr has also spotted this assumption, which he has long called "bean-bag genetics," in Haldane as well. It is far wiser to treat the genome as a whole as the presumptive least "unit of selection" (Mayr 1984).

We have no doubt that Mayr is on the whole more right about this than his opponents. At the same time, Mayr's "physics allergy" is in many ways simply the other side of Fisher's "physics envy." In declaring the "autonomy of [evolutionary] biology" from physics, Mayr may have underestimated a factor in the development of the Darwinian tradition that forms the main theme of this book: the extent to which Darwinian evolutionary biology has maintained its continuity by borrowing dynamical models from physics. To acknowledge this does not imply the reducibility of evolutionary biology to physics. Nor does it undermine the narrative nature of evolutionary explanations considered as natural history. It simply asserts that dynamical models have been used from the beginning to construct evolutionary explanations and that the dynamical models that Darwin used are not those that Galton, Fisher, Wright, Dobzhansky, or Mayr used.

In a famous line in the first edition of *Genetics and the Origin of Species*, Dobzhansky says, "We are compelled at the present level of knowledge reluctantly to put a sign of equality between the mechanisms of micro-evolution and macroevolution" (Dobzhansky 1937, 12). Dobzhansky is saying that evolution at and above the species level is probably an accumulated effect of mechanisms of evolutionary change that are at work within populations, preeminently natural selection attended by drift and migration. Why then does Dobzhansky say "regretfully"? Part of the answer is that the presumed continuity of macroevolution and microevolution offended against filial piety (Burian, in press). It was Dobzhansky's teacher Philipchenko who coined the term *macroevolution* and used it to express his belief that the directional cast of evolution on the grand scale must have causes other than natural selection. No doubt Dobzhansky's willingness to speak less hypothetically, in subsequent editions of *Genetics and the Origin of Species*, about the continuity between micro- and macroevolution is connected with his later success in finding an interpretation of natural selection that offended far less against the cooperative stress of Russian Darwinians, and their allergy to anything that made use of a Malthusian population parameter, than earlier Darwinian theories. The link between Dobzhansky's balancing selection and his attack on eugenics is enough to suggest that. In general, however, it was left to George Gaylord Simpson, a paleontologist, to make a more substantial case than Dobzhansky did in 1937 for the continuity of micro- and macroevolution.

Simpson's job was not entirely an envious one. As the leading paleontologist in America, Simpson had inherited the mantle of Agassiz and his neo-Lamarckian students. His field was still dominated by "orthogenetic" models of macroevolution, that is, by the notion that evolutionary trends reveal progress unfolding in sublime indifference to natural selection. The autonomy of macroevolution had recently been reasserted by Goldschmidt, as well as by Dobzhansky's Russian mentor, showing that it was capable of surviving the transition to genetics. To demonstrate that natural selection has overwhelmingly been the agent of evolutionary change, by showing that it explains the fossil record more plausibly than its competitors, was the burden of Simpson's *Tempo and Mode in Evolution* (1944).

Simpson approached this task with a certain amount of ideological relish. Whereas Dobzhansky saw American democratic culture through the appreciative eyes of one who had escaped from authoritarianism and reveled in diversity, Simpson was preoccupied with the authoritarian religious biases of his American Protestant upbringing. The son of fundamentalist Presbyterian missionaries, he was among a generation of educated Americans who came of age in the Progressive period and who, as a body, were in full rebellion against the narrowness and provincialism of their culture. Such people were scientistic in the strong sense that they saw religion as repressing free inquiry and significantly constraining the pursuit of individual human happiness. For Simpson, Darwinism was a liberation from these things. Where Mayr saw population thinking in

Darwin, therefore, Simpson, who wrote about him just as fervently, saw antivitalistic materialism (Laporte 1990). "Darwin," Simpson said, "destroyed the last stronghold of the supernatural, the providential and the miraculous" (Simpson 1964a). His commitment to Darwinism thus brought with it a vivid sense that religious beliefs were false and degrading. As an undergraduate at the University of Colorado, he reproached his girlfriend, who would later become his wife, for toying with religion. "If I didn't fear I'd do you harm," he wrote, "I'd try to make you an atheist. I really do think that you are a deluded follower of mistaken, superstitious, and cowardly theories" (quoted in Laporte 1990, 505). The titles of Simpson's most popular essays—"The World into Which Darwin Led Us" and "One Hundred Years without Darwinism Are Enough," the second a vigorous attack on creationism—suggest that he saw Darwin the same way Thomas Huxley did. There is a certain Promethean strain of humanism in remarks such as, "A world in which man must rely on himself, in which he is not the darling of the gods but only another, albeit extraordinary, aspect of nature, is by no means congenial to the immature or to wishful thinkers" (Simpson 1964a, 25). The world into which Darwin led us was a purely material world, in which material processes are capable of explaining functional but only apparently purposeful phenomena. By bringing large-scale evolutionary phenomena into the realm of observed microevolutionary causes, thereby denying any role for vitalistic or directional processes in higher-order evolution, Simpson was in effect crusading for materialist humanism.

Simpson could not plausibly reduce macro- to microevolution if he had to assume that large-scale change is gradual in the sense that it unfolds at a uniform or constant rate. An increasingly thick fossil record was revealing nothing of the sort—indeed was supporting what would later be called a "punctuated" pattern. Higher taxa, it was clear, appear rather suddenly in the fossil record and disappear just as quickly. A case in point is the Cambrian explosion. Simple single-celled organisms had been around without much change for what we now recognize as about 80 percent of the earth's history. Suddenly, about 500 million years ago, multicelled life exploded into such a vast number of distinct forms that only a subset of them, albeit suitably diversified, remain today as our major phyla (Gould 1989a). One thing Simpson had to do was to introduce an enriched conception of time, or evolutionary "tempos," as he called them, into genetic Darwinism to accommodate these facts.

It would be no less difficult for Simpson to achieve his aims if he was constrained to work with Fisherian selection. With Wright's and Dobzhansky's work in hand, however, which offered drift, migration, mutation, and recombination as well selection, Simpson was able to construct models of several evolutionary modes, in which the known battery of evolutionary forces could be combined in different proportions to explain his different tempos. In general, Simpson distinguished three tempos and three modes. Tempos could be slow ("bradytelic," from Greek "slow change"), middling ("horotelic"), or rapid ("tachytelic," fast change). To these generally, but loosely, correspond evolution in the phyletic mode, and consequent anagenetic change across grades; speciation, or branching of lineages; and what Simpson called, in homage to the recent revolution in physics, "quantum evolution." Quantum evolution, which involved something like Mayr's "genetic revolutions," accounts for the origin of higher taxa. The mechanism was Wright's genetic drift combined with natural selection and large-scale genetic reorganization to move a lineage quickly up a succession of desirable adaptive peaks.

The explanatory resources of adaptive landscapes are as much at work in Simpson as in Dobzhansky. Macroevolution is represented not only by way of single adaptive peaks but mapped onto whole mountain ranges. Simpson regarded the peaks in these ranges more as large-scale analogues of ecological niches, however, than as harmonious genetic combinations for rolling over cyclical changes, after the manner of Dobzhansky (Eldredge 1989, 27). In his paradigmatic case, for example, the small mammal called eohippus, which was about the size of house dogs, evolved into the modern horse by switching from browsing to grazing, quickly developing dental and other adaptations to do so (Simpson 1944). Ecological zones are viewed as resource-rich attractors, a rather Darwinian point. Strong selection pressures will push lineages toward these zones as successive populations take small steps up the appropriate slopes. That will allow a good deal of anagenesis in addition to cladogenesis.

Simpson knew that he could not carry out his reduction of macro- to microevolution if he had to work with the biological species concept. Although, like Dobzhansky and Mayr, Simpson relied on Wright's genetic drift and associated processes to get things going, Mayr's version of the BSC had scarcely any temporal dimension at all. Mayr himself tends to talk about the origin of higher taxa in terms of "adaptive radiations," and thus in biogeographic terms that call attention to synchronic distributions of populations. One does not get much sense about what is happening to a lineage over time from Mayr's BSC. But it was lineages that Simpson needed to track through time in order to reduce macro- to microevolution. To do so he simply replaced the BSC with his own evolutionary species concept (ESC). A fully articulated version is this: "An evolutionary species is a lineage (an ancestral-descendant sequence of populations) evolving separately from others and with its own unitary evolutionary role and tendencies" (Simpson 1961, 153).

To say that a species is a lineage is not to deny that it is made up of distinct populations. Lineages, like pointillist paintings, are composed of populational dots, each of which is a species, rather than entirely continuous strokes. Simpson is not regressing to Darwin's blur. The point, however, is to say that these populations are related temporally as well as spatially. Seen in these terms, each species will have a history that makes it a node between the two branching events at which it is born and dies, making it easier to construe classification as a map of descent. Accordingly, Simpson's work eventually led to the genealogically based species conceptions of the cladists, which opposed and were opposed by Mayr (Wiley 1981). His "quantum evolution" led by way of similar radicalizing to "punctuated equilibrium" (chapter 14).

The differences we have traced between Dobzhansky, Mayr, and Simpson suggest that there may be no pure, theory-independent fact of the matter about what species are or how they are to be classified. Which definition one chooses is a matter of one's interests, the problems on which one is working, and one's implicit or explicit theoretical commitments. Indeed, even more definitions of species were soon to emerge. Ecologists devised an "ecological species concept," in which a species is defined not by its role in phylogeny but its role in a community (van Valen 1992). Hugh Patterson has proposed a definition that makes reference to "specific-mate recognition systems" (SMRS) (Patterson 1992). These are adaptations that allow conspecifics to recognize each other for mating purposes even where they share ranges and niches with other species. On this view, sympatric speciation does not seem as hopeless as it does on Mayr's model.

A certain pluralism about species concepts and systematics seems, in fact, to be in order, and may be entirely innocent (Mishler and Donoghue 1992; Ereshefsky 1992). There is probably no single species concept or approach to systematics to which all acceptable theories of evolution must conform. The several ways speciation and phyletic evolution can be envisioned show that theories of evolution favor, or even imply, ideas about systematics and what species are. Conversely, competing theories of evolution often require a particular species concept and special criteria for doing systematics. The increasing consensus among biologists that speciation takes place in a number of ways, only some of which will be attended to by any particular evolutionary theory, can only serve to confirm this view. In a world assumed to be as simple and uniform as medieval theologians and classical physicists assumed it must be, this would be a scandalous acknowledgment. Even today there are those who will take such acknowledgments as arguments for a pragmatic or instrumentalistic view of theories: They are means for getting around the world rather than accurately picturing it. Given the essential complexity of the biological world, however, pluralism seems not only ineliminable but desirable, even within a modestly realist framework. For if a plurality of models makes the large-scale generalizations that science has classically hankered after harder to find, it also makes it less likely that particular phenomena in a highly diverse nature are being overlooked or misdescribed if different models and approaches converge on them (Levins 1968; Levins and Lewontin 1985).

In this spirit, it should be recognized that there is enough overlap among the theories of species and speciation we have reviewed, and those to which we have merely alluded, to substantiate the claim that species are groups of populations, whether one considers such groups as genetically, biogeographically, phylogenetically, or ecologically bounded and defined. The typological essentialist conceptions of species that vexed Darwin, that is to say, whose remote background is the Platonic "great chain of being," are no longer a real threat. That is an achievement of the population thinking that was made possible by the probability revolution, which recognizes that many real things in the world are statistically definable arrays, and no less real, individual entities for all that.

As if in recognition of this fact, philosophers and conceptually sensitive biologists have increasingly come to think, in fact, that to treat species primarily as classes is still to retain too many echoes of the old typological and essentialist conception. On their view, the population thinking that informs the modern synthesis now requires treating species not only as distinct natural realities but as spatiotemporal particulars or "individuals." Much as the organism is a particular thing made of many cells, which are its parts, so a species is made of many reproductively linked organisms, which are *its* parts (Ghiselin 1974; Hull 1978). On such a view it becomes easier to think that selection pressure can be exerted on groups that does not reduce to selection pressures exerted on individual organisms, a proposition never contemplated by any of the architects of the modern synthesis, including Wright. Indeed, it even becomes possible to think that selection can select among species in lineages rather than among its component populations or organisms ("species selection").[9]

Thus far we have traced various strands of the modern synthesis as it emerged in the United States. How it developed in the United Kingdom is a slightly different story. Indeed, it is a story that contrasts vividly enough with the American one that we may speak of an American wing or tendency within the modern synthesis and a British one.

Just as Dobzhansky had set out to defend Wright's theory, so in England E. B. Ford led the Oxford school of ecological genetics in undertaking empirical work to confirm Fisher's genetical theory of natural selection (Ford 1980; Turner 1987). The roots of the Oxford School run back to the adaptationist tradition of natural theology. Hence if population genetics was to deserve the name *Darwinism*, it would have to present its adaptationist and selectionist credentials. As it turned out, this was by vindicating Fisher at Wright's expense, in part because genetic drift was generally misinterpreted by English Darwinians as implying speciation entirely through drift, and not at all by means of natural selection:

The tension between Fisherian and Wrightean views of evolution in nature was a central creative factor in the development of evolutionary biology after the early 1930's. . . . The tension between the evolutionary views of Wright and Fisher focused attention upon population size, breeding structure and genetic variation in natural populations, and the relative importance of natural selection and random genetic drift. These were precisely the questions that stimulated Dobzhansky, Ford and a great many others to pursue their research on natural populations. (Provine 1986, 328)

The result was that even though there was a great deal of overlap in terms of concrete results, British versions of the modern synthesis differ in spirit from American versions. Huxley's *Evolution: The Modern Synthesis*, which introduced the phrase, was intended to refute L. T. Hogben's claim that the genetic theory of natural selection is so at odds with Darwin's commitment to blending and Lamarckian inheritance that it does not deserve to be called Darwinism at all, even when prefixed with *neo-* (Huxley 1942, 27; Hogben 1931). To refute that broad claim, one did not have to take sides among particular genetic Darwinians. In spite of its nods toward Wright's work, accordingly, Huxley profusely thanks the hyperadaptationist Ford in the preface to his book. If he does not thank Dobzhansky, Mayr, and Simpson, that is because his book predates Mayr's and Simpson's Jesup lectures, and, as its author acknowledges, was largely written before Dobzhansky's *Genetics and the Origin of Species* appeared (Huxley 1942, 7). The issues raised in these canonical texts do not, accordingly, direct the course of Huxley's book. It is still the bitter quarrel between Mendelians and Darwinians, as well as the old Lamarckian opposition to Weismann, that Huxley tries to put to rest. He tries to do so by showing that genetic Darwinism, a "reborn Darwinism," a "mutated phoenix risen from the ashes," was getting results (Huxley 1942, 8). Just what sort of genetic Darwinism could get the most results is not to the point. Huxley's "modern synthesis" is as much a composite synthesis of modern synthesizers as it is of Mendelism and Darwinism.

To show that genetic Darwinism was indeed worthy of the name *Darwinism*, Huxley tries to demonstrate that the genetic theory of natural selection can answer all the questions a Darwinian theory is supposed to, including the origin of species. To this end, Huxley uses experimental results and observational data from both American and British sources. He assumes an up-to-date populational conception of species that is far more realistic than Fisher's assumed infinite populations (Huxley 1942, 165). He allows what he calls "the Sewall Wright effect," or drift, to operate pretty much on its own at the level of races and subspecies (Huxley 1942, 193). Huxley adamantly maintains, on the other hand, that species are adaptively, and gradually, differentiated from each other and that phyletic evolution by adaptive natural selection across grades is of far greater significance to the long-term course of evolution than the myriad branching events that accompany the adaptive continuum. "Adaptation," Huxley wrote, "is in point of fact omnipresent," even if it is not omnipotent (Huxley 1942, 413).

Huxley took "the Sewall Wright effect" to be equivalent to the fixation of genes in populations entirely by "accident" (Huxley 1942, 194). Huxley's resistance to allowing genetic drift to have much effect in the formation of species, as distinct from races, is based, accordingly, on reading drift as operating entirely without benefit of natural selection. Refuting Wright, or in Huxley's case limiting him, was made a good deal easier because, much to Wright's frustration, this wrong-headed interpretation of his work had already achieved universal currency in Britain. The British tradition, inheriting Fisher's underestimation of population structure, construed drift as a competitor to natural selection. It was a matter of mere sampling error, rather than a mechanism for "spreading the variation" across demes, and so a theory of natural selection in its own right.

Under this inadequate description, Wright's theory made an easy target for people who were even more adaptationist than Huxley. Thus in the late 1940s, Fisher and Ford teamed up to show that the polymorphisms of the moth *Panaxia dominula* could not have have been fixed by drift because the populations in which these polymorphisms were found were statistically too large for accidental drifting of genes. (It was the same sort of argument that Wright had raised against the overly enthusiastic Dobzhansky.) Other members of the Oxford School succeeded, at roughly the same time, in showing that what had been taken to be nonadaptive features, in snails for example, were adaptive after all (Cain and Sheppard 1954). It was in this polemical context that Kettlewell looked more closely than Haldane had at the famous pepper moths, telling a plausible adaptationist story about their light and dark forms (Kettlewell 1955, 1956, 1973). It was in this spirit too that David Lack revisited the Galapagos Islands to show, in homage to Darwin's selectionism, that minute differences in food sources on each island correlate with the different sizes and shapes of the beaks of Darwin's various finches (Lack 1947). Even small, isolated island populations, it seems, are separated by purely adaptive characteristics.

These results were perceived as a decisive refutation of Wright not only because genetic drift was supposed not to be tied to adaptation at all but because in the famous papers of 1930–1932 that had made him Fisher's competitor in the first place, Wright had supported genetic drift by appealing to the then common view that the diagnostic traits systematists use to identify and mark off species are often nonadaptive. "Under the shifting balance process," he wrote in 1931, "complete isolation originates new species differing for the most part in nonadaptive respects" (Wright 1931, 158; Provine 1986, 290). A year later he wrote in a similar vein, "Complete isolation of a portion of a species should result relatively rapidly in specific differentiation and one that is not necessarily adaptive" (Wright, 1932, 363; Provine 1986, 290). Wright had inherited that idea from the American naturalist tradition. It was defended by Robson and Richards as late as 1936 (Robson and Richards 1936). Thus Wright was not claiming something new but simply adding support for his theory by linking it to the conventional wisdom (Provine 1986). It was a bad tactic, for Wright's theory was held hostage to this view. Thus, in 1950, Dobzhansky wrote to his friend from England, with his tongue only partly in his cheek, "You and your works have been annihilated in Oxford by the testimony of several high authorities" (Beatty 1987, 293). Wright responded by denying, apparently even to himself, that he had ever said what he had in fact said about nonadaptive speciation (Provine 1983; 1986).

What Gould has called the "hardening of the modern synthesis" in the 1950s and 1960s must be viewed against this background (Gould 1983; Eldredge, 1985). It is supposed to mean that a transatlantic shift toward the British adaptationist tradition, and a corresponding distancing of the architects of the American synthesis from Wright, is observable. "Perhaps the synthesis was 'hard' in Britain from the first," writes Gould, "and all the change I have documented merely represents a few recalcitrant Americans finally falling into line" after all these paradigmatic adaptationist stories began to roll out (Gould 1983, 86).

That even the American tradition hardened toward adaptationism is a fact there for all to see in successive editions of the classical texts of Dobzhansky and the other Jesup lecturers (Gould 1983; Eldredge 1985), as well as in Huxley himself (Gould 1983, 86–87). References to genetic drift and to its intellectual father wane as assurances wax that adaptive natural selection is the preeminent creative, causal, and explanatory force in evolution at absolutely every level, and as hopes grow that all higher levels can be reduced to natural selection at ever lower ones. Recognizing, however, that the American synthesis was far from monolithic should serve to put the "hardening of the synthesis" in a clearer light. The adaptationism of the American synthesists was never, and never would be, British adaptationism. Wright's theory, as he continually pointed out in frustration, was, in the first place, as much a theory of adaptation by natural selection as Fisher's. Second, when Dobzhansky and Mayr moved away from Wright, it was toward versions of adaptation and natural selection very different from Fisher's and Ford's (Beatty 1987). These shifts would have occurred even if British adaptationism had not gained rhetorical prominence.

Most important, the facade of international consensus that obtained in the years before and after 1959, when much Darwinian horn-blowing could be heard on the occasion of the centennial of *On the Origin of Species,* concealed deep and lasting conceptual tensions between the American and British wings of the synthesis. Dobzhansky's and Mayr's views imply that fitness is so inherently relative to changing environments and to genetic connectivity that the very idea of assigning a fitness level to each gene separately is taken to be either incoherent or only trivially true, a matter of mere "bookkeeping." That very idea, however, was a cornerstone of the British tradition's view about evolutionary causality and explanation. To recall these persistent, if temporarily submerged, differences is not to indulge in Darwinian scholasticism. For, as we will see in chapter 14, conceptual differences suppressed during the 1950s and 1960s resurfaced in force during the 1970s and 1980s, magnifying conflicts that were latent from the beginning, bringing the Darwinian tradition into a crisis only a little less severe than the one that engulfed it around the turn of the century.[10]

In addition, the shift toward hardened versions of the synthesis was given aid and comfort, and was perhaps even pushed along, on both sides of the Atlantic by important international trends in mid-twentieth-century philosophy of science. Throughout this book we have noted that there has been a subtle but powerful interaction between the Darwinian tradition and fashionable theories about scientific method, scientific explanation, and scientific progress. Darwin's relationship with Herschel is a case in point. So is the relationship between Pearson the positivist philosopher of science and Pearson the Darwinian. It is not entirely irrelevant, then, that in the decades when the synthesis was hardening into monocausal adaptationism, "logical empiricist" or "logical positivist" philosophers of science, as well as advocates of the rival "falsificationist" version of empiricism propounded by Karl Popper, acquired great influence in Great Britain and America. In the 1950s and 1960s, in fact, philosophers of science may have played an even more prominent role in dictating criteria for good scientific theories than they did in the time of Herschel, Whewell, and Mill. The new Darwinism interacted with these philosophies of science in a variety of important ways, some of which may have intensified the hardening of the synthesis itself.[11] (By the same token, as we will see in chapter 14, the subsequent "loosening" and "expansion" of the modern synthesis in the 1970s and 1980s coincided with, and was to some extent guided by, the collapse of logical empiricism.) We close this chapter by reviewing some aspects of this interaction.

Views are loosely called positivistic in proportion as they share some of the traits with Comte's proposals in the 1930s for a "positive" philosophy (Hacking 1983). For one thing, positivists are "scientistic," believing that only claims that are validated by something called "the scientific method" can count as pieces of knowledge. Positivists also typically hold that the institutionalization of scientific method is correlated with the emergence of rational, modern societies because science restricts itself to what is knowable about observable phenomena (phenomenalism). By collecting and analyzing "facts, facts, facts," science abjures metaphysics and theology and the alleged irrationalism that accompanies them. Positivists also prize the discovery of empirical laws, for these permit reliable prediction of future phenomena and give hope that humans can control the natural and social processes that such laws guide. They especially prize the highly general laws of the most basic sciences, under which it is sometimes possible elegantly to subsume the less general laws of less general sciences. There is for this reason a "reductionist" streak in the positivist tradition.

The logical positivists who emerged in the 1930s, and acquired considerable influence in England and America from the 1940s to the early 1960s, put their own stamp on these themes by arguing, on the basis of recent advances in mathematical logic, that bits of sense data, or rather the sentences that name them, can (at least in principle) be combined by a logical calculus into sentences referring to larger wholes. Greater control over what we are warranted in asserting was thereby promised because every step in a logical calculus of statements can (at least in principle) be checked. One result of this movement was to assert that metaphysical, religious, and even ethical statements, rather than merely being hard to decide, are actually meaningless, since such statements are (in principle) impossible to verify or falsify. This "verificationist" view of meaning helped shift the emphasis of the positivist tradition from the discovery of scientific hypotheses, and from a psychology-based logic, to the justification of hypotheses by testing under the control of formal or mathematical logic. The main idea of the logical positivists' theory of justification is that any genuinely empirical hypothesis, however or wherever it originates, logically entails certain predictions about future phenomena. When appropriate tests confirm such a prediction, the hypothesis that entails it is, so far forth, partially confirmed. If the prediction is wrong, on the other hand, the hypothesis that entails it is immediately falsified. (Popper's view is that falsification is the best you can hope for, and constitutes the demarcating essence of scientific knowledge, because there is no telling how many successes would actually confirm a law.) From this perspective, laws are general statements or propositions about the world. They serve as premises of explanatory arguments. Explanations, accordingly, are the deductive outputs of such laws, together with statements about specific initial and boundary conditions. Hence explanations are logically identical with predictions, for on this view the only thing that makes an explanation different from a prediction is the time at which the inquirer asserts it (Hempel 1966).

The cultural influence of this way of thinking dominated the postwar and Cold War period in part because it seemed to suggest that the West was a more rational society than those onto which the metaphysical dogmatism of Marxism had fastened itself (Popper 1945). In this atmosphere, practitioners of almost every science rushed to present themselves as engaged in mature, falsifiable, methodologically sophisticated (and fundable) science. This made a great deal of trouble for humanists and some social scientists, who in response soon set out to show that they had comparable, even if not identical, methods by which to work. It also made trouble, however, for the new Darwinians. Verificationist and falsificationist philosophers of science immediately felt moved to declare that genetic Darwinism's key term, *fitness*, is meaningless, that evolutionary biology has no laws comparable in scope and predictive power to those of physics, and that whatever is sound about population genetics was about to be reduced to the more basic science of molecular biology.

The "survival of the fittest," a term Darwin reluctantly took from Spencer, has always caused trouble for Darwinians. The idea seems explanatorily empty because the fit seem to be defined in terms of who actually survives. It is a tautology, like Molière's joke about the quack doctor who explains the effect of a sleeping potion by referring to its *virtus dormitivus* (Latin for "power to induce sleep"). This old conundrum grew more problematic, rather than less, with the roughly synchronous rise of the genetic theory of natural selection and the new, more trenchant versions of empiricism. For genetic Darwinians, fitness no longer meant, even if it implied, prowess of some sort in environment but simply the comparative reproductive success of organisms and genotypes. Those that outreproduce a comparison class are, seemingly by definition, fitter. The rise of logical empiricist and falsificationist philosophies of science turned this puzzle into a scandal, for reproductive success must serve, on that view, not only as a good way of measuring fitness but (since success is the only way to measure fitness) as the source of the empirical meaning of the term *fitness* itself. If the fit are by definition those that are more reproductively successful, the idea can have no empirical content, for it will not be verified or falsified by reference to any facts that might have been otherwise. For this reason, no less an authority than Popper proclaimed at one point that the theory of evolution by natural selection could not be predictive or explanatory because its central concept is empirically nonfalsifiable (Popper 1972, 1974; retracted in Popper 1978, 1980, 1984). Perhaps natural selection was a metaphysical dogma, no better off epistemologically than its creationist opponents.

Under the influence of positivist ideals, some philosophers of biology rose to the bait and tried to present the modern evolutionary synthesis as prepared to meet any criterion a reasonable logical positivist would care to throw at it. What is wrong, they argued, with a little tautology in the axioms of a sophisticated theory? According to logical empiricist philosophers of science themselves, the substantive explanatory work of a theory is done at lower levels by way of "bridge laws" that link axioms to cases (M. Williams 1970; Rosenberg 1983). In a similar spirit, Mendel's laws, and especially their population-level expansion, the Hardy-Weinberg equilibrium formula, could be presented as well-confirmed laws of nature (Ghiselin 1969; Ruse 1979).[12] Finally, if Darwinian evolutionary theory was about to be reduced to molecular genetics, that only proved how good a theory it was in the first place.

Such defenses of Darwinism have attracted ever fewer adherents as the years have gone by, in large measure because much of what they assert is doubtful. New interpretations of "fitness" have been devised that are in fact not tautologous (Mills and Beatty 1979).[13] Neither Mendel's rules nor the Hardy-Weinberg equilibrium formula are laws of nature, as Mayr had already suspected (Beatty 1981). There are, moreover, good reasons to think that population genetics, and even transmission genetics, is irreducible to molecular biology (Hull 1974; Kitcher 1984). In making these points, philosophers of biology do not intend to compromise the scientific maturity of the modern evolutionary synthesis. On the contrary, the status of the synthesis as patently good science has been used to undermine the factitious positivist criteria for good science against which alone modern evolutionary theory seems weak. As a result, some of the conceptual and methodological pressures that catalyzed the hardening of the synthesis were, by the 1970s, alleviated. By then, however, the synthesis had bigger things to worry about, for the issue was no longer whether its principles would be reduced to those of molecular genetics. It was whether molecular genetics would actually refute the modern synthesis. That question opens a new chapter in the evolution of Darwinism.

Reading Guide to Part II

The comparative reception of Darwinism in countries other than Britain has become a scholarly industry in its own right. Glick's anthology, *The Comparative Reception of Darwinism*, first made many of the essential points. The contention that Darwinism could not have been introduced into France was first made by Yves Conry's *Introduction de darwinisme en France au XIXème siècle*. A more tractable way of saying this is that when Darwinism was introduced into France, it was assimilated to an already entrenched Lamarckian-Geoffroyian evolutionary tradition. For more on this, see Moore's "Could Darwinism Be Introduced in France?" Issues surrounding the reception of Darwinism in France, Germany, and Italy are summarized in Corsi and Weindling, "Darwinism in Germany, France, and Italy." A useful overview of the German biological context into which Darwinism was inserted by Haeckel can be found in William Coleman, *Biology in the Nineteenth Century*, and in more detail in Lenoir's *The Strategy of Life*. On the American reception, first read Moore's *The Post-Darwinian Controversies*. A great deal of interesting material has appeared recently on the Russian reception. Especially insightful is Daniel Todes, *Darwin without Malthus*. See also Alexander Vucinich, *Darwin in Russian Thought*, and Scudo and Acanfora, "Darwin and Russian Evolutionary Biology." More work might be done on how the presence or absence of the Newtonian paradigm in various countries affected the reception of *On the Origin of Species*.

In a number of accessible books, Peter Bowler—*Evolution, The Non-Darwinian Revolution*, and *Charles Darwin*—has shown how the evolutionary theories that achieved wide acceptance in the later nineteenth century were not generally Darwinian, even though "the evolution hypothesis" generically was sometimes called "Darwinism." But see Hull's review of Bowler, *The Non-Darwinian Revolution*, which offers an alternative explanation in accord with Hull's view of Darwinism as "a historical entity." We do not find this "non-Darwinian revolution" very surprising. Long before Darwin, evolutionary theory had congealed around a developmentalist core. Darwin's book took the cork out of the bottle and let what was already there breathe fresh air. On Haeckel's developmentalism in guiding the transition to recapitulationist theories like those of the American neo-Lamarckians, read Gould, *Ontogeny and Phylogeny*, and Richards, *Darwin and the Emergence of Evolutionary Theories of Mind and Behavior* and *The Meaning of Evolution*. On the decline of recapitulationism, see Rasmussen, "The Decline of Recapitulationism in Early 20th C. Biology." On Weismann's role in ending Darwinism's studious "pluralism" about hard and soft heredity, see Richards, *Darwinism and the Emergence of Evolutionary Theories*. On Weismann's exaggeration of the sequestration of the germ line, see Buss, *The Evolution of Individuality*. The best guide to what Darwinism looked like at the turn of the century, when it was supposed to be on its deathbed, is Kellogg's remarkably lucid *Darwinism Today*.

One of our central contentions is that Darwinism finally cut its ties to developmentalist theories of evolution because of its encounter with what historians of science now recognize as "the probability revolution," which was spreading across the sciences from the last half of the nineteenth to the first half of the twentieth century. The literature on this revolution is growing rapidly. The idea that probability thinking ushered in something as large as a second scientific revolution has been expressed forthrightly by Stephen Brush in *Statistical Physics and the Atomic Theory of Matter*, "The first scientific revolution, dominated by the physical astonomy of Copernicus, Kepler, Galileo, and Newton established the concept of a 'clockwork universe' or 'world machine' in which all changes are cyclic and all motions are in principle determined by causal laws. The Second Scientific Revolution, associated with the theories of Darwin, Maxwell, Planck, Einstein, Heisenberg, and Schrodinger, substituted a world of process and chance whose ultimate philosophical meaning still remains obscure."

Claims like this have stimulated interesting discussions about whether the probability revolution was a scientific revolution or paradigm shift in a Kuhnian sense. On this topic, consult the interesting exchange among I. Bernard Cohen, Ian Hacking, Lorenz Krüger, and Kuhn himself in Krüger, Daston, and Heidelberger, *The Probabilistic Revolution*. Readers can judge for themselves by reading Hacking's two seminal little books, *The Emergence of Probability* and *The Taming of Change*, as well as Theodore Porter, *The Rise of Statistical Thinking 1820–1900*; Krüger, Daston, and Heidelberger, *The Probabilistic Revolution*; and Gigerenzer et al., *The Empire of Chance: How Probability Changed Science and Everyday Life*. For a more detailed background in statistical mechanics see Brush, *Statistical Physics and the Atomic Theory of Matter* and Stigler, *The History of Statistics*.

The extent to which Darwin himself was already a probability thinker, or even a population thinker, is contested. Mayr, *The Growth of Biological Thought*, assimilates probabilistic thinking to population thinking and under this description does not hesitate to ascribe it to Darwin full blown. Schweber, in "The Origin of the *Origin* Revisited," "The Young Darwin," and "The Wider British Context of Darwin's Theorizing," plays up the analogy between natural selection and statistical mechanics, thereby subordinating what Mayr calls population thinking to statistical and probabilistic reasoning, which he ascribes to Darwin without much hesitation. In "Natural Selection as a Causal, Empirical and Probabilistic Theory," Hodge insists that the appeal to chance and chances in Darwin's Darwinism, and in contemporary Darwinism as well, is rooted in neither biogeographical natural history, à la Mayr, nor statistical mechanics, à la Schweber, but in the autonomous *vera causa* tradition of Herschel, which easily glides over Herschel's own difficulties with statistical laws. Our view is that Darwin is not yet a probability thinker and that this fact weakens even his population thinking.

Perhaps the best way to begin contemplating these issues is by reading the physicist Fleeming Jenkin's review of *On the Origin of Species,* in which Jenkin uses statistical reasoning to confute Darwin. The review can be found in Hull, *Darwin and His Critics.* We believe that the key figure in taking the statistical turn, and freeing Darwinism from its vulnerability to attacks like those of Jenkins, was Galton. Hacking's contention in *The Taming of Chance* that Galton was the first person to show how statistical arguments could be explanatory in their own right seems to support our analysis. So does Gayon's claim, in *Darwin et l'apres Darwin,* which very badly needs to be translated into English, that Galton's Pearson's, and Weldon's statistical approach first made it possible to turn the hypothesis of natural selection into an adequate theory of natural selection.

On Mendel and Mendelism, the indispensable work is Robert Olby, *The Origins of Mendelism,* 2d ed. A good summary of the literature on Mendelism can be found in Bowler's *The Mendelian Revolution.* Gayon argues persuasively that if biometricians like Galton, Pearson, and Weldon had not already invented statistical Darwinism, it would have been a good deal harder to integrate Darwinism and genetics. That it was hard enough as it was is shown clearly by Provine's account in *The Origins of Theoretical Population Genetics* of the "thirty years war" between Mendelians like Bateson and Darwinians like Pearson. Still, Provine's "internalist" account of the integration of Darwinism and Mendelism in this book suffers from two defects. It is, in the first instance, remarkable for its omission of virtually all references to the eugenic background that drove the Darwinian tradition toward integration with genetics. To fill in the blanks, read Daniel J. Kevles, *In the Name of Eugenics,* for the American story, and Mark Adams, ed., *The Well-Born Science,* for a wide, international perspective. (For Muller's brand of test-tube eugenics, nothing is more vivid than Muller's own *Out of the Night.*) Provine also neglects to talk about the wholesale defection of bright, young biologists from the morphological and developmental paradigm, and their move toward statistical studies, around the turn of the century, a movement that affected both Darwinians and Mendelians and contributed to their eventual reconciliation. There has been a considerable debate about the extent of this paradigm shift. Read J. Maienschein, R. Rainger, and K. R. Benson, "Were American Morphologists in Revolt?" and Maienschein, *Transforming Traditions in American Biology*. On Morgan, begin with Allen's *Thomas Hunt Morgan*.

Too linear an account of the development of population genetics, and too unitary a view of the evolutionary synthesis that arose on its foundations, can obscure the fact that the pioneers of theoretical population genetics differed greatly from one another. Recent historical work on Fisher, Wright, Haldane. and Chetverikov makes these differences clear. For Fisher, see J. R. G. Turner, "'The Hypothesis that Explains Mimetic Resemblance Explains Evolution': The Gradualist-Saltationist Schism," "Fisher's Evolutionary Faith and the Challenge of Mimicry," and "Random Genetic Drift," as well as B. Norton, "Fisher's Entrance into Evolutionary Science." On Fisher's life and character see Joan Fisher Box, *R. A. Fisher*, a scientific biography remarkably dispassionate in view of the fact that it was written by his daughter. On Wright, Provine's *Sewall Wright and Evolutionary Biology* is definitive. There are useful summaries of Provine's research about Wright's agreements and disagreements with Fisher in Provine, "The Development of Wright's Theory of Evolution" and "The Wright-Fisher Controversy." (The view of Wright from Fisher's perspective can be found in Turner, "Random Genetic Drift.") Provine is also good on the history of the idea that geographic isolation is a condition for speciation, especially in the school around David Starr Jordan (see also Magnus, "Down the Primrose Path.") For a superb demonstration that Fisher's and Wright's differing philosophical backgrounds and commitments made it virtually impossible, at least in retrospect, for them to agree on anything more than technical issues, see M. J. S. Hodge, "Biology and Philosophy (including Ideology)." On Haldane, there is Ronald Clark, *JBS*. On Chetverikov and the Russian school, Adams, "The Founding of Population Genetics" and "Sergei Chetverikov," are the best guides.

The best source for the evolutionary synthesis of the 1940s and 1950s is Mayr and Provine, *The Evolutionary Synthesis*. For his part, Mayr portrays the synthesis as a triumph of natural history over physics envy in "Prologue," *The Growth of Biological Thought*, "How Does Biology Differ from the Physical Sciences?" and *Toward a New Philosophy of Biology*. How the synthesis was viewed by those who regarded themselves as a party to it on the British side is on display in Ford "Some Recollections Pertaining to the Evolutionary Synthesis." Dobzhansky's reminiscences, "The Birth of the Genetic Theory of Evolution in the Soviet Union in the 1920's," as well as Adams, "Sergei Chetverikov," show how crucially important the Russian tradition of Philipchenko and Chetverikov was to what we have not hesitated (rather more than other authors) to call the American wing of the modern synthesis. Provine, *Sewall Wright and Evolutionary Biology*, contains useful information about Dobzhansky as well as Wright, especially about the research into the genetic composition of fruit fly populations in the wild that Dobzhansky, Wright, and Sturtevant undertook in California. (The papers in which this research is reported have been assembled and republished by R. C. Lewontin et al., *Dobzhansky's Genetics of Natural Populations*.)

Perhaps the most insightful recent work on Dobzhansky's development is Beatty, "Dobzhansky and Drift." More on Dobzhansky is coming out all the time. See the essays in the forthcoming *The Evolution of Theodosius Dobzhansky*, edited by Adams. On Simpson's biography, read Laporte, "The World into Which Darwin Led Simpson." On his work, the best introduction is Stephen Jay Gould, "G. G. Simpson, Paleontology and the Modern Synthesis"; Mayr, "G. G. Simpson"; and Eldredge, *Macroevolutionary Dynamics*. Articles on these figures, and other makers of modern evolutionary theory, appear regularly in the *Journal of the History of Biology*. The scientific fruits of the modern synthesis are summarized in two fine textbooks on evolutionary science, Douglas Futuyma's *Evolutionary Biology*, 2d ed., and Dobzhansky et al., *Evolution*. The first puts things together from the perspective of a student of Wright, the second from the perspective of Dobzhansky's school. Gould, in "The Hardening of the Modern Synthesis," and Eldredge, in *The Unfinished Synthesis*, have supported their contention that the synthesis hardened toward adaptationism by analyzing the pattern of change in successive editions of the four great canonical classics of the synthesis: Dobzhansky, *Genetics and the Origin of the Species*; Huxley, *Evolution, the Modern Synthesis*; Mayr, *Systematics and the Origin of Species*; and Simpson, *Tempo and Mode in Evolution*. The reader will probably learn more about the modern synthesis by repeating this exercise than in any other way.

Historians and philosophers of biology have differed about what sort of thing the modern synthesis was in the first place. Was it an exercise in reductionism under the influence of positivist philosophical programs, as Smocovitis, "Unifying Biology" maintains, or an "interfield theory" that could unify various biological disciplines without reduction, as Darden argues in "Relations among Fields in the Evolutionary Synthesis"? The issue is aired in Bechtel's *Integrating Scientific Disciplines*. We take the view that the synthesis used probabilistic reasoning about distributions of genotypes in populations to solve problems about species, speciation, and systematics. To see how Darwin's views about species contrast with those devised by the synthesists, consult Beatty, "Speaking of Species." To see how species concepts have proliferated, read the essays in Mark Ereshefsky's anthology, *The Units of Evolution*. Mishler and Donaghue, "Species Concepts," seem to have the most tolerant attitude toward this proliferation. M. Ereshefsky's *The Units of Evolution* and Brandon and Burian's *Genes, Organisms, Populations* contain seminal essays about the ontological status of species (are they classes? individuals?). Ghiselin's "A Radical Solution to the Species Problem" and Hull's "Are Species

Really Individuals?" "A Matter of Individuality," "Individuality and Selection," and "Units of Evolution" are seminal. The implications of competing species concepts, and the ontological status of species, for systematics (and vice versa) is a topic scarcely broached in *Darwinism Evolving*. Hull in *Science as Process* gives a dramatic account of how ferocious arguments about these issues can get. Readers are referred to Sober, *Reconstructing the Past*, and to the relevant articles in Sober, *Conceptual Issues in Evolutionary Biology*, 1st and 2d eds, for an introduction to this tumultuous topic.

On the formation of a community of analytic philosophers of biology, and their intervention in various disputes about the modern synthesis, an overview can be found in Ruse, *Philosophy of Biology Today*, which contains a comprehensive bibliography of contemporary philosophy of biology. See also Callebaut, *Taking the Naturalistic Turn*. Sober, *Conceptual Issues in Evolutionary Biology*, 1st and 2d eds., collects many of the essential papers in the field. On philosophical issues surrounding the alleged "tautology of fitness," see, for openers, Gould's "Darwin's Untimely Burial." The "propensity interpretation of fitness," which helps to resolve that problem and at the same time shows the probabilistic nature of Darwinism's fundamental concepts, was first articulated in Mills and Beatty, "The Propensity Interpretation of Fitness," which is reprinted in the first and second editions of Sober's *Conceptual Issues in Evolutionary Biology*. This view has also been taken by Brandon, *Adaptation and Environment* and "Adaptation and Evolutionary Theory"; Burian, "Adaptation"; and Sober, *The Nature of Selection*. Attacks on modern Darwinism for lacking universal laws as solid as those allegedly found in modern physics have been rebutted by philosophers of biology and philosophizing biologists in a number of ways. Ruse, *The Darwinian Revolution*, claims the relevant laws can be found. In "How Does Biology Differ from the Physical Sciences?" and *Towards a New Philosophy of Biology*, Mayr denies it but asserts that laws are not needed anyway and would not help. Sober in *The Nature of Selection* distinguishes between source laws and consequence laws and argues that Darwinism has its fair share of both. Beatty in "What's Wrong with the Received View of Evolutionary Theory?" treats laws as parts of definitions of types of systems, and thus worries far less than other philosophers about exceptions. Thompson, *The Structure of Biological Theories*, and Lloyd, *The Structure and Confirmation of Evolutionary Theory*, take the same view. For a more general account of the philosophical arguments behind this emerging new consensus, see Giere, *Understanding Scientific Reasoning* For competing "rational reconstructions" of the logical and explanatory structure of the modern synthesis, see Sober, *The Nature of Selection*; Rosenberg, *The Structure of Biological Science*; Lloyd, *The Structure and Confirmation of Evolutionary Theory*; and Brandon, *Adaptation and Environment*.

III Molecular Biology, Complex Dynamics, and the Future of Darwinism

The Molecular Revolution

In 1953 a brash American postdoctoral fellow named James Watson and a somewhat older English graduate student, Francis Crick, who were working together at the Cavendish Laboratory at the University of Cambridge, published a brief note in *Nature* entitled "Molecular Structure of Nucleic Acids: A Structure of Deoxyribose Nucleic Acid" (Watson and Crick 1953). This paper transformed the science of biology. Solving the structure of DNA, the information-bearing macromolecule, was unquestionably the most pregnant scientific discovery of the second half of the twentieth century. Together with the contemporaneous invention of the digital computer, this discovery stands at the edge of the information age. The analysis of the structure of this molecule, as Crick and Watson immediately recognized, revealed, in a single stroke, the type of language in which genetic information is encoded. The actual working out of the genetic code was accomplished during the following decade by the application of biochemical techniques in a number of laboratories, especially those of Marshall Nirenberg at the new laboratories of the National Institutes of Health in Bethesda, Maryland.

Crick's and Watson's solution of the structure of DNA was followed by a still unfolding cascade of knowledge about the internal structure and dynamics of the genome. In this avalanche of knowledge, it has become clear that the genome is internally connected in ways that far exceed the "adaptive gene complexes" postulated by the makers of the modern synthesis. It has also become clear that chance plays a demonstrably greater role in the fixation of genes than any of the founders of the modern synthesis suspected, especially at the level of protein evolution. There may well be nothing in all this that contradicts what the modern synthesis assumed about genes. In order to sustain this point, however, defenders of the synthesis have had to stress, in a rather large-scale exercise in revisionist history, that the synthesis was initially much more "pluralistic" about "evolutionary forces" or processes than it came to be in the adaptationist 1950s and 1960s. Sometimes the suggestion seems to be that the processes that were in play in the pluralist early synthesis are the same as those that have now made an appearance, or that they would have appeared of themselves in due course (Ayala 1985).

We do not think that this is the most perspicuous way of viewing the relation between the modern synthesis and molecular genetics. It is, of course, conceivable that interesting new developments in evolutionary theory might have led Darwinians to recognize processes like self-organization and chance fixation of genes by a process of discovery internal to the field. As we will see below, it was a geneticist, Barbara McClintock, who discovered "jumping genes," and a population geneticist, Richard Lewontin, who discovered the vast number of protein polymorphisms in natural populations. Their Darwinian credentials are impeccable. Where pressure has come from outside, however, it is arguble that it has come mostly from molecular genetics. Since the mid-1970s, revelations from molecular biology have formed a sort of agitated field within which debates about the adequacy of the modern evolutionary synthesis have been conducted. Before turning in the following chapter to the response contemporary Darwinians have made to these developments, we will survey some of what molecular genetics has turned up.

An event used to mark the beginning of a revolution usually represents as much a culmination of processes at work within the *ancien régime* as the starting point for a new era. Hence the molecular revolution represents a high point in the much longer history of biochemistry. The first really modern quantitative and experimental studies of the physical and chemical basis of living forms were undertaken by Lavoisier and Priestly. Their pioneering work was conducted under the long shadow of the French Revolution and was abruptly cut off, along with Lavoisier's head, during the Reign of Terror and the British counterrevolution, when Priestly was accused of materialism and atheism and his works were proscribed.

There was some justice to the charge. Materialism, if not atheism, is a philosophical proclivity that has characterized biochemists right up to the time of Crick and Watson. Biochemists have generally exhibited an ardent desire to convince the world that life is merely chemistry and that chemistry is merely physics. It was just such motives, articulated by the physicist Erwin Schrödinger in his seminal *What Is Life?* as well as a self-confessed quest for the glittering prizes, that drove Crick's and Watson's quest to solve the structure of DNA. (Schrödinger 1944; Crick 1966). In an early manifestation of this spirit, Lavoisier had shown that combustion is a material process involving a chemical reaction. It was not due to some abstract theoretical entity or rarefied fluid, "phlogiston," that combined with matter. Lavoisier had heated mercury or lead within a chamber that had a limited air supply, the volume of which could be measured. The ash of the metal weighed more after combustion, and one-fifth of the air (oxygen) had been consumed. It was clear that burning was oxidation, that is, the combining of oxygen with other atoms. At about the same time, Priestly, who had to flee to America because of his revolutionary politics, demonstrated that plants actually produce oxygen and that oxygen is consumed by animals, who depend on it. Armed with this insight, Lavoisier deduced that biological respiration is a process of slow combustion.

The new understanding of respiration showed that living processes rest on physical processes. At the same time, the analysis of fermentation showed that purely physical processes sometimes rely on the action of living things to carry them out. During the next century and a half topics like these sustained an increasingly self-conscious and well-defined research tradition, out of which the science of biochemistry was born (Fruton 1972). We have already noted the larger philosophical and ideological effects of this research tradition on German evolutionary thought in the late nineteenth century, as vitalism and semivitalism gave way to the reductionistic forms of materialism defended by figures such as Helmholtz, Buchner, and Moleshott (chapter 7).

One of the key figures in the rise of biochemistry was Frederick Gowland Hopkins, who came to Cambridge University in 1898. Hopkins founded the biochemistry department at Cambridge, discovered the amino acid tryptophan, and deduced the role of vitamins, for which he was awarded the Nobel Prize for medicine or physiology in 1929 (Weatherall and Kamminga 1992). Hopkins's greatest contributions to the field were, however, conceptual and methodological. He worked on the hypothesis, indeed the conviction, that biological systems are fully subject to the first and second laws of thermodynamics and that for this reason chemical changes occur through a very large number of small changes in structure and energy under the control of enzymatic catalysis (Hopkins 1913). If the number of steps in biochemical reactions is often very large indeed, that is because they have to occur within very tightly constrained thermodynamic limits. Otherwise, cells, and the tissue and organisms made from them, would burn up. Hopkins thus brought to biochemistry the thermodynamic revolution that Maxwell had brought to physics and that Josiah Willard Gibbs, working in isolation in America, had brought to physical chemistry (Needham and Baldwin 1949).

These ideas guided Hopkins's methodological contributions. He was convinced that biochemical changes can and should be studied directly by experimentation on the cell rather than by trying to deduce them, as had previously been done, from test-tube chemistry. That involved a sustained effort to unravel how the cell works. It was no mean feat. From one point of view, the cell is a very small thing. From a more fundamental and more interesting perspective, it is a monster of complexity, a world of its own, in which matter, energy, and information are cycled into energy-transforming pathways of enormous but nonetheless unravelable complexity. By World War II, the chemical transformations involved in fermentation and the related metabolism of glycolysis (from Greek for "sugar splitting") had come to be fully understood. By this time, it was clear that these cellular processes are facilitated by the enzymatic or catalytic action of the complex macromolecules called proteins, whose differing properties depend on the specific sequence of their constituent amino acid chains. It was not long before the proteins that facilitate a series of reactions like those in glycolysis were isolated, largely by the German school of biochemists. The structures of the proteins that catalyze all manner of cellular reactions and pathways were soon being determined by "sequencing" their amino-acid chains, the methods for which were pioneered by the Cambridge biochemist Fred Sanger, and by the X-ray crystallographic methods of Max Perutz and John Kendrew.

By 1961, the pathways of cellular respiration had been largely worked out by English and American biochemists, many of whom had been trained at Cambridge. The function performed by cellular respiration is to transduct energy in the cell, that is, to convert it from one form to another so that it can perform work. The mechanism by means of which this is done involves the capture of the energy produced in metabolism by the molecule adenosine triphosphate (ATP). ATP molecules are stable under the physiological conditions in the cell until and unless they are acted upon by one or more of a myriad of enzymes that can convert the enormous energy stored in them into useful chemical or other types of work. ATP can be viewed as a kind of energy "currency" that connects the various parts of the metabolic "economy" of the cell. Actually, however, it is not ATP per se that holds the energy but the fact that the ratio of ATP to adenosine diphosphate (ADP) is held very far from equilibrium, with ATP constantly being consumed by some metabolic reactions and restored by others. The mechanism by which the energy-yielding reactions of the cell, or the slow combustion of cellular respiration, drives the synthesis of ATP to levels ten orders of magnitude away from equilibrium was eludicated by the Cambridge-educated biochemist Peter Mitchell (Mitchell 1961; see also Williams 1961). Mitchell's proposal, called the "chemiosmotic theory," involves the translocation of protons across the membrane of mitochondria (organelles, or little organs, in the cell that serve as power packs). The coupling of the chemical reactions of respiration and the synthesis of ATP through a flow of protons represents an emergent property of the organization of the membrane system, rather than a process reducible simply to its constituent chemical components. It came as a surprise to some when the Nobel Prize was awarded in 1978 to someone who had solved a problem at the physical roots of life by refusing to abide by what he regarded as overly reductionistic research programs (Weber 1991b).

We may now retrace our steps a bit in order to take notice of the fact that Hopkins's biochemical program first intersected with the genetic theory of natural selection in the imposing figure of J. B. S. Haldane. Hopkins was Haldane's mentor at Cambridge. In fact, Haldane was appointed to the first Sir William Dunn Readership in Biochemistry in 1923, and worked in that capacity in Hopkins's laboratory. Under Hopkins's influence, Haldane acquired his deep conviction that biochemical explanations are more fundamental than morphological ones. In this spirit, Haldane made important contributions to the theory of enzyme kinetics. As far back as 1920, he had begun developing a hypothesis first proposed by Archibald Garrod in 1909 to the effect that a specific gene is responsible for the production of a specific enzyme. Substantiating that idea was a matter of discovering the mechanism by which this is done. The British tradition of "physiological genetics" that Hopkins inspired, and that Haldane began to carry out, was born out of a burning desire to understand the physiological basis and functions of individual genes. Thus, Haldane's own research program in population genetics was focused on finding out why specific mutations of genes lead to alterations in the properties of the enzymes. No wonder that in the face of Mayr's assurances to the contrary, Haldane bluntly proclaimed that "the age of beanbag genetics has hardly begun."

These developments form the remote background of the successful effort to unravel the secret of DNA. This story does not begin where it ends, however, at Cambridge, but at the California Institute of Technology in Pasadena. It begins in 1928 when Thomas Hunt Morgan left his famous "fly room" at Columbia University to set up the Division of Biology at Caltech. Morgan and his Columbia colleagues had already shown that genes are located on chromosomes within the cell nucleus. They did not say precisely where, however, or how, or indeed what genes actually are. Morgan took his time even convincing himself that genes are molecular units, rather than mathematical relationships among cellular and nuclear components. Nonetheless, when Morgan used the free hand he had been given at Caltech to devise research programs in biology, he decided that work on the molecular structure of the gene should form the focus of the new unit. He did so on the understanding that the physical basis of heredity would appear where physics and chemistry and biology meet, and especially where the "new" chemistry and physics that was being articulated at Caltech by R. A. Millikan, R. C. Tolman, J. Robert Oppenheimer, Arthur Noyes, and Linus Pauling, met the new genetic biochemistry (Allen 1975; Kay 1993). By the early 1930s, the program nurtured by Morgan was laying the foundation of what would ultimately become the field of molecular genetics (Fisher and Lipson 1988; Kay 1993). This included the development of research programs in both biochemical genetics and into what were regarded as the simplest of living things, bacterial viruses.

Choice of the proper organism with which to work is often the key to a successful research program (Burian 1992). It was George Beadle who hit on one of the right choices. After receiving his education in plant genetics at the University of Nebraska and Cornell University, Beadle came to Caltech in 1931 as a postdoctoral fellow. There he was introduced to work being conducted in Morgan's own laboratory on the mold *Neurospora*. It was not until he had set up his own laboratory at Stanford University in the late 1930s, however, and had began collaborating with his microbiologist colleague Edward Tatum, that Beadle fully realized the advantages of working with *Neurospora* for achieving the goals of the program of biochemical genetics with which he was also familiar from his Caltech days (Allen 1975). The aim of that program was the same large one that Haldane had set for himself: to determine the relations between genes and enzyme-catalyzed biochemical reactions. In contrast to the organisms used in Haldane's research, as well as to the Columbia workhorse *Drosophila*, *Neurospora* had the advantage that during virtually its entire life cycle it is haploid rather than diploid (it has only one set of chromosomes rather than the usual two). This meant that all mutant genes could be detected phenotypically, for in haploid organisms, recessive genes cannot hide mutations in heterozygotes. Further, the rapid growth of this organism under easily maintained conditions meant that it was easy to identify the biochemical consequences of mutations and ultimately to isolate the mutated enzymes themselves.

Beadle's and Tatum's work supported Haldane's hypotheses that one gene determines one specific enzyme and that mutations in genes will lead to mutant enzymes of altered structure and activity. In the following decade or so, a number of workers were able to show that the work of Beadle and Tatum on *Neurospora* was true more generally. Building on their results, for example, Pauling was able to show that mutant hemoglobins in humans are due to specific amino acid changes in the protein (Pauling et al. 1949). In 1945, Beadle returned from Stanford to become Morgan's successor as director of the Division of Biology at Caltech. He and Tatum received the Nobel Prize for their work in 1954.

It is not only experimenting on the right organisms that leads to successful research strategies. Pregnant ideas are also helpful. The German physicist Max Delbrück, a postdoctoral fellow at Caltech between 1937 and 1939, had just such an idea. After receiving his doctorate in physics at the University of Göttingen in 1930, Delbrück became a postdoctoral fellow with the great Danish theoretical physicist Niels Bohr, who had received the Nobel Prize in 1922 for developing the quantum model of the structure of the atom. By the time Delbrück came to Copenhagen, Bohr was addressing the paradoxical wave-particle duality of electrons that characterized newer quantum models of Prince Louis-Victor DeBroglie (Nobel Prize 1929), Werner Heisenberg (Nobel Prize 1932), Erwin Schrödinger, and Paul Dirac (shared Nobel Prize 1933). It was in this connection that Bohr introduced the famous concept of complementarity, which meant that two mutually exclusive concepts, such as particles and waves, must be applied if we are to get a complete description of many phenomena. By 1932, Bohr was trying to extend the same notion into the biological realm. He suggested that in genetics, the concepts of information content and potency for development form a complementary duality in living things. Delbrück took up this research program. By investigating biological systems with the methods of physicists and by pushing inquiry to the point where paradoxes would emerge, he hoped to find biological complementarity (Fischer and Lipson 1988).

Delbrück first sought to find paradox and complementarity in the descriptions of mutations given by geneticists and how they might be redescribed by physicists. When Delbrück returned to Berlin from Copenhagen in 1932 to work at the Kaiser Wilhelm Institute, he began the process of transforming himself from a theoretical physicist into a biologist and began a research program on the interaction between radiation and the genetic material. He thought he would find that radiation-induced mutation cannot be explained in quantum mechanical terms alone. Quantum mechanical and the genetic explanations would, in that case, both have to be used, even if they could never be conceptually joined, like waves and particles. Delbrück discovered, however, that he could indeed devise a quantum mechanical model of mutation due to radiation that was sufficient to explain the data all by themselves. What was more, his calculations suggested that the size of the target of the radiation that induced a mutation, that is, the size of the illusive gene, must have the dimensions of macromolecules like proteins or nucleic acids (Delbrück, Timofeeff-Ressovsky, and Zimmer 1935). By way of Schrödinger's *What Is Life?* (1944), the "Delbrück picture" of the gene and its mutations influenced a generation of researchers, who came to view the gene as Schrödinger's "aperiodic crystal." This gave the rapidly emerging field of molecular genetics a strongly physical flavor.

When Delbrück arrived in Pasadena in 1937, he was well prepared to profit from a brief collaboration with Emory Ellis, who was working on bacterial viruses. These viruses live by infecting bacteria. Hence, they became known as "bacteriophages" ("bacteria eaters," from the Greek *phagein,* "to eat"), then more simply as "phage." Phage were an answer to a theoretical physicist's wildest biological dreams. It was not only that Delbrück could work on them with simple equipment, and expect within a very short period of time to obtain quantitative data. Phage were reputed to be the smallest possible *living* things, on the boundary between life and chemistry. In fact, phage could be crystallized like mere molecules. Delbrück wondered if this fact would require treating phages under both physical-chemical and the biological descriptions, yielding the complementary he was looking for. He was already generalizing in these terms, once opening a talk at Caltech with the memorable line, "Let us imagine a cell as a homogeneous sphere."

Within two years, Delbrück had laid the foundation of phage genetics in an elegant set of papers that attracted attention from a number of geneticists (Ellis and Delbrück 1939; Delbrück 1940a, 1940b). Delbrück extended this influence by giving phage courses for many summers at the Woods Hole Biological Laboratory (1945–1971). In 1940, he took a position at Vanderbilt University in Tennessee, where the following year he began a long-term research collaboration on phage with Salvadore Luria. Delbrück and Luria demonstrated, by use of electron microscopy, something that turned out to be crucially important: The phage protein shell never enters the bacterium. Yet when the bacterium is lysed, a hundred or so phage particles are liberated (Luria, Delbrück, and Anderson 1943). What kind of magic trick was that? How did the information for making phage get in there? It would be difficult to know unless more was known about bacteria. For this reason, Delbrück and Luria went into bacterial genetics. Their work on the differential sensitivity of mutant bacteria to viral infections was as important as Joshua Lederberg's and Tatum's demonstration of bacterial sexuality (Luria and Delbrück 1943; Lederberg and Tatum 1946). After the war, Delbrück left Vanderbilt to join the faculty at Caltech. Luria stayed at Vanderbilt, later moving to the University of Indiana, where he became research professor for a young graduate student named James Watson. Through Luria, Watson knew Delbrück. Consciously taking up Delbrück's general research program, Watson wrote monthly letters to Delbrück about the progress of his postdoctoral research at Cambridge.

The discovery that phage do not enter bacteria but do come out of them shifted attention to which macromolecule carries genetic information. Since phage consist entirely of only two such molecules, proteins and DNA, it was obvious that one or the other had to be the information-bearing molecule. At the first phage meeting in April 1943 in St. Louis, Delbrück and Luria informed their colleagues about an experiment that had just been performed in Oswald Avery's laboratory at the Rockefeller Institute (Fischer and Lipson 1988). This was an experiment with the bacterium that causes human pneumonia, which in the period after World War I was a leading cause of death. It showed that a nonvirulent strain could be transformed by dead bacteria into a virulent form. Because they could extract dead bacteria of the virulent strain in such a way that proteins, carbohydrates, and lipids were excluded, leaving only nucleic acids as the likely culprit, they concluded that the "transforming principle might be DNA" (Avery, MacLeod, and McCarthy 1944). This flew in the face of the widely held dogma that only proteins are complex enough to hold the information of the gene. It was assumed that DNA and RNA, which are made of only four types of chemical bases, are just too simple.

Unequivocal proof that genes are indeed made of DNA rather than protein soon came in a classic experiment devised and carried out by Alfred Hershey and Martha Chase (Hershey and Chase 1952). Hershey and Chase labeled the protein of bacteriophage with radioactive sulfur and the DNA with radioactive phosphorus, demonstrating by these means that only the DNA enters the bacterial cell. This experiment, and the subsequent work of Crick and Watson, meant that Delbrück would not find the paradox he sought in phage genetics. On the other hand, he had created, and pushed along, one of the most fecund research programs of the twentieth century. For doing this, Delbrück shared a Nobel Prize in 1969 with Luria and Hershey.

Naturally enough, a race ensued after the publication of the Hershey-Chase experiment to determine the structure of DNA (Watson 1968). One of the main contestants was the Caltech group, especially the structural chemist and crystallographer Pauling, who until then had been working on the structure of fibrous proteins such as keratin. Earlier, Pauling had used quantum mechanics to develop a compelling theory of the chemical bond, which described the relationship between chemical structure and reactivity. It was work for which he would receive a Nobel Prize in 1954. True to the deep link between molecular genetics and physics that was the very spirit of Caltech, Pauling used models informed by his type of structural chemistry, together with published data on DNA, to find his way to the structure of DNA.

The other important group of contestants directly employed crystallography. The British school of crystallographic structural analysis at, among others, the Cavendish Laboratory at the University of Cambridge and at Birkbeck College and King's College of the University of London included William Astbury, John D. Bernal, Sir Lawrence Bragg, Max Perutz, Sir John Kendrew, and Maurice Wilkins. Perutz and Kendrew had been deploying X-ray crystallography on protein structure since the mid-1930s, working on the globular oxygen-binding proteins hemoglobin and myoglobin. In Wilkins's laboratory, however, Rosalyn Franklin had begun in 1951 to make careful crystallographic studies of fibers of DNA (Sayer 1975). It was painstaking work of the sort made for constitutional empiricists like Franklin. In 1952, after Hershey and Chase determined that the carrier of genetic information had to be DNA, Franklin's work gained significance and urgency. It was in this climate that Watson arrived at the Cavendish and joined with Crick in the explicit aim of winning the Nobel Prize by discovering the structure of DNA (Watson 1968). Unlike Franklin, who would otherwise have been in a good position to deduce DNA's structure, they were not patient enough to be empiricists. Like Pauling, they were model builders (Judson 1979).

The story of the discovery of the structure of DNA has been told from a highly personal perspective by Watson in his memoir, *The Double Helix* (Watson 1968). A more sober, and slightly less self-important, memoir was later published by Crick (Crick 1988). In defiance of the received image of scientists as selfless high priests of progress, Watson and Crick gloried in competition and quested for recognition. Watson does not seem at all dismayed by his admission that he and Crick extracted crucial crystallographic information from Franklin in ways that were manipulative and chauvinistic. Nonetheless, Watson's and Crick's ability to solve their problem was based on a well-conceived strategy.

Model building was a method that was radical at the time. With help from computers, it is now pervasive throughout science. By constructing elaborate models, which looked like Tinker Toys gone crazy, fitting data to their models and then revising the models, Watson and Crick discovered that all the relevant data fit beautifully into a model of DNA that had the additional advantage of instantly making clear how it could replicate itself and how it could contain enough genetic information to make (in some sense of that word) an organism as complex as any you might wish.

By 1965, an overall picture had emerged of how genetic information in DNA is converted to specified sequences of proteins with enzymatic function, some of it worked out by Crick (1957). Sequences of double-stranded DNA, in the four-letter code of its dioxyribose phosphate base pairs (A-T-C-G) grouped in three-letter "codons," specify particular kinds of amino acids in a particular order in the protein polymer. There are twenty amino acids and sixty-four possible codons (four letters in three-letter codons). The code is, therefore, redundant, even when we allow for codons reserved for "start" and "stop" signals (punctuation marks, as it were). The making of amino acids is accomplished through the intermediary of single-stranded messenger RNA (mRNA), a nucleic acid closely related to DNA. mRNA carries the *transcribed* message from the "master tape" of DNA, as if it were a cassette, to a readout or decoder device called the "ribosome," which in turn translates the nucleotide code into a polypeptide chain of amino acids with the aid of soluble transfer RNA (tRNA) molecules, which recognize codon triplets and carry, through a covalent chemical bond, specific amino acids. When series of amino acids are strung together, the resulting "polypeptide" folds up in regular ways to produce the globular or fibrous entities called proteins. Protein molecules can function as structural elements in the cell. But far and away their most important function is as enzymes that catalyze one of the over ten thousand chemical reactions necessary to sustain a living cell in a multicellular organism.

The molecular view of life is as beautiful and elegant a discovery as has ever graced the history of science. As Nobel Prizes were being handed out almost every year for wave after wave of advances in molecular biology, institutes of molecular biology were established at universities the world around. Newer departments of biology, formed in the 1960s and 1970s, were predominantly devoted to molecular biology. In established universities, traditional areas of biology were marginalized unless they could incorporate molecular biological approaches. ("Whole organism" biologists often felt put upon.) In this heady context, Sanger, working at the Medical Research Council (MRC) laboratories at Cambridge, developed the basic methodology for determining the sequence of the amino acids in a protein. He was duly awarded a Nobel Prize in chemistry in 1958. Later, methods were invented for determining the sequence of the base-pairs in DNA. Sanger, together with the American Walter Gilbert, got a second Nobel Prize in chemistry in 1980 for that. (Sanger is the only person to receive two Nobel Prizes in science. Madame Curie's and Pauling's second Nobel Prizes were for peace.)

As the technology of sequencing continued to improve, it became possible to look directly at the genetic message in the DNA and to envision a sustained effort to sequence the human genome (Wills 1991). The genetic "recipe" for making a human being will eventually fill many large books with small print of sequences written in the genetic code, on which various computer programs will be deployed in the search for patterns. For a while, it looked as if the Human Genome Project would closely resemble the Manhattan Project. The first director of the project was James Watson, playing the role of Oppenheimer. Although the project continues to receive high levels of funding from the U.S. government, there has been a controversial policy decision to privatize large portions of the project. This has raised vexatious problems about private firms' "patenting life," leading to controversies in which Watson was forced to resign because of possible conflicts of interest.

Meanwhile, work on deciphering the mechanism of protein synthesis—in many ways a more complex problem than DNA—had proceeded to the point at which it became possible to think clearly for the first time about the evolution of proteins. Proteins, after all, as well as the things that are made out of them, evolve. Their evolution, moreover, is in some ways more fundamental and telling than the evolution of organisms. As more and more sequences of proteins were determined and interpreted in the light of the genetic code, genealogies of proteins were worked out. Indeed, protein evolution was soon used as an independent source of information about phylogeny and a method of checking conventional morphologically based classifications. This works in the following way.

Proteins are made up of amino acids. Information about which amino acid is to be incorporated at a given place in the protein is what is encoded in structural genes (leaving aside genes that code for various types of RNA itself). With the passage of time, one amino acid can take the place of another in the same protein. If the mutation produces an enzyme with impaired function, or perhaps with improved function, we would expect natural selection to act upon it. It appears, however, that most mutations in the amino acid at a given position do not change the function of the protein of which they are parts. In 1965, Zuckerkandal and Pauling hypothesized that the number of amino acid differences between the same protein in two species is a linear function of the time since those organisms last shared a common ancestor. Basing their work on an examination of the sequences of hemoglobin and myoglobin in a wide variety of species, they proposed that differences between sequences are the result of a regular-ticking "molecular clock." The rate of replacement is more or less constant. A few years later, Walter Fitch and Emmanuel Margoliash used protein sequences to construct a phylogenetic tree showing how amino acids in a very old and very important protein, cytochrome c, differ from species to species (Fitch and Margoliash 1967). These trees are astonishingly convergent with those that had been painstakingly constructed by classical systematists studying morphological differences at the phenotypic level.

There was, however, a difference. Replacements of amino acids seemed to be so regular that they could be used approximately to date branching in actual geological clock time, and not just to date relative degrees of branching. In the early 1970s, the idea of an "evolutionary molecular clock" was championed by Richard Dickerson, who showed that amino acid substitutions in cytochrome c are in fact linear with geological time and that the rate of change in that protein is much slower than in the structurally less constrained hemoglobin (Dickerson 1971; Dickerson, Timkovich, and Almassy 1976; Dickerson 1980). Allan Wilson and Vincent Sarich, who developed a short-cut in which an immunological index was correlated to the amino acid differences between a protein isolated from various species, also saw clocklike behavior (Sarich and Wilson 1966). Indeed, they caused considerable agitation among anthropologists by dating the human-chimpanzee divergence at 5 million years. (The accepted value had until then been 15 million years.)

Twenty years of data collection and analysis have largely vindicated Wilson's claim, and the molecular clock hypothesis for protein evolution more generally (Wilson, Carlson and White 1977; Kimura 1987; Jukes 1987; Zuckerkandel 1987; for limitations and criticisms of the molecular clock hypothesis, see Scherer 1990 and Gillespie 1991). For example, Wilson showed that data from seven proteins, each evolving at a different rate, could be put on a single graph that gave a straight line when the number of amino acid substitutions (or more recently nucleotide base-pair changes in DNA) was plotted against geological time.[1]

A no less interesting discovery than the putative molecular clock is that an enormous amount of the DNA in the genome—perhaps up to 95 percent—turns out to code for no structural proteins at all. This has been dubbed "selfish DNA" by Doolittle and Sapienza (1980). Sometimes it is called "junk DNA." These functionless stretches of DNA are not, by definition, genes. We should not too hastily assume, however, that this extra DNA is in fact functionless. For all we know, it might be chock-full of still undiscovered genetic information. Genes come in two kinds: structural genes, which are functional units that code for the production of proteins, and regulatory genes, which control the expression of structural genes, that is, tell structural genes when to turn on and when to turn off. Perhaps as much as 90 percent of the genome in multicellular organisms is devoted to regulating the expression of genes during embryological development. Developmental genetics is, if not in its infancy, then certainly no more than in its adolescence. Thus even if some stretches of DNA do not code for structural gene products, they might be involved in heretofore unknown regulatory functions or even, as we will much more speculatively note below, in functions related to evolution.

Research into regulatory genes goes back to the seminal work of François Jacob and Jacques Monod (1961). This work was the outcome of a continuous research tradition in microbial regulatory physiology that had been going on at Institute Pasteur in France since the nineteenth century, which unlike much American and British transmission, and even physiological, genetics, had retained a close connection to developmental questions (Burian, Gayon, and Zallen 1988). Jacob and Monod's "operon" model of gene regulation and expression put French genetics on the map, giving French evolutionary thinking a Darwinian cast for the first time. Their model has it that regulatory genes work by coding for "repressors" of a gene whose products would otherwise be produced, or, alternatively, for "inducers" that activate the expression of a gene. Jacob, Monod, and André Lwoff were awarded the Nobel Prize in 1965 for their contribution to understanding the regulation of gene expression in bacteria.

In bacteria, genes so regulated are activated or deactivated in response to changes in metabolism and available nutrients. By analogy, regulatory genes in multicellular organisms, which are large sequences analogous to a computer program, are thought to specify when a particular gene should be turned on or off during development and differentiation. The pattern of gene expression defines the structure and function of the various cell types. The developmental or ontogenetic program as a whole, therefore, specifies (even if does not by itself cause) the distinctive morphology of the organism. Organisms with very similar programs belong to a single species. The more distant the species, the greater are the differences in the program. Even a slight reordering of the developmental program, sometimes through changes as small as a single base pair (point mutation) or in the position of a structural gene that comes under the influence of a different regulatory gene, can quickly and sometimes radically alter the developmental outcome. Sensitivity to initial conditions is apparent. By contrast, changes in structural genes have little evolutionary consequence by themselves. Studies of protein evolution have shown, for example, that human proteins are often identical to chimpanzee proteins. Indeed, comparison of over two thousand positions in the amino acid sequence of six proteins reveals only one amino acid difference between chimpanzees and humans, and there is at most a 1 to 2 percent overall difference between these species (King and Wilson 1975)! This suggests that the putatively great genetic difference between chimps and humans since they last shared a common ancestor, about 5 million years ago, is primarily in their developmental programs, not in their structural genes. This conclusion reinforces the idea that a molecular clock may keep ticking away at the level of protein evolution while more interesting things are happening elsewhere. Comparison of blood proteins shows, in fact, that there is as much variation at the protein level between closely related amphibian species that have not changed morphologically over the past 65 million years as there is between mammals that have diversified significantly over the same time period (Wallace, Maxon, and Wilson 1971). In this case, the reason that phylogenetic trees calculated from protein sequences approximate well to traditional evolutionary trees becomes clear. Protein sequence differences measure time since divergence rather than speciation events per se. It remains a distinct possibility, on the other hand, that speciation events correlate with relatively sudden changes in developmental programs, presumably in small populations (Gould 1982a).

In spite of the intense pace of new discoveries, it is a safe bet that we are still at the early stages of the growth of molecular developmental biology. As more sequences of total genomes, including the human genome, become available and as current and future experiments provide more insight into developmental mechanisms, we should be able to begin deciphering the code in which developmental programs are written. Until then, we are left observing evolution from the molecular perspective as some sort of change in ontogenetic programs and overall genome organization, coupled with regular change in structural genes, which provide a useful method of measuring elapsed time since divergence from common ancestors.

It was their premature fascination with the idea that the action of significant evolutionary change takes place in developmental genetics that led Richard Goldschmidt and C. H. Waddington, who worked before the molecular revolution, to the margins of Darwinian respectability (Goldschmidt 1938; Gould 1982c; Gilbert 1988, 1991). Still, Goldschmidt's and Waddington's insistence that sudden reorganizations of developmental dynamics are of the highest evolutionary importance, as well as their suspicion that an ever more hardened modern synthesis was betraying that fact, was to be proved prescient. The new focus on developmental genetics, informed by recent molecular biology, has finally allowed their insights to resurface and to find analytical terms worthy of them (Gould 1982c).

As information has accumulated about the activity of genes in higher organisms, it has become clear that structural genes come in "pieces," called "exons," which code for portions of proteins. Between exons are "introns," noncoding, intervening base-pair sequences that are spliced out of the mRNA before it leaves the nucleus to direct the synthesis of the protein at ribosomes out in the cytoplasm of the cell. The exon portions correspond to modules of the folded protein structure (called "domains"). A basic function of introns is to provide spacers between domain-functional substructures of the protein. (There are usually somewhere between two and eight of these substructures for each protein.) Introns can do more than that. They may play a catalytic role, for example, in changing nuclear RNA to messenger RNA, a process that involves what appears to be "editing," in which specific bases are added or changed (Rennie 1993).[2]

This experimental and empirical work has provided a more receptive environment for a startling discovery that had been made some time previously but had never been understood, or even, for that very reason, fully accepted. The American corn geneticist Barbara McClintock, working in relative isolation, proved that genes could "jump," or be moved, from one area of the chromosome to another and could even reposition themselves on another chromosome. Just as Mendel's results had been dismissed as pecularities of the organisms with which he worked, so McClintock's work was at first interpreted as a quirk about corn or as an eccentricity of an otherwise highly respected geneticist (Keller 1983). It is true that what was relatively easy to verify in corn, with its capacity to show complex genetic differences in such phenotypes as the color and pattern of kernels, would be difficult to show elsewhere. It turned out, however, that McClintock's results were experimentally reproducible in other systems and in fact were highly generalizable.

That gene transposition can be achieved experimentally, using a virus or a special DNA sequence called a "transposon" as a vehicle of transmission, has not only created an experimental tool of great power and practical significance but has made it possible to see that something like this goes on spontaneously in the genomes of a wide variety of organisms all the time. The transposition of genes from one site on a chromosome to another is possible because specific enzymes can recognize transposons, can cut or cleave the DNA at an appropriate spot, and then can reinsert the gene(s) that are attached to the transposon at another site on the same or a different chromosome. When this occurs, changes are observed in the phenotype, even though there is no change in the gene itself, and no substitution by an alternative allele.

Here is a fertile new source of variation. The mixing and matching of exons allows exploration for new catalytic functions. This can occur without disrupting existing patterns of enzyme-catalyzed metabolism through a process of gene duplication by which some genes make extra copies of themselves. As long as one copy is expressed and "kept" for its original function, the others, which may or may not be expressed, can recombine with other exons to create new types of enzymic activity or (freed from consequences) mutate more freely. Such a cluster of duplicated and diverging genes can form what is called a multigene family.

Hemoglobin, the crucial protein that transports oxygen in red blood cells, provides a good example. There exists a large and diverse family of genes that code for various types of hemoglobin peptide chains. An ancestral gene, a monomer (a single polypeptide chain), has undergone a process of repeated duplications and divergence to produce a tetramer (an association of four polypeptide chains), which is composed of two related types of monomers. This tetramer has emergent properties that give it an altered and cooperative affinity for oxygen, as well the ability to regulate that affinity. In addition, chains used to make the tetramer have become specialized to the various needs of the embryo, the fetus, and three types of adult chains, two of which have only recently diverged from each other. In the hemoglobin gene family, there are also some nonexpressed genes, called "pseudo-genes," which show a greater degree of sequence difference from the myoglobin gene than can be seen in hemoglobin genes that are tied down to specific functions in the life cycle. These may represent "evolution on the fly," as it were, or at the very least variation that is available for selection at a later time. In the case of the family of genes that code for antibody proteins, it is difficult to avoid the conclusion that the ability to generate significant variation has itself been selected for, for this family of genes shows an even more complex pattern of gene duplication than hemoglobin, as well as a great deal of exon shuffling, in which a few hundred genes can be recombined in specialized somatic cells to produce antibodies for over 10 million different possible antigens, any of which can be manufactured on site when the organism has been challenged by a particular antigen.[3]

While some multigene families develop divergent functions, others maintain extraordinary homogeneity through rectification and appear capable of evolving in concert (Campbell 1987). Sometimes a group of different genes that code for a complex phenotyptic trait lie next to each other and are linked. Such gene clusters are called "supergene complexes" (Wills 1989). The clustering of genes in a supergene complex arises when, through a process of transposition, scattered genes are brought to one location. Supergene complexes are involved in many cases of insect mimicry, especially in butterflies. Because there has been selection for a concert of genes that increases the rate of production of mutants, such complexes are able to evolve quickly in both the model organisms and the mimics. They are both engaged in what has been called a "designer arms race." We can anticipate that such supergene complexes will show, when sequenced, some sort of functional organization that will resemble that seen in multigene families.

The possible evolutionary significance of these newly discovered genetic mechanisms is startling. They bring into view a whole world of readily available, and easily maintained, variation. Moreover, that variation can go to fixation without passing through Mendel's laws and their extrapolation to populations by way of the Hardy-Weinberg equilibrium formula (Dover 1982). Sometimes called "molecular drive," some speak of this phenomenon as "non-Mendelian." Insertion of a transposable element into a gene that regulates cell growth has been shown to inactivate the gene and thus to transform the cell into a cancerous one (Rennie 1993). Extrapolation of this observation suggests that transposons might act as a type of regulator themselves, suppressing or activating regulatory genes, thereby changing the developmental pattern of the ontogenetic program. In this way, duplication, transposition, and exon shuffling may well provide mechanisms for rapid evolutionary change in regulatory programs, speciation in small populations, and punctuated patterns in phylogenesis. In this case, a three-tiered picture of evolutionary tempo and mode comes into view, but on discontinuationist assumptions that would have bothered Simpson: rapid or "tachytelic" evolution that bypasses Mendel's laws and severs macroevolution from microevolution; "horotelic" evolution at the level of organisms in populations by means of anagenetic change; and slow or bradytelic evolution by means of the routine, clocklike replacement of amino acids in evolving proteins (Rennie 1993).

The rich, complex structure and dynamics that has been revealed about genes through molecular biology is a far cry from the simple processes of classical genetics or their straightforward correlation to stretches of DNA sequence as envisioned in the early days of molecular biology. Campbell and Wills have speculated that the genome's capacity to change rapidly by the various routes described above is itself an adaptation, designed to enhance a lineage's capacity to evolve. Wills speaks of the mobile genome as an "evolutionary tool box," which facilitates the evolutionary process itself (Wills 1989). Campbell even speculates about the capacity of mobile genomes to anticipate the future (Campbell 1985). In either case, something is being claimed that goes beyond Dobzhansky's extension of adaptation by talking about selection for variation-maintenance. The capacity to evolve further by means of the production of variation and its rapid incorporation into genetic programs, if it is adaptive, must be an adaptation for a range of temporal cycles and spatial parameters wider and deeper than anything Dobzhansky had in mind, or even could have had in mind. If any of this is even partially true, we stand at a watershed in the history of evolutionary theory.

The rise of molecular genetics and biochemistry has been attended from the outset by philosophical interest. Indeed, some of its founding fathers have been eager to put themselves forward as the bearers of glad tidings about the philosophical implications of molecular biology. As if he really feared vitalism were still vital, for example, Crick argued that with the rise of molecular biology, the old specter of supernaturalism had finally been laid to rest and that not only the cell but the origin of life and ultimately the whole nervous system are understandable in terms of physics and chemistry (Crick 1966). Jacob is even more explicit. "The aim of modern biology," he has written, "is to interpret the properties of the organism by the structure of its constituent molecules" (Jacob 1973, 9). For Jacob, the paradoxes of living systems disappear with the understanding that heredity is *nothing but* a molecular program. "Nothing but" is the reductionist's war cry.

The implications of the molecular revolution for evolutionary theory are a far more interesting theme than the death of an already dead vitalism. The decoding of DNA seemed at first good news for Darwinism. The picket-fence of Weismann's barrier, on which the genetic theory of natural selection was based, was suddenly transformed into the Berlin wall of the central dogma of molecular biology: *Information in biological systems flows unidirectionally from nucleic acid to protein.* To say that information cannot flow from protein to nucleic acids was not simply to repeat in molecular terms the principle of Weismann's barrier. It was to place the prohibition much deeper down in the roots of biological process. So well, in fact, did molecular genetics seem to confirm Mendelian genetics that in the late 1950s and early 1960s there was much talk among philosophers of science about "reducing" Mendelian to molecular genetics. This talk was encouraged by the "unity of science" program promoted by the reductionistic enthusiasms of then-ascendant logical empiricist philosophers of science, who held that scientific progress involves the deduction of less powerful theories from more powerful ones by way of laws operating at more fundamental levels. Some biologists, it must be admitted, were wary. In particular, Mayr proclaimed the "autonomy" of evolutionary biology from molecular biology in the well-placed suspicion that too reductionistic a spirit about genetics would undermine the knowledge evolutionary biologists had acquired about the historical and ecological processes that are indispensable for any realistic understanding of evolution. In addition, Mayr thought that molecular reductionism would impose misguided and crippling criteria on the further development of evolutionary biology (Mayr 1985, 1988).

Mayr need not have worried quite so much. The consensus emerging from a vigorous discussion about the putative reducibility of Mendelian to molecular genetics orchestrated by philosophers of science is that the dissimilarities between this case and the paradigmatic reduction of the phenomenological gas laws to statistical mechanics far outweigh their likenesses. If there is to be any reduction at all of population genetics to molecular genetics, classical transmission genetics must serve as an intermediary. It is perspicuous, however, to regard the classical transmission genetics of the Morgan school as an "interfield connection" between two research traditions, the Mendelians of the early twentieth century and cytology (Darden 1991; cf. Darden and Maull 1977 for the notion of interfield connection). It is unlikely that the theoretical terms, concepts, and background assumptions used by either of these intersecting research programs could ever fit neatly enough into the other to allow talk of reduction to be anything other than loose. For example, the Mendelians viewed an organism as composed of unit characters transmitted transgenerationally by way of germ cells. Morgan's school thought that organismic characters are caused by genes. What became the chromosomal theory of heredity connected these views. This theory stated that genes are located on chromosomes and that, through the process of meiotic division, Mendelian segregation is achieved. The conceptual shift to thinking about genes on paired chromosomes in turn suggested the possibility that genes could have a molecular structure. As we have seen, this hypothesis became the basis of research programs at Caltech and Cambridge. After the successful analysis of DNA, however, the molecular gene itself eventually was divided, like the atom, into overlapping parts with different functions, such as mutation, recombination, and coding (Benzer 1955). These functions have only multiplied, like subatomic particles, with the discovery of introns and exons. The shift from the highly theoretical conception of the gene that was postulated at the beginning of this sequence of theories to the highly concrete, but still disseminating, conception at its end will not yield a sufficently univocal sense of the term to permit reduction. This is not the case in classical statistical dynamics, where it was indeed possible to say that heat equals molecular motion.

Kitcher summed up the case against reductionism by arguing in "1953 and All That: A Tale of Two Sciences" that molecular and Mendelian genetics do intersect at several points but that each remains essentially connected to entities, processes, propositions, background assumptions, and scientific practices that cannot be mapped onto the other, even when definitions of common terms are extensively fiddled with (Kitcher 1984; cf. Hull 1974; Ruse 1984a, 1984b, 1988a, 1988b, for background; but cf. Waters 1990, 1994).[4] This is not, however, a proclamation of pragmatic antireductionism, based on an instrumentalistic view of theories (*pace* Rosenberg 1989). It reflects something about the complexity that is in the grain of nature and how it is organized (Gasper 1992). Failure to reduce transmission genetics to molecular genetics is not due to still imperfect knowledge but, in the end, to the fact that living nature has different levels of organization. The diversity of scientific communities and research traditions in biological sciences roughly reflects this fact (Darden 1991; van der Steen 1993; Bechtel 1993; Burian 1993). Thus the practice of transmission geneticists, who focus on the cellular properties and mechanisms of chromosomes rather than on Mendel's laws, is not reflected very well in the "law-talk" that the logical empiricist philosophy of science imposes on working scientific communities. Similarly, transmission geneticists show little interest in the biochemical description of genes and their molecular mechanisms unless these affect the Mendelian ratios, as they do in the case of transposition. Conversely, molecular geneticists are not even trying to derive the principles of transmission genetics from molecular genetics. Kitcher suggests that in view of these practical facts about communities of inquiry, as well as the theoretical equivocity of the relevant terms and concepts, there is no reasonable expectation that bridge principles will ever be found between DNA sequence segments and the "gene" of transmission genetics, or that molecular accounts will bring out features of transmission genetics that are interesting (Kitcher 1984).

At the same time, rejection of reductionism does not, at least for Kitcher, entail rejecting the unity of science (Kitcher 1981). (It entails only rejecting the logical empiricists' *conception* of the unity of the science.) Molecular genetics, says Kitcher, answered a worry about a presupposition of classical genetics—how gene replication and mutation occur. In allaying that concern, molecular biology affirmed the presupposition rather than reducing the old theory to the new. In the process, molecular genetics offered refinements of what counts as a gene. Each new conception of "gene" was not a reduction of an old one to a new but a sequence of replacements along a continuous line in which a dialectic between innovation and tradition is maintained (Kitcher 1982b; Burian 1985; Depew and Weber 1985).

The retreat from reductionism in genetic and evolutionary theory is part of a general retreat from the positivist tradition in the philosophy of science since the late 1960s. Many philosophers, reflecting the post-Kuhnian shift to viewing scientific change as a social process, have been finding other alternatives to reductionism by looking at the practices of communities of inquirers, and their cross-fertilization, rather than solely at the logical structure of bare theories (Darden and Maull 1977; Darden 1986, 1991; Bechtel 1986; Grene 1985). In this new context, it has became increasingly clear that what was wrong was not the continued fidelity of evolutionary theory to the natural history tradition but, as Mayr had suspected, the philosophical criteria that had been used to disparage that fidelity. The waning of the spirit of logical empiricism has also contributed to making reductionism, in biology at least, what Marjorie Grene (Grene 1971) has called a "side issue." Arguments showing the limitations and biases of reductionist research strategies in evolutionary biology soon acquired wider significance (Wimsatt 1980; Sober and Lewontin 1982). By showing that logical empiricist criteria did not hold well even in some well-developed natural sciences, the new philosophers of science helped to trim back the universalizing pretensions of the positivist tradition more generally. They thereby gave new breathing room to interpretive forms of inquiry in the human sciences, such as cultural anthropology, sociology, and history, which could never, and probably should never, meet positivist criteria.

Whether they supported or opposed reductionism, however, philosophers of biology and theoretical biologists seldom expected that there might be developments within molecular genetics that could undermine, or at least limit the applicability of, the genetic theory of natural selection itself. Crick and Watson's discovery certainly did not do this. Subsequent developments in molecular biology, however, began to suggest just this worrisome possibility. The molecular clock of protein evolution, for example, raised the possibility that the role of genes in at least some basic evolutionary processes does not depend much on natural selection. More ominous, transposition of mobile genetic elements suggested that genes can bypass the Mendelian mechanisms that mediate between molecular genetics and the modern synthesis. In fact, since the mid-1970s, the upshot of the molecular revolution has been to encourage a volcano of new theorizing about evolutionary processes, unmatched in intensity, diversity, and speculative character since the turn of the century. All in all, the present period bears more than a little resemblance to the situation at the turn of the century when Mendelism, in its de Vriesian and Batesonian dispensation, seemed about to displace the Darwinian tradition altogether. It was inevitable that the modern synthesis would be affected by these developments. In responding to these challenges, old tensions within the modern synthesis resurfaced and were intensified. These form the subject of the next chapter.

14 Expanding the Synthesis: The Modern Synthesis Responds to the Molecular Revolution

Earlier we noted that the conceptual differences among Wright, Fisher, Haldane, Dobzhansky, Mayr, and Simpson were played down when the synthesis "hardened" and when positivist philosophers of science seemed to require that the practitioners of an allegedly mature science must be parties to some blandly harmonious set of agreements (chapter 12). In this chapter, we suggest that the revolution in molecular biology has created an environment in which these suppressed tensions have come back both to haunt and to enliven contemporary Darwinism. The current atmosphere of challenge, defensiveness, reconsideration, and sheer excitement in evolutionary theory should be understood largely as a magnification of these old conflicts under the new conditions created by molecular genetics. What is true of the substance of evolutionary theory is also true of disputes about method. The molecular revolution has led some contemporary Darwinians to talk once more about universal laws and reductionistic ideals. It has led others in the precisely opposite direction.

In pointing out this pattern, we need not take a stand on whether molecular biology does, or at some future time will, actually contradict the modern synthesis. Whatever the disposition of that issue, it remains true that the rise of molecular biology has generated a series of challenges to which the modern synthesis has had to make plausible responses. In the course of doing precisely that, contemporary Darwinians often invoke, and then transform, the basic ideas of one or another of genetic Darwinism's founding fathers. As a result, differences among the makers of the modern synthesis have in recent years come to the fore far more than their agreements. The revered founders, moreover, often come back dressed in clothes in which they would scarcely recognize themselves. This pattern can be seen in each of the three recent controversies that form the subject of this chapter: debates about how, and why, selectively *neutral mutations* are fixed in populations; disputes about the *"selfish gene hypothesis"*; and discussions of proposals for an *"expanded synthesis."*

The "neutralist debate" was stimulated by the revelation that amino acids making up proteins go to fixation in populations in ways that have no particular selective value and are for this reason "invisible" to natural selection. This discovery led the Japanese population geneticist Motoo Kimura to present himself as a latter-day champion of Sewall Wright, and at the same time, of Muller's theory of "purifying" selection. Kimura's way of interpreting this data has met with stiff resistance from Dobzhansky's heirs and disciples. Indeed, Dobzhansky's quarrel with Muller has been reenacted by a new generation of defenders of balancing selection who, unlike their intellectual fathers, have been schooled in the ways of molecular geneticists.

The idea that genes are "selfish" arose when George C. Williams, and, in a more strenuous voice, Richard Dawkins appealed to properties of the gene discovered by molecular geneticists to reaffirm Fisher's atomistic approach to genetic Darwinism. In *Adaptation and Natural Selection*, Williams defended the perspective of the individual gene over the whole organism in terms of the positivistic contention that the most parsimonious among otherwise more or less equally powerful theories is always to be preferred (G. Williams 1966). Dawkins, for his part, has defended Fisher's assumption that context-free additive fitness value can be assigned to single genes less on methodological grounds than by appealing to properties of DNA or of any other automatically more-making entity that is able to serve as a "replicator" (Dawkins 1976, 1989). In this way, a proposal that was on the defensive during the heyday of Dobzhansky's and Mayr's genetic relativity has been refreshed by appeals to contemporary molecular genetics.

The third dispute to which we call attention is about the suggestion that Darwinian selection can be "expanded" to accommodate selection on a variety of levels and entities, including groups and species, and at different rates (Gould 1980, 1982b; Eldredge 1985). On this view, genic selection, rather than being incoherent, may well be one of several levels at which natural selection, or something closely analogous to it, operates. If genic selection is possible, however, so must be selection on groups, and even on species (Stanley 1979; Gould 1982b; Vrba and Eldredge 1984; Eldredge 1985; cf. Hull 1980, 1981). Advocates of a synthesis expanded this way, pluralists as they are about units and levels of selection, often look back fondly to the openness and pluralism of the "prehardened" synthesis of Wright and the early Dobzhansky and to prophetic heretics like Goldschmidt. They do so to rekindle their hope that Darwinism will continue to thrive by remaking itself into a more flexible theory of selective processes at any and every level of scale (Gould 1982c, 1983). By deflecting the selectionist activity that stands at the core of the Darwinian tradition to levels above and below the organism, Darwinians of this tolerant stripe intend to be ready should molecular biology show how rapid speciation can occur by means other than classical selection (Gould 1982a).

Studying these controversies allows us to paint a fairly comprehensive picture of the condition of the Darwinian tradition on the eve of the complexity revolution. What contributes most to the excitement and even the *pathos* of these and other contemporary evolutionary debates is the possibility that the knowledge we are acquiring about the roles of chance and self-organization at every level of biological reality will eventually reveal the inherent limits of the entire Darwinian tradition. For their part, those who are inclined to believe that the molecular revolution reveals the bounds of Darwinism are eager to see in the accommodations of Darwinian expansionists signs of appeasement. They regard the expanded synthesis as an expression of a more or less desperate desire to keep nonselectionist processes at bay by broadening, and hence weakening, the very notion of natural selection (Brooks and Wiley 1988; Ho and Saunders 1984; Ho and Fox 1988; Goodwin, 1988; Goodwin and Saunders 1989; Salthe 1993).

In reply, advocates of pluralistic Darwinism view the various programs of non-Darwinian evolution that have been put forward as suspicious recrudescences of nineteenth-century developmentalism dressed up in the clothing of the new dynamics and thermodynamics. Proponents of an expanded synthesis are certainly as anxious to welcome complexity as their opponents. In their case, however, respect for complexity means abandonment of the demand for too much explanatory unity in evolutionary biology. They hope instead to honor the contemporary perception that biological complexity must be respected by reconstructing Darwinism as a family of explanatory models that can be applied flexibly to a wide and varied range of phenomena. Their resistance to explaining evolution in terms of a single law-governed process, including the "laws of form" that the new Geoffroyians hope to see precipitating out of complex systems dynamics, is thus continuous with Mayr's attack on the putative role of general laws in evolutionary biology.

In all three of these debates, the voice of Dobzhansky's student Richard Lewontin has made itself heard. It was Lewontin who most strenuously opposed Kimura's neutralism by defending the viability of balancing selection at the molecular level (Lewontin 1974). Lewontin has also rejected genic selectionism, and the sociobiological proposals associated with it, by offering an even more radical defense of the primacy of organisms than Mayr. He argues not only that selection begins with the interaction between organisms and environments but that the activity of organisms in creating their own environments shows up the passivity that hyperadaptationists attribute to organisms as an ideologically generated distortion of reality (Lewontin 1983; Levins and Lewontin 1985). (For Lewontin ideology shows up where the reach of scientists exceeds their real, concrete grasp.) In the early 1970s, finally, Lewontin defended the possibility of selection on groups or demes, and so helped lay the foundations for expanding the synthesis to countenance multiple units and levels of selection (Lewontin 1970). In all of these contexts, Lewontin, working with his Harvard colleague, the Marxist population ecologist Richard Levins, has insisted that reductionism and overly mechanistic conceptions of evolutionary dynamics can be countered only by adopting a "dialectical" philosophy of biology, which allows one to recognize when ideology is distorting real hard science by suggesting false explanatory ideals and claims. We will pay a good deal of attention throughout to Lewontin's and Levins's polemics.

In 1968, the Japanese population geneticist Motoo Kimura advanced what he called "the neutral theory of protein evolution" (Kimura 1968). He was followed a year later by the American molecular biologists Jack King and Thomas Jukes, who, in the title of a coauthored article, rebaptized neutralism "non-Darwinian evolution" (King and Jukes 1969). Neutralists claim that the rate at which amino acids are replaced in the evolution of proteins is too regular, and maintenance of function when genes are replaced too continuous, to be affected by natural selection. Neutralists concluded that the most likely interpretation of the regular ticking of the molecular clock is that protein evolution is for the most part selectively neutral. Confirmation of the molecular clock hypothesis has aided their case.

Mutations can be considered equifunctional when one amino acid (for example, lysine) replaces another (such as arginine), without affecting the functional property of the latter that is required in that position for the function of the protein, in this case, chemical basicity. There is no reason for natural selection to scrutinize equifunctional amino acids. Indeed, there is no way that it can. Such substitutions, called neutral mutations, are selectively neutral. Nonetheless, over time genes with neutral mutations can go to fixation with predictable frequency, and so have long-run consequences for the evolution of proteins. Kimura worked out an elegant mathematical formula that predicts much of protein evolution without any need to know particular sequences of biotic and environmental change. As in Wright's shifting-balance theory, the most important predictive factors in this formula are population size and rate of mutation. He also demonstrated that for diploid organisms, the overall rate of molecular evolution is a direct function of the rate of appearance of neutral mutations averaged over all alleles over a long time, regardless of population size. Computer modeling also shows that neutral mutations are randomly fixed at an overall rate that is linear with geological time (chapter 12).

For neutralists, deviations from the molecular clock arise when natural selection is at work. Perhaps the most fascinating aspect of Kimura's account is that neutral mutation represents the expected maximum rate of allelic substitution, while the force of natural selection in removing deleterious genes and building or shoring up adaptations can be measured as a departure from this expected maximum. This is significant for the historical trajectory of Darwinism because, as we have seen throughout this book, changes in computational baselines mark major reformulations of the Darwinian tradition and, in the process, create new Darwinian research programs. For Darwin, the computational baseline was Malthusian reproduction. For the biometricians, it was the normal distribution of phenotypes. For the genetic theory of natural selection, it was the Hardy-Weinberg equilibrium. Advocates of neutralism are proposing yet another change in the baseline from which evolutionary change can be measured. "These equations," Gould says, "give us for the first time a base-level criterion for assessing *any kind* of genetic change. If neutralism holds, then actual outcomes will fit the equations. If selection predominates, then results will depart from predictions" (Gould 1989b, 20).

In spite of their provocative use of the term *non-Darwinian evolution,* accordingly, the larger significance of the neutralists' proposal is not that they want to replace Darwinism but that they want to replace the Hardy-Weinberg formula as the null hypothesis against which selection can be measured. From this perspective, "non-Darwinian evolution" is a misleading name for what neutralists advocate. Kimura's theory is a kind of Darwinism. Only when a certain conception of Darwinian natural selection is taken as equivalent to the very idea of natural selection will one be tempted to see in Kimura's theory an anti- or even a non-Darwinian theory of evolution. This is, nevertheless, a substantial change in the Darwinian tradition. The Hardy-Weinberg baseline predicts that no evolutionary change will take place unless selection or other forces intervene. If such forces are at work but are balanced, no net change will take place (Sober 1984a). The neutral hypothesis, on the contrary, predicts that protein evolution at least, and perhaps much more, will occur at a regular, and maximal, rate unless it is opposed or altered by selection. This makes the fixation of selectively neutral genes the creative aspect of evolutionary change and, in the spirit of the "classical" view of natural selection advanced by Muller, assigns to selection, as Kimura was quite aware, the negative role of executioner, purifying a presumably adapted population of maladapted genes. King and Jukes put Kimura's view as follows: "Natural selection is the editor, rather than the composer, of the genetic message" (King and Jukes 1969; Kimura 1983).

If this theory is generally true, natural selection can no longer be presumed to be even heuristically the primary agent in evolutionary processes, and genetic drift, or something like it, can no longer be blithely treated as a trivial or merely annoying secondary evolutionary force. Already aware of this aspect of the matter, Kimura came to the United States soon after the end of World War II with the explicit aim of doing postdoctoral work with Wright. In stressing the explanatory role of parameters such as population size and mutation rate, Kimura saw himself following in Wright's footsteps. He saw neutralism, moreover, as a way of vindicating the role Wright had ascribed to drift (Provine 1986). At the same time, Kimura transforms the general notion of predictable chance fixation of genes into something rather different, and decidedly less selectionist, than Wright. Kimura proposes to push the probability revolution further down into the structure of evolutionary theory by making the fixation of genes in populations a purely stochastic process. On his view, a mutation is not something that climbs up a hill but, in the felicitous analogy of Christopher Wills, is more like a pebble thrown into a pond (Wills 1989). For most mutations, even slightly favorable ones, there is a brief rippling in the gene pool that rapidly dies out. Every so often, however, one of the ripples sweeps the pond. A fluctuation in gene frequency is amplified until it is fixed. Something closer to sheer chance, or even quantum mechanical effects, may be at work in the fixation of alleles coding for amino acids. With this degree of sensitivity to initial conditions, no interesting causal stories may be down there to tell at all. The result of this changed perspective is drastic. By sacrificing Wright's attachment to the primacy and creativity of natural selection, Kimura turns Wright's anti-Fisherian stress on population parameters and numbers of generations into an argument in which it is claimed that drift can generate significant evolutionary change in protein lineages without help from natural selection at all. Drift, accordingly, is not primarily a mechanism for making variation available for selection in demes. It is a major evolutionary player in its own right. With friends like Kimura, Wright scarcely needed enemies in his long and prickly campaign to restore his selectionist credentials in the eyes of British adaptationists. It is not surprising that Wright distanced himself from his self-proclaimed disciple from the start (Provine 1986).

It is possible to play down the radical implications of neutralism, while admitting much of what Kimura and other neutralists say about protein evolution, by taking a hierarchical view of evolution. Hierarchical versions of Darwinism, in which selection and other evolutionary processes may be combined in presumptively different proportions at different levels of the biological hierarchy, have become prominent in the last twenty-five years (Gould 1982b). We suspect, in fact, that the neutralist debate provided the initial stimulus for the development of contemporary hierarchical forms of Darwinism. Only if one assumes reductionist criteria, in which it is presumed that in mature theories what goes on at the lowest level must account for what goes on at all higher levels, does neutralism make real trouble for Darwinism. On reductionist assumptions, and neopositivist criteria for verifying theories, protein evolution, and molecular evolution generally, will have to be selectionist if higher levels are to be, on pain of falsifying the entire theory of natural selection. Conversely, if the most basic level of evolution is governed by a nonselectionist molecular clock, reductionism implies that natural selection must be weak even at higher levels. Those who see in neutralism a falsification of Darwinism usually show themselves sooner or later to be reductionists of this sort. The quickest way for Darwinians to slip out of this straightjacket, accordingly, is to point out that reductionist ideals in complex sciences are no longer in good order, products as they are of physics envy and simplistic ideas about the unity of science. The quickest way in turn to enforce this general point about evolutionary biology is to assert that autonomous processes are at work at the various levels of biological organization, from genes to organisms to populations. By holding that amino acid sequences can change freely as long as the proteins they code for maintain the same organismic function, Kimura and other neutralists have made it possible for Gould and other defenders of an explicitly hierarchical Darwinism to cling, with palpable relief, to a view like the following:

In the domain of organisms and their good designs, we have little reason to doubt the strong, probably dominant influence of deterministic forces like natural selection. The intricate, well-functioning forms of organisms—the wing of a bird or the mimicry of a dead twig by an insect (for instance "walking sticks")—are too complex to arise as long sequences of sheer good fortune under the simplest random models. But this stricture of complexity need not apply to the nucleotide-by-nucleotide substitutions that build the smallest increments of evolutionary change at the molecular level. In this domain of basic changes in DNA, the neutralist theory, based on simple random models, has been challenging conventional Darwinism with marked success during the past twenty years. (Gould 1989b, 16)

From a causal point of view, enhanced degrees of freedom at the level of protein evolution can arise because function is expressed in the three-dimensional folded structure (tertiary structure) of the protein, on which neutral substitutions of amino acids have equal or neutral effects. The biochemist Harold Morowitz has called this arrangement "hierarchical embedding" (Smith and Morowitz 1982). Processes at a lower hierarchical level can tolerate the free play of variations as long as the structure and function of the higher level is maintained. Hierarchical embedding was an idea with a future in front of it. It figures prominently in calls for an expanded synthesis.

Even if all this proves sound, neutralism still poses severe challenges for the modern synthesis. For one thing, even if it leaves natural selection at the organismic level untouched, the presumed universality of the modern synthesis will have been impugned. From the perspective of Darwinism, whether old or new, it is simply astonishing that there should be clocklike behavior at the level of protein evolution and that such a large amount of evolutionary change should occur without selective significance. It was explicitly presumed by the founders of the synthesis that adequate adaptationist explanations of protein evolution would be forthcoming and that those explanations would reflect the jerks, starts, and stops of adaptation and speciation. Simpson stated this expectation categorically: "To an evolutionary biologist, it seems highly improbable that proteins, supposedly fully determined by genes, should have non-functional parts, that dormant genes should exist over periods of generations, or that molecules should change in a regular but nonadaptive way. . . . [Natural selection] is the composer of the genetic message, and DNA, RNA, enzymes, and other molecules in the system are successively its messengers" (Simpson 1964b, 1538). Similarly, the protein chemist Emil Smith, who first sequenced cytochrome *c*, wrote: "Each of these amino acids must have a unique survival value in the phenotype of the organism—the phenotype being manifested in the structure of the proteins. This is as true for a single protein as for the whole organism" (Smith 1968, 249).

By failing to anticipate the empirical discoveries that vindicated the molecular clock hypothesis, neutralism cast at least subjective doubt on the predictive prowess and explanatory fecundity of the modern synthesis. This gave rise to understandable suspicions that subsequent attempts by defenders of the synthesis to come to terms with neutralism were ex post facto and ad hoc. Ayala, for example, has tried to show that empirical findings supporting the neutral theory are consistent with the synthesis. He has, however, repeatedly shifted his ground, initially arguing, in fact, that the regularity in accepting mutations in proteins over time was only apparent, an artifact of measuring over inappropriate time frames; then that it was real but irrelevant; and finally that it provided yet another source of variation. (Ayala 1974; Ayala 1982; Stebbins and Ayala 1985). There ought to be, and usually is, an uneasy feeling when facts not anticipated by a research program are blithely explained or explained away as if they could have been, but simply were not, uncovered by that program until a rival theory discovered their existence and gave a plausible rival account of them. At the very least, the clash between neutralism and selectionism elevated the explanatory burden borne by the modern synthesis. It shifted the burden of proof.

Lewontin has defended Dobzhansky's legacy, and in particular his theory of balancing selection, with greater subtlety than Ayala. Lewontin's response to Kimura's 1968 paper was set out in his Jesup lectures of 1969, published in 1974 as *The Genetic Basis of Evolutionary Change*. This work precipitated a spectacular quarrel about neutralism that involved not only technical but ideological considerations. Kimura's 1983 book, *The Neutral Theory of Molecular Evolution*, was in large measure a reply to Lewontin.

Lewontin recognized that Kimura's reworking of Wright's genetic drift had been purchased in the currency of Muller's old theory of "purifying" selection, according to which the primary role of selection is to prune harmful mutations from presumptively well-adapted, and not very heterogeneous, populations. That idea had long before formed the theoretical basis of eugenics. Lewontin was deeply distressed, therefore, by the threat to Dobzhansky's pluralistic, democratic values that he saw lurking in Kimura's explicit indebtedness to Muller. Lewontin's book, which he piously offered as the homage of a devoted "epigone" to a great man and a great scientist, thus came to occupy in the new world of the molecular revolution the niche previously occupied by Dobzhansky's own Jesup lectures, *Genetics and the Origin of Species* (Lewontin 1974, ix). As Dobzhansky's seminal work had been directed against Muller, so Lewontin now addressed himself to someone he cast as Muller's avatar, Motoo Kimura.

Lewontin was in a position to speak authoritatively. The data on which Kimura had built his interpretation were to a large extent *Lewontin's* data. Lewontin and his coworkers had used gel electrophoresis to verify that there are vast seas of heterozygosity in natural populations, mostly in the form of polymorphic proteins that serve as enzymes (Lewontin and Hubby 1966; Lewontin 1970, 1974).[1] The average frequency of alternative alleles for humans, for example, is 6 percent and varies between 2 and 17 percent. This translates to about six thousand heterozygotic pairs of alleles per person. We have seen that estimates of the amount of genetic variation in natural populations have been going up steadily throughout the century. "Balancing" views of selection, accordingly, have generally displaced "classical" theories. Yet the degree of polymorphism revealed by electrophoretic analysis of proteins had never been anticipated even by Darwinians like Dobzhansky, who were otherwise very lavish about natural selection's tolerance for variation. Although most of this diversity is carried within recessive heterozygotes, the number of mismatches in a breeding population would presumably have a marked effect on fitness, or at least would throw the burden of proof once more onto those who cleaved unto Dobzhansky's expansive, but insufficiently evidenced, confidence that large "genetic loads" are nothing to worry about. Contemporary evolutionary debates began when it had to be decided what to say about all this unexpected variation being carried around in natural populations.

Kimura's interpretation was that there was indeed nothing to worry about. Selection is blind to all this variation at the protein level. Thus it constitutes no genetic load. Lewontin put the claim in the following way:

If we take it as given [as Muller and Kimura do] that balancing selection is rare and that natural selection is nearly always directional and "purifying," how can we explain the observed polymorphism for electrophoretic variants at so many loci? We can do so [they say] by claiming that the variation is only apparent. That is, we can suppose that the substitution of a single amino acid, although detectable in an electrophoretic apparatus, is in most cases not detectable by the organism. . . . [On this view] amino acids are "genetic junk," revealed by the superior technology of the laboratory but redundant physiologically (Lewontin 1974, 197–98).

What bothered Lewontin was that Kimura's interpretation left open the possibility that if all this polymorphism were visible to natural selection, that is, if it served as variation in a selectionist context, it would constitute, as Muller would have had it, an intolerable genetic load. It was on precisely this assumption, in fact, that Kimura responsored Muller's theory of negative or purifying selection and rejected Dobzhansky's balance theory, explicitly proclaiming that "it has been known since the great work of Muller in the early days of *Drosophila* genetics that negative selection is the most common form of natural selection" and that "most of the experimental evidence on which Dobzhansky thought to support his balance hypothesis turned out to be invalid" (Kimura 1983, 19, 51).

Lewontin challenged the logic of the argument. He saw the pervasive polymorphism he had turned up not as an anomaly for Dobzhansky's theory of balancing selection but as a way of reconfirming and even radicalizing its theoretical and moral messages. The increased level of polymorphism revealed by molecular genetics sustains Dobzhansky's insistence on the tolerance of natural populations for variation and the great resources that variation presents to selection. Selection itself favors the maintenance of equivalent polymorphisms because populations having wide diversity are most fit for the changing environments they encounter, which they are constantly creating, destroying, and recreating by their own resource-utilizing activities. By showing that the amount of enzymatic polymorphism within a population is on the same order of magnitude as variation between populations, Lewontin sought to reaffirm Dobzhansky's social and political pluralism. The genetic basis of racism, he thought, is simply refuted by this evidence (Lewontin 1974; Lewontin, Rose, and Kamin 1984). Nature is not only tolerant of differences, and prudently stores much variation, but embraces humans within one large family, in which between-group differences are no more pronounced than within-group differences. For this reason, between-group conflicts cannot be rationally defended in racialist or other biologistic terms. That, says Lewontin, is an ideologically contaminated misuse of biology (Lewontin 1992).[2]

It was not only worries about Kimura's invocation of Muller, however, with its latter-day eugenicist overtones, that caused Lewontin's juices to flow. His fire was soon being trained on what he regarded as the faulty assumptions and ideologically distorted conclusions that the Fisherian wing of the synthesis was reading out of, or into, the molecular revolution in genetics. One of these developments was "genic selectionism," the idea that the unit of selection, and its chief beneficiary, is the single gene, considered as a chunk of DNA large enough to survive recombination over many generations. To this issue we now turn.

Genic selectionism arose in the context of the notion of kin selection, the idea that genotypes gain greater representation in future generations by favoring cooperative or altruistic adaptations at the level of related organisms. Many of my genes are, after all, in my cousins and nephews. If I care for them and they care for me, the genes will more successfully be passed along, albeit in different combinations and in different bodies. The idea has been around long enough for Haldane to have expressed it in his typically epigrammatic way by asserting, "I will die for two brothers, or eight cousins." When looked at from the genic perspective, therefore, natural selection has reason to favor cooperative phenotypes. The good news for Darwinians in this idea is that an explanation for altruism could at last be fashioned in Darwinian terms without falling back, as Darwin himself did, on "group selection," that is, selection for traits that are ascribable to individuals only insofar as they share in group-level properties (Darwin 1871). The bad news is that Darwin's stress on individual organisms might be subverted by attributing mysterious powers and interests to individual genes, for kin selection seems most readily explicable by making genes, rather than organisms, the "units," and beneficiaries, of selection.

Contemporary genic selectionism builds on Fisher's preference for doing business at the genotypical level. At that level, Fisher hoped to find additive variance in fitnesses, and so predictive laws. After 1953, the putative reducibility of Mendelian and transmission genetics to molecular genetics encouraged the further development of such a "gene's eye" view of natural selection. As it happened, moreover, this interanimation between molecular genetics and gene-centered versions of population genetics occurred just when logical empiricist philosophers of science, who prized the reducibility of theories to the laws of more basic sciences, were at the height of their influence. For both scientific and philosophical reasons, accordingly, Fisher's preference for doing evolutionary biology in genotype space, and for assigning average fitness to each separate gene, was less exclusively a British preoccupation by the early 1960s than it had been. Thus in 1966 a philosophically *au courant* American biologist named George C. Williams argued that each gene can in principle be assigned an independent fitness value, even when it works in tandem with others, like rowers in a racing crew (as Dawkins puts it), and that in consequence we ought to assume that the gene is the "unit of selection" in the interests of explanatory parsimony, enhanced predictive prowess, and other positivist philosophical desiderata (Williams 1966).

In treating the gene as the "unit of selection," Williams meant that genes, considered as coding segments of DNA that survive the shuffling of meiosis, are the stable entities that survive into the future, while the individual organisms that bear them arise and perish evanescently. Since it is the sole business of genotypes to replicate themselves by the meiotic process so beautifully explained by the structure of DNA, those genes that are represented in greater numbers in the next generation are the beneficiaries of the selective process. Accordingly, genes that code for phenotypic traits which enhance their own replicative success will be favored by natural selection. Some of the phenotypes in question may well be cooperative behaviors.

Chief among Williams's whipping boys, and the explicit stimulus and target of his book, was E. C. Wynne-Edwards, a Scot whose ecologically centered view of evolution had led him to affirm that groups, especially groups that develop means of regulating their own numbers, are sometimes units and beneficiaries of selective processes (Wynne-Edwards 1962). Wynne-Edwards was bravely, if somewhat haplessly, struggling with one of the Darwinian tradition's oldest and biggest headaches, as had the Russian Darwinians and American field ecologists before him. How, in a purely competitive world, where each organism is assumed to be in a Hobbesian relation to every other and is living under the threat of Malthusian limits, could cooperation and self-sacrifice, or what is generally called "altruism" (coined from Latin *alter*, "other," as in other-centeredness) take hold, evolve, or even have arisen in the first place? Wynne-Edwards was following Darwin in thinking that under such conditions, altruistic behavior could arise and become an adaptive product of natural selection only when cooperating groups are the entities on which selection works. In particular, he argued that group selection is at work in creating mechanisms that allow populations to regulate their reproduction in the face of scarcity.

Williams opposed this idea by arguing that the mitigation of individual self-interest required by this group-selectionist hypothesis is, in the first place, inherently implausible. Indeed, in a strictly Darwinian world, it is well-nigh impossible. Cooperative behavior is vulnerable to a few cheaters, or even to one. In a widely varying and competitive world we can expect that noncooperators will abound and will prosper. In any case, an alternative explanation of altruism that conforms much better to well-entrenched Darwinian assumptions was already available. Taking a gene's-eye perspective allowed Williams to affirm W. D. Hamilton's ascription of the causal priority of kin selection over other forms of group selection and to reduce kin selection to individual selection at the genic level (Hamilton 1964). Hamilton, like Haldane before him and Williams after him, noted that if genes rather than organisms are treated as the primary units of selection, it matters not a whit how many bodies are used to transmit genetic information to the next generation. Cooperation may not be the best deal for each individual organism. Indeed, it is positively harmful to the self-sacrificing ones. But it may be the best deal for a genotype that, by its very nature, seeks to get as many copies of itself as possible inserted into the chromosomes of the next generation by using the bodies of genetic relatives as carriers.

The argument is this: Relatives share genes in proportion as their degree of propinquity increases. That is the point of Haldane's quip about the two brothers and the eight cousins, for eight cousins contain, on average, an equal amount of one's genetic composition as two brothers. Using the mathematics of games, in which each player must decide what to do to maximize its benefits in a world in which all other players are also assumed to be acting exclusively in their own self-interest, Hamilton showed that cooperative behavioral phenotypes will, under a wide range of conditions and parameters, maximize the representation of a gene in the next generation (Hamilton 1964). Accordingly, there will be selection pressure for such phenotypes. By showing how kin-selectionist ideas could explain the obvious and pervasive fact of cooperation in nature without sacrificing Darwinism's ontology of self-interested entities and without nonparsimonious recourse to group selection, Hamilton, Williams, and others whose work lies at the foundation of the research program called sociobiology appeared to have solved one of Darwinism's most nagging problems. The resulting enthusiasm for sociobiology has its roots in the understandable air of triumph that followed these developments. [3]

This enthusiasm was intensified by the fact that Hamilton had provided startling empirical evidence for this hypothesis by using game theory to explain how and why cooperation arises among social insects. Many social insects among the *Hymenoptera*, a group that includes bees and ants, possess a particular kind of genetic system that is favorable to the evolution of cooperative behavior, which apparently has emerged among them at least eleven separate times. *Hymenopteran* females share a greater percentage of their genetic makeup with each other than they do with their brothers.[4] If it could be arranged, daughters would thus pass on more of their genes by having their mother produce more sisters than if they themselves reproduce. Deflection of reproduction onto a single queen, economic cooperation among sterile castes, and the marginalized status of male drones in insect colonies can readily and persuasively be justified by computing the game-theoretical consequences of the facts about *Hymenopteran* genetics, given the assumption that cooperative adaptations will be proportional to the degree of genetic relatedness. The result will be an explanation of the social structure of insect societies on the basis of possibilities and payoffs inherent in their genetic system, and the concomitant suggestion that genetic systems play a greater causal role in evolution than organismic selectionists have assumed. Soon kin-selectionist scenarios were being generalized by Darwinians who favored a gene-centered point of view beyond species with lopsided replicative systems. By assuming that cooperation in species is correlated to degrees of genetic relatedness and that altruism is a function of someone or something's long-range self-interest, kin selection seemed to explain the pervasive fact of cooperation in nature.

In *The Selfish Gene* (1976, 1989), Dawkins integrated these results into a version of genic selectionism that went well beyond the concern with methodological parsimony that had motivated Williams. "The gene's eye view of Darwinism," Dawkins says in the preface to the second edition of this popular work, "is implicit in the writings of R. A. Fisher and the other great pioneers of neo-Darwinism in the early thirties, but was made explicit by W. D. Hamilton and G. C. Williams in the sixties. To me their insight had a visionary quality. But I found their expression of it too laconic, not full-throated enough" (Dawkins 1989, ix). For Dawkins, Fisher was "the greatest biologist of the twentieth century," (Dawkins 1989, 124). By "full throated" Dawkins, however, seems to mean that Fisher, and Williams and Hamilton after him, all full of positivist delicacy, were far too reticent to make a realistic commitment to the literal truth of the fact that, by their very nature, genes are selfish mechanisms whose sole concern, as modern molecular genetics has shown, is to make copies of themselves. In contenting themselves with accounts of genic selection whose chief advantages were theoretical parsimony and predictive adequacy, Dawkins implies that his predecessors unnecessarily exposed themselves to the objections of Mayr and fellow-traveling philosophers, who argued that genic selectionism is just genetic "bookkeeping," a way of representing evolutionary change devoid of (much) causal significance (Wimsatt 1980; Sober and Lewontin 1982; Brandon 1990). What was needed, Dawkins thought, was a more causally oriented restatement of genic selectionism.

Dawkins does not deny that organisms, which he regards as vehicles for the transmission of genes, must successfully compete at the phenotypic and environment level and that in this sense organisms are targets of selection. Nor does he deny that phenotypes and genotypes are connected in as complex a web of many-to-many relationships as Mayr insists (Dawkins 1982). However, if Dawkins is not a genic reductionist in these respects, he is still a genic causalist, for he gives a realistic twist to genic selection by asserting that genes are causally effective in creating, through natural selection, just those phenotypes that will help them get represented in greater numbers in the next generation. Accordingly, Dawkins has proclaimed that genes are "selfish," not only because it is the inherent business of replicating macromolecules to make more of themselves but because they actually build adapted organisms in order to enhance their own replicative prowess. They are the beneficiaries of selective processes, where benefit is measured by increased representation in the next generation. By transferring the site at which adaptive selection works from the organismic to the genic level, the "selfish gene hypothesis" does more, therefore, than innocently look at natural selection from Williams's parsimonious point of view. It explicitly opposes both Mayr's principle of genetic relativity, and his related stress on the causal primacy of environmental selection, in which the interaction between the organism and its environment is the first step in the process of natural selection.

In order to sustain his claim, Dawkins has had to have some help from a revisionist ontology. By "unit of selection" Dawkins means whatever survives in greater numbers after the selective process. Genes are the only units of selection in Dawkins's sense because they alone possess properties that make them less transient than the organisms that they pass through. Dawkins defines a gene as "any portion of chromosome lasting enough generations to be a unit of selection" (Dawkins 1989, 32). This does not necessarily mean that genes must be made of DNA or of any other particular substance. It does mean that whatever substance performs the work of replication, if it is to be the most prominent agent in and beneficiary of the evolutionary process, must be the most *real* entity in the process. Here is where the ontology starts doing most of the heavy work. Dawkins tacitly treats longevity as an index of ontological primacy. "Genes, like diamonds, are [potentially] forever," says Dawkins, while "individuals and groups are like clouds in the sky or dust storms in the desert" (Dawkins 1989, 34–35). Genes, in the sense of "naked replicators," were there at the beginning of life, he claims, floating around in the primeval soup. They multiplied because they were, by definition, entities that make more of themselves by the process explained by Crick and Watson. Nevertheless, some are more successful in sending copies of themselves across generations than others. Replicators will differ from one another in their "longevity, fidelity and fecundity" because some are more successful than others in coding for adaptive phenotypes that allow the gene breathing room to blithely make more of itself. For this reason, Dawkins says that organisms are the "survival machines" of genes (Dawkins 1989, 20). He claims that if Darwin had known modern molecular genetics he would have taken the same view (Dawkins 1989, 20).

An apparent advantage of Dawkins's theory is that it can more easily explain the proliferation of "selfish DNA," that is, the massive amount of that substance in each chromosome that apparently codes for no gene products, than our speculation that some function might eventually be found for all this stuff. For Dawkins, replicating is precisely what DNA can be expected to do unless something stops it. "The true purpose of DNA is to survive, no more and no less. The simplest way to explain the surplus DNA is to suppose that it is a parasite, or at best a harmless but useless passenger, hitching a ride in the survival machines created by other DNA" (Dawkins 1989, 45). Similarly, the discovery that genes come in multiple copies and in large families, many of which can be put to new uses while functional work is carried out by others, is consistent with this more-making tendency. Thus, Dawkins asserts that the "selfish DNA" hypothesis is a subcase of the selfish-gene hypothesis (Dawkins

1989, 45, 275). That is not because selfish DNA is a gene. It is because it is *selfish*.

Dawkins's version of genic selectionism leads him not only to favor the reducibility of Mendelian theory to molecular genetics but to take a reductionist view of the relation of molecular genetics to physics. He asserts that "the differential survival of replicators is a special case of a deeper, more universal physical law governing 'the survival of the most stable'" (Dawkins 1989, 13). In saying this, Dawkins is repudiating Mayr's "autonomy of biology." He is not, however, arguing against the autonomy of biology from physics merely on grounds of methodological parsimony. His reductionism is not only about the reducibility of one theory to another. Dawkins is an entity reductionist as well, who wants to say that the entities mentioned by a more basic science are more real than those that figure in a less fundamental one. Dawkins's effort to give genes causal efficacy depends on reifying them, that is, on treating them as distinct centers of causal power.

Dawkins might reply that this interpretation pushes his metaphors too hard, that he is simply attempting to express Fisher's and Williams's preference for parsimony by using vivid analogies and metaphors. Sometimes he says things like this. Metaphors, however, are not rhetorically so innocent. As readers of this book will by now be aware, it is just because metaphors play roles in explanations that one is not entitled simply to say, "Oh, that's just my way of putting it." Even when they perform little or no explanatory work, moreover, metaphors carry a good deal of metaphysical and epistemological freight. Indeed, whenever there is a deficit between theoretical reach and empirical support the difference is usually made up by invoking ontology to do the missing work. Similarly, epistemological and methodological ideals are sometimes used to intimate on highly general grounds that the theory in question must be true. In such cases, Lewontin and Levins argue, we are entitled at least to suspect that ideology may be involved (Levins and Lewontin 1985).

In this respect, genic selectionism has been seen as an attempt to shore up competitive individualism by shifting it from the organismic to the genic level (Sahlins 1976). On this interpretation, Dawkins proposes to preserve Darwinism's connection to competitive individualism, which may be among its least attractive and most dispensable feature, and in large measure an artifact of the simplistic dynamical models with which it was forced to work, by replacing Darwin's selfish organism theory of natural selection with a Hobbesianism of selfish genes. Just when it might be possible to pry natural selection loose from the simplistic interpretive frameworks in which it has hitherto been encoded, Dawkins transfers the competitive struggle of individuals from the organism above to the genes below. In doing so, Dawkins clearly wants to reaffirm the continuity of the British tradition in genetic Darwinism that traces back to Fisher and Ford. This does not mean that ideology is not also in play. It is nonetheless useful to note that Dawkins's defense of genic reductionism is less explicitly tied up with the individualism of capitalist economics than with attempts to use reductionist materialism as a weapon against Christian spiritualism. Dawkins systematically invests his metaphors with disturbing semantic reverberations that harken back to Enlightenment themes and turn his theory into a direct competitor of Christian creationism. Only radical materialist reductionism, it seems, can block religious dogmatism, even if the cost is making us pawns of our genes rather than of a tyrannical Calvinist God. Thus, in a stunning passage whose rhetoric mimics and mocks Christian theology, Dawkins writes:

Was there to be any end to the gradual improvement in the techniques and artifices used by the replicators to ensure their own continuance in the world? There would be plenty of time for improvement. What weird engines of self-preservation would the millennia bring forth? Four thousand million years on, what was to be the fate of the ancient replicators? They did not die out, for they are past masters of the survival arts. But do not look for them floating loose in the sea; they gave up that cavalier freedom long ago. Now they swarm in huge colonies, safe inside gigantic lumbering robots, sealed off from the outside world, communicating with it by tortuous indirect routes, manipulating it by remote control. They are in you and me; they created us, body and mind; and their preservation is the ultimate rationale for our existence. (Dawkins 1989, 21)

In his 1975 book *Sociobiology: The New Synthesis*, E. O. Wilson, a Harvard entomologist and population biologist, used the applicability of game-theoretical mathematics to social insects to argue for the generalizability of processes like kin and genic selectionism and for their relevance to human evolution. In the sociobiology debate that followed the publication of this book, echoes of genetic determinism, harking back vaguely to social Darwinism and the eugenics movement, were put on the table. No one was more instrumental in introducing these unwelcome themes than Wilson's colleagues, Lewontin and Levins. Lewontin and Levins had worked with Wilson at the University of Chicago, where, together with Robert MacArthur and others, they had waged a common effort to integrate population genetics with ecology. Now these former colleagues divided on a range of issues connected with Wilson's sociobiological extension of kin selection theory and genic selectionism (Segerstrale 1986).[5] Against the agitated background of the Vietnam War, in protest against which Lewontin resigned from the National Academy of Science, Levins and Lewontin formed Science for the People, and later the Dialectics of Biology Group, to oppose genetic reductionism (= mechanism), atomism (= individualism), and determinism (= social and political passivity) in any and all of their forms. Lewontin found it especially suspicious that these ideas, as well as the roughly contemporaneous refloating of creationism and the notion that IQ differences are heritable traits correlated with different races, were being popularized just as the United States was turning, as it had in the 1920s, toward political conservatism and away from the Progressive legacy that had been so well defended by Dobzhansky's genetic pluralism (Levins and Lewontin 1985; Lewontin 1992; cf. Jensen 1969). Throughout the 1970s and 1980s, debates raged as old friendships shattered. Indeed, the resultant press battle was one of the most vivid evolutionary controversies to be displayed before the public since Cuvier confronted Geoffroy or since Huxley butted heads with the bishop of Oxford.

One notion that consistently emerges from Lewontin's and Levins's polemics is that by giving scientific cover to the view that only individual entities in competitive relations are real and that actual successes and failures faithfully reflect the inherent capacities and deficiencies of individuals, genic selectionists are, consciously or not, projecting capitalist competition onto genetic atoms. These allegedly scientific ideas thus serve to perpetuate and justify antiprogressive social policies by justifying laissez-faire property and exchange relations as the only conditions in which true ability can show itself (Levins and Lewontin 1985; Lewontin 1992).[6] Attention is thereby drawn away from the particular economic and political preconditions that must hold if that atypical view of the world is to be turned into what looks like common sense. Genic selectionism is thus for Lewontin and Levins a successor ideology of the social Darwinism, eugenics, racism, and sexism that had for a while been rendered recessive in politics by the horror of the Holocaust, by the Progressive tradition in American politics, and by Dobzhansky's genetic pluralism and ethic of toleration, for all of these movements share similar ideas about the conditions in which innate talent can manifest itself.

Lewontin and Levins can be easily misunderstood on this score. They are not claiming that genic selectionists are social Darwinians, or that they favor eugenics, or that the new sociobiology is a disguised version of these older theories. They are saying instead that all such theories have been made out of the same conceptual materials and that ideology is at its most effective when it expresses conceptual biases rather than when it is shilling for some particular claim or policy. Adopting rhetorical flourishes as highly colored as Dawkins, Lewontin, Levins, and their fellow "dialectical biologists" have, accordingly, followed Marx in ridiculing the "topsy-turvy" ontology of reified genes as a reflection of what human relations look like under the individualist capitalism that they help justify and perpetuate.

As in his controversy with Kimura, Lewontin had a mixture of empirical, theoretical, philosophical, and ideological points to make. Empirically, he was able to point out that he and colleagues in his laboratory had already advanced proof that at least one case of group selection exists that does not reduce to kin selection. The case had nothing to do with the vulnerable sort of group selection proposed by Wynne-Edwards, in which organisms are said to regulate their own numbers in accord with the needs of the group. It concerned so-called t alleles in house mice, organisms that live and breed in small demes (Lewontin and Hubby 1966). Males that are homozygous for t alleles will die. The demes in which such males spread their genes around, with the help of females, will go extinct. But this gene also has a tendency to cheat Mendel's laws by making multiple copies of itself. It thereby gets itself overrepresented, in Hardy-Weinberg terms, in the next generation. The continued existence of the house mouse thus rests on an equilibrium between these two processes. In establishing that equilibrium, the conclusion cannot be avoided that the fitness of each mouse, in the now canonical sense of expected reproductive success, so deeply depends on which deme it is in that otherwise identical individuals will be fit in one group and unfit in another. Subsequently, other cases of interdemic group selection have been found in flour beetles and other organisms, and the general conditions for interdemic selection were clarified (Wade 1976, 1977, 1978; D. S. Wilson 1975, 1983; Sober 1984a; Lloyd 1988).

Interdemic selection confirms Wright's early insistence on the importance of demic structure. It thereby points to a different aspect of Wright's legacy than Kimura did. It suggests that if the gene below must be taken into account, so must the group above (Sober 1984a; Brandon 1990). A hierarchy of "units of selection" thus began to take shape. No less important, interdemic selection strongly reaffirms the principle of genetic relativity against efforts by Fisher's latter-day disciples to appeal to the structural properties of DNA to defend context-independent additive genetic fitness. Indeed, Lewontin defends the principle of the relativity of fitness even when group selection is not a factor, showing himself to be Dobzhansky's heir and Mayr's ally. "The fitness at a single locus ripped from its interactive context," Lewontin wrote at the end of the *Genetic Basis of Evolutionary Change*, "is about as relevant to real problems of evolutionary genetics, as the study of the psychology of individuals isolated from their social context is to an understanding of man's sociopolitical evolution. In both cases, context and interaction are not simply second-order effects to be superimposed on a primary monadic analysis. Context and interaction are of the essence" (Lewontin 1974, 318).

Lewontin proves this by demystifying the causal efficacy that genic selectionists attribute to coding stretches of DNA. Population geneticists can identify and isolate the effects of genes, he points out, only by plotting how well organisms reproduce and develop under different conditions. This kind of work does not result in attributing causal properties to genes but in statistical "norms of reaction" whose relativity to different circumstances is ineliminable:

Each genotype has its own norm of reaction, specifying how the developing organism will respond to various environments. In general a genotype cannot be characterized by a unique phenotype. In some cases the norm of reaction of one genotype is consistently below that of another in all environments. . . . [For example], the number of light receptor cells, or facets, in the compound eye of the fruit fly *Drosophila* is usually about 1000, but certain gene mutations severely reduce the number of facets. For example, flies carrying the mutation *Ultrabar* have only about 100. However, the number of eye cells also depends on the temperature at which the flies develop. Flies of the normal genoyptye produce about 1,100 cells at 15 degrees C, but only 750 cells at 30gC. . . . So, for example, we can say unambiguously that *Ultrabar* flies have smaller eyes than normal fies because that is true at every temperative of development. However, another mutation, *Infrabar,* also have fewer cells than the norm, but has an opposite relation to temperature. (Levins and Lewontin 1985, 90–91)

Discoveries like these point to the fact that genes are only one of a large number of factors that specify a phenotype:

It is a fundamental principle of developmental genetics that every organism is the outcome of a unique interaction between genes and environmental sequences modulated by the random chances of cell growth and division, and that all these together finally produce an organism. . . . We are not determined by our genes, although surely we are influenced by them. . . . *Even if I knew the complete molecular specification of every gene in an organism, I could not predict what that organism would be.* (Lewontin 1992, 25–26, italics added)

These various influences are radically interactive. Regulatory genes, for example, determine when genes are turned on and off. But signals from the environment are also among the factors that tell a regulatory gene when to go to work. These environmental signals in turn contain an element of chance, or what Lewontin calls "developmental noise," and are highly sensitive to both random and purposive acts of developing organisms themselves. In sum,

The system that interprets environmental signals and determines the response is so complex it must be described in terms of a large number of strongly interacting variables. *Such a system is likely to be dynamically unstable,* showing complicated fluctuations of state, and is unlikely to reach a resting state (stable equilibrium) even in the absence of all external signals. Thus spontaneous activity arises in organisms out of the complex evolution of responses to the environment. (Levins and Lewontin 1985, 44, italics added)

We reach by this route the heart of Lewontin's and Levins's vision of nature. It is a vision of irreducible complexity, and of a beauty that is intellectually and ethically betrayed by the reductionist, mechanist, and individualist legacy. In order to recognize this complexity, we have no choice but to adopt the perspective of the active, interpretive organism itself. "It is not that organisms find environments and either adapt themselves to environments or die," Lewontin writes. "They actually construct their environments out of bits and pieces," both by internalizing the information that enables them to deal with their world, and by changing their world to meet their needs and desires (Lewontin 1992, 112). From that perspective, we not only see the primacy of the agents' own causality in its environment but are prevented from objectifying, reifying, or, as in Dawkins's case, deifying, genes.

All this may sound like Lamarck's "active materialism." In a way it is, since, through Marx, Lewontin and Levins have reconnected with the revolutionary "active materialist" tradition in whose earlier phases Lamarck played a prominent role. They have correspondingly distanced themselves from the mechanistic or Cartesian materialism in which Darwinians have tended to encode natural selection, which Levins and Lewontin believe marred their own scientific educations (Levins and Lewontin 1985, 267). This does not mean, on the other hand, that Lewontin and Levins are Lamarckians. In their view the flexible capacities that enable organisms to change their world have been accumulated by natural selection working to create and preserve the capacities for dealing and changes and riding over cycles that Dobzhansky so clearly discerned (chapter 12). Indeed, Lewontin says that

whereas Lamarck supposed that changes in the external world would cause changes in the internal structure, we see that the reverse is true. An organism's genes, to the extent that they influence what that organism does in its behavior, physiology, and morphology are at the same time helping to construct an environment by allowing the organism to act on inherited information and to respond to new information on the basis of inherited capacities. If genes change in evolution, the environment of the organism will change too. (Lewontin 1992, 112)

In stressing the responsiveness of active, problem-solving organisms to changing environments, Lewontin and Levins are not only carrying forward Dobzhansky's legacy but, we believe, more or less unreflectively recovering an even more long-lived and protean tendency in American Darwinism which, because it first flourished during the Progressive period and served the aims of Progressive politics, might well be called "Progressive Darwinism."[7] Although Progressive Darwinians, such as James Mark Baldwin, William James, John Dewey, and Lester Ward, differed a great deal from one another, they all opposed what Dewey called "the ordinary biological theory of society," by which he meant Spencerian social Darwinism (Richards 1987; Dewey 1894). Social Darwinians maintained that if humans are to continue to adapt and evolve, the Malthusian competitive conditions that drive adaptive natural selection elsewhere, and until recently have dominated human societies, must continue to be enforced. This was to be done by institutionalizing laissez-faire capitalism. In opposing this view, some American Progressives thought it was necessary to reject evolutionary theory altogether. William Jennings Bryan, for example, assumed that if we are under evolution's control, it must be in the grim way the social Darwinians say we are. In speaking against any sort of evolution, however, as he did in the famous Scopes Monkey Trial in Tennessee, Bryan played into the hands of fundamentalists. Other Progressives, notably educated liberal Protestants, subscribed to non-Darwinian theories of evolutionary progress. Their ideas fell into disfavor with the decline of the developmentalist views of evolutionary progress. The people we are calling Progressive Darwinians differed from both of these parties by affirming evolution *and* Darwinism, but denying that social Darwinism is good Darwinism in the first place. They argued that Malthusian conditions are not ubiquitous enough to establish any inference from Weismann's refutation of the inheritance of acquired characteristics to the idea that organisms are passive pawns of a cut-throat organic and inorganic environment that predetermines what every player must do just to survive. This being so, natural selection will tend to select for the problem-solving capacities that Dewey calls "intelligence," namely, the ability to anticipate the consequences of behavior and thereby to guide that behavior toward useful ends, primarily by molding the environment to one's needs. So construed, intelligence is not a push-pull mechanism but an anticipatory capability that is most highly developed in humans but is shared to one degree or another by all organisms. "Intelligence," Dewey writes, "is indirection, checking the natural direct action, and taking a circuitous course" (Dewey 1894, 408). If natural selection creates the capacities for problem solving, it leaves the particulars to learning. It thereby creates selection pressure for the sociality that allows learning to be an effective agent.[8]

This tradition did not die with the coming of genetic Darwinism. On the contrary, it preformed the contours along which the American wing of the modern synthesis ran, and into which the contributions of famous émigrés like Mayr and Dobzhansky were received. It can be felt, for example, in Mayr's, Lewontin's, and Levins's common insistence on the causal primacy of what organisms *do* in their environments. This insistence has been renewed by the contemporary rediscovery of what Progressive Darwinians already knew: that "the same behavior which is the product of evolution also is a partial cause in the process of evolution" (Plotkin 1988, 1). Such agent-centered views of evolutionary process provide resistance to eugenics and other forms of genetic determinism by asserting, with Wright and Dobzhansky, that nature is just too complicated and protean to tolerate reproductive tinkering. If this is so, we may well wonder why Lewontin and Levins insist on encoding their thoughts in the language of Marxist dialectics rather than the good old American problem-solving talk in which these ideas have traditionally been phrased. Our guess is that in the wake of the positivist ascendancy, only the heavy artillery of dialectical thinking can do critical work that in simpler times Dobzhansky was able to leave to the Deweyan pragmatists' tolerant vision of democratic problem solving through an experimental method oriented to human needs.[9]

We have seen that Darwinians who are Wright's and Dobzhansky's heirs see in genic selectionism and sociobiology a latter-day resurgence of

Fisher's genetic atomism. They also see in Kimura's neutralism a no less disagreeable recurrence of Muller's "purifying selection." Proposals for expanding the modern synthesis into a more general, and explicitly hierarchical, theory, in which genes, organisms, groups, and even species can be units of selection, are one way of meeting both of these challenges. Whatever else one might say, this proposal is a clever way of admitting what Kimura, Williams, Hamilton, and Dawkins were saying, while at the same time resisting the reductionist implications that seemed to follow. There certainly is a level at which selection for genes might occur. Similarly, the fixation of selectively neutral alleles is true at the level of protein and nucleic acid evolution. All this can admitted without prejudice to classical Darwinian selection at the level of organisms in populations, however, or even to interdemic group selection, for the very same conceptual considerations that countenance genic selection also countenance some forms of group selection. Perhaps, chimed in Steven Stanley, Elizabeth Vrba, Niles Eldredge, and Stephen Gould, there can even be some sort of higher-order selection still, in which the spatiotemporally individuated species of the modern synthesis are said to stand to the overall direction in whole clades (branching phylogenetic lineages, from Greek for "branch") as variation among individual organisms stands to the adaptedness in populations (Stanley, 1979; Gould 1980, 1982b; Vrba and Eldredge 1984; Vrba and Gould 1986; Eldredge 1985). In this event, direction in a clade will be a result of a sorting process in which characteristics of the clade itself have the statistical edge. These might include higher speciation rates, lower extinction rates, or other "traits" that cannot be reduced to the adaptive prowess of individual members of the component species of a lineage. In this case, macroevolution will be irreducible to microevolutionary processes even within a roughly Darwinian frame of thinking.

The idea that whole species may be units of selection in their own right was first rendered at least thinkable by several respectable voices, biological and philosophical, who argued that if species are to play the role assigned to them even in orthodox Darwinism, where they appear as outcomes of a selection process and hence "units of evolution," they must be treated not as classes (a tenacious remnant of the old, preevolutionary essentialist ontology) but as "individuals," or more precisely as spatially and temporally bounded particulars held together by the exchange of genetic information (Ghiselin 1974; Hull, 1976, 1978, 1980, 1981). From the original synthesists' redefinition of species as populations and their rejection of the typological species concept, it was only a small step to the ideas that species are spatiotemporally distinct particulars or individuals. If so, however, there seems to be no conceptual reason that prevents species from being units of selection as well as units of evolution. As individuals, they are the sorts of things that can be selected; and if, as individuals, they happen to have properties that are more likely to survive a sorting process than other individuals at the same level, they will, in all probability, be selected.

It was probably Gould's interest in species selection that prompted him to announce that "a new and more general theory of evolution is emerging": "When a proper hierarchical theory is fully elaborated, it will not be entirely Darwinian in the strict sense of reduction to natural selection acting upon organisms. Yet I suspect that it will embody the essence of Darwinian argument in a more abstract and general form. We will have a series of levels with a source for the generation of variation and a mode (or set of modes) for selection among individuals at each level" (Gould 1982b, 104; see also Gould, 1980).

A hierarchically expanded Darwinism is possible only if Darwinism's two-step process of variation and differential retention is separated from the hypothesis according to which adaptive selection working on organisms is the primary causal agent of differential retention of genetic variation over generations at every mode and level. Gould's idea amounts to a proposal to save the selectionist core of the Darwinian tradition (*sensu* Lakatos) by abandoning the dispensable auxiliary assumptions of adaptationism and gradualism that for so long tied down Darwinism to whole organisms alone. It is against these auxiliary assumptions that Gould most vigorously deploys his considerable argumentative skills (Gould 1980, 1982b).

Like all other such proposals, this one runs the risk of further eroding the tradition it was designed to save. Darwinism without gradualism or adaptationism might not be Darwinism at all. It certainly would not be to Charles Darwin, who explicitly held that if these two axioms were taken away, his entire theory would fall immediately to the ground (Darwin 1859, 194, 199). For Gould, though, the risks were worth running. For an additional, and perhaps compelling, advantage in moving toward a hierarchically expanded theory is that it allows chance and self-organization, whose evolutionary roles had been greatly augmented by the molecular revolution, to serve along with selection as significant agents of change. Neutral mutations will go to fixation when they are let loose to operate at levels from which the heavy hand of selection pressure has been deflected. Even where it operates, moreover, natural selection will labor under strong evolutionary constraints that determine ontogeny by way of connectivity in the genome, especially its regulatory sector. Hierarchical thinking not only permits Kimura to be right about the role of chance at the level protein evolution, therefore, but developmental biologists to be right about the pervasive role of constraints in the ontogeny of whole organisms: "The constraints of inherited form and developmental pathways may so channel any change that even though selection induces motion down a permitted path, the channel itself represents the primary determinant of evolutionary direction" (Gould 1982b, 383).

Finally, an expanded synthesis had implications for macroevolutionary issues that were of particular interest to Gould and his fellow paleontologists. Restricting Darwinism to organismic selection driven by gradual adaptation has tended, in their view, to prejudge two crucial macroevolutionary issues. Simpson in particular linked speciation too closely to adaptation and, without much evidence, foreclosed the issue of whether the evolution of clades is a gradual, linear function of adaptive evolution at the organismic level (Gould 1980, 1982b). What evidence there was for the latter claim was for Gould and Eldredge too closely tied to the dogmatic assertion that the gaps that appear in the fossil record simply reflect ignorance, an assumption that goes back to Darwin's own gradualist convictions about "missing links," and beyond that to Lyell's uniformitarianism. But Eldredge and Gould had argued as far back as 1972 that

significant evolutionary change arises in coincidence with events of branching speciation, and not primarily through the *in toto* transformation of lineages (classical anagenesis). [What Gould and Eldredge called "punctuated equilibrium"] maintains, speaking of tempo, that the proper geological scaling of speciation renders branching events *geologically instantaneous* and that, following this rapid origin, most species fluctuate only mildly in morphology during a period of *stasis* that usually lasts for several million years. (Gould 1982b, 83, italics added; cf. Eldredge and Gould 1972)

The thesis of punctuated equilibrium, according to which significant genetic change is normally concentrated around speciation events, exhibiting thereafter little morphological change that adds up to anything, seemed to Gould to square with new work in developmental genetics (Gould 1982b). Study of regulatory genes now makes it likely that speciation is related to sudden reorganizations of developmental programs, which are then locked into place. Such events may well depend for the most part on spontaneous processes, such as genetic drift and genomic self-organization, and therefore may occur without the significant adaptation that doubtless precedes and follows them. Since such events will predictably occur with much higher frequency in small, isolated populations, Gould even had the cheek to treat this conclusion as a natural inference from Mayr's "peripatric" theory of speciation by isolation and subsequent genome reorganization, at least when that theory is freed from Mayr's adaptationist prejudices (Eldredge and Gould 1972; Gould 1982a, 1982b). Goldschmidt's hopeful monsters, suitably redescribed, live again (Gould 1982c). The problem of evidence then takes care of itself. On this theory, speciation will be a common occurrence. Since it is improbable, however, that we will ever find the remains of more than a fraction of these evanescent little populations, direct (as opposed to inferential) evidence for this claim will be lacking for statistically respectable reasons rather than for the dogmatic reasons that Gould and Eldredge ascribe to Lyell, Darwin, and Simpson.

Lewontin played an important if a sometimes reluctant, role in articulating the theoretical basis for an expanded synthesis. It was Lewontin who, in arguing for the possibility of interdemic group selection in a 1970 article, first used the term *units of selection*. He went on to speculate that natural selection might operate on such entities as genes, cells, and demes in addition to organisms, and suggested that if it does, its effects may well differ widely from those we associate with the adaptedness of organisms, organic parts, and behaviors, and therefore with adaptation-*ism* (Lewontin 1970). This idea was subsequently taken up by several philosophers of biology (Hull 1980, 1981; Sober 1984a; Brandon 1990). They saw in the notion of "units of selection" a way of clarifying the conceptual structure of evolutionary theory without being too closely tied to the (merely empirical) notion that selection operates on organisms to produce adapted populations. They also found in the idea of multiple units and levels of selection relief from overly reductionistic ideals in the philosophy of science from which they were trying to free themselves. The idea was then reimported into theoretical biology by scientists who were in touch with these philosophers. The expanded synthesis idea is, for this reason, a scientific research program perhaps more closely tied to the work of philosophers than most others.[10]

It was, in fact, the philosopher David Hull who first caught and ran with the ball Lewontin had thrown out. In a seminal pair of essays, Hull argued that in spite of himself Dawkins had made an important contribution to a hierarchically expanded Darwinism. In order to claim that genes use organisms as instruments of their own differential fecundity, and so are the only acceptable units of selection, Dawkins had distinguished "replicators" from "vehicles" (Dawkins, 1976, 1982, 1989; Hull 1980, 1981). To avoid begging the question in favor of the ontological primacy, and causal efficacy, of genes, Hull in turn distinguished between replicators and interactors, that is, between "entities that pass on their structure directly in replication" and "entities that produce differential replication by means of directly interacting as cohesive wholes with their environments" (Hull 1981, in Brandon and Burian 1984, 150). Replicators are entities that with varying degrees of fidelity make more entities like themselves, and thereby conserve and transmit information through an intergenerational process of selection. Interactors, by contrast, are entities that deal immediately with their environments in ways that inhibit or foster the differential retention of replicators. Replicators form lineages and have genealogies. Interactors are causally at work in environments within a single generation. In most versions of genetic Darwinism, organisms are interactors and genes are replicators. More generally, we may say that replicators influence interactors by feeding them the information that gives some interactors a competitive edge over others in dealing with their environment. Interactors in turn influence replicators by determining what information gets transmitted and what does not. Which replicators survive and which do not depends causally on which interactors reproduce (Brandon 1985, 1990). That independent fact cannot be screened off as causally irrelevant except where the question in favor of the causal role of genes in development and organic behavior is begged, as it tends to be by Dawkins.[11]

It follows from these considerations that a hierarchically expanded synthesis requires two hierarchies: a genealogical one and an ecological one. These hierarchies appear forthwith when Hull's replicator-interactor distinction, itself an expansion of Dawkin's replicator-vehicle distinction, is writ even larger by practicing biologists who advocate an expanded synthesis. Replicators then array themselves into a genealogical hierarchy consisting of entities whose more-making results in more or less long-lived lineages. Genes, genomes, cellular lineages, chromosomes, organisms, demes, species, and the higher taxa are members of the genealogical hierarchy. The ecological hierarchy, meanwhile, will be composed of interactors that serve as nodes of energy transduction and dissipation at various levels and rates (Eldredge 1985; Salthe 1985; Eldredge and Grene 1992). Cells, organisms considered as somatic wholes, colonies, populations in niches, ecological communities, regional biota, and at a limit the entire terrestrial system, whose behavior is strongly linked, we now recognize, to the ebb and flow of organic life—all of these can be interactors. Note that organisms show up twice. Perhaps the traditional restrictions of evolutionary theory to organismic selection derive from the fact that organisms seem real to us for at least two reasons. We have an ontological prejudice in favor of middle-sized commonsense objects like ourselves. Organisms, moreover, at least the relatively large ones we see, seem more causally efficacious to us, and hence more real, because they are both replicators and interactors.[12]

According to advocates of an expanded synthesis, evolution occurs when an item picked from the ecological hierarchy is matched with an item from the genealogical. Questions then arise about the processes by which entities on both sides of the dual hierarchy are interrelated. Presumably, selection of some sort plays a major role. At this point, however, we reach a fork in the road. Not all of these units can plausibly be connected by way of natural selection, at least when it is working to produce adaptations and at least in the same way organisms in populations do. One sort of expanded synthesis arises, therefore, when the tie between these units is restricted to processes that do result in adaptive natural selection or very close analogues of it. Brandon proposes such a narrowly expanded synthesis (Brandon 1990). Species selection is conceptually possible on this view but empirically unlikely. A different sort of expanded synthesis arises, on the other hand, when units along the two hierarchies are connected through processes that, while they involve selection, may not produce *adaptation*.[13]

Lewontin has expressed little enthusiasm for either the highest or the lowest reaches of an expanded synthesis. His doubts about genic selection seem matched to his indifference to species selection. One reason might be that if real causality is transferred to the species or the genic level, Lewontin's and Levins's stress on the agency of organisms working within groups will be dissipated into abstract sorting processes in which evolutionary outcomes are passively accumulated, as they are in the mechanistic forms of Darwinism that Lewontin and Levins so deeply oppose. It is ironic, accordingly, that some of Lewontin's own ideas have been seminal in the articulation of nonadapationist, but still selectionist, versions of the expanded synthesis, in which genic and species selection figure prominently. The most important of these is Lewontin's attempt to dissociate natural selection from adaptationist research programs.

Natural selection, Lewontin claims, will occur whenever there is (1) "variation in morphological, physiological and behavioral traits"; (2) heritability of at least part of this variation "so that individuals resemble their relations more than they resemble unrelated individuals"; and (3) differential reproduction, in which "different variants leave different numbers of offspring" (Lewontin 1980, in Sober 1984a, 244). Adaptation does not appear on this list, because, Lewontin says, adaptation is not necessary for natural selection. It was a "postulate added by Darwin as an *explanation* of the mechanical cause of the phenomenon of differential reproduction and survival" (Lewontin 1980, in Sober 1984a, 245, italics added). Lewontin's dissociation of natural selection from adaptation (an idea that Darwin would have found incoherent) has made it possible for Gould and others to argue that natural selection, *sensu* Lewontin, is not tied to sorting processes in which adapted entities must result.

Lewontin's efforts to separate natural selection from adaptation, and a fortiori from adaptationism, are driven by his attempt to separate Darwinism from genic selectionism and sociobiology. These programs assume that adaptationist reasons can be given for every trait. Without that assumption, the idea that genes build survival machines would make no sense, for survival machines are *adapted* machines. Thus adaptation*ism*, according to which "the world, both living and dead, [is nothing but] a large and complicated system of gears and levers," is one effect, Lewontin thinks, of the distorting influence of Cartesian mechanistic ideals on Darwinism (Lewontin 1992, 12). Whatever the merits of this way of looking at the relationship between adaptation and natural selection may or may not be, it is undeniable that Lewontin has used it to make a number of telling points against the adaptationist program of sociobiology and to reveal degrees of complexity in the biological world that strong adaptationists tend to miss or misconstrue.[14] When the difference between natural selection and adaptedness has been acknowledged, natural selection can be seen as working in ways that are sometimes nonadaptive, or even, as American neo-Lamarckians held, overadapted and hence maladaptive.

The relativity of fitness, for example, entails that the adaptive status of a trait can change as its frequency goes up or down in a population.

Mimics that avoid predation by building adaptations that disguise them from prey will, for example, lose their advantage as their disguise grows more common. It is possible that in some conditions "frequency-dependent" selection will result in maladaptedness. Similar effects can attend "density-dependent selection." For example, "a mutation that doubles the egg-laying rate in an insect, limited by the amount of food available to the immature stages, would very rapidly spread through the population. Yet the end result would be a population with the same adult density as before but twice the density of early immatures and much greater competition between larval stages" (Lewontin 1980, in Sober, 1984b 248; Sober 1984a, 171–211). Moreover, like Gould, Lewontin thinks that adaptationism blinds sociobiologists to the explanatory role of constraints. Evolution, says Lewontin, is an organism's way of getting around constraints (Lewontin 1983). The same idea is implicit in Jacob's idea that evolution is a "tinkerer" (in French, a *bricoleur,*) constructing Rube Goldberg–like machines out of spare and used parts under tough conditions, rather than sleek, efficient BMWs that roll fresh off the assembly line (Jacob 1977, 1983). In such cases, as Lewontin and Gould teamed up to argue in a now legendary article colorfully entitled "The Spandrels of San Marco and the Panglossian Paradigm," it is the constraints that do the explaining rather than some adaptationist story (Gould and Lewontin 1979; cf. Selzer 1993). What they say is worth recounting.

The spandrels of San Marco are architectural features of St. Mark's in Venice. These spaces have been decorated so beautifully that it is easy to think that they were designed to provide a place for an artist to work. In fact, they are merely artifacts left over when two other real architectural features came together, leaving an odd space that needed clever mosaic to "make it look good." Pangloss, as you may recall, was the philosopher in Voltaire's *Candide* who justified every rotten thing that happened to the other characters, and even to himself, as occurring for the best because this is the "best of all possible worlds," as the rationalist philosopher Gottfried Leibniz was reputed to have held. Gould's and Lewontin's point is that if you look at the biological world through rose-colored lenses like these, as Paley and pan-adaptationists after him, including contemporary sociobiologists, have tended to do, you are likely to think that a feature is an adaptation when it is not. You are then likely to tell some story about it that is false. One source of this mistake is the tendency to cut organisms up into parts on the assumption, often unjustified, that the traits you pick out are really there, and so must be there for a reason. One can easily imagine a long and tedious adaptationist story about the evolution of chins. But as Gould and Lewontin point out, there is no such thing as a chin. What looks to us like a chin, and gets named as one in natural languages, is from an anatomical point of view the result of changes occurring in two distinct morphological fields, as in the case of the spandrels. There are no genes for chins. Another error is that if you are determined to think that there is a reason for everything, as Pangloss does, the failure of one adaptationist story, when someone gives evidence against it, simply launches you into another. That leads to explanatory emptiness because you are arbitrarily ruling out a highly likely alternative: that the trait has a nonadaptive cause. Perhaps it is a structural feature of the materials out of which the organism is built. Perhaps it was built up by natural selection for some other adaptive role and then was put to a new use (what Gould calls an "exaptation") Something very basic, in any case, is being assumed in many sociobiological "just-so stories," as Gould calls them (after Rudyard Kipling's book about such things as how the leopard got its spots). What is being assumed is that every feature has actually been built by natural selection to serve a function. Sociobiologists love to tell such stories. By reminding them that adaptation, when it occurs, occurs under severe constraints, Gould and Lewontin are attempting to show sociobiologists that citing constraints rather than telling adaptationist stories often provides the best explanation of an evolutionary question. Explanation is a context-sensitive notion (Garfinkel 1980). In the cases considered by Gould and Lewontin, what is needed for an explanatory "ah-ha experience" is a particular piece of information about the structural properties of materials or the embedded developmental facts about particular lineages that closes them to further selection, and not another just-so story.

The enhanced array of evolutionary units, agents, forces, processes, and levels accessible to expanded forms of Darwinism, including genetic drift and inherited constraints, countenances a wide range of possible evolutionary scenarios with which inquirers can work in particular cases.[15] Explanatory Procrustean beds, usually the consequence of over-generalizing a scenario that fits only a limited number of cases, will be discouraged. Since this temptation often arises from misplaced attempts to find general laws and deductive arguments, advocates of an expanded synthesis have found allies among philosophers of biology who, in resisting the methodological biases of reductionism, have placed stress on the explanatory role of models and scenarios in evolutionary problem solving (Beatty 1981, 1986; Lloyd 1988; Depew and Weber 1985). It is even possible to argue that if philosophers of biology had not already pulled away sharply from the explanatory ideals of logical positivism in depicting unexpanded versions of synthesis as mature science, expanded versions would never have been formulated, or at least would not have gained the degree of support they now enjoy. The idea of an expanded synthesis depends, in fact, for its methodological health on trends in the new philosophy of science, according to which problem-solving heuristics are more important than parsimonious theories, and an ever-widening arsenal of models, brought to bear on the study of particular problems, is more important than an ever-decreasing set of laws. Just as the hardening of the modern synthesis was facilitated by the ascendancy of logical positivist philosophy of science, so the idea of an expanded synthesis has been facilitated by the decline of positivistic ideals among most contemporary philosophers of biology. Accordingly, one can profitably view the explanatory pluralism of the expanded synthesis, and its recommendation to concentrate on particular cases, as a radicalization of Mayr's views about the specificity of evolutionary explanations (Beatty 1981; Brandon 1978, 1990; Sober 1984a; Depew and Weber 1985).

Nowhere is the epistemological temper of the expanded synthesis more apparent than in the role it assigns to narratives or stories in evolutionary explanation. In stressing structural and phylogenetic constraints, it might superficially appear that Lewontin's and Gould's argument against sociobiological just-so stories is an attack on story-telling generally in evolutionary biology. It is important to recognize, however, that Lewontin and Gould are not arguing against the ineliminable role of evolutionary scenarios or narratives in evolutionary biology. On the contrary, it is just because they value narratives as the proper, and perhaps only, way of explaining unique sequences of events that Gould and Lewontin have spoken so ardently against the explanatory weaknesses of *adaptationist* narratives.

Appeal to story-telling as a distinct form of explanation, which can throw explanatory light on particular, nonrepeated historical sequences, has long been a weapon used by humanists, especially historians and philosophers of history, to show that this discipline should be given immunity to the "law-covered" forms of explanation favored by positivists (Gallie 1964; Mink 1979; Hexter 1971; cf. Depew 1985). The point of these arguments is usually that modes of explanation proper to natural sciences are improper for inquiry into human affairs. It comes as something of a surprise, accordingly, that in recent years the same arguments have been taken up by some practitioners of evolutionary biology, a natural science. Lewontin now says, for example, that the notion that "science consists of universal claims as opposed to mere historical statements is rubbish," and that in fact "a great deal of the body of biological research and knowledge consists of narrative statements" (Lewontin 1991, 142–43). Gould too has argued that events strung together into causal chains are neither sufficiently necessitated, reversible, nor predictable to make anything other than inherently retrospective narrative reconstructions possible (Gould 1989a).

Gould's and Lewontin's narrativism underscores their rejection of sociobiology because not everything in a narrative is action. In fact, actions and interactions, whether in historical or biological inquiry, or even in fiction, can be understood only against a background of constraints. Accordingly, one cannot get a good explanation of an organism's traits merely from what would, a priori, be the most efficient, most desirable way for an organism to function. One cannot do this for roughly the same reasons that one cannot get a good historical narrative, or even a good fictional one, merely by mentioning what an agent intends to do, tacitly supposing that it will be done. Life just is not like that. Just-so stories in biology are, in this respect, like hero worship in history: neither can produce a convincing, or even a real, narrative.

Drawing the line between narrative and nonnarrative explanations *within* natural science, rather than at the boundary between the natural and cultural sciences, is certainly one way of defending the autonomy of biology. In suggesting that evolutionary biology's failure to resemble physics is due to its resemblance to narrative history, however, Gould goes beyond anything Mayr ever claimed on behalf of evolutionary natural history. One reason for this rather startling shift is that in their retreat from adaptationism, even of Mayr's modest sort, proponents of an expanded synthesis recognize that chance has a better hand in evolution than Darwinians have hitherto suspected. Darwin's Darwinism, we can now see, may have opened up the tidy world of Victorian Christianity to vast and sometimes terrifying processes of change. But the comfortable rationalism of adaptationism still bound classical Darwinians to natural theologians like Paley. The expanded synthesis undercuts these assumptions by admitting degrees of chance several orders of magnitude beyond Darwinism's customary levels, both at the level of the very small and the very large. At one end, adapted organisms besport themselves on a shifting ground laid down for the most part by the chance fixation of neutral mutations (Gould 1989b). At the other, if the birth and death of species stands to the evolution of clades as genetic variation stands to the adaptedness of organisms, what enters into the misleadingly inevitable logic of our taxonomic systems is for the most part a matter of accident (Gould 1989a). Evolutionary theory, and philosophical reflection on it, are, from this perspective at least, drifting deeper and deeper into a world that is certainly full of novelties, complexities, and propensities, which sometimes congeal into some sort of shape and pattern behind us but whose future, and even present, course we cannot even in principle anticipate. *Narrativism in biology is one way of acknowledging ineliminable complexity.* To recognize this condition is to render it risky to make inferences from the mere existence of selection pressures to evolutionary outcomes, or to treat natural selection as an explanation of first resort, appealing to other forces only when selection fails (Dyke and Depew 1988). It is true that laboratory setups can reduce the number of variables sufficiently to verify, to almost anyone's satisfaction, the existence of evolution by natural selection. But the one thing the natural world is not is a laboratory.

For some time, philosophers of biology have been conducting discussions about precisely where the statistical element in evolutionary theory lies and how deep it runs (Sober 1984a; Rosenberg 1989; Horan 1994). The answer to this question depends on what kind of evolutionary theory, and what kind of Darwinism, you have in mind. When the contrast is between early Mendelism and Fisher's equations, which show how frequencies of genes in populations will predictably change under calculable selection pressures, the statistical element in genetic Darwinism lies in features of models that it shares with statistical mechanics (Rosenberg 1985, 1989). When the implied contrast is between Fisher's deterministic equations and Wright's genetic drift, the statistical element becomes a random or stochastic process at work in small populations, where evolution's wild cards can be found (Sober 1984a, 110). When the implied contrast is between Dobzhansky's balancing selection and Kimura's neutralism, the statistical element points to a stochastic element at levels where natural selection has been constrained from operating freely. When the implied contrast is between a fixed environment and the relativity of fitness to changing environments, genetic drift, density dependence, frequency dependence, developmental noise, multilevel selection, and other simultaneously changing variables, as Lewontin and Levins say, the statistical element suggests that deterministic, and in principle predictable, processes that can be found only within an ocean of chancy events that cannot be predicted even from moment to moment (Richardson and Burian 1992, 353). When the contrast is between the adaptedness of organisms and subtle patterns that may or may not slowly emerge from the differential birth and death of species, the statistical element shows up as slight directional biases that appear in a world governed largely by continency (Gould 1989a). It seems, then, that the statistical element in contemporary Darwinism cannot be found in a single place. The more comprehensive, expansive, and realistic one's Darwinism is, the more it seems to be everywhere.

15 Developmentalism *Redivivus:* Evolution's Unsolved Mysteries

A general theory of biological evolution should include within its domain a number of problems that have hitherto resisted solution within the broad confines of the Darwinian, or indeed any other, research tradition. These problems include how life evolved from nonlife; how developmental programs evolve; what impact, if any, developmental dynamics have on the evolution of species; the relation between ecological dynamics and species diversification; and what is the best way of conceiving the mix between pattern and contingency in phylogeny. The purpose of this chapter is to convey some understanding of the history of these questions and of the direction of inquiry into them now.

Our list of questions is not entirely haphazard. The origins of life, development, ecology, phylogenesis—these are the big questions that people think of when they hear the word *evolution*. It is answers to these questions that people want from evolutionists. That is why they so often feel put off when Darwinians confine themselves to talking about changing gene frequencies in populations and to throwing cold water on ideas about evolutionary direction, meaning, and progress. The reason people feel this way has to do with something more than the intrinsic interest of these questions. It has a historical dimension as well.

Concern with developmental, ecological, and phylogenetic dynamics, and even with the origin of life, became the premier evolutionary problems at a time when developmentalist or transformational models and metaphors, as distinct from Darwinian variational and selectionist models and metaphors, dominated evolutionary theorizing (Levins and Lewontin 1985; Sober 1985). As we saw in chapter 7, that was in the later nineteenth century, before genetic Darwinism pushed these concerns to the periphery. Under the developmentalist dispensation, what evolutionary theories were supposed to explain was relativized to what developmentalists countenanced as evolutionary phenomena in the first place. These were isomorphically structured phenomena found at different levels of scale ranging from individual development to ecological succession to phylogenetic differentiation. The very terms *ontogeny, ecology,* and *phylogeny* were coined by the recapitulationalist Haeckel. On this account, the steady complexification of ecological systems toward a "climax community" was supposed to connect the regularity of individual ontogeny with the larger unfolding of the grand tree of life or phylogeny. The idea that it is phenomena (rather than the facts or data that help us to establish these phenomena) that science sets out to explain suggests that the processes to which our list of problems point were recognized as phenomena because it was already assumed, or at least anticipated, that evolution is nothing but development writ large (on phenomena as explananda, chapter 1; Bogen and Woodward 1988). The big questions about evolution were thus nurtured in, and remained conceptually tied to, the developmentalist tradition. So did the presumptive answers to these questions. General developmental laws were assumed to be the key to explaining phenomena that were perceived to be developmental in the first place.

Spencer, whose views about evolution were as influential as Haeckel's, shared most of Haeckel's expectations about how evolving systems were expected to behave. Spencer spoke of Darwinism as "the developmental hypothesis" in ways that left Darwin's head spinning (chapter 6). Admittedly, Spencer was not as strongly recapitulationist as Haeckel, for even before Darwin's work had appeared, Spencer had generalized von Baer's laws, which eschew strict recapitulationism, into a universal theory about the transformational dynamics of all natural and social systems. Nonetheless, Spencer's dynamics were dominantly transformational. He defined evolution as "integration of matter and dissipation of motion" and concluded that "from the lowest living forms upwards, the degree of development is marked by the degree in which the several parts constitute a cooperative assemblage" (Spencer 1864, 286, 328).

This account implies that the questions we ask about phenomena, as well as the answers we expect to receive about them, are tied closely to different scientific research traditions. This link acquires even greater importance when another of the themes of this book is brought to mind. Scientific traditions are often no less long-lived and capable of revival than are social, political, and religious traditions. Languishing traditions in science are always out there, waiting for a chance to reassert themselves when a reigning paradigm seems vulnerable. It should not be surprising, then, that whenever the Darwinian tradition encounters difficulties, developmentalists of all stripes, whose tradition became recessive with the consolidation of the modern synthesis, come out of the woodwork, ready, willing, and eager as ever to declare Darwinism dead once again. Ours is just such a time.

These reflections suggest that we must revise what we said at the outset. It is not quite true that the Darwinian tradition has *failed* to solve problems about phenomena such as the origin of life, developmental genetics, ecological dynamics, or phylogenetic order. Rather, ever since Darwin, Darwinians have tended to set aside such questions for the time being or, more often, to deny that there are any such phenomena to explain. Darwin himself set aside the problem of the origin of life for both scientific and religious reasons. The makers of the modern synthesis tended to set aside developmental biology until genetics had matured. Developmental geneticists like Goldschmidt and Waddington, who were unwilling to do this, were marginalized. Modern Darwinians are often reluctant to admit that anything like ecological or phylogenetic complexification even exists, particularly when the implication is that there is evolutionary "progress." They take as much delight as Lyell in pointing out that at least as many species have evolved into less complex forms as have evolved toward greater complexity and that measures of complexity are arbitrary in the first place (Williams 1966; cf. Nitecki 1988). Controlling what counts as phenomena is a real function of research traditions.

Still it is not clear that the consensus against the developmental tradition will hold much longer. Even Darwinians are once again flirting with evolutionary direction and complexification, even if they have cast off talk about progress as nineteenth-century baggage (Gould 1988; Ruse 1988b; Wimsatt and Schank 1988). We think that the new willingness of Darwinians to consider evolutionary direction and complexification, as well as to grapple with development and ecology, reflects the steadily increasing power of the molecular revolution to unsettle the assumptions on which the modern synthesis was built. The molecular revolution, in addition to providing occasions for transfigured versions of the old developmentalism to reassert its claims against the Darwinian tradition, is also stimulating Darwinians to acknowledge and try to explain evolutionary phenomena that have for a long time fallen within the orbit of the developmentalist tradition. As the twentieth century ends, the most vital, and for that reason the most confused and confusing, debates about evolution are taking place in the contested territory where expanded Darwinians and transformed developmentalists meet. The remainder of this book will be largely concerned with mapping these confrontations.

In order to see why the stirring of the developmentalist tradition reflects the rise of molecular biology and how that tradition is being revitalized by complex systems dynamics, we must look back over some of the ground we have traversed. The Darwinian tradition was reborn in this century when Weismann's sequestration of the germ line was joined with Mendelism to seal off questions about ontogeny from the causes of replication, variation, and evolution. This had the effect of putting distance between twentieth-century Darwinians and figures like Haeckel and Spencer. It also created a situation in which twentieth-century Darwinians were inclined to conjure up a founding father in their own image. Their Darwin was ostensibly neutral about inheritance and held views about development and evolution that were further from those of Haeckel or Spencer than the real Darwin's actually were. The real

Darwin, as we saw in part I, thought that variation, the raw material of natural selection, arises from disturbances to the growth process. So far forth, he continued to cling to the epigenetic tradition that went back to Aristotle, who saw nutrition, growth, and reproduction as parts of a single, continuous cycle (Hodge 1985; Bowler 1990). Darwin agreed, moreover, that the complex interaction between environments and embryos will result in some sort of parallel between individual development and phylogeny. What was most novel about Darwin's work was that he used natural selection to defend a weak recapitulationist version of the parallel between ontogeny and phylogeny, according to which the tree of life is ramified after the fashion of an expanding capitalist economy by the action and interaction of competitive individuals rather than by the unfolding of developmental laws. In this respect the differences between Darwin and Spencer were not nearly as great, accordingly, as some twentieth-century Darwinians have assumed (Richards 1988).

Weismannian Mendelism displaced this epigenetically oriented set of issues with an updated version of preformationism, in which the underlying information that guides reproduction is largely unaffected by what happens in and to a developing embryo. At first, molecular genetics seemed to support this view. The central dogma of molecular biology, according to which information flows solely from DNA to RNA to protein, seemed to underwrite the isolation of evolutionary from developmental biology. It has become clear, however, that "reverse information flow" does in fact occur. Moreover, genomic DNA, "once thought to be a static and unchanging 'information store,' turns out to be extremely fluid. Amplifications, deletions, rearrangements, and mutations occur frequently during development and in response to environmental stimuli" (Ho and Fox 1988, 10; chapter 13). As experiments in transposing genes show, these newly discovered sources of variation are triggered by genomic stress. Perhaps there is something after all to Darwin's old assumption that variation is caused by developmental disruptions. Perhaps it is even true that this is a functional process, a product of adaptive evolution that enhances the chances of continued evolutionary success by producing variation just when a species or a lineage is in evolutionary trouble (Campbell 1987; Wills 1989). If variation leads autonomously and suddenly to reorganized genomes and gene pools, and these changes form the nodal points of phylogenesis, it is even possible to suspect once again that phylogeny and ontogeny are more strictly parallel than orthodox Darwinians have assumed (Oyama 1985, 1988). Screened off from the transient effects of shifting environments, which are evanescently and reversibly tracked by natural selection operating at the microevolutionary level, phylogeny will record structural changes. Developmental programs will record, at differently embedded depths, aspects of phylogenetic history. The spirit of Geoffroy will once more preside over evolutionary theory in the grand sense that really interests people.

The notion is once more abroad, then, that a general theory of evolution that embraces and resolves evolution's larger mysteries will be found only when the Darwinian tradition ceases (once again) to be the obligatory framework for productive inquiry. Books and papers have been regularly appearing bearing titles such as *Beyond Neo-Darwinism: Introduction to the New Evolutionary Paradigm*. In these books, authors from a range of disciplines wonder aloud whether the developmentalist tradition might once more become ascendant in a new form, or they solemnly pronounce that a "new evolutionary paradigm" has in fact made its appearance already (Ho and Saunders 1984; Ho and Fox 1988; Salthe 1993; Swenson 1995). From the perspective afforded by such books, Darwinians who have advocated an "expanded synthesis," or who speak of an "unfinished synthesis," are viewed as bailing out a leaking ship. Gould and Eldredge are right, new developmentalists will say, to think that an adequate theory of evolution must recognize that living nature is hierarchically organized and that evolutionary processes and forces take many forms. In assuming, however, that every level of the biological hierarchy must be governed by selection or its analogues, Darwinism expanded turns into Darwinism weakened. In any case, expanded Darwinism provides little or no insight into *why* biological nature should be hierarchically arranged in the first place.

New developmentalists generally applaud advocates of an expanded synthesis for seeing that "if there are multiple levels at which selection works, and higher levels evolved later than lower ones . . . a kind of progressive change . . . seems inevitable" (Wimsatt and Schank 1988, 232). They also commend pluralistic Darwinians for seeing that the distinction between replicators and interactors can be generalized until all biological entities and processes are shown to fall into two interacting hierarchies, a genealogical hierarchy of conserved information and an ecological hierarchy composed of energy processing units (Eldredge 1985; Salthe 1985, 1993; Eldredge and Grene 1992). Admittedly, authors sympathetic to a new evolutionary paradigm often differ among themselves about whether evolutionary causality falls more on the genealogical or the ecological side of the hierarchy (Brooks and Wiley, 1986, 1988; Wicken 1987). They are united in affirming, however, that there are autonomous dynamics at work both within and between the ecological and genealogical hierarchies and that significant evolutionary change depends in considerable measure on these dynamics rather than on the weak forms of selective sorting that Darwinians invoke. The point is not simply that evolution occurs within an "ecological theatre," as the ecologist G. Evelyn Hutchinson says (Hutchinson 1965). It is that the theater itself changes by way of the autonomous dynamics of self-organizing systems that utilize high-grade and dissipate low-grade energy and that, together with similarly autonomous processes governing informational change within genomes and gene pools, these processes account for major evolutionary phenomena. This means that the new developmentalists are united in denying the assumption that allows advocates of the expanded synthesis to retain their filial piety toward the Darwinian tradition: that units along these two hierarchies are woven together by natural selection or some analogue of it. For this reason, new developmentalists conclude that "we do not need a revised Darwinism, but a new conceptual framework" (Webster 1989, 13).[1]

Since structural transformations, self-organizational phenomena, and energetic dynamics are closely tied to mathematics and physics, the new developmentalists tend to place Mayr's proud proclamation of the autonomy of biology from the physical sciences in an unflattering light. Far from being a justifiable demand that the modern synthesis be given world enough and time to develop (Mayr 1985, 1988), the autonomy of biology turns, in their mind, into a defensive plea arbitrarily sealing off the relevance of more basic processes from problems falling within the domain of the synthesis (Wicken 1987). New developmentalists suggest that under the management of such luminaries as Mayr, Gould, and Lewontin, Darwinism has been moving steadily away from links with the sciences that have always breathed new life into science: "Biological science as a whole has suffered the severing of a long-standing relationship with the so called 'exact sciences,' physics, chemistry, and mathematics. . . . While one grand architect of the neo-Darwinian synthesis may rejoice in the final emancipation of biology from physics and chemistry [Mayr 1982], a great deal of current emphasis is placed on integrating biological explanations with contemporary physics, chemistry, and mathematics" (Ho and Fox 1988, 4).

From this perspective, the narrativist turn taken by recent Darwinians can very quickly become an object of ridicule as well. The admission by many contemporary Darwinians that evolutionary theory does not use universal laws as premises but instead constructs "vast genealogical narratives comparable in form . . . to the medieval *chansons de geste* or the Icelandic sagas," can be construed as a confession of scientific defeat (Webster 1989). When the polemical juices really begin to flow, the inadequacies of Darwinism are ascribed to its ideological biases. Mae-wan Ho, for example, argues that Darwinism "originates from, and in turn reinforces, a particular perception of life which emphasizes competition and strife at the expense of all other human qualities, a distortion of reality that carries obvious political dangers" (Ho and Fox 1988, 3). Stanley Salthe, an apostate Darwinian developmentalist, asserts that Darwinism must be wrong in part because it is morally unacceptable (Salthe 1993).

Darwinians, both orthodox and expansionary, can respond to such charges by claiming that, like their predecessors, the new developmentalists oscillate uncertainly between vitalism and reductionism. If developmentalists assume that there is an overall unfolding direction to life, it will be said that they must be ascribing a mysterious, crypto-vitalist purposiveness to evolutionary processes. If, on the other hand, new developmentalists repudiate such claims by asserting that evolutionary complexification is merely a matter of the ineluctable laws of physics, chemistry, and mathematics, they open themselves to the charge of reductionism. These are charges that new developments should not try to evade. Nor do they have to. The revolution in complex systems dynamics is now making it possible to hope that complex, self-organized systems, including those investigated by evolutionary biology, can be more closely linked to physics and chemistry without reductionism or vitalism. The hope that complex systems dynamics will be helpful in recognizing and explaining evolutionary phenomena like those pointed to by nineteenth-century evolutionists is already evident in titles such as *Theoretical Biology: Epigenetic and Evolutionary Order from Complex Systems* (Goodwin and Saunders 1989).

Having shown some sympathy for ideas like these, we must now record that we are less inclined than many new developmentalists to assert that appeal to self-organizational dynamics means the death of Darwinism or its confinement to tightly constrained microevolutionary ranges (Weber et al. 1989). Even where arguments to this effect do not depend on caricatures of Darwinism, they usually assume that Darwinism is inseparably tied to the Newtonian or Boltzmannian models that have nurtured and revitalized it in the past but have not defined or exhausted it. Whereas new developmentalists believe that natural selection, tied as it is to old dynamical models, cannot explain significant evolutionary change, we think, like Kauffman and Wimsatt, that it is possible for natural selection to be effective in causing evolutionary change precisely because it works in and on a field of self-organized entities subject to nonlinear changes (Kauffman 1993; Wimsatt and Schank 1988). It is possible that the complexity revolution, far from revealing the limits of the Darwinian tradition will serve instead to show the limits only of the background assumptions on which that tradition has hitherto relied. We suspect, in fact, that by recasting itself in terms of new dynamical models, Darwinism can fruitfully address such thorny issues as the origins of life; the relations between ecology, evolution, and development; and the disputed issue of pattern and progress in phylogeny. For the present, however, the task is to say more about these phenomena themselves and how biologists in both the Darwinian and developmentalist traditions have been dealing with them.

For Darwin, the problem of the origin of life was largely bracketed off from the problem of life's subsequent evolution. In *On the Origin of Species*, Darwin spoke of life having been "breathed in to one or several ancestral forms" (Darwin 1859, 484, 490). Privately, however, in a letter to Hooker in 1871 he wrote, "But if (and oh what a big if) we could conceive in some warm little pond with all sorts of ammonia and phosphoric salts, light, heat, electricity and etc., present, that a protein compound was chemically formed, ready to undergo still more complex changes" (DAR 94, 188–89).

Until recently, the problem of the origin of life has remained almost as peripheral among Darwinians as it was for Darwin himself. The molecular revolution, however, has given increased saliency to the issue and has suggested experiments that have uncovered some intriguing facts. Indeed, it is now overwhelmingly likely that life, under some reasonable definition, will be produced and replicated under laboratory conditions within a fairly short time. When that happens, the origin of life will cease being conceived as a one-time event, after the manner of Genesis I, and will begin to be treated as an expected phenomenon that has emerged from and within a wider range of evolutionary processes.

The best estimates indicate that the earth was formed about 4.5 billion years ago. It took close to 500 million years after that for its surface to cool sufficiently for water to condense and for oceans to begin to form. Until this time, the earth was too hot for molecules such as amino acids and nucleotides, the building blocks of proteins and nucleic acids, to survive in it. During the next 200 million years, continents formed, volcanic activity fell off in intensity, and the frequency of meteoric impacts decreased, in part because a thickening atmosphere protected the earth. The remarkable fact is, however, that cellular life appeared as early as 3.8 billion years ago and all the major metabolic features appear to have been present by 3.5 billion years ago (Schopf 1983). Life's advent seems, in fact, to have occurred as soon as it was chemically possible on the primitive earth (Morowitz 1992).

The first experimental exploration of the origin of life was done in 1953 at the University of Chicago. Stanley Miller, a graduate student working in the laboratory of the Nobel laureate chemist Harold Urey, passed a spark through a chamber filled with ammonia, methane, hydrogen, and water, all of which are believed to have been components of the early atmosphere. Remarkably, the input of energy into this chemically reducing atmosphere produced a substantial number of the amino acids that are found in the proteins of modern organisms, including some with complex structures. It is much easier, of course, to produce a rich variety of amino acids than to make the constituents for nucleic acids. Experiments during the 1960s by Leslie Orgel at the Salk Institute showed, however, that chemical routes were also available under prebiotic conditions that would spontaneously produce large quantities of adenine, one the four bases that make up the nucleotides found in nucleic acids. Orgel also showed that the sugar component of nucleotides can be readily synthesized from formaldehyde, which is also believed to have been available early in the earth's history (Miller and Orgel 1974). Darwin's hypothetical "warm little pond" thereupon became something more than a thought experiment. Chemical conditions likely to have obtained on the primitive earth would spontaneously have produced large quantities of a number of amino acids, nucleotides, and even primitive-like replicating macromolecules.

The question now became whether proteins, made of chains of amino acids, or nucleic acids came first in the emergence of a living thing, or occurred simultaneously. Darwinians have, on the whole, been attracted to a "replicators first" strategy, even though this way of attacking the problem has a disconcertingly chicken-and-egg aspect to it. DNA will not self-replicate unless it is coupled with specific enzymatic proteins. In turn, these enzymes depend for their synthesis on information encoded in RNA and DNA.

Recently workers in this research program have been cheered by the discovery that RNA has self-catalytic properties. It has been shown that the heteronuclear RNA of eukaryotic cells self-splices out introns, or the noncoding intervening sequences in a gene whose exons code for the structure of the protein, without the involvement of any protein (Cech 1990). Thus, there has arisen of late an excited emphasis on an "RNA world," in which the origin of self-replicating RNA molecules would mark the crucial event in the emergence of life. If catalytic peptides of nonrandom, but not yet specified, sequences were also available, they could help catalyze not only the reactions that produce nucleotides but also participate in "polymerizing" or stringing together such nucleotides. Perhaps evolutionary bootstrapping is possible after all.

Those who take this view must then face a further issue: Should they regard the origin of life, construed as the emergence of "naked replicators," as an improbable accident that needed to occur only once to produce what we see around us, or should they demand that it be the result of a more or less predictable process? Jacques Monod, one of the fathers of the "operon" model of regulatory genes, is the most famous adherent of the view that life on earth originated in a low-probability event. "We must conclude," he has written, "that the emergence of life on earth was probably unpredictable before it happened. We must conclude that the existence of any particular species is a singular event, an event that occurred only once in the whole of the universe, and therefore one that is also basically and completely unpredictable, including that species we are, namely man" (Monod 1974, 23).

When Dawkins says in *The Selfish Gene* that "at some point a particularly remarkable molecule was formed *by accident*," he too is professing adherence to the "frozen accident" theory. "A molecule which makes copies of itself," Dawkins says, "is not as difficult to imagine as it seems at first, and it only had to arise once. If it had affinities for its own kind, the building blocks available in the primeval soup will be pulled into a replicate of it" (Dawkins 1989, 16). This is, of course, just what one would expect of someone who makes genes the primary causal agent of every evolutionary process.

The view that life on earth originated in a low-probability event resembles Darwin's official, published line, with talk about "frozen accidents" replacing Darwin's hand-waving remark about "life having been breathed" only once into a primitive organism. It also bears some resemblance to the views of contemporary physicists, who think of our universe as the result of a "singularity," the explanation of which depends not on prediction based on laws—after all, laws apply to events that result from singularities—but on working backward from the undeniable fact that we are, in any case, here. (This is the so-called anthropic principle.)

Darwinians do not, however, have to rely on "frozen accidents." If they are expanded Darwinians, they can increase the probability that life would predictably have emerged from the primeval soup by considering that entities that can be affected by natural selection, that is, living things, emerge out of more basic physical and chemical processes in which other forms of selection play a role. Selection of the reproductively *fit* rides on the back of, and emerges out of, selection of the physically *stable* and the chemically *efficient*. Chemical selection rewards reaction pathways and cycles that remake the reactive intermediates required for the synthesis of products. These are called "autocatalytic cycles." Self-catalyzing chemical kinds become the object of a selective process in which self-organization gives a boost to the emergence of entities among which natural selection proper can occur.

This kind of thinking animates the research program of Nobel laureate chemist Manfred Eigen. Eigen has mathematically demonstrated that an integrated set of autocatalytic cycles, which he calls "hypercycles," will reward increased catalytic efficiency and increased specificity in directing the production of a particular peptide sequence (Eigen and Schuster 1979, 1982). Thus, even before any other metabolism had arisen, a type of selection would have produced molecules capable of replication, with the aid of proteins for which they coded. It is easy to imagine that such systems could function in the autocatalytic cycles of the hypercycle model and would develop specific sequences prior to the incorporation of proteins into the cycle. Since polynucleotides and peptides interact in nonrandom fashions, it is likely that whatever RNA happened to be around would, on this account, facilitate the production of a peptide that would have a specific catalytic ability to make more RNA. (The discovery of autocatalytic RNA has also led to a more recent variation of this scenario in which RNA acts alone, as a "ribozyme," to catalyze its self-replication [Cech 1990; Gesteland and Atkins 1993].)

Molecular Darwinians of both Monod's and Eigen's stripes usually assume that the self-replicating RNA ancestors of living systems arose as naked replicators in dilute solution in the oceans. Harold Morowitz and several colleagues have shown, however, just how physically unlikely it is that life could have begun in that way (Morowitz, Heinz, and Deamer 1988; Morowitz 1992). Morowitz is among those biochemists who have long recognized that no account of the emergence of life will be satisfying unless it explains how cell membranes, as well as replication, arose, since lacking such membranes, replicators would have no stable home that would allow concentration of the chemical components. People of this persuasion have followed in the footsteps of Russian biochemist Alexander I. Oparin, who first proposed scenarios for the origin of life that emphasized the emergence of metabolism before replication (Oparin 1938).

Sidney Fox significantly advanced this research program by demonstrating that amino acids will self-organize into little proteinoid microspheres that are resistant enough to a watery fate to provide a site at which the information that directs catalysis and replication can accumulate (Fox 1965, 1984, 1988.) Since it is chemical biases and affinities, as well as the exploitation of energy gradients, that drive the emergence of life, rather than natural selection, which operates only when living things are already up and running, Fox has dissociated himself from efforts to push natural selection proper into the prebiotic realm, and more recently from Darwinian views about evolutionary dynamics generally (Fox 1988).

Those who take a "protein-first," and a "cell-first," approach to the origin of life tend to make greater use of self-organizational dynamics than do molecular Darwinians like Eigen (van Holde 1980). In this spirit, Jeffrey Wicken builds on the notion of a compartmentalized chemical space by appealing to the effect of energy fluxes far from equilibrium to drive the internal organization of emerging proto-cells (Wicken 1985). In this way it is possible to produce models for life's emergence in which autocatalysis, replication, and phase separation through membrane formation arise together as a single system. Chemical selection will favor autocatalytic cycles whose reliability in producing products useful for the next round of functioning is itself enhanced by their ability to capture and store information in macromolecules. We will return to such scenarios for the origin of life in chapter 17.

If thermodynamic considerations are as crucial in explaining the emergence of life as Fox, Morowitz, and Wicken think they are, it seems reasonable to assume that considerations governing energy flow will remain important in the subsequent history and differentiation of living things. In that case, we might also expect that the science of ecology, which explicitly places life into the context of energetics, would be highly relevant to the study not only of prebiotic but of biotic evolution. In this spirit, Morowitz writes, "Sustained life is a property of an ecological system rather than of a single organism or species. . . . A more ecologically balanced point of view would examine the proto-ecological cycles and subsequent chemical systems that must have developed and flourished while objects resembling organisms appeared" (Morowitz 1992, 54).

The fact of the matter is, however, that although ecology was once closely linked to evolutionary thinking, evolution and ecology have for the better part of this century been in tension with each other. The original ecology emerged within a developmentalist vision of evolution. It was displaced with the rise of modern Darwinism. The beginning of wisdom about the history of ecology is to recognize that this word now designates three distinct, but overlapping, disciplinary matrices. Ecology first bifurcated into "systems ecology" and "community ecology." In turn, community ecology gave birth to "population ecology," which has taken up bits and pieces of community ecology into the framework of population genetics. At each point in this process, the study of ecological "succession" or development has acquired less and less evolutionary significance, and its analogy to ontogeny has seemed less and less persuasive.[2]

We hear today calls for reunion between ecology and evolutionary theory. These arise in large part from the growing perception that, in a cultural environment in which the two great nineteenth-century ideologies of free-market capitalism and authoritarian state socialism both attract blame for fostering an exploitative view of nature, ecology is fast becoming an architectonic discipline within which a new ethics of care for the earth and a "green politics" is being nurtured. It seems crucial to have an evolutionary theory that interfaces well with this vision of nature. Many ecologists who share this noble vision have been disconcerted to discover that modern ecology has been rejoined to Darwinian evolutionary theory mostly by way of mathematical population genetics, kin selection, game theory, and selfish genes, all of which have competitive overtones, and that their own hopes resonate better with the macroevolutionary theory of a bygone, seemingly irrecoverable, era. Those who call for interanimation between evolutionary biology and ecology should not assume, therefore, that what they are asking for will be easy to achieve. It will require radical changes in both evolutionary theory and in ecology. In particular, it will require the resources of the complexity revolution to revitalize, and demystify, the old developmental ecology, and to defend the claim that autonomous ecological dynamics are a powerful causal factor in evolutionary change.

We have already noted that the term *ecology* was coined by Haeckel. The Greek root he used to form the term was *oikos* (household), the same root from which the word *economy* is derived. Originally *oikonomia* referred to the small-scale economics of the household, as it was studied by Aristotle in *Politics* I. Haeckel's coinage suggested that a community of organisms is like a household. Ecology first flourished in the 1890s, when developmental and directional theories of evolution prevailed.

Thus, early ecologists not only saw communities of species as organically interrelated wholes, indeed as "superorganisms" in their own right, but as systems that changed and grew. In particular, they thought that ecosystems exhibit the same patterns of endogenous change and differentiation that they thought they saw in ontogeny and phylogeny. In this matter they were influenced by Spencer's generalization of von Baer's laws into developmental trajectories that every natural system, including communities of species, could be expected to follow as it moved from the incoherent and homogeneous to the coherent and heterogeneous. Because Spencer saw all entities that exhibited these patterns of endogenous development as *evolving*, there seemed no reason for ecologists not to think of what became known as ecological succession as a form of evolution at a level lying between individual ontology and phylogeny (Odum 1969).

Ecology was born out of biogeography, the study of the causes of the spatial distribution of populations. Darwin himself was a biological geographer. The first explicit step from biogeography to ecology was taken, however, when German-trained physiologists used inorganic variables to analyze the geographic distribution of species. It was a Danish botanist named Eugenius Warming who made the most progress in developing what became ecology. Warming plotted the distribution of plants against temperature and moisture. Rather than considering what would happen to different genotypes when they share the same environment (as population ecologists later would), Warming considered instead what would happen when organisms with a single inherited makeup are inserted into different environments. What would happen is what Warming called "epharmonic convergence": Different organisms, such as the ancestors of South African and North American cacti, would make the same "move" in the evolutionary game. In response to lack of water, they would develop succulent, fleshy, water-retaining stems, and thin down the leaf to a spine (Worster 1977, 199).

Warming also amended received wisdom about another matter. Darwin thought of competition within species as mitigating competition between them by allowing related species to occupy different niches. Warming had an even rosier view of this process. He thought that competition is concentrated within a species because only members of the same species struggle over precisely the same resources. Real interspecific cooperation, of the kind devoutly wished for by anti-Malthusian Russian Darwinians like Kropotkin, can develop because of what is now called niche exclusion: Representative communities of different species divide up the environmental pie in such a way that different species occupy not only nonconflicting places at nature's table but places that are mutually helpful in what have come to be called food chains. Finally, there is a temporal dynamic in Warming's work. Warming made it a principle that each community, unless otherwise inhibited, moves toward a "final community," or "climax formation," in which cooperative diversity is maximized.

American biologists, especially those studying the Great Plains, liked Warming's work in part because it resonated well with their Spencerian belief that competition gives birth to cooperation. As a result, it was in America that the discipline of ecology first became institutionalized. The translation in 1909 of Warming's *The Ecology of Plants* (1895) stirred Henry Cowles's at the new University of Chicago to study ecological succession on the Michigan dunes. An ecological society was founded in 1915. One of Cowles's students, Charles Christopher Adams, became the first editor of the journal *Ecology*. Meanwhile, Asa Gray's student Charles Bessay had taken up a not-too-prestigious job at the University of Nebraska. His student, Frederick Clements, became the next great figure in American ecology. It was Clements who fully developed Warming's notion of a "climax community."

Clements entered the University of Nebraska in 1890 at the age of sixteen. Having grown up in the sparsely settled grasslands of Nebraska, he was already a keen observer of its biota and their apparent resilience in the face of natural, if not man-made, disasters. Clements's *The Phytogeography of Nebraska*, together with his *Plant Succession* (1916), represents a turning point in early ecology. Like Warming, Clements sought to reconcile Darwinian selection with cooperation by arguing that selection on the basis of competition favors the eventual development of a highly stable cooperative web of relationships in a diversified community. Just as organisms move from homogeneity to heterogeneity, Clements reasoned that a community of species reaches a steady state of maximum diversity and maximum cooperation. Clements took the analogy between communities and individuals seriously. He regarded an ecosytem as an organism. "The unit of vegetation," he wrote, "the climax formation, is an organic entity." Like an organism, therefore, a community "arises, grows, matures and dies. . . . The climax formation is the adult organism, the fully developed community, of which all initial and intermediate stages are but stages of development"(Clements 1916, 124–25).

This vision was a ripe for disillusionment. There is, moreover, some truth in Donald Worster's claim that it was the Dust Bowl that served as the catalyst for this expected fall (Worster 1977, 219–53). The idea that succession moves toward a single, mature, resilient climax community was blown away in the vast and sudden ecological catastrophe that engulfed the Great Plains in the early 1930s, disrupting the lives of millions of the people whose monocultural practices were a principal cause of their misery. For his part, Clements preserved his faith in nature's developmental ways throughout the debacle. He did so, however, only by subtracting human predators from the picture, except for hunters and gatherers like the native Americans, who presumably lived in harmony with nature's ways. This way of reacting to the collapse of the old ecology eventually led through Aldo Leopold's "land ethic" to the contemporary "green movement" and to the philosophy of deep ecology, which ceaselessly inveighs against the ideological sin of human "speciesism" and recommends an immediate and radical drop in human populations until we are fit once more to share the earth with our fellow species (Naess 1983).

The discipline of community ecology represented an alternative way of responding to the crisis of the old systems ecology. Although questions about ecology as evolution, and even about the mystical properties of "climax communities," were generally left aside, community ecology maintained a fairly strenuous holism. William Morton Wheeler, for example, a University of Chicago entomologist, saw ant colonies as "superorganisms" and thought of ecological systems in something like the same terms. Typically, however, community ecologists were more interested in the notion of steady states and populational equilibria than in the Spencerian idea of progressive differentiation or succession. Theirs was a more synchronic than diachronic view of ecological communities. That is in part because community ecologists had been charged with protecting wild populations and species from human predation, as well as with managing animal populations in national parks. They did so guided by Charles Elton's *Animal Ecology* (1927), from which they acquired a standing interest in mathematical models and empirical studies that throw light on how stable numbers of linked populations of predators and prey are to be maintained.

Still the rise of community ecology did not mean that systems ecology disappeared. "Reformed" systems ecologists, we will call them, continued to believe that ecological succession bears some similarities to ontogeny, even if ecologies have multiple "climaxes," including some that are disastrous insofar as they promote the dominance of a single species, thereby opening the entire system to collapse (Johnson 1988, 1990). For reform systems ecologists, the dynamic tendencies of ecosystems come into view in proportion as ecosystems are viewed in terms of physical, especially thermodynamic, laws and imperatives, in which organisms and populations serve as nodal points for energy use, transfer, and dissipation and in which life itself emerges and differentiates in accord with thermodynamic imperatives. From their perspective, ecosystems are not perspicuously viewed as loosely integrated superorganisms, as Wheeler believed.[3] On the contrary, reformed systems ecologists tend to view organisms as very tightly integrated ecological systems (Odum 1953; Odum and Odum 1982; Ulanowicz 1986; Johnson 1992; Wicken 1987; Weber et al. 1989).

The work of Alfred Lotka provided the foundations of reformed systems ecology. The whole of European culture seems to have gone into the making of Lotka. Born of American parents in 1880 in what is now Ukraine, Lotka received the bulk of his education in France and Germany before completing degrees in chemistry and physics at the University of Birmingham in 1901. He then went to the University of Leipzig to study the newly emerging field of physical chemistry (Servos 1990). It was there that in 1901–1902 Lotka attended a course of lectures presented by Friedrich Wilhelm Ostwald at the newly established Physical-Chemistry Institute. Taking to his breast Ostwald's Helmholtzian proclamation that energy is the fundamental reality, and the central organizing and integrating concept of both the physical and biological sciences, Lotka vowed to create a discipline of "physical biology," which he conceived as analogous to physical chemistry. In 1902, Lotka came to the United States, where he held a variety of jobs in the chemical industry, government laboratories, and scientific journalism. Seeking an academic post, Lotka obtained a master's degree in physics from Cornell in 1909. Working in his spare time, he began publishing papers on the application of physical principles to biological organisms. In 1912, he submitted twelve of these papers to the University of Birmingham for the D.Sc. degree, hoping thereby to qualify himself for an academic career. Lotka's friendship with the influential biologist Raymond Pearl led to an honorary-fellow title from Johns Hopkins, and an opportunity to use Hopkins as an academic address. It did not, however, lead to an appointment. Under these marginalized conditions and in the shadow of the emergence and consolidation of genetic Darwinism, Lotka wrote his unique masterpiece, *Elements of Physical Biology* (1924). Neither biologists or biochemists paid it much mind. The only people who seemed to be interested in the book were ecologists. Lotka, however, was not very interested in them. Charles Adams had to inform Lotka that his work was of considerable interest to ecologists and to beg him to let him publish a review of *Elements of Physical Biology* in *Ecology*. That was, however, the end of that. In frustration, Lotka finally took a job as a mathematical demographer with the Metropolitan Life Insurance, where he worked until 1947.

Ecologists were interested in Lotka's work because he emphasized the complex interactions among animals, plants, minerals, and the physical environment and considered these processes as an interconnected and integrated process in which organisms and communities of species exchange matter and energy with their surroundings. Lotka saw an evolutionary element in these processes. It was not the evolution of isolated populations and species, however, but rather of whole systems made up of interacting species, each of which, by degrading energy, is affected by and affects all others in the system. Lotka also believed that selection is the key concept needed to understand how evolution occurs in a complex interacting network like this. He was not trying to develop an ecological theory of natural selection, however, but rather to relativize the conditions under which natural selection can be effective to more fundamental forms of what we call chemical selection ("the survival of the efficient") and physical selection ("the survival of the stable"). Unlike the reductionistic materialists of Helmholtz's generation, however, Lotka did want to contract biology to fit current physics and chemistry. Rather, he was trying to expand physics and physical chemistry to account for energy flows in and through biological systems, including organisms.

Lotka was keenly aware that biological systems are not thermodynamically isolated, as Clausius's statement of the second law requires. Although the earth as a whole can be considered approximately closed, since it receives only energy (rather than matter and energy) from the sun, nothing on earth, and certainly not organisms, is ever at equilibrium. On the contrary, the entire terrestrial system is maintained at a steady state very far from thermodynamic equilibrium. Organisms themselves are the clearest examples we have of systems stabilized far from thermodynamic equilibrium. Their continued existence depends on a continuous influx of matter and energy and an increasingly elaborate process of degrading it into sinks. These sinks are, for the most part, other organisms in a complex hierarchical chain. (We now recognize more clearly than Lotka that the entire terrestrial system, including its abiotic as well as its biotic parts, is regulated by the expansion and contraction of biological processes [Lovelock 1979].)

These insights led Lotka insightfully to conclude that successive states of biological systems at any level of scale cannot be predicted by a simple physical maximum principle like the second law, which merely provides a boundary condition for physical, chemical, and biological activity. Any law of evolution will have to take into account the various kinetic pathways by and through which energy is accumulated and distributed in the organic world. What Lotka used for this purpose was a purported "law of evolution" that he formulated as follows:

Evolution proceeds in such direction as to make the total energy flux through the system a maximum compatible with the constraints. (Lotka 1924, 357)

With his vision of the unity of physics, chemistry, and biology, Lotka proposed this as a fourth law of thermodynamics (Kingsland 1985). If this law were to prove universally valid, evolution could not be explained by tracking a single species, or even pairwise sets of predators and prey. What evolution could and would produce under any given set of circumstances would be a function of properties of whole communities. The problem was how to track the pathways by which systems like these evolved. According to Lotka, this was to be done by using a principle of selection in which it is assumed that selective advantage will always go to matter-and-energy cycles that can utilize their own products, or autocatalyze, more effectively than competitors.[4] The units of selection in Lotka's system, accordingly, are not entities that outreproduce competitors but cycles of energy-and-material flow that have differential autocatalytic properties. On this view, the effects of reiterated selection are to increase the rate at which matter cycles through the system, driven by ever more intense energy flows, increasing the total biomass; to favor those organisms that are not only more efficient in their utilization of energy but are capable of tapping previously unused energy sources, increasing total diversity; and to favor species that, in entering into complex cooperative webs with other species, allow the ecosytem as a whole to conform to the evolutionary laws that govern it. In this way, Lotka gave analytic substance and causal mechanisms, for the first time, for the cooperative vision of selection that had been intuitively adopted by Russian and American field biologists, as well as to the directional and diversifying tendencies that Spencer, Warming, Clements, and others had generalized out of von Baer's laws.

Lotka's vision was generally pushed aside by the rise of genetic Darwinism in the 1930s. Nonetheless, he had a number of admirers and followers who have articulated his research program. Much that became known as general systems theory was inspired by Lotka, a fact insufficiently acknowledged by its founder, Ludwig van Bertallanffy. It was the ecologists Eugene and Howard Odum, however, who most faithfully and creatively continued Lotka's work. Early among Eugene Odum's observations was that ecosystem homeostasis could be maintained through the existence of compensatory flow pathways when more than one concatenation of transfers links two species in a community. If one pathway is disrupted, its loss can be compensated by augmented flow through other, unperturbed pathways. Odum built on this concept to organize a wide range of empirical observations about ecosystems dynamics (Odum 1969). As a result of ecological succession, Odum argued, mature communities tend to possess increased energy flow, greater variety of species, narrower trophic specialization by their members, enhanced amount of cycling, and longer retention of media in the system.

At the heart of this ordering process lies the type of positive feedback inherent in autocatalytic cycling, which defines the field on which selection processes of all kinds work (Odum and Odum 1982; Odum 1988). This positive feedback can be envisioned as the result of a concatenation of influences in which one item in a chain catalyzes another. In ideal cases, the causal links in the loop are the only ones at work. Real loops, however, will always be embedded within larger networks of causalities. If one observes only some of the elements of a loop, they will appear to function in a nonautonomous fashion at the behest of the boundary conditions that drive them. Once the scale of observation is enlarged to include all members of the loop, however, its overall autonomy emerges as an attribute of the system as a whole. Thus, an increase in the activity of any element in the loop will augment the activities of all loop members, including itself.

In such systems, selection will favor subsystems that have the ability to incorporate within their boundaries whatever information is relevant to directing the next autocatalytic cycle (Weber et al. 1989). From this perspective, we can plausibly define organisms, and some colonial approximations to them, as *informationally informed autocatalytic systems* (Wicken 1987). No doubt the spontaneous emergence of replicating macromolecules would have been of immense signficance in the initial emergence of such entities. The imperatives of chemical or autocatalytic selection will give a tremendous edge to any entities that can commandeer the information-storing capacities of such molecules to guide metabolism and to carry information across generations. Chemical selection gives rise in this way to natural selection proper, as the blurred boundaries between prebiotic, proto-biotic, and fully biotic systems are crossed. Increasingly sophisticated, complex, informationally rich, and interconnected entities of this sort can emerge only as parts of increasingly diversified and integrated ecological systems. The evolution of individual species is causally conditioned, therefore, on ecological processes that cannot be reduced to lower-level entities or interactions. In these processes, cooperation and competition are inseparable aspects of a single process.

The large vision of reformed systems ecologists was never taken up by community ecologists, or by the population ecologists who followed them. Yet this is not to say that Lotka's work had no influence on them. On the contrary, Pearl, Adams, and others immediately saw in Lotka's work an important contribution to the problems of population dynamics in which they were professionally interested. In particular, community, and later, population ecologists took from Lotka's *Elements of Physical Biology* the now famous "logistical equation" for population growth. Based on a direct analogy with autocatalytic chemical reactions, the equation takes the form: $dN/dt = rN(K - N/K)$, where N is the population size at time t, r is the maximum growth rate, and K is the upper limit of the population that can be sustained by a given environment. (At about the same time Lotka was working on the logistic growth equation, the Italian mathematical physicist Vito Volterra published a similar approach based on classical mechanics and the principle of least action [Volterra 1926]. The equation has since been known as the Lotka-Volterra equation.) Lotka's theory of population growth differs from Malthus's because it took into account what other species were doing. It also portended far less disastrous consequences to each, since Lotka's equation predicted that any two species would, after a period of "logistic growth," settle into a stable oscillation (although Lotka was aware that even when a constant environment and a constant gene pool are assumed, several solutions, and hence several outcomes, were usually possible).

In order to apply the Lotka-Volterra equation to ecological and evolutionary problems, a new conception of the niche had to be developed. The niche would no longer be just a role in a community, as it was for community ecologists like Elton, but a multidimensional description of all the variables that affect and are affected by a given organism. Such a conception, it turns out, had already emerged in the work of a school of Russian ecologists, notably Vladimir I. Vernadsky. Building on the earlier work of Russian ecology, Vernadsky developed a theory of biogeochemical ecology that shared many of Lotka's concerns and assumptions. During the 1930s Georgii F. Gause extended Vernadsky's concept of a multidimensional niche to a general "principle of competitive exclusion" in order to account for his laboratory studies on predator-prey studies of microorganisms. Gause's work was in turn taken up by G. Evelyn Hutchinson and David Lack. From there it became central to the genetic population ecology of Hutchinson's student Robert MacArthur, whose program in population ecology was aimed at bringing the community ecology into the modern synthesis. Working with Levins, MacArthur developed the idea of an evolving Hutchinsonian niche and integrated it with Dobzhansky's evolution in changing environments. With E. O. Wilson, MacArthur developed the idea of a preferred change from "r-selection" (selection for fecundity and colonizing) to "k-selection" (selection for the efficiency and adaptedness) as Malthusian population limits are approached. Game theory was soon being used by participants in this research program to analyze predator-prey relationships. In the course of these inquiries, however, complexity has reared its head, for it has been shown that predator-prey relations not only fail sometimes to reach stable points or limit cycles but often exhibit chaotic dynamics (May 1974; Schaffer and Kot 1985a, 1985b).

The complicated story of ecology suggests that those who call for a new synthesis between evolution and ecology would do well to remember that ecology and evolution have been synthesized once already. They were integrated, however, within the bosom of the developmentalist tradition, whose analytic tools were not powerful enough to keep various bits and pieces of its comprehensive theory from being yanked out of their original context and assimilated piecemeal to quite different research programs and traditions. Nowhere is this more true than in the case of Lotka, whose relational and interactive vision of nature was lost even as some of his contributions were taken up into community and population ecology. As population ecology follows the logic of its own development, however, it is discovering once again that all evolution is in fact coevolution and that coevolution is a relational process distributed over a large dynamic system. A second synthesis between ecology and evolutionary theory lies further along this path.

The relationship between developmental biology and evolutionary theory, or more grandly between ontogeny and phylogeny, has as convoluted a history as the relationship between ecology and evolution.[5] We have already noted that von Baer became the father of modern developmental biology, as Aristotle was the father of ancient developmental biology, when he put forward a set of developmental laws (von Baer 1828). Von Baer concluded that the most general features of a group of animals appear earlier in embryological development than special features; that less general characters develop from more general; that the embryo of a given species does not pass through the stages of adult forms of lower species; and that the embryo of a higher species is never like the adult forms of lower species but only like their embryos. Although Darwin accepted these laws of development, von Baer's opposition to transmutation and common descent eventually prompted evolutionists to take a more strongly recapitulationist view of ontogeny than von Bear in order to support their already strongly recapitulationist views about phylogeny. Von Baer proved to be (more or less) right about development (Hall 1990; Wimsatt 1986). In the strongly recapitulationist context of late nineteenth-century evolution, however, that fact simply led developmental biologists to distance themselves from evolutionary theory. Nor did this tendency dissipate when genetics began to define itself as distinct from embryology by concentrating on gene transmission rather than gene expression at the time of Morgan. On the contrary, the increasingly successful integration of genetics with evolutionary theory left developmental biologists more isolated than they had been before. Developmental biology played virtually no role in the formation of the modern synthesis (Mayr and Provine 1980). Nor has it ever been fully integrated into genetic evolutionary theory.

Calls for reintegrating development, genetics, and evolution were raised as early as the 1940s by biologists as eminent as Waddington. Today such calls are heard as often as calls for reintegration between ecology and evolution. They arise, moreover, against a similar background. Ontogeny and phylogeny were once as integrated as ecology and evolution. These syntheses took place, however, in a conceptual matrix that proved less and less explanatorily fecund even as genetic Darwinism and molecular genetics were going from strength to strength. Developmental biologists, grasping this fact only too well, pulled away from the evolutionary matrix in which their field emerged without fully crossing over to the Darwinian tradition.

Developmental biology reached a crucial moment in the late nineteenth century. By then, its central problem was to understand how sequences of cell divisions generate a differentiated yet fully integrated organism. A methodological dispute broke out between the two most eminent embryologists of the day, Wilhelm Roux and Hans Driesch. Roux was a mechanistic materialist in the tradition of Haeckel. He spoke of his research program as *Entwicklungsmechanik* ("developmental mechanics"). The idea was that each stage in the development of an embryo is supposed to be triggered by the one before it, like cars crashing into each other on a highway. Driesch, by contrast, took a more teleological or end-oriented approach and was, in consequence, something of a vitalist.

Roux's and Driesch's philosophical disagreement was connected with different views about where the information that guides ontogeny is to be found. Roux's methodological mechanism was complemented by cellular holism. His "mosaic" theory of development asserted that each cell has only some of the information needed to instruct cell division and epigenetic differentiation. Integration and individuality could, in his view, hardly be explained on any other terms. Driesch, on the other hand, thought that every cell has all the required information to specify an organism's cell division and developmental differentiation. Driesch proved that he was right when he experimentally demonstrated that even isolated cells (of sea urchins) will produce developmentally normal organisms. Driesch was supported by Morgan, who had become his friend when they worked together as young men at a Biological Research Station in Naples (Allen 1978). In fact, Driesch's results set up Morgan's research program in genetics. He would henceforth try to find out where the "equipotential" information in each cell is located. It would be found, he thought, in the cell nucleus, and specifically in the chromosomes.

Although Morgan was an embryologist, he held that the problems of (transmission) genetics could be studied apart from questions about the embryological expression of genes (Allen 1978; S. Gilbert 1978, 1991). As a result, genetics developed toward its union with molecular biology and evolutionary theory without leaving an important role for development. The falsification of "mosaic development" and the success of Morgan's research programs did not, on the other hand, deter Roux and others from continuing to look for laws in which the causes of development and inheritance are distributed throughout the cytoplasm. Those who opposed Driesch and Morgan were inclined to believe that in most cases the germ line is not sealed off early in ontogeny. Indeed, they were convinced that the Weismannian and Mendelian assumptions behind Driesch's and Morgan's "magic molecule" thinking represented little more than a recrudescence of the old preformationist tradition that epigeneticists like von Baer had already put to rest. This critique of Weismannism forms a continuous tradition running from Charles Manning Childs's opposition to Morgan through N. B. Berrill's resistance to the modern synthesis to Leo Buss's recent work, *The Evolution of Individuality* (1987) (Gilbert 1992).

By the 1940s, developmental geneticists like Richard Goldschmidt and C. H. Waddington were responding more creatively to the marginalization of developmental biology by genetic Darwinism. Neither put as much stock as did Childs or Berrill in cytoplasmic, as opposed to genetic, inheritance. In this respect, they were genetic Darwinians themselves. Instead of concentrating on transmission genetics, however, Goldschmidt and Waddington made pioneering attempts to study gene expression. Both were convinced that the silence of the modern synthesis on this topic meant that important sources of evolutionary change had been overlooked.

Goldschmidt was born in 1878 in Frankfurt of a wealthy and cultured Jewish family. He was trained in embryology in Munich before moving to Berlin in 1913 to become the director of the Kaiser Wilhelm Institute for Biology, a position he held until he fled Nazi persecution in 1935. Eventually he settled at the University of California at Berkeley. As a geneticist, Goldschmidt was less interested in genetic transmission than in how genes can affect the rates of the chemical processes they specify. In his 1938 book *Physiological Genetics,* he argued that an orderly pattern of development was possible because changes in the position of what he called "rate genes" altered the timing of their expression. This suggested that a single, small genetic change early in ontogeny could cause a subsequent cascade of phenotypic effects. "The mutant gene produces its effect, the difference from wild-type, by changing the rates of the partial processes of development" (Goldschmidt 1938, 51).

Goldschmidt worked out the larger implications of his view for evolution in *The Material Basis of Evolution,* a work that, for better or worse, appeared just as the classics of the modern synthesis were rolling off the presses (Goldschmidt 1940). Goldschmidt attacked a central assumption of the emerging synthesis by arguing that macroevolution is not merely microevolution working itself out over a longer period of time. Instead, rapidly appearing discontinuities, or "macromutations," caused by changes in the expression of the regulatory genes in early development, suggested a saltationist account of speciation in which what Goldschmidt colorfully called "hopeful monsters" play the crucial role. Although Goldschmidt's particular claims were shown to be false, molecular genetics has since shown how the repositioning of a gene into a different regulatory regime can indeed alter its timing of expression. In all probability, it is partly through developmental changes that many new species are created. Goldschmidt is now getting a more respectful hearing for his basic ideas by advocates of punctuated equilibrium and other antigradualist theorists of speciation, even although he was totally ignorant of the concepts and mechanisms that have in retrospect cast him in the role of a prophet (Gould 1982c; cf. chapter 13).

Conrad Hal Waddington fought somewhat more successfully than Goldschmidt against the marginalization of developmental genetics in the modern synthesis and sought doggedly to introject embryology into evolutionary theory. Waddington was a member of the Theoretical Biology Club (or Gathering, as it called itself) that was assembled by the philosopher J. H. Woodger in Cambridge in the early 1930s. Its members included Joseph Needham, soon to be famous for his history of science in China, and J. D. Bernal, one of the Cambridge crystallographers whose work led to understanding the structure of DNA. Among the aspirations of this group was to move biology away from the static biases of the taxonomic and morphological tradition to a dynamical picture, in which what D'Arcy Thompson called "the growth of form" is modeled mathematically (Thompson 1917, 1942). With these concerns in mind, the group adopted the "process philosophy" of Alfred North Whitehead, which replaces a traditional metaphysics of static substances with an ontology in which what we think of as things are actually emerging processes. This research community competed for funding with the reductionist programs in molecular biology of the gene then getting underway at Cambridge and Caltech. It lost that competition in part because of the antireductionist tendencies it had picked up from Whitehead. This background helps explain why Needham, in recalling the glories of the original Theoretical Biology Club, later remarked: "Waddington always used to say that molecular biology and the study of the properties of DNA and RNA was all very well, but that people would have to come back to embryology in the end if they wanted really to understand the properties of the higher living systems" (Needham 1984, vii–viii).

Waddington's holistic and processive tendencies are evident in his insistence that causality in biology is always at least a two-sided affair. He deeply distrusted models in which genes are portrayed as directly causing development or directly acted upon by natural selection. Like Goldschmidt, Waddington argued that regulatory genes, purely theoretical entities that he called "evocator genes," control development through the production of specific substances at specific times. He was aware, accordingly, that genes would do nothing except where a potential for expressing them is also present. Even then, it is not simply a matter of triggering a gene into action. There is, for one thing, much interaction between cytoplasm and nucleus during development. Waddington was particularly alive to the fact that gene expression is sensitive to environmental disturbances, especially early in ontogeny.[6] Two organisms sharing the same genotype might, accordingly, end up with different phenotypes, just as two organisms sharing the same phenotype might have different genotypes. The plasticity between phenotype and genotype implied that new genotypes, and hence new evolutionary directions, could be established in ways that were causally dependent on selection among phenotypes. Waddington, who had a gift for putting evocative names onto theoretical processes and entities that neither he nor any one else fully understand, called this "genetic assimilation" (Thom 1989). He demonstrated it experimentally by inducing mutations by thermal shock into the eggs of *Drosophila*, subjecting these to selection pressures, and producing lineages of mutant flies.

Waddington's most influential attempt to integrate development with genetics is to be found in his concept of epigenetic landscapes. Waddington portrays development as taking place on a range of hills and valleys very much like Wright's and Dobzhansky's adaptive landscapes. In Waddington's models, however, hills and valleys do not represent distributions of genotypes—in Waddington's models a crisscrossing network of genes lies *below* the whole landscape, like the wiring below a radio chassis—but possible developmental pathways. Rather than trudg-

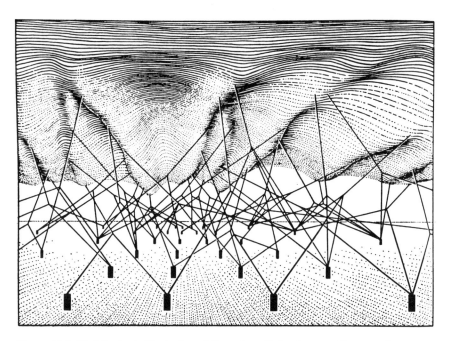

Figure 15.1 Waddington's illustration of the epigenetic landscape with a complex system of interactions underlying it. The pegs represent genes, and the strings represent the chmical tendencies produced by the genes. As reproduced in Gilbert (1991) from Waddington (1956). Reprinted with permission.

ing up the sides of hills in pursuit of higher fitness, moreover, as evolving populations do, embryos develop by rolling down one or another of the valleys. Early in development, at the top of the epigenetic landscape, an embryo might be pushed into any of a number of pathways. Whatever valley it goes down, however, the developing embryo is buffered against disruptions because the valleys themselves have been carved out by selection, often by way of the process of genetic assimilation, into pathways that are resilient to disturbances. Waddington called this "canalization" (canals = channels). The chief method by which canalization keeps an embryo on track is "homeorhesis" (from Greek *homeo* + *rheo*, "the same flow") which restarts an embryo down the right path when something goes wrong.

The epigenetic landscape idea was supposed to be a heuristic guide rather than to represent an actual process. In proposing it, however, Waddington was attempting to do more than guide developmental genetics toward fruitful hypotheses by way of suggestive images. By consciously conflating traditional embryological diagrams, in which successive cell divisions are represented as moving down a series of branching tracks, with the already much transformed models of adaptive landscapes that Dobzhansky inherited from Wright, Waddington was declaring that development and genetics are, and should be, integrated (Haraway 1976).

Developmentalism *Redivivus*

The memory of the original Theoretical Biology Club did not fade. In 1967, Waddington assembled an international cast of biologists, mathematicians, and philosophers to judge the prospects of theoretical developmental genetics once again. The meetings took place in Serbelloni, Italy. By then, several things of great importance had happened. First, the purely hypothetical notion of regulatory genes had become a reality in work initiated by Monod's and Jacob's "operon" model, about which Waddington had waxed enthusiastic in his *New Patterns in Genetics and Development* (1962) (in spite of the reductionistic enthusiasms of its codiscoverers). Another was the use that the cybernetic genius Alan Turing had made of "reaction-diffusion" equations to model morphogenesis (Harrison 1988). In retrospect, Turing's model appears as the first concrete hint that the emergence of biological form is not dependent solely on what natural selection has haphazardly assembled but is instead generated, guided, and facilitated by dynamical properties that will apply to any system with an appropriately large number of interacting parts. In such systems, sudden bifurcations spontaneously appear. On this view, the canalized valleys of Waddington's epigenetic models do not have to have been carved exclusively by natural selection. René Thom's "catastrophe theory" proved useful in making mathematical models of the conditions under which sudden bifurcations will (probably) occur (Thom 1972).

Inspired by the 1967 meeting, the English embryologist Brian Goodwin went on to champion the use of structuralist and field theoretical approaches to morphological transformation. Goodwin and his colleague Garry Webster have postulated a "morphogenetic field" that can model the cellular cleavage patterns of early ontogeny up to 128 cell divisions, when the symmetry breaking of gastrulation sets in (Webster and Goodwin 1982; Goodwin 1984). Goodwin and Webster are aware that they are walking not only in Waddington's footsteps but in the long shadow cast by Geoffroy. They are also aware that their work is expanding what D'Arcy Thompson demonstrated in his remarkable book *Growth and Form*, which uses traditional geometry to show just how systematic and structural are the morphological transformations of body plans at every level of the taxonomic hierarchy (Webster 1984; Goodwin 1989b, 91; Thompson 1917, 1942). One might even reconstruct "theoretical biology" as the latest set of research programs within a continuous research tradition going back to the early nineteenth century. It is worth at least passing note, accordingly, that the members of this research community often come from countries in which the original evolutionary synthesis of Lamarck and Geoffroy was never fully displaced by Darwinism and that, like Geoffroy's and Lamarck's original British followers Grant and Knox, British and commonwealth members of the community often teach in universities other than the most elite, and sometimes project an appropriately revolutionary aura to match.

In 1987, Goodwin and his colleagues held another meeting of their far-flung research community. This time the meetings took place in Mexico. In addition to commemorating Waddington and the twentieth anniversary of the 1967 sessions that he had convened, there were many new developments about which to talk (Goodwin and Saunders 1989). The transposable genetic elements first postulated by McClintock had given way to a picture of the mobile genome, capable of responding to environmental challenges with a bag of tricks scarcely dreamed of by Waddington. In addition, the emergence of nonlinear dynamics had made it possible to explore morphogenetic fields by introducing concepts like sensitivity to initial conditions, fractal geometry, and self-organization. In this spirit, the American microbiologist and developmental biologist Franklin Harold has asserted that there is now an emerging consensus according to which biological self-organization is a gradual, cumulative, epigenetic process in which sequential, linear molecular information at the scale of nanometers in the genome provides a recipe by which three-dimensional cellular structures up to five orders of magnitude larger will spontaneously, but predictably, emerge:

Small initial differences (sometimes random or nonspecific, environmental cues) are progressively amplified, generating spatial fields of one kind or another; the nature of these fields may vary from one case to the next. These fields direct the localization of molecules and forces that actually shape the visible structure, and therefore serve as obligatory intermediates in all developmental pathways (except for those that are wholly explicable as the result of molecular self-assembly). (Harold 1990, 415)

Recognition that development runs along dynamical contours has been of particular help in modeling interesting phenomena such as the life cycle of the slime mold *Dictystelium discoideum* (Garfinkel 1987). The slime mold is not a real mold at all but a single-cell amoeba that feeds on bacteria. When there is a scarcity of food, the individuals aggregate, forming colonies of thousands of cells. These colonies can migrate as a unit over relatively large distances. Over time, the homogeneous assemblage of cells differentiates in such a way that part of it becomes a base rich in cellulose, while the other part becomes a "fruiting body" rich in polysaccharides. The fruiting body then bursts, scattering spores, which yield mobile cells when food is again available. The cycle thereupon starts over again with the individual amoeba. A number of attempts to model this pattern of self-organization and symmetry breaking by making reductionist assumptions have met with little success. While some oscillating behaviors can be modeled, important global behavior, includings waves of aggregation, are not typically captured. That is because these structural changes are due, in Goodwin's view, to self-organizing amplifications of stochastic fluctuations (Goodwin and Trainor 1980). Use of nonlinear models of global behavior promises to show how and why slime molds behave the way they do and to yield more general insight into the structural dynamics of organismic growth and evolutionary change.

We now turn to phylogeny. If natural selection working at microevolutionary levels is the primary cause of phylogenetic differentiation, phylogeny will reflect little more than the transient, noncumulative properties of changing local environments. Simplification will be almost as commonplace as complexification. What complexification exists will be due to the fact that natural selection tends in the long run to reward organisms, populations, and species that devise new tricks to outwit their opponents. These adaptations will exhibit greater complexity because, like the tendency toward technological improvement within a capitalist economy, that is the general direction in which competitive advantage lies in evolutionary arms races. There is little or no reason, on this view, to see any isomorphism between individual development and the larger course of evolution, and certainly no reason to postulate an inner drive toward complexity that operates independent of natural selection. This conclusion, constrained at first by Darwin's own modest developmentalism, has been vigorously defended by modern Darwinians (Nitecki 1988).

By contrast, the basic idea of new developmentalists of every stripe is that macroevolutionary change reveals an intelligible pattern of structural transformations. Advocates of this view can point to a number of phenomenological regularities that seem modestly robust (Cope's rule, for example, according to which body size tends to increase in lineages). The problem is that directional ideas about evolution have always come trailing clouds of transcendent and teleological glory. This suspiciously vitalistic aura has led tough-minded Darwinians to downplay observations about directional phenomena in evolution. Indeed, the old developmentalism gave way to Darwinism in part because the former seemed too hopelessly purposive and vitalistic to make evolutionary theory a fully natural science. If this tradition is to make headway, it must separate whatever inherent tendencies toward complexification can be justified on dynamical (or perhaps thermodynamic) grounds from any sense that this tendency is indefeasible or purposive.

The notion that speciation is correlated with sudden, if subtle, changes in developmental programs, which then lock onto new stable patterns, has breathed new life into this old debate by changing the terms on which expansionistic Darwinians are willing to discuss it. Gould and Eldredge have argued that it is very nearly a consequence of contemporary orthodox views about species and speciation that macroevolution will show a pattern of "punctuated equilibrium" rather than gradual change, for peripatric speciation in small, isolated populations, if correlated with genomic organizations that soon settle into a new regime, will produce a statistical pattern in which many fossil remains of organisms with similar (but not identical) constructions will be found, while the remains of unsuccessful species will seldom turn up.

Orthodox Darwinians might well think that with friends like Gould and Eldredge, they do not need enemies. According to Gould and Eldredge, evolutionary trends do not result from adaptive advantage at the level of organisms and populations in the context of evolutionary arms races. Precisely because speciation is not well correlated with anagenetic adaptive change, phylogenetic pattern, such as it is, will more probably result when properties irreducibly ascribable to clades themselves, such as higher speciation or lower extinction rates, cause differential retention of individual species, and hence the differential persistence and fecundity of the clades to which these species belong. They will result, that is, from *species selection,* the biases of which will eventually make themselves apparent to the eye of the statistically trained paleontologist.[7]

To make matters more interesting still, however, Gould does not think that species selection applies at the taxonomic level of phyla, where the fundamental body plans on which organic diversity are laid down. To Gould's mind, there is no getting around the fact that, in deciding which phyla were to be fruitful and to multiply, an element of sheer contingency and arbitrariness has overwhelmed whatever ordering tendencies may have been at work. This suggestion is vividly expressed in Gould's gracefully written account of how interpretations of the extraordinary ensemble of fossils from the Burgess Shale in British Columbia have changed over time (Gould 1989a). The Burgess Shale was formed 530 million years ago, shortly after a mass extinction marked the change from the Precambrian to Cambrian. After this major culling, the so-called Cambrian explosion occurred: rapid evolution and an extraordinary, even chaotic, diversification of multicellular organisms ensued. Yet by the start of the Devonian period, 395 million years ago, only our (relatively few) modern phyla are to be found in the fossil record. Among the Burgess fossils are remains of some species that clearly belong to these phyla, particularly those falling into the arthropod body plan. Until the careful work of a new generation of paleontologists, who used mathematical techniques first developed by X-ray crystallographers, it was assumed that nearly all of the disparate fossils could be "shoehorned," as Gould puts its, into these few extant groups. The intellectual drama that Gould traces in *Wonderful Life* involves a paradigm shift, however, in which paleontologists have now recognized the existence of many radically different body plans in the Burgess Shale that do not correspond to modern organisms. Gould contends, in fact, that the Burgess Shale contains representatives of no fewer than fifteen to twenty phyla that are now extinct. It seems inescapable that a major decimation—a reduction not in species diversity but in the disparity between organisms with fundamentally different body plans—took place about 500 million years ago to account for this. That would have been, on the geological time scale, shortly after the formation of the Burgess Shale.

Why, Gould goes on to ask, did only a few phyla make it through the highly constricted bottleneck out of which modern life emerged? Was there anything special about the phyla that did? In our day, unlike Darwin's, when catastrophism, shorn of its old theological overtones, is making a more than a modest comeback, we can readily imagine that mass extinction had something to do with this. So does Gould. At least five such "great dyings" have occurred, the most notorious being the Cretaceous-Tertiary extinction that took place 65 million years ago, when the dinosaurs rapidly died out. In that case, there is strong evidence that one or more asteroids or a fragmented comet hit the earth, throwing enough debris into the atmosphere to greatly stress the photosynthetic food chain that supported dinosaurs and other taxa (Alvarez, et al. 1980; Alvarez and Asaro 1990; Kerr 1991; 1992a, 1992b). Volcanic activity can also explain mass extinctions to the satisfaction of the new catastrophist sensibility (Courtillot 1990).

Gould speculates, in fact, that major decimations or extinctions are responsible for the overall shape of macroevolution, in which periods in which life forms expand alternate with massive contractions. To validate and generalize this hypothesis, he draws on a quantitative study of the fossil record of marine invertebrate genera conducted by a research group under his direction (Gould, Gilinsky, and German 1987). The group produced 708 "spindle diagrams," in which time is represented as moving upward on the vertical axis, while diversity within genera is mapped onto the plus and minus sides of a horizontal axis. The pattern that is repeatedly seen in these diagrams tells the same story as the Burgess Shale at a lower taxonomic scale. After the occurrence of a decimation or mass extinction, there follows a period of rapid diversification and "experimentation," which is then greatly reduced by the time half of a given time period has gone by. During and after that time, the number of species within the surviving genera continues to increase.[8]

Some of this reduction is due to selection of a rather Dobzhanskyan sort. Gould argues that criteria for adaptive survival can differ during times of mass extinction and "normal" times. Not every lineage has as many resources to ride through these rough times as others. Those that do tend that make it. Those that do not perish. Diatoms are a good example of the former sort. They alternate between a period of rapid growth and reproduction during the seasonal upswelling of the ocean, which brings fresh nutrients into their niche, and a period of metabolic shutdown and sporulation as they exhaust resources. But in the hard times of an extinction, diatoms can survive by staying in the spore form for a more extended period of time. This does not mean that sporulation was specifically selected as a trait for survival in a major catastrophe; it does mean that it is an adaptation aimed at riding over the cycle time of the annual pattern of the ocean's response to the seasons and that different adaptive strategies will prevail when cycles have high troughs and valleys, or chaotic frequencies, than when they are more moderate. By similar reasoning, it seems likely that some lineages may be better able than others to survive oscillations between ice ages and warming periods, or even changes in the earth's magnetic field. By contrast, species and clades that come into existence when times are good, and whose bag of tricks is fixed at that time, will not be as well buffered when wide cycles recur.

No organisms can be very well buffered, however, when events are not cyclical at all, or are so infrequent that adaptive response cannot possibly be tuned to them. Gould hypothesizes that this is the case whenever extinction rates reach high proportions of life-forms, as in the Permo-Triassic extinction of about 230 million years ago, when 96 percent of all species present on earth went extinct. In such cases, it is hard to give credence to the belief that the survivors were any more fit or quick than the dead. They were probably just luckier.[9] For this reason, Gould does not see any criteria by which the roughly twenty Burgess phyla that failed to survive were any less qualified than the few that did. There is no reason to believe that the ancestor whose line ultimately led to us, *Pikaia*, the first chordate, was superior to those of its contemporaries that went under. It is surely the case that if our earliest ancestor had died out during decimation we would not be here. Nor would any type of creature remotely like us. That affords, however, no reason to think that we are the culmination of an intelligible ascent. Gould exclaims, "Our own evolution is a joy and a wonder because such a curious chain of events would probably never happen again, but having occurred makes eminent sense" (Gould 1989, 285). That is precisely why life is *wonderful*.[10]

Still, there is, says Gould, a logic of contingency, albeit a narrative logic. Contingency is manifest in an unpredictable sequence of antecedent states without which a present state would not exist. No link in the chain has to be precisely the way it was. Any change in a preceding state might in fact so alter the ensuing cascade of consequences that things might have turned out quite differently. Of the possible causal chains, however, at least one does in fact survive. That is what we can, and do, tell stories about. But in constructing such stories we cannot, without doing violence to the phenomena, make use of predictive or deductive laws any more than of overarching purposes that guide the entire affair. Gould writes:

The resolution of history must be rooted in the reconstruction of past events themselves—in their own terms—based on narrative evidence of their own unique phenomena. . . . The issue of verification by repetition does not arise because we are trying to account for uniqueness of detail that cannot, both by laws of probability and time's arrow of irreversibility, occur together again. We do not attempt to interpret the complex events of narrative by reducing them to simple consequences of natural law; historical events do not, of course violate any general principles of matter and motion, but their occurrence lies in a realm of contingent detail. (Gould 1989a, 278)

We have encountered narrativist approaches to complexity and contingency before (chapter 14). What is clever about Gould's argument, however, is that if there are macroevolutionary patterns and trends, they can reliably appear only when ideas about progress and directedness notions are totally abandoned, for it is only then that the specific logic of historical sequences shows itself. This insight radicalizes Mayr's and Lewontin's narrativism.

Nonetheless, when biological evolution is viewed in the wider context of chemical and ecological evolution in which it occurs, Gould's argument may not be quite as strong as it seems. It depends on looking at what is happening to individual organisms and species rather than at the larger ecological and geophysical world to which they belong, which, if Lotka and his followers are right, sets their evolution in the context of complexifying changes in the systems of which species and other lineages are parts. Perhaps the developmentalist tradition can mount a reasonably good rebuttal of Gould's analysis in these terms. If so, it would go in something like the following way:

Gould himself admits that the origin of life is a likely emergence from the laws of chemistry under the appropriate initial and boundary conditions (Gould 1989a, 289). But he sees physical principles receding quickly thereafter as causally relevant in the dynamics of evolution and the kind of contingency that calls for narrative explanations taking over. To establish this, Gould denies that life has grown steadily more complex since its origin and even that it would have grown more complex if it had not been interrupted by catastrophes. In support of this view, Gould cites the fact that from about 3.8 billion years ago prokaryotes dominated the earth until the more complex eukaryotes more or less suddenly became dominant somewhere about 2.2 to 1.5 billion years ago. (This major change probably resulted from the inclusion of prokaryotes as symbiotes in ancestral eukaryotes as organelles called mitochondria and chloroplasts [Margulis 1981].) It took another 900 million years before the Cambrian explosion of multicellular forms. Gould interprets such long lag times as an indication that there is no lawlike drive toward complexification by arguing, in effect, that during the intervening times, nothing happens (Gould 1989a).

Admittedly, nothing much may have been happening to the morphology or behavior of individual, or even colonial, organisms. In fact, however, a great deal was happening between 3.5 and 1.5 million years ago in the larger world of which they were part. Bigger and more complex eukaryotic cells, utilizing ATP, could not have appeared until molecular oxygen reached a certain level in the atmosphere. But for free oxygen to be present in the atmosphere, it was first necessary for photosynthetic bacteria, which uses nonoxygenated compounds like hydrogen sulfide as their source of electrons, to evolve into forms that could "split" water, producing and releasing molecular oxygen as a metabolic by-product and for the oxygen so produced to react with elements such as iron. After all the iron, for example, had been oxidized, it was possible for the steady-state concentration of oxygen to rise. This probably occurred around 2 to 2.5 billion years ago. It then took probably another 500 million years to produce enough oxygen to saturate the oceans and to complete a variety of oxidation reactions until there was a surplus of oxygen in the atmosphere. When the concentration of free oxygen started to rise beyond a certain point a major mass extinction of the anaerobic life forms that had earlier ruled the earth took place. To them oxygen was poison. (It was the first and perhaps most dramatic so far of many cases in which living things have undermined the conditions of their own life, as humans may well be doing today.) This situation created enormously strong selection pressures. Anaerobes that survived at all retreated to oxygen-poor niches. The ancestors of contemporary aerobic organisms, on the other hand, were under intense pressure to adapt metabolic machinery that could detoxify oxygen. (It was probably achieved through what Gould calls "exaptation," since the machinery to do this may well first have been used for some type of early photosynthesis [Dickerson 1980].) A consequence of this achievement was that new aerobic organisms could extract a far greater source of useful energy than anerobic organisms and primitive aerobic organisms by synthesizing much more ATP. The accessibility of these new energy gradients produced an explosion of life-forms—the Cambrian explosion.

Interestingly, oxygen concentration in the atmosphere did not rise even to about 1 or 2 percent of current levels until about 800 million or so years ago, about the time multicellular organisms first appeared. By the time of the Cambrian explosion, which probably began about 530 million years ago and lasted a mere 5 or 10 million years, there had been a dramatic rise to 10 to 20 percent of present levels (Bowring et al. 1993). From these considerations, it appears that life on earth not only arrived as soon as it possibly could but went multicellular and complex as soon as the thermodynamic field had been charged up and amplified through newly available kinetic pathways as soon, that is, as physical and chemical conditions allowed (Runnegar 1982; Loomis 1988). Was this not as expectable as the origins of life itself, and on roughly the same sorts of grounds? There is more pattern in the system as a whole than in its isolated parts. Can we say for sure, moreover, that ecosystemic developments of this sort have not continued ever since, interrupted no doubt by catastrophes and mass extinction but facilitated by them as well?

Positive answers to these questions will be much more likely if evolution is seen as coevolution, and if coevolution is viewed as a process of partitioning a dynamic ecological system under thermodynamic imperatives, and not as morphological and behavioral tinkering. About the latter, it is possible to tell only higgledy-piggledy narratives. From the former point of view, however, one might expect ecosystems to complexify and hence diversify due to increased entropy production, which uses autocatalytic kinetic systems, including informationally informed autocatalytic systems, or organisms, as preferred dissipative pathways (Swenson 1989; Swenson and Turvey 1991). Clearly this propensity to growth and complexification can be, indeed has been, interrupted by catastrophes. Nothing guarantees that life on earth might not have been, and still might be, utterly extinguished, or that it might not have to start all over again from some fairly impoverished beginnings. Even so, periods of rapid diversification following decimations are also to be expected, working on whatever genomic and morphological architectures happen to be around. Even when such expansions are predictably followed by prunings, life on earth tends, ceteris paribus, to grow more diverse. As Darwin believed, individual species may grow less complex. Complexity can even average out over all taxa into a more or less steady mean. Still, ecosystems, regional biota, and the entire earth system might at the same time be growing more complex under the impact of thermodynamic imperatives.

Consider the hardest case of all. Gould has always wanted to free evolutionary theory from any lingering thought that human evolution was inevitable, or that consciousness and intelligence were bound to emerge in the course of biological evolution (Gould 1977b, 1989b). If our chordate ancestor in the Burgess Shale had become extinct, he says, our type of complex brain might never have evolved. It is sobering to think that the only reason this ancestor made it was dumb luck. Admittedly, humans are improbable and fragile evolutionary entities, and all the more valuable and interesting for that. But even as we should not expect that our line of descent has anything inevitable about it, neither should we assume that only this line of descent could lead to intelligence or to complex cultures, or that only our species could have had self-consciousness. Appeals to contingency make that possible as well. If brains had not evolved our way, can we say for sure that they would not have evolved in another way, particularly if we were not around to occupy the relevant niche? Do we know enough about other actual, extinct, or possible body plans and genomic wiring diagrams to rule that out? Such creatures might be nothing like us, but they would still be the product of a natural development.

The argument might continue along the following lines: If species come into existence by reorganizations of the architecture of the genome, as Gould says they do, rather than by adaptive natural selection under the control of changing and different environmental necessities, will not changes in developmental programs constitute a record of life, with earlier modules safely embedded or "generatively entrenched" within later ones? Freed from the ups and downs of adaptation, might it not be possible to discern a logic in these shifts in developmental programs, a logic that may well reveal laws of form like those dreamed about by Geoffroy? Might ontogeny not be recapitulating phylogeny, and internalizing ecological complexification, after all?

In this chapter we have considered a number of phenomena that were once integrated into a synthesis by developmentally centered visions of biological process—the origins of life, ontogeny, ecology, and phylogeny. This vision was dimmed by the rise of genetic Darwinism. Is it possible for Darwinism to acknowledge and explain at least some of these phenomena? We will suggest in the following chapter that if the answer is yes, it will be because Darwinism has used new dynamical models.

16 New Models of Evolutionary Dynamics: Selection, Self-Organization, and Complex Systems

If there is one thing advocates of an expanded Darwinism and their new developmentalist counterparts have in common it is antipathy to adaptation*ism*. So great is this shared antipathy that in both cases it sometimes spills over to the notions of adaptation and adaptedness themselves, and to natural selection to the (considerable) extent that it is tied to these concepts. For this reason, both expanded Darwinians and new developmentalists make chance and constraint play more prominent explanatory roles in evolutionary explanations than natural selection. Still, it remains at least logically possible to reconceptualize and reform, rather than marginalize or junk, selection and adaptation. Perhaps there is some way in which adaptive natural selection, chance, and self-organization might be integrally related instead of tugging constantly at each others' sleeves. Perhaps a new conception of adaptedness can even do productive work on the large problems we outlined in the previous chapter, bringing Darwinism, broadly construed, to bear on issues about the origins of life, ontogeny, ecology, and phylogeny.

This chapter is a case study of at least one attempt to do precisely this. The developmental geneticist Stuart Kauffman has appealed to some relatively accessible aspects of complex dynamical models in order to shift the concepts of adaptation, fitness, and natural selection into a higher key, where they range over objects as distant from sense perception as the autocatalytic systems that led to life, evolving ontogenetic programs, and coevolving species within a complex and constantly changing adaptive landscape with many interacting agents. With his stress on the generic order that spontaneously emerges in systems as complex as these, and on how limited the hand of natural selection often is in the face of self-organizational constraints, Kauffman may look to traditional Darwinians very much like a member of the developmentalist opposition. To developmentalists, on the other hand, Kauffman seems suspiciously Darwinian. Is it not Darwinian to expand the scope of Dobzhansky's adaptive landscapes, to insist that selection is a powerful evolutionary agent when it can select among an array of self-organized systems, and to speculate that natural selection keeps complex systems within a range where this can occur? The fact is that Kaufmann has inherited both traditions, learning personally from evolutionists as different from one another as Waddington, Goodwin, and Maynard Smith. To the extent that Kauffman is a Darwinian, his Darwinism is what Darwinism might look like when it operates against a new set of background assumptions taken from the study of complex systems. "We must," he writes, "build a larger theory which marries Darwin's idea of natural selection to the self-organized properties of the entities that selection was privileged to operate upon" (Kauffman 1989, 87). Our claim in this chapter is that any such theory—it need not be Kauffman's—must rely on a new set of dynamical assumptions, the dynamics of complex systems.

Every so often in science, as in other human endeavors, a new way of looking at the world and the phenomena it contains emerges. Many people now argue that science is undergoing such a shift as it crosses what Heinz Pagels has called the "complexity barrier" (Gleick 1987; Kellert 1993; Lewin 1992; Pagels 1988; Stewart 1989; Waldrop 1992). Perhaps it is just because this is happening that in our own day we can now see clearly that another such shift occurred when the sciences underwent a probability revolution during the second half of nineteenth and the first half of the twentieth century (Gigerenzer et al. 1989; Krüger, Daston, and Heidelberger 1987; Hacking 1990) Perhaps it is part of what people mean by the "postmodern condition" that with the probability revolution behind us and the complexity revolution ahead of us, the Newtonian world in whose last days Darwin lived seems more distant to us than ever before, and the Aristotelian world it displaced positively archaic.

We have learned in this book that the probability revolution, whose paradigm cases are statistical mechanics and thermodynamics, facilitated the consolidation of a revitalized Darwinian research tradition. Kauffman's hope (and ours as well) is that as we understand more about complex systems dynamics, something that grand can happen to the Darwinian tradition again. We are less concerned, accordingly, with technical questions about whether Kaufmann's models succeed or fail to solve particular evolutionary problems than with the ways in which his work points toward a new conception of natural selection, and beyond that to new ways of envisoning how law, necessity, chance, and historicity are related in evolutionary dynamics. We do not seek to resolve these issues here. We do want to get them formulated perspicuously, however, and to suggest that when this has been done, it is plausible to believe that Darwinism can indeed revitalize itself within a new dynamic framework.

Stuart Kauffman, born in California, received a B.A. in philosophy from Dartmouth in 1961. He then spent two years at Magdalen College, Oxford, where he added physiology and psychology to the study of philosophy. Kauffman returned to California to take premedical courses at the University of California, Berkeley. By 1965, at age twenty-four, he was a second-year medical student at the University of California, San Francisco, a bit bored with routine medical studies. What he was interested in was whether natural selection alone could account for the abundant biological order that he was learning about in his embryology courses. Perhaps he could do research in developmental biology. Kauffman had an intuition that there had to be a natural, spontaneous, lawlike source of order beyond that provided by natural selection. He knew that D'Arcy Thompson and others had already suggested that, but their work had failed to incorporate genetics. Like Waddington, Kauffman was deeply impressed by the model of gene regulation in bacteria that Jacob and Monod had recently published (Jacob and Monod 1961; Waddington 1962). He realized that genes would interact in parallel in regulatory systems rather than sequentially and that what would matter would be a stable pattern of gene activity. Kauffman recalls thinking at the time about how that stability is achieved and maintained. "Self-organization," he surmised, "is a natural property of complex genetic systems. There is 'order for free' out there, a spontaneous crystallization of generic order out of complex systems, with no need for natural selection or any other external force" (quoted by Lewin 1992, 24).

Kauffman decided to see how far he could get making a mathematical model based on a regulatory gene that turns on and turns off a structural gene. The mathematics he chose was fitted to articulate his intuition about parallel processing. In Boolean networks, named after the nineteenth-century English mathematician and logician George Boole, a string or array of symbols, or "digitules," is transformed by a set of simple rules. In the model, each element of an ensemble of N elements has two possible states (on or off in the case of gene expression) and receives K inputs from other elements. Applied to hypothetical genes in a regulatory system, rules like the following might be specified: If any signal is positive, the gene will be turned on. Another might be: All signals must be positive for the gene to be turned on. Whatever the rules are, the idea is that at any instant the entire network is in a state whereby each element receives input from those elements to which it is connected and becomes active or inactive in accord with the rules that govern the system. These interactions produce the next state of the system, and so on.

Working with relatively simple computers Kauffman found that Boolean networks often settle down to a set of states that are repeated over and over again. Boolean networks, that is, can become spontaneously ordered. They do so, moreover, in ways that are surprisingly impervious to perturbations. These patterns are called *state cycles*. State cycles are emergent properties of computational systems in which sets of "local" rules, such as Boolean functions, give rise to global order. With these thoughts in mind, Kauffman attempted to see whether his model could explain why the approximately 100,000 genes of *Homo sapiens sapiens* direct the development of only 250 or so cell types, when 10^{30000} activity patterns are mathematically possible. He first took on a simpler calculation, asking what would happen in a model system with just 10^{30} possible activity states. To Kauffman's surprise, in just sixteen iterations the model settled down to a stable pattern of just four state cycles (figure 16.1).

Kauffman then varied K, that is, the number of inputs per element. He found that the number of state cycles at $K = 2$ is approximately equal to the square root of N, the number of elements in the system. For $N = 100,000$, for example, $N^{0.5} = 317$. That is approximately the number of cell types in multicellular organisms. Perhaps, then, this range of cell types had been determined not by natural selection, or even by the properties of particular materials that cells are made from, but from the inherent mathematical properties of systems with a large array of elements and connections among them.

The binary mathematics of Boolean networks are naturals for experimenting and playing on computers, for the computer itself is a binary calculating machine. Of course, calling the strings and arrays of symbols "genes," as Kauffman did, and the combinatorial dynamical behavior induced by the relevant rules "gene regulation," is not enough to turn this into real biology. Nor, accordingly, does making such models imply that biological systems, including brains, are enough like computers or computer programs to be called computational devices. What can emerge from the kind of experiments Kauffman was conducting are only simple models that may or may not capture an aspect of the complex behavioral dynamics of biological systems. Still, it would be remarkable if so simple a mathematical model could show properties that do resemble the dynamics of real biological systems. Kauffman's model of cell types seemed to him just such a result.

Subsequently, Kauffman used Boolean models to deduce or explain other characteristics that are isomorphic with developing biological systems. He showed, for example, that cells can switch in development to only one or two other kinds of cells. Here was an evolutionary constraint that natural selection would have a hard time getting around. He also proved that modifications to early embryological stages, when they take hold at all, will be less open to subsequent change than later; why something like von Baer's laws must generally be true; and why, accordingly, ontogeny will exhibit the same pattern across many phylogenetic taxa. Kauffman later summarized these results as follows:

Complex systems, such as the genomic regularity networks underlying ontogeny, exhibit powerful "self-organized" structural and dynamical properties. The kind of order which arises spontaneously in such systems

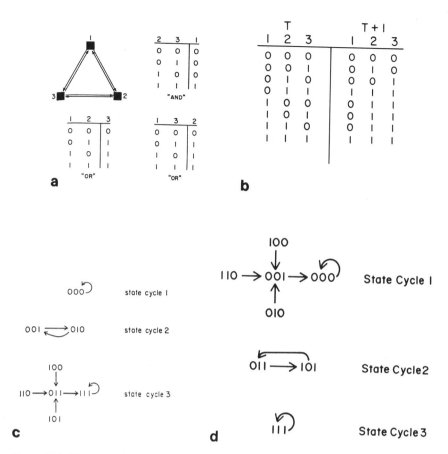

Figure 16.1 This example is of a simple case of a Boolean network with only three elements. *a:* The wiring diagram of the Boolean network that contains three binary elements, one of which is governed by the Boolean "And" function, the other two of which are governed by the "Or" function. *b:* A rewriting of the Boolean rules to show all eight possible states. *c:* The state transition graph of *a* and *b,* showing the three state cycles to which the system can settle down. *d:* The consequences of changing the rule governing element 2 from "Or" to "And." From p. 466 (Fig. 12.8) of Kauffman (1993). Reprinted with permission.

is strikingly similar to the order found in organisms. This raises the plausible possibility that the spontaneous order found in such complex system accounts for some or much of the order found in organisms. (Kauffman 1989, 67)

This sounds a lot like the "laws of form" tradition that we dubbed the new Geoffroyism in the previous chapter. Kauffman himself says that his opus magnum, *Origins of Order,* "is an effort to continue in D'Arcy Thompson's tradition" (Kauffman 1993, 644). Early on, these links were tightened when the MIT neurophysiologist Warren McCullough, whose pioneering application of Boolean networks to neural networks had led Kauffman to seek him out, introduced the third-year medical student to the British developmental biologist Brian Goodwin. Through Goodwin,

New Models of Evolutionary Dynamics

Kauffman was invited to participate in the 1968 meetings on theoretical biology convened by Waddington at Serbelloni, Italy. There he presented his models of cell types (Lewin 1992).

This encounter with the "laws of form" tradition does not mean that Kauffman gave up on Darwinism. On the contrary, he responded positively when Maynard Smith, doyen of British Darwinians, challenged him at Serbellini and later to integrate his models of generic order with natural selection, and taught him the mathematics of fitness landscapes that would be required to achieve that (Lewin 1992). After completing medical school, Kauffman took his first academic job at the University of Chicago, where, in a spirit open to both Darwinism and developmentalism, he worked on the developmental genetics of that workhorse of population biology, *Drosophila*. Kauffman continued this work when he moved to the Department of Biochemistry and Biophysics at the University of Pennsylvania Medical School in 1975.

To meet Maynard Smith's challenge, Kauffman tried to fuse his NK models with models of fitness or adaptive landscapes. "The fitness landscape image," he writes, "is a powerful, basic, and proper starting point to think about selection" (Kauffman 1989, 69) We know by now that this is a metaphor with a varied and plastic history. We have seen how Wright transfigured Fisher's idea of fitness as a peak, itself an inverted image of thermodynamic gradients, into a picture in which there are many peaks and valleys in an adaptive landscape (Wright 1932). We know too that Dobzhansky mapped adaptive landscapes onto real ecologies. More recently, we have seen how Waddington transformed adaptive landscapes into epigenetic landscapes. Now Kauffman defined the adaptive landscape when $K = 0$, that is, when there are *no connections* among N elements in an array, as *fully correlated* and as yielding a single global fitness peak. This is what will happen if we assume that all traits are "atomistic" or fully independent and that fitnesses are, accordingly, additive. This is, in short, Fisher's dream world. By contrast, when $K = N - 1$, that is, when each gene's fitness contribution depends epistatically on its connections with all the others to do its work, then the fitness landscape is defined as fully uncorrelated. Here there is no global optimum among a very large number of local optima that on average are no higher than the statistical mean fitness possible for the ensemble, its generic properties. Between these extremes, the fitness landscape is said to be tunably rugged, with adaptive peaks higher than the statistical mean. If K remains small, say around $K = 2$, then as N increases, the contribution of each trait to overall fitness becomes less. As K increases beyond $K = 2$, however, traits grow more strongly coupled. Conflicting "design constraints" arising from this coupling mean that only lower fitness peaks will be available. This is the world of the evolutionary trade-offs made familiar by genetical game theorists like Maynard Smith. Finally, if N is a large array and K increases toward the limit of $K = N - 1$, a complexity catastrophe will ensue. The peaks of the fitness landscape fall toward the generic ensemble properties, that is, the mean fitness the system will have solely in virtue of the values of N and K.

Mutations provide a way of walking through fitness landscapes. What happens to mutations at each point along the continuum between $K = 0$ and $K = N - 1$ differs. When there is a single optimum, for example, each mutational step (other than those that are selectively neutral) will induce the system to climb up or down the optimal surface. In a moderately rugged fitness landscape, natural selection will use mutation, and the variation that arises from it, to keep a system on a local optimum. Even when selection is strong, and the system is held to a single peak, however, the height and shape of the peak will distantly reflect the generic properties of the ensemble as a whole. If selection is weak, on the other hand, the system will wander through the fitness landscape in a less goal-oriented way. Genes will drift to fixation by chance. Generic ensemble properties will begin to show themselves more powerfully. As we reach fully uncorrelated landscapes, so rugged that they make it difficult for organisms and environments to remain tuned to each other at all, selection, mutation, drift, and other evolutionary forces will be held in check by generic ensemble properties. That is a complexity catastrophe.

So likely are complexity catastrophes in systems as complex as regulatory programs that we can regard the fully uncorrelated landscapes of $K = N - 1$ as a null hypothesis against which the work of natural selection and other conventional evolutionary "forces" is to be measured. We see here the recurrence of a phenomenon that we have had occasion to note often before in the history of Darwinism—a new zero state, baseline, or null hypothesis against which to measure evolutionary change, and so the stirrings of a new research program within the larger Darwinian research tradition (Burian and Richardson 1992). This proposal departs in at least one crucial respect, however, from the otherwise heterogeneous proposals of predecessors like Galton, Fisher, and Wright. The problem with all previous forms of Darwinism is that they have assumed simple, or what Herbert Simon calls "decomposable," systems as their null hypothesis (Simon 1962). In systems that are simple enough to block the formation of generic ensemble properties, and even to screen them off from view, selection and other forces can plausibly be measured against the normal curve or the Hardy-Weinberg equilibrium or other such stable backgrounds. Working within these assumptions, Darwinians have gone on to take account of an ever larger array of evolutionary forces in addition to selection, such as drift, density dependence, frequency dependence, and developmental noise. As they have done so, they have become correspondingly incapable of predicting successive states of even so simple a genetic system as two alleles at two loci (Burian and Richardson 1991; Lewontin 1992). Rather than rejecting the simplistic background assumption that causes the trouble, most Darwinians have preferred to give up the ideal of predictive sufficiency and to deal with the complexity that they now acknowledge by substituting retrospective narrative reconstruction as an ideal that should be taken over into evolutionary theory from natural history. In systems as complex as genetically regulated organisms clearly are, however, it is more realistic and parsimonious, and now more computationally feasible as well, to treat the generic properties that spontaneously arise in complex systems as the expected baseline against which selection must make its way. Predictions deducible from lawlike behavior can then be ventured. By mapping the language of adaptive landscapes onto his NK models, Kauffman makes it possible to reverse one of the deepest presumptions of the Darwinian tradition.

"The establishment of a null hypothesis of this sort," write Burian and Richardson, "is a major accomplishment. . . . Deviations from observed genomic architectures. . . . could be used to detect the perturbing effects of selection and other 'agents' of evolutionary change" (Burian and Richardson 1991, 269). Burian and Richardson admit, however, that this requires a major "gestalt switch" (Burian and Richardson 1991, 282) because Darwinians have to pay a steep price for Kauffman's advances. They must be prepared to admit that in many cases natural selection cannot be expected to do all or even most of the work, that as explanatory models become more realistic natural selection ceases to be an explanation of first resort, and that when selection operates, it does so in a fairly narrow range of possibility space, since it selects among entities that are already self-organized modules and that are in the process of spontaneously forming into still higher levels of self-organization (Dyke and Depew 1988).

In 1985, Kauffman was invited to join the Sante Fe Institute for the Study of Complex Systems. The Santa Fe Institute, catalyzed into existence by such well-known and respected scientific figures as the physicist Murray Gell-Mann and the economist Kenneth Arrow, provides a site for interdisciplinary research into complex systems. The thrust of work there has been to use computer modeling not only to solve problems in such areas as meteorology, economics, ecology, and evolutionary biology but to produce, in effect, a new dynamical theory, or at least a new dynamical language, for the study of complex systems (Lewin 1992; Waldrop 1992). Kauffman's Santa Fe connection, facilitated by his selection as a MacArthur Fellow, has enabled him to move beyond his earlier work in at least three ways. He has reframed that work within a larger theoretical setting; he has suggested a way of preserving the primacy of natural selection within the Darwinian tradition by showing its role in the dynamics of complex adaptive systems; and he has deepened and extended the range and scope of his explanatory models by bringing the traditional problems of the developmentalist tradition, from origins of life to phylogeny, under the sway of his revised Darwinian framework.

By the time he arrived in Santa Fe in the mid-1980s, the power of computers had increased by many orders of magnitude over the primitive setups Kauffman had used to produce his first models of cell types. It had become clear by then that all this increased computational power allowed one to see dynamics in action, as graphic displays on computer screens showed the results of reiterating transformation rules over and over again. The old days of pen-and-paper calculation were gone for good. So many surprising results were showing up, in fact, that it seemed reasonable to people at the Santa Fe Institute to reconceive the very notion of a dynamical system, just as it had been reconceived when calculus was invented in the seventeenth century. It was within the terms of this new, more general dynamical language that Kauffman was soon redescribing his NK models and adaptive landscapes.

Classical dynamical systems are described by mathematical functions in differential equations. Until the advent of the modern computer, only the simplest kinds of dynamical systems could be studied: those described by linear differential equations under relatively rigid boundary conditions and parameters (West 1985). In practice, therefore, classical dynamics has been restricted to calculating changes in position and momentum on the part of entities possessing the generic properties of physical objects, such as mass, force, weight, charge, or their close analogues. (That is why it is not entirely unreasonable to think of most of what has passed for science, both natural and social, in the past few centuries as metaphors in search of differential equations.) Systems that do not fit well into this model must be laboriously approximated. The new dynamics differs from classical dynamics in these respects as well as others. It is a more abstract, formal, mathematical, logical, and symbolic science. It does not drag around, if only metaphorically, residues of physical concepts like mass, momentum, or charge. Moreover, instead of treating complexity by extrapolating and approximating from multiple layers of simplicity, it seeks to describe the dynamics of complex systems in their complexity.

What, then, are complex systems? Complex systems are not just complicated systems. A snowflake is complicated, but the rules for generating it are simple. The structure of a snowflake, moreover, persists unchanged, and crystalline, from the first moment of its existence until it melts, while complex systems change over time. It is true that a turbulent river rushing through the narrow channel of a rapids changes over time too, but it changes chaotically. The kind of change characteristic of complex systems lies somewhere between the pure order of crystalline snowflakes and the disorder of chaotic or turbulent flow. So identified, complex systems are systems that have a large number of components that can interact simultaneously in a sufficiently rich number of parallel ways so that the system shows spontaneous self-organization and produces global, emergent structures.

This usually means that the sets of equations used to track the dynamics of complex systems contain nonlinear terms. The sum of two solutions for a nonlinear equation does not, in turn, yield a solution. In classical dynamics, the existence of nonlinearities can have a devastating effect on predictability. Approximations become necessary. Enhanced computational ability now makes it possible however, not only to do that, but to sidestep the whole approach by varying the parameters at which equations are run and seeing what patterns turns up. (That is what Kauffman is doing when he runs his models at various K and N.) Rather than forcing us to give up on predictions, manipulating parameters in this way yields a new sort of prediction, what Kellert calls *qualitative predictions* (Kellert 1993). One may not be able predict the state of a complex system, or its components, at any given point in past or future time. One can, however, identify what patterns recur under particular parameter values.

Trajectories of dynamical systems are described in the language of phase spaces. A phase space is a geometrical model that describes states of an object, a system of objects, or an ensemble (a system composed of a large number of items whose trajectories can be summed and averaged, like Maxwellian gases) in terms of a number of variables, or degrees of freedom, that define it. At any instant in phase space, how things stand with the variables of an object, system or ensemble can be represented by a single point. In the course of time, this point will move, describing a trajectory in phase space. The trajectory of an object or system through the phase space of classical dynamics is produced by one or more differential equations. The trajectory of a baseball, for example, is sufficiently described by a set of differential equations involving just two variables, momentum and direction, each in three dimensions. That means we can use a six-dimensional phase space to create a visual, yet purely mathematical, picture of the dynamics of such an object— a phase portrait. This is useful as soon as more objects, variables, and parameters are added, for at that point our low-dimensional imagination is bound to fail us. It already fails us, in fact, even in our baseball example. Actually the motion of three baseballs, say during infield practice, can be described as a point in an eighteen-dimensional phase space. We cannot visualize what eighteen-dimensional space looks like. We can, however, mathematically follow the trajectory of the point that defines such a system in phase space, even if not in real space.

Complex systems dynamics, like classical dynamics, also uses the language of phase spaces. It recognizes, however, that as systems travel through phase space, they eventually settle down, in one of three possible ways, into regions where they are relatively impervious to further perturbation. These regions of phase space are called *attractors*. When a system moves to a stable point, we say that it has a *point attractor*. A pendulum that ultimately stops swinging, due to friction, has reached a point attractor where no further change occurs in phase space. Some systems, by contrast, settle down to a closed loop. They are said to exhibit the periodic behavior of a limit cycle attractor. A pendulum, for example, could avoid eventually settling down to a point attractor if we used a magnetic or mechanical device to give it a boost to overcome friction and to maintain a minimal swing at the natural frequency of the pendulum. Finally, there are chaotic, or what are sometimes called strange, attractors. They are what is most novel and intriguing about the dynamics of complex systems (Gleick 1987; Stewart 1989; Kellert 1993). In classical dynamics, there are no chaotic attractors. There are either point or limit cycle attractors, or else mere disorder. Owing to the enhanced capacity of computers to track patterns in phase space, however, we can now see that phenomena that have hitherto been written off as orderless, or, worse, have been inappropriately compressed into point attractors or limit cycles, often have a good deal of order in them, albeit of a nonrepetitive sort. Even a pendulum will exhibit a complex and nonrepetitive pattern, or chaotic dynamics, if given the right kind of tweaks and shoves (Tritton 1986).[1] Within the house of chaotic attractors, moreover, there are many mansions, many different kinds of order.

Those who have been studying the dynamics of complex systems have been inclined to take the chaotic dynamics of complex systems as paradigmatic and to treat point and limit cycle attractors as special cases that occur when particular boundary conditions and parameter values are held rather tightly in place. This is a wise move in virtue of the fact that the first discovery of the new dynamics was that even the most deterministic equations, of the sort dear to the heart of classical physics, do not always yield predictable trajectories as they move through phase space. The meteorologist Edward Lorenz discovered this when he used a set of three deterministic differential equations to model aspects of the weather (Lorenz 1963). In two of these equations there were nonlinear terms (such as xy and xz). Such sets of equations generally have to be solved numerically with the aid of a computer. Still, it had generally been assumed that, due to the averaging properties on which people had been relying since Maxwell and Boltzmann, the nonlinearities would not compromise the predictability one might otherwise expect from deterministic equations. Much to his surprise, however, Lorenz discovered that very slight differences in the initial conditions for the system of equations he employed gave widely divergent trajectories in phase space when the same program was run over and over again (Gleick 1987). Perturbances that were expected to wash out did not. What is more remarkable is that, rather than falsifying his hypothesis about weather patterns, Lorenz discovered a pattern in the resultant phase portraits that modeled the weather pattern better than what he had originally hoped for. In this case, sensitivity to initial conditions produces a bi-lobed plot in the x-y plane that looks rather like butterfly wings (Lorenz 1993). (In three dimensions the plot looks more like a kidney.) This strange attractor is now called a Lorenz attractor (figure 16.2).

The Lorenz attractor was the first example of a chaotic attractor to be found in a phase space with more than two dimensions. Systems that appear to be structureless in space and time can show very beautiful structures in phase space. Such systems seem to be all around us. Time intervals between drips of a faucet, for example, turn out to be described by a strange attractor in a three-dimensional phase space (Shaw 1984). Perhaps we ourselves live, move, and have our being at unpredictable points within chaotic attractors.

In Santa Fe, Kauffman began to describe his NK Boolean networks in the language of the new dynamics. In this effort, he was aided by the work of Chris Langton, a Santa Fe colleague who was trying to produce computer models of living systems, or to create "artificial life" (Langton 1989,1992). In the course of devising computer programs called cellular automata, to which Kauffman's Boolean networks are related, Langton distinguished four kinds of rules that define programs and four kinds of behavior in phase space to which they give rise. Three of these were already known. One class of rules gives rise to a type of order in which there is no change in functions over time ("frozen dynamics"). Another gives rise to order in which there is periodic change ("oscillatory dynamics") that is relatively stable to perturbations. A third class produces chaotic dynamics with sensitivity to initial conditions. Langton discovered, however, that there is a fourth set of rules governing a fourth region in phase space lying in a narrow band between order and chaos. Here, at "the edge of chaos," a population of simple elements interacts under local rules in ways that give rise to emergent global structures and phenomena. The membranelike character of the edge of chaos may, however, be a reflection of the particular boundary conditions chosen, and it may be more perspicuous to think of a *region* that shows self-organizing and adaptive behavior. (Mitchell, Hraber, and Crutchfield 1993). Arrays that show such behavior are said to exhibit complex dynamics (Langton 1986).

In describing what happens at or near the edge of chaos, Langton and Kauffman have also made use of a notion that Per Bak, a physicist at the Brookhaven National Laboratory, called the "power-law diagnostic." Bak discovered that large, interactive dynamical systems develop quite naturally to a "critical state" at or near the edge of chaos (Bak and Chen 1991). At this critical state, the system shows a range of responses to perturbations. For a given perturbation, the smaller responses are most frequent, and major change is rare. To help us visualize this distribution, Bak would have us imagine a stream of sand falling onto a plate. As the pile gets higher, it becomes unstable. Avalanches occur. Small avalanches often occur, but once in a while a really large avalanche happens. The

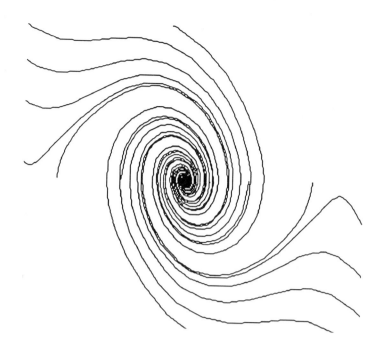

LORENZ MAP

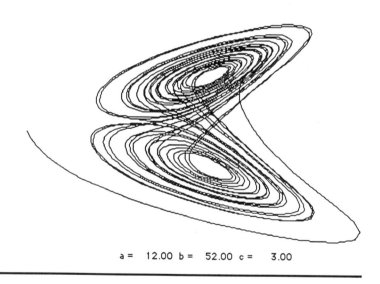

Viewing the X–Z plane   a = 12.00 b = 52.00 c = 3.00

Figure 16.2 An example of a point attractor *(top)* and an example of a Lorenz attractor *(bottom)*. With the point attractor, the same end state is reached regardless of the starting point. In contrast, with the Lorenz attractor, the trajectory does not settle down to a point but rather keeps moving and never exactly repeats itself. Drawn with *Chaos*, Dynamics Software, 1989.

New Models of Evolutionary Dynamics

equation that describes this relationship of frequency and magnitude corresponds to mathematical powers, such as 2^2 and 2^4. Hence it is called a power law.

It was because his Boolean networks are like cellular automata that Kauffman could assimilate his own work to Langton's and Bak's. Kauffman soon realized that all the marvelous isomorphies with biological phenomena that he had found when he set $K = 2$ were possible just because when $K = 2$ the system is sitting in the ordered regime near the edge of chaos, where it is subject to perturbations in accord with probabilistic schedules set by Bak's power law. At $K = 1$ and below, systems are too tightly bounded to display the properties characteristic of complex systems. Their behavior is ordered. The ordered state is analogous to the frozen state of a crystal. There exists a fixed or oscillating pattern that can be summarized by a simple algorithm. If $K = N - 1$, on the other hand, all sorts of interactions are possible, but systems have an essentially random structure, and hence do not do anything interesting. Their behavior is chaotic. Over time, the system will settle onto and cycle through attractors with vast numbers of states, taking billions of times longer than the history of the universe to traverse these enormous attractors. At about $K = 2$, near the transition between order and chaos, however, interesting things can and do happen. The behavior of systems in this range is complex. It might move toward one or several different attractors. The amount of information needed to describe this regime is much greater than for either of the other two regimes. There are no easy algorithmic or statistical shortcuts. This region of phase space is analogous, then, to a biological membrane rather than to a crystal or a gas, for it has elements of order, fluidity, and disorder. Transient islands of ordered structure can arise in a sea of chaos, only to melt away as new order appears elsewhere. At the edge of chaos, moreover, and especially in a narrower region that lies in the ordered regime close to the edge, complex systems exhibit the distinctive dynamics of self-organization and adaptability (Langton 1986, 1992; Lewin 1992; Waldrop 1992; Ulanowicz 1994).

Soon Kauffman was arguing that adaptation can in fact take place best in systems that are near the edge of chaos. Only at, or near, the boundary between order and chaos are attractors ordered but loose enough to allow the system to show significant variation or even to evolve from one attractor to another (through processes like genetic recombination, for example). Adaptation, considered as a process, becomes a walk-in parameter space near the edge of chaos in search of a better fit to the fitness landscape or even to find better attractors, while an essential subset of the system's variables is kept within tight bounds. Such attractors are located by means of selective feedback processes in which differentially retained variants are stabilized and amplified. It is hard for a system, however, to stay in this region. Kauffman was soon arguing that since natural selection considered as a phenomenon in its own right rather than something that explains other phenomena, can evolve in this region of phase space, and thus bring systems that are *near* the edge of chaos *to* the edge of chaos and hold them there. This implies that the most important adaptive properties of complex systems are adaptations that optimize the ability of such systems to persist in a regime that is fluid enough to allow them to evolve further:

Ordered systems encounter relatively correlated landscapes, chaotic systems adapt on very rugged landscapes. Landscape structure governs both the evolvability and sustained fitness. . . . Further, *natural selection* itself may be the force which pulls complex adaptive systems into this boundary region. If so, we begin to have a powerful tool with which to examine the collaborative interaction between self-organization and selection. . . . At the boundary between order and chaos, the frozen regime is melting and the functionally isolated unfrozen islands are in tenuous shifting contact with one another. It seems plausible that the most complex, integrated and also the most evolvable behavior might occur in this boundary region, near the edge of chaos. (Kauffman 1993, 218–19)

In our view, this way of situating the phenomenon of natural selection in dynamical space throws considerable light on why research programs inspired by Dobzhansky have been more productive than those initiated by Fisher. Dobzhansky's stress on variation maintenance as a condition for further adaptation in changing environments trades implicitly in the area of phase space that Kauffman identifies as most productive for further evolution. Accordingly, Dobzhansky's intuitive vision of adaptations for adaptability acquires here, perhaps for the first time, an adequate theoretical, or perhaps meta-theoretical, backing. The sorts of adaptations Dobzhansky had in mind were not simple morphological parts but complex life-cycle traits aimed at allowing organisms and populations to ride over environmental cycles of various amplitudes and frequencies. Dobzhansky's theory of balancing selection was intended to capture how such traits emerge and are maintained by selection (Beatty 1987). The subsequent development of Dobzhansky's research program made it possible, by Lewontin's time, to see that even when variation is fixed in populations by nonselective means, such as neutral mutation, adaptive capabilities of this sort will in the long run be enhanced. More recently still, the self-organizing properties of mobile genomes have further enlarged the notion that adaptability is the best adaptation by making it possible to think of adaptations as enhancing the capacity for evolvability itself.

Kauffman's theory of the evolution of adaptive systems at the edge of chaos provides a good dynamical backing for these developments and a good explanation of the problem-solving productivity of research programs developed in Dobzhansky's shadow. It does so because it suggests that complex adaptive behavior will develop only in systems whose range of elements and connections is such that wide variation, a feedback-driven selection process, and self-organization are integral aspects of a single process. Fisher's model, by contrast, operates in too well correlated a landscape to produce the properties that living systems need most, as Wright immediately saw. "We need a true theory of the biological embracing self-organization, selection, and historical accident," Kauffman wrote in 1989 (Kauffman 1989, 87). By 1993, he had articulated just such a theory by using the dynamical language that he had learned and helped formulate at the Santa Fe Institute.

Kauffman seems to regard the explanatory power of evolutionary theories to be significantly measured by their capacity to solve evolution's grand mysteries. His encounters with the laws of form tradition sometimes even seem to have led him to identify and describe these phenomena in ways familiar to the developmentalist tradition. It is indicative of Kauffman's large design in *The Origins of Order*, accordingly, that he rather systematically brings his models to bear on topics ranging from the origins of life and the evolution of developmental programs to ecological coevolution and to phylogeny. In each case, the trick is to assign N and K to the relevant objects and connections at different levels. Kauffman seems convinced that at every level of the biological hierarchy, his models will show how the generic properties of ensembles spontaneously produce self-organization, which in turn provides a platform for the selection that keeps systems within a range where they can evolve further.[2] In consequence, Darwinian processes are found in places where the developmentalist tradition has been unable, and unwilling, to see them, while Darwinism recognizes and explains phenomena hitherto associated exclusively with the developmentalist tradition.

Kauffman's account of the origin of life expresses a decided preference for theories in which life's emergence is regarded as an expectable phenomenon and in which autocatalytic properties drive the coevolution of proteins and nucleic acids. "We can think of the origin of life," he writes, "as an *expected emergent collective property* of a modestly complex mixture of catalytic polymers" (Kauffman 1993, xvi). For Kauffman, life began whole and integrated, not disconnected and disorganized. What is unique about Kauffman's model of the origin of life is that his NK models make it clear that generic properties will self-organize a set of proteins in such a way that some significant catalytic functions will arise and stabilize without putting all the weight, or at first even most of it, on replicating ribonucleic acids, or even their vaunted self-catalyzing powers. If anything Kauffman's is a "protein-first" model, in which prebiotic selection favors the properties that drive systems toward better autocatalysis and emergent metabolism. Once generic proteins begin to acquire specific catalytic functions, their need to be "remembered" if more coherent, reliable, effective and heritable metabolism is to be possible makes further selective integration with similarly self-organizing and evolving replicative macromolecules an expectable development.

Kauffman defines a protein sequence space in which N equals the number of amino acids strung together to make the proteins. For such an ensemble of protein molecules, the size of sequence phase space is $20N$. A given sequence is represented as a point in this space. It has $19N$ point mutation neighbors. K, meanwhile, represents the number of functionally important interactions that can obtain among the constituent amino acids. Kauffman then introduces a concept he calls "catalytic task space," "an abstract representation, or mapping, of all chemical reactions which can be catalyzed onto a space of 'tasks'" (Kauffman 1993, 122.) (Overall, Kauffman estimates that there are about 10^8 unique catalytic tasks.) He then imagines mapping the catalytic task space onto protein sequence spaces. The fitness of proteins will in this case be measured by their capacity to perform some catalytic task and to become sufficiently stable to persist under the obtaining conditions. (This model immediately explains why, once a protein settles on a catalytic task and has evolved optimal properties, there can be a random walk within the allowable volumes of task space due to neutral mutations, as Kimura had shown [Kauffman 1993, 108–12]).

To model the emergence of life, consider an ensemble of generic proteins of random sequence, produced by polymerization of abiotically formed amino acids. These ensembles can catalyze various chemical reactions, including autocatalytic reactions that produce more of themselves. It is likely that a sufficiently large ensemble of N proteins (of varying length) would be able to cover enough of catalytic task space that a loosely connected network of collectively autocatalytic chemical reactions would spontaneously self-organize. This is due to K interactions between the catalytic proteins via mutual reactants and products. Such a reaction network would start in the chaotic regime, but because proteins can collectively make copies of themselves by autocatalysis and peptide interactions, enough variation can arise to permit chemically more efficient sequences to drive out those less efficient at collective autocatalysis. Over time, prebiotic ensembles reduce K and move toward the edge of chaos, where greater stability and articulation of reaction networks is possible. At this point, a proto-metabolism will emerge *even in the absence of any genetic information*. "If the model is correct," Kauffman writes, "then the routes to life in the universe are broader than imagined" (Kauffman 1993, 330). All that is needed to get a connected web of catalyzed chemical transformations as an emergent property is a sufficiently complex set of catalytic proteins (or RNAs) (Kauffman 1993, 337). Several investigators have shown, in fact, that sets of proteins or RNAs of random sequences can produce new catalytic activities (Johnsson et al. 1993; Bartel and Szostak 1993). Replicating nucleic acid templates for these collectively autocatalytic sets provides a more reliable "memory" of those sequences that have greater chemical efficiency. Once there is a true genetic informational system, this proto-metabolic system produces a genuine metabolism, and life as we know it emerges. With this emergence arises not just chemical effiency but biological fitness. Natural selection itself emerges from prebiotic selection in reaction networks. While prebiotic selection is strongly constrained by the properties of the molecular species involved, biological selection can act more strongly on a less constrained genetic regulatory system.

We turn now to the evolution of regulatory genes, the evolutionary level for which Kauffman developed his NK models in the first place. Here N refers to genes coding for proteins, and K to regulatory connections among them. Regulatory genes, it is now recognized, come in two kinds, cis-regulatory and trans-regulatory genes. Cis-regulatory genes are stretches of DNA lying next to structural genes, to which diffusable products of a trans-regulatory gene can bind. In the regulatory system of bacteria that Monod described, the trans gene was called the r-gene and the cis gene the o-gene. In eukaryotes, the trans genes are not necessarily packaged as an operon, as they are in bacteria. The product of the trans gene can act as a cis gene locus remote on the same chromosome or on another chromosome altogether. Further, the cis gene can affect the activity of any trans-gene that is contiguous with it. A vector with its base in the trans gene and its head in the cis gene, and vice-versa, represents the network connectivity or "wiring diagram" of the regulation. For purposes of computer simulation, accordingly, assume that there are sixteen trans-cis-structural genetic units distributed over four chromosomes (figure 16.3)

If the trans and cis genes are packaged together in functional units, the regulation is like that of bacterial systems, and the diagram represents sixteen independent feedback loops with trans-1 affecting cis-1, which in turn can affect trans-1. If we group the trans-1 with cis-16 and the trans-2 with cis-1, and so forth, we get a circular wiring diagram composed of all the elements. If we permit only mutations of the type of transposition and gene duplication to occur randomly in the computer, we can generate on this circle very complex wiring diagrams (figure 16.4).

If the number of genes (N) is considerably greater than the number of regulatory connections allowed (K), the scrambled genetic circuits are small and relatively unconnected with each other. As K increases relative to N, however, crystallization of large, connected genetic regulatory circuits takes place at about $N = K$. These circuits grow more complex up through about $K = 2N$, where large connected circuits generally emerge. As the number of connections K increases beyond that point, the wiring diagram undergoes a phase transition toward something like the generic complex pattern, *even in the face of continued selection* (figure 16.5).

When the complexity of the regulatory connectivity becomes large enough in the model, selection cannot override the drift toward the generic ensemble properties. This has convinced Kauffman that in real

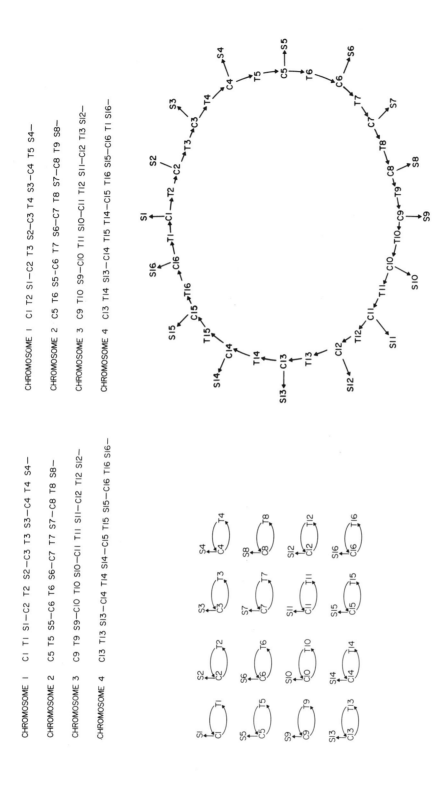

Figure 16.3 *Left top:* Hypothetical set of four haploid chromosomes with sixteen kinds of cis-acting (C_1, C_2, . . .) and trans-acting (T_1, T_2, . . .) along with the structural genes (S_1, S_2, . . .) whose expression is regulated. *Bottom left:* Regulatory regulations for the left-top set. *Top right:* Same as top left except that the T genes have been repositioned by one location, i.e., $T(x+1)$. *Bottom left:* Regulatory interactions of the top-right set. Notice how different the regulatory patterns can be due to a small rearrangement of the regulatory elements. From pp. 174–176 (Figs. 1a,b and 2a,b) of Kauffman (1985). Reprinted with permission.

New Models of Evolutionary Dynamics

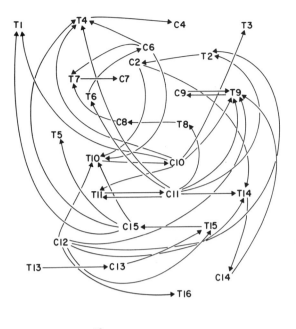

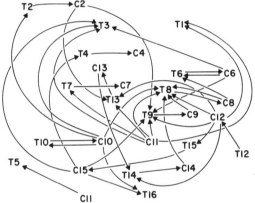

Figure 16.4 *Top:* Regulatory interaction of the chromosome set on the left of figure 16.3 after 2000 transpositions and duplications have occurred, but structural genes and fully disconnected regulatory genes are omitted. *Bottom:* Similar to top but based on the chromosome set on the right of figure 16.3. From pp. 178–9 (Fig. 3a,b) of Kauffman (1985). Reprinted with permission.

regulatory programs, there can be basic features that were never selected for. On the contrary, they are there in spite of selection (Kauffman 1985).

Can living systems realistically sustain a connectivity as large as $N = K$ and still survive the resultant chaotic change in the regulation of gene expression? For $K = N$ networks, the actual behavior will be chaotic. The length of state cycles will grow exponentially by the function ($2^{N/2}$). For a network of 200 elements and 200 connections there will be 2^{200} or 10^{60} different states possible, and the state cycles will have an average length

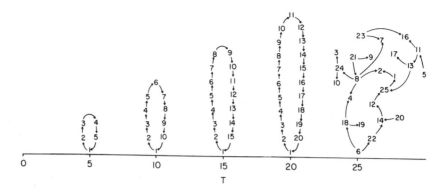

Figure 16.5 Maximally adapted wiring diagram achievable in a population by selection for a single closed regulatory loop, where T is the number of regulatory connections. By $T = 25$ the single loop cannot be maintained, and the maximally adapted wiring diagram achievable falls toward the ensemble generic properties. From p. 190 (Fig. 8) of Kauffman (1985). Reprinted with permission.

of 10^{30}. (To put the size of this number in perspective, recall that the estimated age of the universe is just 10^{17} seconds.). Such networks exhibit extreme sensitivity to initial conditions. At the same time, they possess an element of order. The number of possible state cycles is $2^N/e$. This means, for example, that if $N = 200 = K$, say for a small organism, the number of attractors is just 74. Further, Kauffman demonstrates that about two-thirds of the states will fall under the influence of just a few attractors. While most attractors possess only a few states, some will contain a large number. The greater the number of states that lead to an attractor, the more immune it is to the effects of perturbations. Thus, although the system is in the chaotic regime, there will be a rather small number of attractors to which the system would settle down. When we consider contemporary biological genomes, which typically have values of N in the thousands, the number of attractors at high values of K would still be very large and result in too unstable a situation for the survival of living systems.

The chaotic behavior of $N = K$ Boolean networks persists as the number of connections between elements is reduced to about $K = 3$. But at $K = 2$ there is a dramatic emergence of order. In $K = 2$ networks, the number and the length of the state cycles are approximately equal to the square root of N. For a genome of 100,000 genes (the estimated size of the human genome) analysis predicts that there are only 317 possible state cycles or attractors. Kauffman notes with great interest that this is about the number of cell types (254) observed in humans. This suggests that these attractors could define cell types as products of self-organization. Indeed, across the phylogenetic range, from bacteria through humans, there seems to be a rough correlation between the number of calculated attractors and the observed number of cell types.[3] In this antichaotic regime at or near the edge of chaos, in contrast to the chaotic regime, there is a fair amount of insensitivity to initial conditions. Kauffman notes that "most mutations in such networks alter the attractors only slightly. The ordered network regime is therefore characterized by a homeostatic quality; networks typically return to their original attractors after perturbations. And homeostasis . . . is a property of all living things" (Kauffman 1991, 81). The advantages of such stability would provide a strong selection acting on newly emergent regulatory systems to bring them toward the edge of chaos.

For developing embryos, the NK model predicts that mutations early in development occur on a relatively uncorrelated fitness landscape. Hence the chances are low that the mutant will be more fit. Late in development, on the other hand, the fitness landscape can be presumed to be more correlated. Chances are better that a mutation will be fit rather than deleterious. Thus, Kauffman's model generates a a redescribed von Baer's law (Kauffman 1993, 75; cf. Wimsatt 1986). The von Baerian process of differentiation itself can be understood, in fact, as a response to perturbations carrying a cell into the basin of an attractor for another cell type. Only a few of the possible attractors are in the region that can be accessed. Thus, the overall process of development from a fertilized egg must follow a branching pathway. The trajectory of development is such that once a cell has begun to differentiate along a certain pathway, it is no longer able to differentiate along all possible pathways. Here, Kauffman believes, are the dynamics behind Waddington's epigenetic landscapes. Indeed, Kauffman comes close in this analysis to formulating the synthesis of genetics, developmental biology, and evolutionary theory called for by Waddington and the Theoretical Biology Gathering.

When cell types are viewed as attractors, one can readily speculate that attractors in the ontogenetic-NK space, if expressed early in the developmental trajectory, can account for stable ontogenetic programs. It will be easier then to reconceive the process of speciation as a shift in attractors acting late in the developmental trajectory. Because of possible looseness in attractors near the edge of chaos, and the tendency of complex systems to restabilize, it might not take many mutations for an ontogenetic program to come under the influence of another attractor that could produce an emergent evolutionary novelty in a geologically relatively short period of time-giving rise to a punctuated paleontological record. A small number of mutations, for example, apparently no more than about half a dozen, was sufficient to transform teosinte grass into corn (Culotta 1991).

Perhaps this conception of development provides theoretical backing for views of speciation such as those favored by Gould (Gould 1982a). The idea that developmental programs are full of structural constraints, and contain many features that are shielded from the action of selection, is to be expected on dynamical grounds. So is the idea that periods of rapid change alternate with periods of long stasis. Indeed, from this deepened theoretical background, pan-adaptationist and gradualist programs cannot give any plausible explanation of speciation at all. It is no wonder, then, that Fisher's tradition never developed any such theories. It could not. A Fisherian conception in which fully correlated adaptive landscapes, where K approaches 0, rather than being the conditions most favorable to speciation, are among the least favorable, for selection is a more powerful evolutionary agent on more rugged landscape near the edge of chaos, where a limited number of attractors allows for selection on ensembles of regulatory genes. Indeed, natural selection may have its greatest impact near the edge of chaos where the possibility of adaptive evolution is greatest. It follows from this that "the ability to take advantage of natural selection would be one of the first traits selected" and that the most favored traits will be those "life cycle traits" that enable organisms to deal with their changing environments and to ride over disruptions (Kauffman 1991, 82.).

With these thoughts in mind, we turn to ecology, and in particular to coevolution among many species linked, primarily as predators and prey, in ecological communities. Ever since Lotka, ecologists have seen oscillatory behavior in the dynamics of communities construed this way. Much appears to be simply periodic. Since the 1950s, however, it has become apparent that there are also chaotic patterns in data for changes in natural populations over time. Until computational methods became available, such chaotic behavior could not be subjected to rigorous analysis. It was the English zoologist Robert May who first demonstrated that data on the population density of insects as a function of time could be modeled only by using the newly emerging mathematics of nonlinear dynamics and chaos theory (May 1974). A number of subsequent studies have shown just how widespread such phenomena are (Schaffer and Kot 1985). NK models can throw additional light on both the phenomenology and the causes of such phenomena.

Kauffman's approach to topics in population ecology is to treat N as the total number of traits per species within an ecosystem, with K epistatic, interactive traits. The interactions among the species are then modeled as interacting adaptive landscapes that, in addition to changing their own environments, mutually deform each other's. The background, then, is adaptive landscapes treated ecologically, after the fashion of Dobzhansky, but linked to one another in such a way that the fitness landscape of each species depends upon the other species with which it interacts. Species are thus nodes in a complex network. The null hypothesis in this case, accordingly, is one in which selection and other evolutionary forces are challenged to operate against a generic ensemble of linked fitnesses for interacting coevolving populations, all of which deform constantly each other's fitness landscape. It is the kind of problem that could not even be envisioned, let alone solved, without the representational and analytical devices of complex systems dynamics.

Kauffman uses as a simple example the interaction between frogs and flies. As frogs zap out their tongues, they will catch a certain number of flies. Those flies that are slower or tend to stick to tongues will be less adaptive, and their numbers will decrease. But if, through mutation or migration, flies appear with slipperier bodies, fewer of this new phenotype will be eaten. Not only has the adaptive landscape of the flies changed. So has that of the frogs. If some of the frogs should develop stickier tongues, again through mutation or migration, the adaptive landscape of both species changes once again.

In computer simulations of such coupled, coevolving adaptive landscapes, the effects of changes in the physical environment are included as random, external perturbations of the adaptive landscape of the species. The effect may be to make the species as a whole less fit. Over time the species can ascend one of the new adaptive peaks through the traditional process of variation and selection. But in that process, they may well change the adaptive landscapes of other species with which they interact. This leads Kauffman to assert: "In coevolution, organisms adapt under natural selection via a *metadynamics* where each organism myopically alters the structure of its fitness landscape and the extent to which that landscape is deformed by the adaptive moves of other organisms . . . the entire ecosystem coevolves to a poised state at the edge of chaos" (Kauffman 1993, 261).

When an ecosystem evolves to near the edge of chaos, where adaptive natural selection is accessible, it will be more stable than when K was larger. Change is not impossible, however, in such a system state. What happens is that the number of changes and their size follows the "power law" diagnostic that Kauffman, following Bak, sees at work on the edge of chaos, where ecological systems are presumably located and maintained.[4] If the number of species interactions is low, the effect of a perturbation will likely disappear with little effect on the community at large. There is also an intermediate state, where some external perturbations cause various ripplings of small changes, but only occasionally launch a large cascade or "avalanche" of massive changes. If the number of interactions, or the connectedness, is high, however, as when an ecosystem is away somewhat from the edge of chaos toward the chaotic regime, even small perturbations can more readily sweep through the whole system and change it. Sometimes massive changes like these are dynamically equivalent to the mass extinctions observed in the fossil record. It does not always take a major catastrophe, then, to cause massive change. It all depends where the system is at.

Kauffman's approach to macroevolution combines his conception of the branching pattern of ontogeny with the power-law dynamics of ecology to produce what amounts to a punctuated pattern of evolution.

Just as cell types differentiate as they move from one attractor to another in the epigenetic landscape, so phylogeny occurs, in highly interconnected, ecological networks, when from one ontogenetic attractor a species gains access to one or another of a few nearby attractors. This gives rise to a branching pattern in which certain attractors become effectively inaccessible. (Wings evolved from limbs, but horses, having their forelimbs committed to running, are unlikely to evolve wings.) This does not mean that evolution is development writ large, or that ontogeny mechanically recapitulates phylogeny. Rather, the two processes exhibit similar trajectories due to the fact that both are expressions of the same dynamics working on ensembles of genes.

Such patterns can be facilitated, as well as interrupted, by mass extinctions, for such extinctions are often followed by rapid proliferation of new forms that occupy newly empty ecological niches. The most dramatic event of this type was the Cambrian explosion, when many new themes, or body plans, were laid down. This maximal disparity was soon diminished, however, and evolution contented itself with producing maximal diversity on a relatively small number of body plans (Gould 1989a). What needs to be explained is why the same pattern did not characterize recovery from the Permian extinction of later times (Kauffman 1989, 1993). Kauffman's coevolutionary model of interacting adaptive landscapes can be used to argue that the regulatory genetic programs of newly evolved multicellular organisms in the early Cambrian were still on a rugged landscape, where there was more chance for innovation and for large-scale avalanches in regulatory genetic reorganization, resulting in big gains or big losses. As the the ontogenetic programs of individual species evolved to entrench the earlier phases of development, the later phases would have less rugged landscapes. By the time the Permian extinction and recovery took place, the genetic regulatory programs for multicellular organisms were probably stabilized at, or near, the edge of chaos. Thereafter, change would have occurred by way of small alterations in the latter parts of the ontogenetic programs of fairly secure developmental regimes. This would have resulted in a greater diversification of species after the extinction to fill recently the vacant niches, but in little or no increased disparity (Kauffman 1993, 76–83). In our opinion, if this analysis is correct, two remarkable facts come into view. First, the larger contingencies of life's history occur against a background in which developmental biology moves further toward the edge of chaos, where stable developmental programs and enhanced adaptations can be found. Second, what happens under the influence of potentially catastrophic events depends on how sensitive the system is to initial conditions.

To Darwinians, Kauffman's ideas often seem reductionistic. "Kauffman's approach," write Burian and Richardson, "is that of a physicist studying complex materials and complex systems. . . . Kauffman holds that a sort of statistical mechanics governs ensembles of complex systems, depending only on formally characterizable properties of underlying entities" (Burian and Richardson 1992, 271). As a result, there is nothing distinctively biological about Kauffman's principles or much that is particularly relevant to the precise problems that evolutionary biologists are trying to answer (Burian and Richardson 1992).

In spite of the fact that Kauffman himself sometimes talks in ways that invite responses like these, this does not seem to us the best way of characterizing either the aims of his project or its preliminary results (Kauffman 1993, 310, 340, 403–4, 470–72, 487–88). Even less does it characterize what his methods portend for the future.[5] The bare fact that Kauffman uses the language of dynamics to describe natural selection and other evolutionary processes cannot in itself be said to betray a desire to assimilate evolutionary biology to physics. If there is one thing this book has demonstrated, it is that treating evolution in terms of dynamical models has been going on from the start. Indeed, because the models of dynamical systems in which Kauffman trades do not depend on or refer to specific properties of physical systems, he actually runs less risk of undermining the autonomy of biology than, for example, Fisher. If there is nothing distinctively biological about Kauffman's models, neither, it should be said, is there anything distinctively physical about them. Accordingly, it seems jejune to characterize Kauffman's search for biological order as a reduction to a statistical mechanics. It was Fisher who did that. Fisher to the contrary notwithstanding, the dynamical rules of statistical mechanics, where $K = 0$ or $K = N$, are precisely what evolving biological systems, as biological systems, do not obey.

It is, moreover, an achievement to show, from a position well within the Darwinian tradition, that long-held assumptions to the effect that adaptive selection is at its most powerful when adaptive landscapes are highly correlated, and when K approaches 0, are wrong. It would even be an achievement if Kauffman had done no more than reverse the burden of proof in this matter. The relevance of his analysis to particular cases and problems, accordingly, while it is undoubtedly high, is less important than his virtual demonstration that the conditions under which natural selection is most powerful are also those in which we can expect much self-organized order and at the same time much that is chancy and historically contingent.

Misunderstandings on these points seem to have arisen in part because in Kauffman's earlier work, the generic properties of ensembles were often used to insist that biological order is due primarily to the effects of self-organization and to emphasize the limits to adaptive natural selection. That is still a prominent theme in *The Origins of Order*, where to some extent it competes with Kauffman's explorations of the dynamical conditions under which natural selection can arise and can be an effective evolutionary force. One source of this tension is that generic properties of large ensembles actually perform three overlapping, but slightly different roles, which are not always discriminated. The generic properties of regulatory genetic ensembles provide, in the first place, a plausible null hypothesis against which the effects of selection and other evolutionary forces are to be measured in systems as presumptively complex as regulatory genetic networks. They also serve as what Levins calls "sufficient parameters," that is, variables that reduce complexity without sacrificing explanatory or predictive power (Levins 1966, 1968; Levins and Lewontin 1985). Finally, they sometimes seem to be presented as universal laws from which evolutionary explanations and predictions can be deductively derived after the fashion of the received philosophy of science.

Kauffman's tendency to treat generic ensemble properties as general laws may well have been stimulated by his resistance to those recent Darwinians who have been confronting complexity by stressing the role of narrative reconstructions rather than of laws in explaining the products of evolution's historically contingent tinkering (Chapter 14). Kauffman's resistance to this trend manifests itself in his tacit fidelity to fairly conventional views about the role of universal laws of nature in biological explanations. If such laws cannot be found in distinctively contingent, and far from universal, products of evolutionary contingency like Mendel's laws, the Hardy-Weinberg equilbrium, or Fisher's fundamental law of natural selection, perhaps, he seems to suggest, they can be be found at a more basic level (Depew and Weber 1985). To the extent that Kauffman talks this way, the objection that Burian and Richardson raise may be valid. In not relying on specifically biological premises, Kauffman cannot pretend to explain specifically biological phenomena, and continues to rely on conceptions of natural law tilted toward physics.

If Kauffman's stress on the lawlike character of generic properties does not seem sufficiently distinguished from the idea of "covering laws" it might be because, in the received philosophy of scientific explanation and in the world of classical physics that it reflects, there is not much difference between a law of nature, a sufficient parameter, and a null hypothesis.[6] In a fully decomposable and deterministic world, laws of nature serve both as a boundary conditions for every possible state of affairs and as rules for moving from state to state within those boundaries (Dyke 1988). In an essentially or presumptively complex world, however, boundary conditions and transformation rules are seldom the same thing. The most universal boundary conditions, such as the second law of thermodynamics, for example, tell us virtually nothing about the kinetic pathways that systems use to obey it. Concrete transformation rules, moreover, allow for reliable prediction only under very well-defined boundary conditions. Finding what those are is usually more significant for getting robust explanation, and more difficult, than finding the rules that apply within the boundaries. It is not too far wrong to suggest, in

New Models of Evolutionary Dynamics

fact, that what used to be called laws of nature should, in a complex world, be treated as a set of "closure conditions," which, in foreclosing other possibilities, allow us to deal only with the degrees of freedom that remain within the system as defined (Dyke 1988). In such a world, the worst thing a scientist can do is to increase his or her ability to control those degrees of freedom by taking for granted too simple a set of boundary conditions. It happens, however, all the time.

When they are viewed as null hypotheses and as sources for finding sufficient parameters, on the other hand, Kauffman's dynamics seem to us to be not only very promising but far from reductionistic. Rather than enforcing, or trading upon, a severe contrast between what is invariantly law governed and what is historically contingent, this way of putting the matter seems to us to have the salutary effect of breaking down just that opposition. The generic properties of ensembles describe shapes, curves, and warps in the phase space within which evolutionary forces move. On this view, the distinctly biological emerges in a range of boundary conditions, described in the dynamical language of phase spaces, which presumptively couple chance, self-organization, and selection. Explanatory hypotheses that take highly particular conditions into account are thereby licensed and encouraged rather than discouraged or ruled out.

It is, of course, entirely possible that Kauffman himself does not take this view of the matter. It is entirely possible that Kauffman subscribes to a philosophy of science that does not do full justice to his own accomplishment. He may interpret his work as encoding discoveries about laws of nature. He may even want to find universal laws in order to reduce biology to physics, and to sacrifice the autonomy of biology on the altar of the unity of science. It is possible, moreover, that a penchant for reductionistic theories results from a reductionistic and mechanistic ontology. This very objection to Kauffman's work has been raised, in fact, by Robert Ulanowicz, who argues that the reductionist bias in Kauffman's work derives from his use of cellular automata as modeling devices (Ulanowicz 1994). Cellular automata are, after all, mechanisms; and mechanisms, no matter how complex and self-organizing, have a bias toward decomposability and atomism. More generally, Robert Rosen has argued that modern biology is still pervaded with and limited by machine metaphors that undervalue the relational complexity of living beings and systems (Rosen 1991).

This may miss the main significance of Kauffman's work. What Kauffman has done is to show that many phenomena that have become well accepted in contemporary evolutionary science flow rather easily and directly from background assumptions taken from complex dynamics. These phenomena run from neutralism about protein evolution to intrinsic constraints in developmental programs, from the inherent probability of the emergence of life to punctuated patterns of macroevolution, and from evolutionary trade-offs and ecological coadaptedness to Dobzhansky's stress on variation-maintaining mechanisms. So great is the contrast between these phenomena, many of which could be discovered only after the molecular revolution, and the expectations of classical or statistical dynamics that many contemporary Darwinians are prone to cast aside the role of dynamics in Darwinism altogether and to content themselves with narrative reconstructions that sacrifice generality to explanatory specificity. Kauffman's models, however, need not imply that the current stress on particular cases, contingency, and retrospective narrative reconstruction is wrong. It is part of our recognition of the intrinsic complexity of the biological world that these methods should prove both unavoidable and useful. What Kauffman's models enable us to do, however, is to situate this complexity against a dynamical background that renders it expectable, tractable, and comprehensible. It thus serves the aim of unifying biological science with physics without reductionism, while avoiding at the same time too strenuous a proclamation of the autonomy of biology. This may not be Kauffman's own way of looking at the implications of his models. Indeed, it probably is not. It is, however, a possible interpretation, and in the light of the history traversed in this book, a useful way.

This enhanced problem-solving power reduces adhocery about propositions already known to biologists. In addition, Kauffman's principles enable him to make headway on a number of large problems that have thus far resisted integration within Darwinism and that have for this reason remained within the orbit of the developmentalistic tradition. This is a major accomplishment, showing the continued fecundity of the Darwinian tradition, or at the very least, reducing the gap between Darwinism and its developmentalist rivals.

None of these virtues depend on whether Kauffman's particular models, cellular automata, will eventually prove to be the best way of tracing and modeling the complex systems dynamics of living systems. They may well be overly mechanistic. If, however, Kauffman's results can be accomplished with such simple and even mechanistic models, new models of complex dynamical systems will presumably do even better in their problem-solving power. It is highly unlikely that we will ever be in a position where it is reasonable to go back to simple dynamics. The Rubicon has been crossed.

17 The Thermodynamics of Evolution

Wherever we look in nature, we see complex physical, chemical, or biological systems in which matter tends to organize itself into complex patterns and structures. It does this in the face of the dissolution that is ultimately guaranteed by the second law of thermodynamics. We will see in this chapter that such structure building is expected rather than contradicted by thermodynamics. This process occurs in a wide range of phenomena, from convection cells in a heated pan of water to tornadoes and hurricanes to "waves" of chemical activity in certain types of reactions to analogous waves of activity in the excitable cells of a human heart and in the patterns of complexification of a developing embryo. As we suggested in the last chapter, we are quickly acquiring the means to track the distinctive dynamics of processes like these in phase space. Nonetheless, dynamics is not everything. Thermodynamics, and the various kinetic pathways that systems take in order to abide by its ineluctable laws, are no less important. Speaking about one of his computer models of artificial life, Langton says, "The whole system represents a dynamical pattern with energy being dissipated through it . . . Take away the energy, and the whole thing collapses" (quoted in Lewin 1992, 190). In a similar spirit, we suggest in this chapter why the sorts of dynamical models Kauffman is exploring are displayed first and foremost in characteristics that living systems have in virtue of being energy-dissipating systems. Whatever patterns evolution exhibits are rather closely related to this fact.

We begin with some history about the troubled but changing relationship between thermodynamics and Darwinian evolutionary theory. When Boltzmann called the nineteenth century the Century of Darwin, what he had in mind was that Darwin introduced "deep time" into biology. If our earlier analysis is correct, Darwin did not quite do that. He still had one foot in the eighteenth century, since he was trying to extend the Newtonian science of Lyell and Herschel to biology. We are more inclined, in fact, to describe the nineteenth century, or at least its second half, as *Boltzmann's* Century, and to regard figures like Fisher, Wright, and

Simpson as giving us in the twentieth century, perhaps for the first time, a Darwinism fit for deep time, for statistical thermodynamics, and processes like those its models, are more radically temporal, irreversible, and directional than Darwin's Darwinism. Indeed, the advent of thermodynamics in the mid-nineteenth century was generally perceived to be so contrary to what Darwin's theory of evolution by natural selection required as virtually to refute the latter. Darwin's theory implied the possibility that systems could increase in complexity, even while denying that such increases were necessary. The second law, on the contrary, seemed to imply that as entropy increases to a maximum, the energy differences in the universe will smooth out until no more work is possible. Heterogeneity, structure, and life itself do not seem able to follow this arrow of time.

William Thompson (Lord Kelvin), one of the most influential thermodynamicists of the nineteenth century, called Darwin's attention to this paradox in a vivid way. He applied the new methods of thermodynamics to data on the dissipation of energy from the sun and the earth in order to estimate their age. His analyses suggested that the earth was rather young, in the range of 20 million to 400 million years, and probably was only between 98 million and 200 million years old (Smith and Wise 1989). This was a far cry from Darwin's vision, in which a single geological era alone might well last 300 million to 400 million years. In response to Kelvin, Darwin conceded in the sixth edition of *On the Origin of Species* that a time scale as short as Kelvin demanded was insufficient for the type of organic change that he envisioned. He acknowledged that this constituted a grave criticism of the theory of evolution by natural selection. He appears to have tried to take the heat off (literally) by shifting a bit more of the work from selection to the inheritance of acquired characteristics. Darwin also had the good sense to say that not enough was yet known about the processes and dynamics of the earth to obtain a reliable estimate of the age of the earth, and hence to close the matter (Darwin 1872).

Darwin turned out to be right. What was missing from physics of Kelvin's day was knowledge about radioactivity and nuclear processes. With the enriched physics of our time, the age of the earth is now estimated to be about 4.5 *billion* years, an age with which Darwin could happily live, and that he might even be said to have anticipated. Even as concerns about the age of the earth abated, however, the fundamental problem remained. From a thermodynamic point of view, a view presumably more basic than the biological one, the universe appeared to be running down toward a state of maximum entropy, while living systems seemed to maintain and even increase their order and organization. So deep was this tension felt by later nineteenth-century intellectuals and artists that, even if Kelvin's and Darwin's time scales had been reconciled, it is far from clear that this would have removed the depressive cloud that hung over late nineteenth-century writers like Henry Adams when they contemplated the anticipated "heat death" of the universe. Even Huxley began to share the gloom (Huxley 1894). The tension between a physics that ran down and an evolution that was assumed to progress was another reason that many late nineteenth-century evolutionists preferred spiritualistic theories of evolution to any sort of selective theories; it did not seem that nature could do this on its own. This ambience also explains why later those who remained thoroughly Darwinian, like Fisher and Wright, attempted to show how physical and natural selection could work together in a "two tendency universe" (Turner 1985; Hodge 1992a).

Considerable sweetness, as well as light, was at last thrown on this subject in a seminal little book that appeared in 1944, *What is Life?* by the quantum physicist Erwin Schrödinger. The problem is not as bad as it seems, Schrödinger argued. The second law requires only that the universe *as a whole* must show an increase in entropy. Eddies of order, or what Schrödinger called "negentropy," could be sustained in the great flow of ever-increasing entropy. Accordingly, a living cell, an organism, even an entire ecosystem, might maintain its internal structure if it could be coupled to its surroundings in such a way that the entropy of the environment remains greater than the internal "negentropic" decrease within the boundaries of the system in question. This could happen only so long as the system remained far from equilibrium. Schrödinger wrote:

It is by avoiding the rapid decay into the inert state of "equilibrium," that an organism appears so enigmatic. . . . Everything that is going on in nature means an increase of the entropy of the part of the world where it is going on. Thus a living organism continually increases its entropy— or, as you may say, produces positive entropy—and thus tends to approach the dangerous state of maximum entropy, which is death. It can only keep aloof from it, i.e. alive, by continually drawing from its environment negative entropy. . . . What an organism feeds upon is negative entropy. Or to put it less paradoxically, the essential thing in metabolism is that the organism succeeds in freeing itself from all the entropy it cannot help producing while alive. (Schrödinger 1944, 71–72)

What Is Life? inspired a good deal of research into physical, chemical, and biological systems that exhibit, under specific conditions, stability far from equilibrium, irreversibility, order, structure building, and, of late, chaotic and edge-of-chaos dynamics. One of those who has explored how entropy increase and order go hand in hand in systems stabilized away from thermodynamic equilibrium is the Russian-born physical chemist and Nobel laureate Ilya Prigogine, who has worked primarily in Belgium and the United States. In Prigogine's view, the second law is not just consistent with evolution, as Schrödinger maintained. It helps explain it.

Self-organization, the dynamical trajectories of which can be modeled on computers, arises in the real world only when specific physical and chemical conditions permit certain kinds of behavior. Some of these conditions have been described by Prigogine. Many systems that show nonlinear dynamics are, or are closely linked to, systems that are stabilized far from thermodynamic equilibrium. They can be so stabilized because they are open systems: They maintain their internal structure by pumping energy and matter into themselves, using it to form and maintain structure and do work, and then dissipating it in a more disordered or degraded state to their surroundings. Open systems include things as different from one another as tornados, cells, organisms, ecological communities, and economic systems. Many of these things are the entities that belong to, and help define, the ecological, as distinct from the genealogical, hierarchy dear to expanded Darwinism. It is precisely by building internal structure that open systems pay their debt to the second law. That is because structured pathways are often the preferred means for dissipating degraded energy. Prigogine calls these *dissipative structures*. Dissipative structures in effect collapse energy-matter gradients. If the gradient were not replenished, the structures themselves would collapse. A cell deprived of nutrients dies. Given the continued availability of gradients, however, dissipative structures have various sorts of abilities to pull in the resources they require for their maintenance, producing in consequence a steady state that remains far from thermodynamic equilibrium. The inherent tendency of open systems to increase, even to maximize, their dissipative rate or some other similar quantity is, accordingly, linked to the capacity of such systems to build better dissipative pathways in the form of more efficient internal structures. To the extent that dissipative structures can be influenced by selection, there will be selection pressure for them to do just this, and to do it better than other dissipative structures that are competing for resources with them.

Dissipative structures are usually built up by some form of self-organization. We can take it for granted, therefore, that some of the energy that a dissipative structure captures is used to build structure within the system itself, in effect, creating an internal sink and facilitating increased entropy production in the external sink. That is what environmental degradation is. It is a big problem because it is an indispensable condition of life itself. The most effective way of building structure and dissipating entropy is by means of autocatalysis. A chemical reaction that produces a substance that can help the production of another reaction just like it will show a rapid amplification of the concentration of the substance in question. This is called an *autocatalytic cycle*. This is already a reason dissipative structures can be expected to show nonlinear dynamics (Ulanowicz 1989). Dissipative structures, by the very nature of the self-organizing processes that produce them, are highly sensitive to changes in initial conditions. Depending on what initial (and boundary) conditions obtain, they are capable of generating dynamics that, displayed in phase space, produce order, chaos, or complex organization at the edge of chaos. The equations Lorenz used for his dynamical model of the weather, for example, which we considered in the last chapter, were originally devised to describe dissipative phenomena like convection cells. In any case, dissipative structures are not particularly at home in the gradualist world of Lyell or Darwin. Instead, they do things like "explode" and "crash."

Convection cells are, in fact, a good place to begin thinking about dissipative structures. They were first studied in a classical experiment conducted by Henri Benard in about 1900. A contemporary version of this experiment involves heating a thin layer of oil from the bottom (source) and keeping the top of the liquid in contact with a transparent cooling plate (sink). As the temperature from below rises, more energy flows into the system. This is reflected in a higher kinetic energy of the liquid's molecules and a larger difference in temperature or thermal gradient across the liquid. The result is increased energy flow through the system and increased dissipation of energy, or entropy production, to the sink. At a critical size of the gradient, or the thermodynamic force field, when the effect of viscosity is exceeded by heat transport, the intensified random motion of the molecules becomes insufficient to dissipate the energy flux. Large-scale macroscopic streams of hot liquid then flow to the top, forming convection cells that are observable because of the different index of refraction between the hotter and cooler liquid. We observe convection flows when we do something as ordinary as boil a pan of water. The honeycomb pattern we see toward the bottom of the pan is the result of coupling buoyancy, thermal diffusion, and viscous forces, creating a macroscopic structure that is perceptible because it is at least 100 billion times the scale of molecular dimensions in water itself. Self-amplifications like these are "circularly causal" in the sense that they take the form of an autocatalytic cycle, in which the incipient formation of structure—here a convection cell—reflexively reinforces the formation of more of the structure until the system settles down to a new steady state (Swenson and Turvey 1991). This is an example of self-amplification of a stochastic fluctuation that breaks the symmetry of the linear regime (Garfinkel 1987). Such symmetry breaking takes the system into the realm of nonlinear dynamics (Swenson 1989, 1991; Swenson and Turvey 1991).

When a system is stabilized far from equilibrium, there is a thermodynamic potential, which, like an electrical potential, provides the drive to move the system to equilibrium (Schneider and Kay in press). For the system to reach equilibrium, however, the thermodynamic potential must have a pathway of change. Henceforth we will term this a *kinetic pathway*. In an electric circuit, no current will flow through a wire unless the switch is closed to complete the circuit, even though there is an electromotive potential. If the circuit includes a motor, the electric potential can be converted into mechanical work. If there is a short circuit, the electrons will move down the path of lower resistance, that is, the most favored kinetic pathway that happens to be available. Something like that is true for thermodynamic systems as well as for electric ones. In systems maintained far from equilibrium, the state of equilibrium is denied to the system as long as energy is provided by the source. We can say, then, that the input of energy either drives the system away from equilibrium or, when the gradients become large enough, that the system is "constrained away from" the normal kinetic paths that would otherwise lead it to equilibrium. If they are physically available to the system, new kinetic pathways will be accessed, which, in forming autocatalytic structures, provide novel and more efficient paths toward equilibrium. Like Sisyphus forever rolling his stone uphill but never achieving his goal, equilibrium will not be reached as long as an energy gradient is maintained.

When a system is constrained far from equilibrium, macroscopic order arises not as a violation of the second law of thermodynamics but as a consequence of it. Creationists are wrong, accordingly, when they still tout the contradiction between the second law and evolution that entangled nineteenth-century minds (Kitcher 1982).[1] There would be a contradiction if living systems were so simple or so random that, like atoms in a gas, they hover around and quickly access thermodynamic equilibrium. If anything is clear by now, however, it is that living systems are virtual paradigms of systems that are constrained away from equilibrium and that pay what they owe to the second law by building internal kinetic pathways that send things in their environment, instead of themselves, to thermodynamic equilibrium.

In phase space, steady states far from equilibrium normally appear as point attractors. With a larger gradient, such systems can show periodic behavior, oscillating between two different steady states. As the gradient is increased further, however, a critical point will be reached at which the system becomes unstable before it settles down into a more complex dissipative structure. At this critical point, the system is said to bifurcate. In this case, its dynamics are defined by a chaotic attractor. Finally, a system can be so stressed that no attractors at all can be recognized even in phase space. At its outer limit, such a system will become completely random, or ergodic. In ergodic systems, all possible positions in phase space will be explored and occupied over time without making much difference. Kelvin's heat death of the universe is the ultimate ergodic attractor.

We may now use phase portraits of these attractors to talk about how selection processes of various kinds are related to autocatalytic dissipating systems. The Benard cell is an example of physical selection, or selection of the stable. There is, in the first instance, selection for the optimal dimensions of the hexagonal form. The hexagonal form itself represents the best way of getting "order for free," for given the boundary conditions under which Benard cells form, hexagonal shapes are the arrangement that will give the best packing in the available space. The optimal size of this form is also selected because it is the most stable; it provides the best balance between hydrodynamic stability and ability to dissipate excess energy through convection. Swenson has shown that the hexagonal cells initially generated in a Benard cell do not at first display the ordered array of cells of the same size that we finally seen when the system settles down to the steady state (Swenson 1989). In the early stages of convection cell formation, cells arise that are larger or smaller than the optimal final size, along with those that are just right. In time, smaller cells fuse and large cells split until they all achieve the optimum size (Swenson 1989). That process is physical selection. It encompasses all other forms of selection only in the sense that physical stability is a necessary condition for all physical systems, not because other forms can be reduced to it. In natural selection, for example, selection for fitness cannot be reduced to selection for stability, at least when radically reductionistic forms of genic selectionism are set aside (Dawkins 1989).

Autocatalytic chemical reaction systems also give rise to macroscopic dissipative structures. They reflect selection not only for physical stability, however, but also for efficiency in catalytic processing and in energy dissipation. Chemical selection finds and selects dissipative pathways. It is a kinetic theory, accordingly, bounded by thermodynamic laws. Periodic or clock reactions are simple, or at least clear, examples of such chemical systems (Tyson 1976; Field and Györgyi 1993). The first instance of such a reaction was published by the Russian chemist Boris Belousov (Belousov 1958). It was later refined by the Russian biochemist Anatoly Zhabotinskii (Zhabotinskii 1964). Hence this phenomenon is now known as the Belousov-Zhabotinskii reaction or, for short, the BZ reaction. In the BZ reaction, a complex reaction among citric acid, sulfuric acid, potassium bromate, and an iron salt produces a dramatic color change that alternates suddenly and repeatedly between blue and red. This color change in a stirred solution is rapid, global, and quite periodic. If the solution is spead in a thin layer, it is possible to produce stunningly beautiful circular and spiral waves of color. In either case, macroscopic structure, both temporal and spatial, is observable.

Belousov originally developed his complex reaction system to model the reactions of the Krebs cycle, the metabolic pathway in aerobic cells by which carbon dioxide is formed and energy is captured for the synthesis of ATP. The problem was that random reactions at the microscopic level were not expected to give such sharp, periodic color transitions or waves of chemical reaction. In fact, Belousov had considerable difficulty publishing his results because the referees and the editor argued that chemical reactions could not possibly happen this way. It took Belousov two years to find a sufficiently obscure journal in which to publish his work. There it languished until its significance was recognized and championed by Zhabotinskii.

In the BZ reaction, an energy gradient exists between the chemical bonds of reactants and products. The reaction pathway between them involves autocatalysis. A chemical intermediate is produced that facilitates the production of more of itself, while at the same time it generates more of the product. The resulting nonlinearity produces the observed dissipative structure. When the reaction is pushed by using larger gradients, the BZ reaction shows chaotic patterns. In a phase portrait, these map as a strange attractor—a "Rössler attractor" in this case (Field and Györgyi 1993). We have, accordingly, a dynamical system that progresses from initial homogeneity to periodic inhomogeneity in its chemical waves and ultimately to chaotic inhomogeneity. Along the way, selection occurs for the efficiency of the autocatalytic cycle itself. This sort of selection goes beyond the basic requirements of physical stability because the kinetic pathways of the possible autocatalytic cycles are selected for their capacity to handle the matter-energy gradient and for the speed with which they can accomplish the dissipation of the gradient. This contributes to the overall efficiency of the system in which these kinetic pathways are contained (Wicken 1987; Weber et al. 1989; Swenson 1989, 1991a, 1991b, 1992, 1995).

The general point is that, under appropriate initial and boundary conditions, physical systems, such as convection cells or tornados, and chemical systems, such as oscillating reactions, show spontaneously but perfectly understandable self-organizing characteristics. That is why hurricanes, for example, occur during seasons of maximal disparity between the temperatures of tropical and temperate oceans. The structures produced by self-organization keep systems from accessing or overloading the carrying capacity of the kinetic pathways that normally allow these systems to reach their equilibrium attractors. Instead, these systems utilize alternative kinetic pathways that result in the formation of macroscopic dissipative structures. The energy gradients that drive self-organization are a fact of nature from the cosmic scale down to the biochemical scale within cells. These self-organizing processes all involve one or another form of selection. Eventually, we will want to situate natural selection into this picture.

With this aim in mind, let us shift from chemical autocatalysis to the autocatalytic aspects of living systems. Lavoisier likened the process of cellular respiration to slow combustion. In modern terms, we can say that metabolism burns with an enzyme-catalyzed "flame" (Mitchell 1962). (This analogy is apt, for a flame is itself a dissipative structure.) A cell, or any other biological entity, is an open thermodynamic system that processes matter and energy flowing through it, thereby stabilizing its internal structure and more rapidly increasing the entropy of its surroundings. While they live, all living entities maintain themselves at a steady state far from equilibrium. The only time they achieve thermodynamic equilibrium is when they are dead. Indeed, that is probably the best definition of biological death. We are certainly entitled, then, to consider biological systems, from cells to ecosystems, as dissipative structures.

Clearly, however, living things differ in crucial ways from convection cells and tornados. Although metabolism is a chemical process that exhibits self-organizing properties, we can distinguish it from a BZ reaction, for example, because the information that guides metabolism is internal to the cell. Cells do not have to reinvent metabolic pathways of chemical transformation in each cycle or generation. Instead, they carry within themselves the information that encodes the organization of metabolism and morphology. In particular, they carry a more or less accurate record of the kinds of structures that have in the past been required for the survival and functioning of entities like themselves and for the ability to pass this information along to a new generation. Each cell and each organism carries within itself a history of its lineage in two books. The first book is the book of structural genes. It is about the information needed to make proteins and various RNA molecules. The second is the book of regulatory genes, which encodes information that regulates when structural genes are to be expressed. In complex organisms, some of the regulatory genes are devoted to defining an ontogenetic program that sets down the basic rules and patterns for the development of specialized tissues and the overall blueprint of the organism.

This view of the matter allows us to answer Schrödinger's question, "What is life?" in a definite way. We assert that a system is alive if and only if it exhibits the structures and behavior of informed, self-replicating, dissipative autocatalytic cycles (Wicken 1987). That is about as good a definition of life as you will get, as long as it is also understood that a living cell must have a membrane that provides physical phase separation from the rest of the world. It is, in fact, across this membrane that the matter-energy flows occur which power metabolism (Mitchell 1961; Morowitz 1992). Call living things, then, *bounded, informed, self-replicating, autocatalytic, dissipative structures.*

Holding this definition of what life is clearly in mind, let us speculate a bit about how life emerges from chemically autocatalytic systems in a thermodynamic context. Assume a prebiotic soup of a variety of chemical compounds. It is predictable, given the buildup of internal potential that can drive the system further from equilibrium, that there will come a point when a prebiotic system, at least one that compartmentalizes membranous enclosures, will reach a critical complexity. A phase transition will then occur, triggering a connected web of reactions that in turn crystallizes catalytic closure so that a system of polymers becomes self-reproducing. In this context, drawing too severe a contrast between proto-proteins and nucleic acids, and then wondering about who pulls what up by whose bookstraps, quickly becomes unhelpful. In a thermodynamic, dissipative, and autocatalytic context, proto-proteins are not entirely devoid of self-replicating, or at least self-reproducing, capacities themselves. Nor are nucleic acids devoid of catalytic properties, as RNA research has recently learned. Metabolism would have begun to emerge, accordingly, at the same time as ensembles of generic polymers of peptides and nucleotides in a context of autocatalytic, and more than likely cross-catalytic, self-replication (Odum 1971; Wicken 1987; Kauffman 1993). These considerations suggest that a model of the origin of life in which the coding functions of nucleic acids can appear after membranes and catalytic peptides make metabolism possible is at least as plausible as the prevailing view. This picture is not unlike the one painted by Kauffman (Kauffman 1993). It too declines to put all the stress on self-replicating RNAs that subsequently decorate themselves with proteins and phospholipids (Eigen and Schuster 1979; Cech 1990; Dawkins 1989). Although the template properties of DNA and RNA are now essential for life, they were not necessarily essential for its emergence under the action of thermodynamic and kinetic imperatives (Wicken 1987; Williams 1991; Morowitz 1992). Indeed, peptide-catalyzed proto-metabolism in some sort of chemically segregated space is probably essential for the emergence of replicating and coding nucleic acids themselves. This picture places Kauffman's dynamical models within a thermodynamic context that puts some causal fire under what Kauffman himself expects on purely dynamical grounds.

Whether nucleic acids capable of autocatalytic replication preceded or followed the appearance of catalytic proteins and metabolic transformations, it is in any case likely that the critical chemistry necessary for the emergence of life occurred in a small, segregated, phase-separated space. This was likely provided by an "amphiphile bilayer" vesicle, which would have arisen as a stable structure due to the chemistry of its chemical components. Such a vesicle would have been composed of bifunctional, or amphiphilic, molecules that have long hydrocarbon "tails" that are hydrophobic (water hating) and "heads" that are hydrophilic (water loving). The hydrophobic portions would have self-associated in order to avoid the water while projecting the hydrophilic portions toward the water. This would have given a thermodynamic driving force that increases the overall entropy of water plus amphiphiles through formation of a two-layered hollow sphere. Such vesicles would functionally resemble the lipid bilayers of contemporary cells (Deamer 1986; Morowitz, Heinz, and Deamer 1988; Deamer and Pashley 1989). Morowitz argues that the amphiphile bilayer vesicle is preferable to alternative containers, such as protein microspheres, not only because the hydrophobic core of the bilayer provides a thermodynamic barrier separating the inside of the vesicle from the environment but also because such an arrangement is closer to what we see in cells today (Morowitz

1992, contra Fox 1984). The amphiphile molecules that Deamer and Morowitz postulate to make up the membranes of protocells have been observed in carbonaceous chondrite meteorites, which are believed to have been formed by the same chemical processes that obtained on the primitive earth (Deamer 1985).

Such phase-separated entities would not be living things, but they would provide good cradles within which life could emerge. It would have been through interactions between generic proteins and nucleic acids within a proto-membrane and the context of a proto-metabolism that living systems acquired their special characteristics (Carter and Kraut 1974). What would most likely have driven this emergent self-organization would have been an energy source like sunlight, trapped perhaps by a dye molecule in the bilayer (Deamer 1992). Such a source could sustain a proton gradient across the bilayer, which in turn could drive the formation of polyphosphate (Morowitz 1992; Deamer and Harang 1990). The formation of polyphospate is important because the kinetic pathways by which the thermodynamic imperative might be met in living systems are highly constrained both by what is required for the physical selection of the stable, and by the "Goldilocks" or "just right" mix of chemical properties and efficient reactions indispensable for basic metabolism in a watery (aqueous) environment. There are very few alternatives to polyphosphate as a molecule to capture and transfer chemical energy for metabolism. Perhaps there are none (Westheimer 1987). Presumably, then, polyphosphates helped drive the polymerization of amino acids and nucleotides and indirectly helped the proto-cells take up these monomers from the environment. Later in the evolution of living cells, polyphosphate became modified into the ATP that now provides the "energy currency" for all living systems on earth.

The system of macromolecules and metabolic processes we have described at the dawn of life are dissipative structures. Accordingly, they increase the entropy of both the system of proto-cells itself and the environment. Entities like these, in which nucleic acids and proteins would have become coupled synergistically in autocatalytic cycles, can be expected to have progressively acquired more precisely defined metabolic functions over time (Wicken 1987; Weber et al. 1989; Weber 1991a). This potentiality would have been realized and fostered through competition with other such autocatalytic systems for energy fluxes. As successful patterns of dissipation emerge under these competitive conditions, we would expect progressive tightening of the nucleic acid-protein relationship as proto-cells evolve into true cells. Competition of this sort is neither physical selection, however, nor natural selection. It is thermodynamic or chemical selection of the efficient rather than physical selection of the stable or biological selection of the fit, in which the relevant units of selection are energy-capturing and energy-utilizing cycles. Self-organization provides much order without selection in autocatalytic cycles; chemical selection acts upon these cycles to produce increased efficiency.

The emerging ensemble of proteins that catalyze the reactions constituting primitive metabolism would originally have been "generic" proteins. Over time they would have acquired more specific catalytic functions. In a world in which autocatalytic cycles compete for efficiency in finding, utilizing, and dissipating energy sources, however, there would have been keen selection pressure for any entity that could increase these efficiencies by storing the information needed for autocatalysis and for expanding autocatalytic prowess by using these information-storing capacities in new ways. The close coupling of replicating macromolecules to autocatalytic proteins would have been highly prized in this context, and much to be expected. It would have been of enormous competitive advantage to such catalytic units if they were able to "remember" information that enhances autocatalytic activity by coding it in the polymers of nucleic acids that are formed by chemical selection itself. It is by this route that the inorganic flows smoothly into the organic. Genetic information accumulates under thermodynamic selection for stable patterns of entropy production.[2] We assume that the properties required for the emergence of life will be maintained by living systems thereafter. That is why we view living things as *informed* autocatalytic systems that sustain themselves by efficient environmental energy exchanges, yet vary under the drive to configurational randomness in such a way that new information, guiding new catalytic functions, can be selected from this variation (Wicken 1987).[3]

Natural selection of the reproductively fit is emergent from chemical selection of the autocatalytically efficient; it is a process that can be ascribed only to the autocatalytic dissipative structures that capture information within strongly defined boundaries and use it to guide efficient autocatalysis. Such entities would have to be able to pass the information they have to successor entities, for without that additional property, the whole point of internalizing information would be lost. Among such entities themselves, therefore, what would have been even more crucially contested than storing and deploying metabolic information would have been the ability to reproduce themselves and the information they possess. Fitness is a measure of that ability. It cannot be reduced to the efficiency of chemical selection, any more than the efficiency of chemical selection can be reduced to the stability of physical selection, for the relevant processes and entities capable of engaging in them do not exist at those levels. To say that these processes cannot be reduced to lower-level ones is not, however, to say that they are not part of a single, coherent process. They are. It is to say that evolution exhibits emergent levels and emergent properties. (The use of brains and symbolic communication to store and transmit culturally acquired information, and to keep finding more of it, is a remarkable adaptation that confers on species that have it very sophisticated means of enhancing autocatalysis and making other species pay their entropic debts. It signals another evolutionary level beyond natural selection, for it results in, and relies on, an autonomous form of cultural selection that cannot be reduced to natural selection, just as natural selection cannot be reduced to chemical selection, or chemical selection to physical selection.)

The idea of natural selection, as well as the notion of fitness used to measure it, is itself poised on the edge of chaos. The fitness of various sorts of organisms is not necessarily, or even probably, enhanced by superiority in a single trait. What is fit for the goose, and perhaps for the gander, may not be fit for the duck, or even fit for the goose next year. In fact, the emergence of the ability to take advantage in resource competitions of an indefinite number of often infinitesimally small differences creates degrees of freedom, in both the technical and the ordinary senses, well beyond what can be achieved by merely chemical and physical systems. It also creates more variables and interactions among them than can be tracked. It is impossible, then, to reduce the components of fitness to any single language or system of variables.

This situation has given rise to the notion that fitness is a supervenient property (Kim 1978; Rosenberg 1985; Sober 1984). The notion of supervenience is inspired, distantly at least, by the conceptual framework of statistical mechanics. A property is supervenient when the same macrostate can be accessed by any number of microstates. The very idea illustrates how the probability revolution is affecting conceptual as well as empirical issues as it makes its influence ever more broadly felt in culture (Gigerenzer et al. 1989). In many ways, discussions involving supervenience rerun old nineteenth-century debates about vitalism with the sophisticated weapons of the twentieth. Supervenience differs from the notion of "emergence" because it does not imply, or even suggest, what Donald Campbell calls "downward causation" (Campbell 1974). On the contrary, supervenience has been of interest to cognitive psychologists even more than to evolutionary biology because it seems to achieve at least one of the goals of reductionism without reductionism's discomforts. No mental state without a physical embodiment, it says, even though it is not possible to redefine a mental state as a physical state, and in that sense to reduce the mental to the physical.

If fitness is a supervenient property, then the same level of reproductive ability (in properly defined comparison classes) can be achieved by quite different arrays of physical properties. Some evolutionary biologists have argued that the magical properties of supervenience do not do much real work in this connection (Lloyd, in Callebaut 1993, 150–154). Others have enthused over it (Rosenberg 1985). Whatever its positive merits may or may not be, however, supervenience has probably already paid its way in discussions of fitness by doing some important, if unwelcome, deconstructive work.

Ever since the Russian ecological Darwinians began measuring energy flows in communities of animals, ecologically minded Darwinians have hoped that energetics could provide a common coin for measuring fitness. "An ecological system," says Levins,

was reduced to a system for the transformation of energy—a flow of energy. The notion of energy as the fundamental thing to look at, as the universal medium of exchange, is clearly brought into biology by analogy with economic exchange. . . . There was a hope . . . that we could ignore all the complexity of interacting species, the heterogeneity of populations, the complexities of competition and symbiosis, of mutualism and predation, and reduce everything to a single medium of ecological exchange, which was designated "energy." (quoted in Callebaut, 1993, 263–64)

The complexities that separate fitness from the commensurable language of energetics have provoked a number of reactions on the part of ecologists. Those who have already given up the earlier aspirations of community ecology will find nothing to mourn. They will already have become Darwinian population ecologists, who use fitness measures to describe relationships among species in a community, and have little use for the language of energy flow. Others may remain faithful to the older ecological tradition by restricting themselves to the study of succession and other such processes that are indeed measured in terms of energy but seem to have nothing directly to do with evolution by means of natural selection. Finally, there will be those who cling to ecology's founding vision by seeing in ecological succession an instance of or a cause of evolution by rejecting the relevance of fitness altogether. Such theories drift quickly out of Darwinism into the developmentalist orbit (Salthe 1993).

Yet the fact that fitness cannot be definitionally reduced to energetics need not lead however, to an artificial separation between ecology and evolution, as it sometimes has in the context of Darwinian population ecology's effort to tear itself away from the older community and systems ecology (see Kingsland 1985). The fact that the meaning of the concept of fitness cannot be reduced to the language of energetics does not imply in the least that energy flow is causally irrelevant to evolution, even to evolution by means of natural selection. Nor does the fact that fitness admittedly supervenes on a host of heterogeneous properties mean that the fit traits created by natural selection are not causally relevant to how energy flows through ecological communities. Representatives of species in local communities (sometimes called "avators") are the crucial nodes of energy transduction within the complex web of relationships in which they participate (Wicken 1987; Weber et al., 1989). The traits they have developed to enhance their own survival and reproduction are also, accordingly, means by which energy circulates in ecological systems. At the same time, the flow of energy through the nodes of an ecological system, while it would not happen except through the competitive agency of more and less fit organisms, acquires a certain autonomous shape of its own that involves, as Russian Darwinians and followers of Lotka have long suspected, cooperation among the species of a community to enhance, even if not necessarily to maximize, the rate of energy flow through the system as a whole. Cooperation and competition are jointly rewarded (Lotka 1924; Wicken 1987; Weber et al., 1989). A certain component of fitness, then, goes to organisms that enhance these webs better than their competitors. Those populations are fittest that best enhance the autocatalytic behavior of the matter-energy loops in which they participate (Ulanowicz 1986). The effect of such competition would be an emergent tendency during the later stages of ecosystem development toward more highly articulated networks of flow, wherein those pathways that foster more efficient transfers flourish at the expense of less effective routes. In saying there things, we are neither reducing fitness to energetics, nor conceding that the supervenience of fitness means that energetic considerations are not components of fitness.

The interpenetration between ecological and evolutionary processes we have been sketching comes into view most clearly when organisms are defined as informed autocatalytic dissipative structures. According to that definition, the entities that emerged at life's beginning, and have multiplied and diversified ever since as they accessed and parceled out the free energy in environments, are both energetic and informational entities. Indeed, organisms are the point at which the ecological and genealogical hierarchies are joined together because organisms are uniquely both replicators and interactors (Eldredge and Grene 1992). Moreover, the evolution of species comes into proper perspective only when the dynamics of interacting species are placed within the rich context of coevolution in ecological communities. The evolution of living systems takes places within what Hutchinson calls an ecological theater, in which the evolution *of* species, like the development of characters in a play, occurs through a process of coevolution *among* species (Hutchinson 1965). We saw in the last chapter that, on Kauffman's models, communities linked together attain average fitness levels. The very fact that they are stabilized at or near the edge of chaos means, however, that they can be destabilized and eventually reintegrated by self-organization. In that process, speciation is most likely to occur. This dynamical pattern is realized in a thermodynamic and kinetic field in which pathways of energy flow exhibit and guide this pattern of stability, disruption, and reintegration.

One of the more remarkable things about ecological succession is that its trajectory bears an uncanny resemblance to the life cycles of individual organisms. Ecological systems, like developing individuals, move from the homogeneous to the heterogeneous, as von Baer, following Aristotle, claimed. During succession, under the right kinds of boundary conditions, there is a continual, hyperbolic increase in "complicatedness," that is, an increase not only in total biomass but in the number of different types of components and their connectedness or articulation. As this process proceeds, there is an increase in internal stability as the rate of development slows. When succession has been reached in one or more climaxes, there is an increase in vulnerability to perturbations, the specific entropy production (entropy per unit time per unit biomass) is minimized, and the system becomes senescent (Salthe 1993, Ulanowicz 1986; Weber et al. 1989). In this respect it was rather prescient of Haeckel to situate ecology between ontogeny and phylogeny.

The successional pattern of ecology has been most fully analyzed in terms of thermodynamic requirements and preferred kinetic pathways by Ulanowicz, who has introduced the term *ascendancy* to quantify the relationship between the total energy transfer through an ecosystem and the level of interconnectiveness among the components of that ecosystem over time. Increase in ascendancy reflects the effects of positive feedback in successive restructurings of ecosystems. On the whole, Lotka turns out to have been on the right track: Ecosystems favor species that, in funneling energy into their own production and reproduction, also increase the total energy flow through the system. The effect is to increase the dissipation of energy as entropy production to the surroundings. This greater entropy production produces autocatalytic energy and matter flow cycles that not only become sinks in themselves but also facilitate greater flow of energy through the system as a whole. This process results in a decrease in the intensity of energy flow as the biomass increases, after an initial "charging" period when the intensity of energy flow is increased. Individual nodes become more efficient.

The parallels between ontogeny and ecology have led people in the developmental tradition to think of ecologies as ontogenies written large and of phylogeny as ecology written larger still. Guided by our definition of living systems as bounded, informed, autocatalytic dissipative structures, we think the opposite way of looking at this matter is more explanatory. An organism is a very tightly integrated ecological system in which the thermodynamic patterns seen in ecological ascendancy achieve high boundedness, stability, and predictability. During the process in which a fertilized egg divides to form an embryo or an organism grows and develops after birth, there is an increase in the energy flow through the system, with a concomitant increase in biomass. A maximum of that energy flow goes into creating organization, a minimum into metabolic uses. After an initial increase, the specific metabolic energy, that is, the energy flow divided by the biomass, then begins to decrease.

Throughout this process, there is a continual, hyperbolic increase in complicatedness in the size and number of an organism's components and their interactions. As a result, there is an increase in internal stability as the rate of development slows. Finally, as aging takes place, the organism is more and more easily disrupted by fluctuations in its environment. Viewed as a whole, this process suggests that the stored information curve is a kind of inverse of the energy intensity curve. While the organism is changing most rapidly, it has the greatest specific energy flow through it. On the other hand, increasing complexity results in less flexibility, curtailed metabolic rates, and an increased vulnerability to insults. The organism's ability to heal or regenerate decreases until a limit is reached and the organism is "recycled"—another name for death (Salthe 1993).[4]

What makes an organism different from an ecological system is not this pattern itself but the ability that it has to internalize the relevant information for processing energy and matter. That is precisely what an ecosystem lacks. Organisms employ informational macromolecules to achieve stable, homeostatic (stabilized state of a single parameter), and homeorhetic (stabilized flow or trajectory) metabolic pathways that can never be achieved by entities that depend on external signals. Let us repeat and expand at this point our definition of an organism. An organism is an informed autocatalytic system possessing, in virtue of information stored in macromolecules, an internal organization of kinetic relationships such that it is able to maintain itself by pulling environmental resources into its own production and faithful reproduction and dissipating unusable energy to appropriate sinks. This view suggests that organisms originated, and may continue in part to evolve, by incorporating, storing, and deploying information about the environment in developmental programs, which information describes patterns of energy flow in more general kinds of autocatalytic cycling systems (Weber et al. 1989). The adaptations that would have been most favored by natural selection are those that enable organisms to process information from the environment that allows them, and their descendants, to roll over cycles, disruptions, catastrophes. These are the "life-cycle" traits on which students of evolution since Dobzhansky have been most interested. Ecologies, let us emphasize, are not superorganisms. Organisms are highly integrated ecosystems.

If that is how ontogeny looks from the perspective of ecology, the view of phylogeny from the perspective of ecology is quite different. Phylogeny reflects the complexification of the earth in some dimensions. In this respect, it bears some resemblance to the ecological and ontogenetic patterns we have described. Whereas ontogeny encodes ecology processes that have been more or less successfully incorporated into internalized information, phylogeny is as much a record of failures as well as successes. What survives phylogenetic sorting is open to enormous contingency. Speciation itself occurs in the context of highly unstable dynamics, in which sensitivity to initial conditions carries with it the threat of descent into ecological chaos without hope of getting back to the adaptive edge. Macroevolutionary pattern thereafter is as highly punctuated and full of historical quirks, as Gould and Eldredge have asserted and as Kauffman's models predict. For these reasons alone, neither ontogeny nor phylogeny recapitulates the other. On the contrary, phylogeny's pattern *contrasts* with that of ontogeny by giving chance a better hand than either selection or self-organization.[5] We should on these grounds be wary of too close a phylogeny-ontogeny parallel. Still, we need not be overly skeptical about evolutionary direction and complexification, or even of a teleological aspect of the evolutionary process as the following reflections suggest.

Traditionally, the subject of evolutionary complexification has been tied to the notion of evolutionary direction through the old idea of biological teleology. Darwinians have often felt that the only way to reject the second idea is to reject the first. When evolution is seen in the perspective of complex dynamics and far-from-equilibrium thermodynamics, however, it is much easier to countenance the existence of a genuinely teleological aspect of biological adaptedness. Should we do so? And if so, in what sense of teleology?

Traditionally, biological teleology has referred to three different things: overall direction of the phylogenetic process toward greater complexity or some other end-state, the end-directedness of ontogeny and the functional adaptedness of organisms to environments and organs to organisms. The last is what Paley had in mind, and what Asa Gray meant when he said that Darwinism is "teleology wedded to morphology" (Gray 1876). In recent years there have been calls to rehabilitate "teleology" in the second and third senses even among distinctively sober philosophers of science who will have no truck at all with the first (Dobzyhansky et al. 1977, 497–516; Mayr 1982; Wright 1973; Brandon 1981). This new receptivity to analyzing adaptationist arguments in terms of the concept of teleology has arisen, we think, because the probability revolution has made it possible to give accounts of functional and goal-oriented processes, which, although they are genuinely goal oriented or end directed in some sense, do not depend, even tacitly, on the assumption that underlies the argument for the existence of God from design: that whatever has an end, goal, or function is the result of someone or something's *intention* or *purpose*. Wright, for example, has analyzed teleological claims as falling under the form "X is there *because* it does Y" (Wright 1973). Such claims answer "what for" questions and yield "in order to" answers. Their general form, accordingly, is common to both design and natural selection arguments. The former make reference to the intentions of an artificer as the reason an entity has some feature; the latter refer only to the accumulated effects of natural selection. Thus, Brandon argues that natural selection explanations cite "the effects of past instances of A (or precursers of A) and show how these effects increased the adaptedness of A's possessors (or the possessors of A's precursors) and so led to the evolution of A" (Brandon 1981, 103). There is no reason, Brandon says, to deny that such arguments are teleological.

A problem with this otherwise reasonable view is that adaptationist accounts of an acceptable sort can easily be replaced by disguised or tacit design arguments. Gould and Lewontin have been highly effective in pointing out how easily sociobiologists, for example, slide down the slippery slope into disguised design by retailing just-so stories (Gould and Lewontin 1979). This slide from selectionist reasoning to intentionalist cryptoteleology occurs in proportion as the intentional form of teleology is treated as paradigmatic, and the natural-selectionist form is regarded as a derivative analogue. One way to drive a wedge between natural selection and design forms of teleological argumentation is to anchor natural selection in natural processes more deeply, and so to distance the sort of functional arguments that refer to natural selection from arguments referring, however tacitly, to intentional design by a nonnatural agent. We believe that the ecological picture of natural selection does this by anchoring selective processes in energetics (Weber et al. 1989).

The premises that generate these arguments, however, also seem to countenance talk about direction of evolution or "teleology" in the grand sense. If kinetic pathways are naturally selected means whereby a necessary and inevitable entropic debt is dissipated, these pathways will have a propensity for complexification and organization even as they are discharging their entropic debt. This propensity is precisely what distinguishes living systems from machines. It is also what distinguishes what is teleological about biological adaptedness from paradigms grounded in design. Ironically evolutionary direction, albeit contingent, is what keeps adaptedness from regressing to design.

On this account, shifting to complex systems dynamical background assumptions as a framework for the further development of the Darwinian tradition leads to the expectation of adaptedness, and so relieves Darwinian explanations of temptations to disguised teleology of the design sort. No longer need Darwinians argue that nature has been able to achieve the same kind of effect that an engineer-god could attain by using different means. It is telling in this connection that Paley's watchmaker does not completely disappear in Dawkins's version of evolutionary theory (Dawkins 1986). He is said only to be a "blind watchmaker." From our perspective, however, there is no watchmaker, blind or sighted, for the simple reason that there is no watch. Natural organization is not an artifact, or anything like it, but instead a manifestation of the action of energy flows in informed systems poised between order and chaos. Directionalities, propensities, and self-organization in a thermodynamic perspective actually exclude the notion that evolution is oriented toward an end in the intentional or design sense. The thermodynamic perspective allows biological adaptedness precisely by excluding design arguments. Directionality excludes directedness.

18 Natural Selection, Self-Organization, and the Future of Darwinism

We have been considering ways in which biological theory might benefit from the analysis of complex, self-organizing systems. Most recently, we have been interested in natural systems that are maintained far from thermodynamic equilibrium. With their sensitivity to initial conditions, their nonlinear changes in phase space trajectories, their self-organizing properties, and their ability to adapt on rugged landscapes, far-from equilibrium systems are the kinds of systems in which the dynamical models we have pointed are most readily instantiated. Interestingly, they are also the kinds of systems we most convincingly call living and evolving. Thus, we have also tried to suggest how these models, placed in a thermodynamic setting, can illuminate phenomena like origin of life, the dynamics of development, community and systems ecology, and phylogenetic pattern and direction, all of which have thus far been incompletely assimilated into the modern synthesis and thus remain under the spell of the once dominant, but freshly stirring, developmentalist tradition.

Whatever else may happen, we are reasonably certain that evolutionary theory will remain incomplete as long as self-organizational and dissipative phenomena are kept at a distance. This still leaves open, of course, whether complex dynamical models and nonequilibrium thermodynamics will testify in favor of the developmentalist or of the Darwinian tradition—or indeed will stir into being what might at some future date become a new scientific research tradition altogether. Can self-organization and dissipative structures be brought into the present evolutionary synthesis or some expanded version of it? Alternatively, will assimilation be so challenging that it will require a change of background assumptions in the Darwinian tradition comparable to that which produced the modern synthesis itself? Or, finally, are self-organization and dissipative structuring so foreign to Darwinism's core concept, natural selection, that giving them an important place in evolutionary theory will put an end to the Darwinian tradition itself?

By framing the history of the Darwinian tradition in the way we have, we have in effect been arguing for the second alternative. We concede that it is far too early to be entirely confident about this. Nonetheless, what can surely be done at present is to set out a range of logically possible relationships that can obtain between natural selection and self-organization, to surmise how the Darwinian tradition would be affected in each case, and to show why the way we choose to look at this relationship is preferable. On this note we close this book.

Just as there are many niches in an ecological community, so there are many niches in the possibility space that determines how self-organization and natural selection can be related. We discriminate seven such niches:

1. Natural selection and self-organization are not related at all.

2. Self-organization is auxiliary to natural selection.

3. Self-organization constrains natural selection, which drives evolution.

4. Natural selection constrains self-organization, which drives evolution.

5. Natural selection instantiates self-organization.

6. Natural selection generates self-organization.

7. Natural selection and self-organization are aspects of a single process.

Most of these niches have been at least tentatively occupied by early explorers of this terrain. Each niche, moreover, is spacious enough to accommodate a number of competing theoretical positions. Locating individuals in the same niche does not, accordingly, imply total agreement. On the contrary, since niche exclusion is as contested a process in discursive as in ecological space, cohabitants are often more critical of each other than they are of theorists from whom they are more widely separated.

The first position is that self-organization and natural selection have little or nothing to do with each other. This is not merely a possible position, but, until recently, has been the received wisdom. It may be still be more right than wrong. Nonetheless, the developments we have traced in part III of this book probably suffice to change the presumption that has for so long existed in favor of this view. When the argumentative weight has been shifted in this way, a second niche comes into view—a conservative position. On this view of the matter, the self-organizational properties of biological entities like genomes, developmental programs, and ecological communities represent simply one more alternative to natural selection among the causes of evolution. Self-organization, that is, can be added to any list of approved "forces," such as genetic drift, mutation pressure, gene flow, and natural selection, all of which might be at work in particular evolutionary phenomena, at least in any respectably pluralistic version of Darwinism (Sober 1984a).

Alternative forces have always been downplayed, of course, by Darwinians of a pan-adaptationist turn of mind, even when they pay lip-service to such processes as genetic drift and mutation pressure. Doubtless self-organization will meet a similar fate at their hands. Heirs of Wright, on the other hand, who admit a significant role for chance processes, especially in important events like speciation, will recognize that appeals to self-organization are parallel to appeals to chance and can greatly enrich Darwinism's arsenal of explanatory tools. Indeed, self-organization is demonstrably at work in many complex systems in which chance also plays a significant role.

Even on such an admirably pluralistic view, natural selection is often taken to be the explanation of first resort (Dyke and Depew 1988). Other "forces" are added to take account of presumed perturbations in a fundamentally selective process operating on relatively decomposable systems with uncorrelated fitness landscapes. At the very least, natural selection is accorded heuristic primacy: Look for selection first, the admonition goes, and other "forces" only when selection fails to yield a persuasive explanation. In a synthesis expanded to include it as an orthodox "force," self-organization would, on this view, be accorded a role as presumptively subservient to natural selection as genetic drift. This would hardly amount to a scientific revolution. That is why the second niche represents a conservative position on the relationship between self-organization and natural selection.

It is, of course, possible to elevate a force other than selection to a heuristically central position within the same conceptual rationale. It is even possible to reduce natural selection to an explanation of last resort. Kimura's neutral theory of protein evolution, for example, takes the more or less regular rate at which mutations become fixed as the expectable cause of evolutionary change. Natural selection is treated as a perturbation. A theory in which self-organization plays the same heuristic role that genetic drift does in Kimura's theory of protein evolution is, it seems, at least logically possible. By this route, we reach a third conceptual niche. Self-organization, on this conception, is not an auxiliary evolutionary "force" but a fairly substantial and expectable constraint on the freedom of action of such forces, preeminently natural selection.

This position seems to be the majority view in an important essay on evolutionary constraints coauthored by a group of contemporary evolutionary thinkers whose variance in matters theoretical is as wide as it could possibly be, including as it does Maynard Smith, Richard Burian, Stuart Kauffman, P. Alberch, John Campbell, Brian Goodwin, Russell Lande, David Raup, and L. Wolpert (Maynard Smith et al. 1985). The central concern of the authors of this paper is the relation between developmental constraints and natural selection. Developmental constraints are defined as biases on the production of variant phenotypes, or limitations to phenotypic variability, caused by the structure, architecture, functions, and dynamics of developmental systems. Such constraints are often viewed less as prohibitions than as opportunities, since they allow exploration of certain regions of genomic, morphological, or behavioral space in a nonrandom manner.

Highly adaptationist versions of the synthesis will hear about such notions as reluctantly as they will hear any other sort of talk about constraints on natural selection. For they (or at least Panglossian caricatures of them) assume that selection has potentially unlimited and directionally unbiased variability upon which to work. But this assumption is highly implausible. The stasis recorded in the fossil record, on which paleontologists have been insisting, is most plausibly taken as evidence of structural or developmental constraints that limit the effectiveness of natural selection and bias its direction (Gould 1982a). It is true that stasis can also be explained in Dobzhanskyan terms by an analogue of balancing selection called stabilizing selection (Charlesworth, Lande, and Slatkin 1982). It is also true that the potentially complex interaction between developmental constraints and natural selection will make it at least as difficult to distinguish and measure the contribution of each as it is to distinguish the contributions of drift from natural selection even in orthodox accounts (Beatty 1984). In spite of these epistemological and methodological problems, however, the consensus has now shifted in favor of the widespread occurrence of constraints on selection, and thus in favor of something closer to niche 3 than niche 2.

The focus of the joint paper we are considering is on what its authors term "local constraints." These are developmental constraints responsible for the persistence and stability of particular features of particular taxa. It is constraints of this sort on which Gould, for example, places high value as sources of (narrative) explanation. This stress sidesteps the more interesting question of "universal constraints," which may be imposed not by the quirks of phylogenetic history, or even by the concrete laws of physics and chemistry, but by inherent mathematical or formal properties of complex systems architecture. To say that this topic is sidestepped in the essay we are considering is to say that the voice of Stuart Kauffman is not heard very loudly in it. When Kauffman argues that the self-organizational properties of sufficiently complex genetic regulatory programs can be affected by selective forces to at best a limited extent, he seems to have universal constraints in mind (Kauffman 1985, 1989, 1991, 1993). Kauffman argues that universal, as distinct from local, constraints are mathematically expectable consequences of interconnections among large genetic ensembles. Self-organization is to that extent assigned a different role than Maynard Smith and his colleagues accord to it. Self-organization constrains natural selection, as well as other evolutionary forces, in the quite different sense that it is the expected background against which selection and other evolutionary processes are to be measured. In this view there is latent a potentially rich alternative to the conservative reading of the role of constraints embedded in the essay we have been considering (Dyke and Depew 1988).

Yet as constraints on natural selection become less local and more universal, however, it becomes more and more unclear whether we are remaining within the permissible, if generous, boundaries of the Darwinian tradition at all. Addressing this issue draws us into toward new possibilities. Self-organization can be construed, for example, as the driving force of large-scale evolutionary dynamics rather than as a constraint on natural selection's role as evolutionary driver, or as the baseline against which forces like selection are measured. Here a fourth possible niche comes into view. Dan Brooks and Ed Wiley have made a controversial proposal about the causes of phylogenetic branching that illustrates this idea (Brooks and Wiley 1986, 1988). On their view, the self-organization of ontogenetic programs is coupled with the idea that over time species, viewed as discrete or individuated information-exchanging entities, dissipate something these authors call "informational entropy." Natural selection is assigned a role that harks back to Muller. Its job is to prune down the novelty that is constantly *and autonomously* being generated in this way. On this view, macroevolution might in principle occur without any microevolutionary natural selection at all. Even when it is admitted that this autonomous process is to some extent constrained by natural selection, we judge this position to lie somewhat beyond the Darwinian pale. It supporters think so too. Its non-Darwinian character is in their eyes one of its most attractive features (Salthe 1993).

This proposal has met with considerable resistance on both conceptual and empirical grounds (Wicken 1986; Depew 1986; Olmsted 1988; Hariri, Weber, and Olmsted 1990; Morowitz 1992). Nonetheless, the notion that natural selection constrains an autonomous process of phylogenetic diversification has been congenial to biologists like Salthe and Goodwin, who have argued that macroevolutionary dynamics and large-scale biological order owe less to natural selection than to autonomous developmental dynamics (Salthe 1993; Goodwin 1989a, 1989b). This structuralist and self-organizational conception represents a renewal in the age of molecular biology and nonequilibrium thermodynamics of the long displaced, but once extremely powerful, Geoffroyian research tradition according to which phylogeny runs on its own, and records its progress in ontogeny. Since the whole point is to screen off natural selection from macroevolution, this can hardly be Darwinism of any sort.

Even more problematically Darwinian is a view in which natural selection becomes an instantiation of the most basic physical processes under a specific set of initial and boundary conditions. Under this rubric, a fifth niche comes into view. Swenson, for example, has argued that selection, generically considered, will be a derived consequence of self-organization whenever a system is moving toward a state of maximum entropy production (Swenson 1989; Swenson and Turvey 1991). "Selection is entailed by self-organization," he writes, "and Darwinian selection ('natural selection') is a special case where the components are replicating. Thus self-organization (order production) is necessary (onto-logically prior) to natural selection, but not the other way around by any means" (Swenson, personal communication). Natural selection provides on this account additional pathways for entropic dissipation. This conception exhibits a reductionistic tendency that downplays what is novel about specifically natural selection, and the explicitly biological objects on which it works, preferring instead to see natural selection as a manifestation of deeper, directional laws governing all systems that select, in the long run, in favor what is physically stable (Swenson 1995). (Charles Sanders Peirce might have favored this idea.)

A sixth niche is in some ways the opposite of the fifth. Some theorists have speculated that natural selection is in some ways the author, or at least the shaper, of self-organization. This view depends for its plausibility on treating natural selection and its products in sophisticated ways that do not take isolated and mechanical morphological items as paradigmatic of what natural selection can create. Natural selection is viewed instead the way Dobzhansky, Lewontin, and Levins view it. Its most challenging task is to produce a range of life-cycle traits that allow populations, or perhaps species and lineages, to respond to changing environments. Adaptability, on this view, is itself a paradigmatic adaptation. So construed, adaptability depends on maintaining a great deal of variation in populations. It also depends on a highly mobile genome that can respond to problems. The fact that chance, in the form of protein polymorphism, and self-organization, in the form of genomic responsiveness, are now empirically validated phenomena, and often work in tandem, adds considerable power to this view. In addition to balancing selection, or perhaps as an extended form of it, natural selection itself can be construed as favoring self-organized genetic arrays as a source of variation. This view runs considerably lower risks of undermining the autonomy of evolutionary biology than those we have most recently been considering. It is also quite Darwinian in spirit.

Wimsatt's models of the evolution of developmental programs illustrate this approach. His ideas about the "generative entrenchment" of traits in ontogeny resonate well with Kauffman's except for the fact that, whereas on Kauffman's view ontogenetic stability is first and foremost a product of self-organization, Wimsatt believes that selection stabilizes developmental programs by selecting for their self-organizing properties (Wimsatt 1986). His reasoning is that mutations and other changes occurring early in development have a greater chance of being harmful than later ones. There is selection pressure, then, for self-organized "locks" that keep ontogeny proceeding down Waddington's canalized valleys. In articulating this view, Wimsatt makes some use of newer dynamical resources. It is unclear, however, what he regards as the general null hypothesis against which evolutionary change occurs. He does not share

Kauffman's conception that self-organizing properties of genetic arrays are to be taken as the relevant background assumption.

At least one more niche seems possible. Self-organization and natural selection might be seen as mutually entailing, complementary aspects of a single process. Neither can get very far without the other in producing significant evolution. When Kauffman complements his stress on self-organization as constraining natural selection by arguing that selection acts to keep living systems at the crucial interface of order and chaos, the "edge of chaos," he seems to us to be moving into this niche (Kauffman 1991, 1993). In the important region of phase space between too much order and too much chaos, self-organization presents to natural selection entities that can play significant roles in evolutionary change. At the same time, selection maintains systems in precisely the region of phase space that affords them the greatest opportunity to evolve further. Similar ideas have been presented by John Campbell and by Christopher Wills, who argue that natural selection will favor traits that enhance the possibility of further evolution, and so reveal evolvability to be the greatest adaptation of all (Campbell 1987; Wills 1989).

Similar assumptions are at work, albeit in a more thermodynamic than dynamical mode, in a school of thought that has its roots in systems ecology. On their view of the matter, natural selection emerges from more basic forms of selection that favor autocatalytic cycles. In rewarding more effective and better informed autocatalysis, and so improved capture and utilization of resources, selection among such cycles results in the creation and maintenance of self-organizing networks of informed energy utilization and dissipation. Natural selection of the fit would have emerged from chemical selection for the energetically efficient with the appearance of informational macromolecules in which the information guiding such processes, and enhancing autocatalytic prowess, is stored and transmitted. Organisms will, on this account, be construed as informed patterns of thermodynamic flow. Those populations will be fittest that best enhance the autocatalytic behavior of the reward loops in which they participate. One advantage of this notion is that it makes it possible to contextualize natural selection to the wider array of processes in which it occurs, and to project a vision of ecological communities in which cooperation becomes as characteristic as competition, or indeed inseparably linked to it.

Attempts to use the self-organizational aspects of thermodynamics and dissipation to situate natural selection within a framework that regards ecological energy flow as paradigmatic have been made by various "ecological Darwinians" (Lotka 1922; Odum 1988; Johnson 1981, 1988, 1992; Ulanowicz 1986; Schneider 1988; Weber et al. 1989; Wicken 1987). We have counted ourselves among their number. There is reason to believe that theories falling into this niche are Darwinian enough to accord a central explanatory role in specifically biological matters to natural selection. That explanatory role is not predicated, however, on pushing other processes to the margins or downplaying their importance. On the contrary, the point of natural selection is to allow alternative processes to do as much work as they can precisely so that natural selection can be as effective as possible.

As we have been considering successively more dynamically enriched ways in which natural selection and self-organization might be related, we have at the same time been steadily moving toward a deeper appreciation of the pervasively probabilistic and statistical character of the world in which organisms, including ourselves, live. The autonomous power of statistical arrays to produce order hardly appears at all when self-organization is treated as just one more "force" to be added to those recognized by a "pluralistic" evolutionary synthesis. Indeed, the very notion of force, as it operates on the metaphorical field assumed by this conservative view, implies an inertial conception of natural order that requires something else to make it change. At the other end of our spectrum of seven positions, probabilistic thinking about large statistical arrays generates a vision of an autonomously created natural order in which natural selection is both an expected phenomenon and a powerful evolutionary agent.

This vision of nature is not accessible to dynamical models that Darwinism inherited from modern science's two previous revolutions: the mechanical revolution of the seventeenth and eighteenth centuries and the statistical revolution of the nineteenth and twentieth. It can come into view only when theoretical biology internalizes the nonlinear dynamics of complex systems. This turn may produce some sense of discontinuity between Darwinism's past and future. It requires, as Burian and Richardson have admitted, "a gestalt switch" (Burian and Richardson 1992). Yet it is entirely possible that appeal to complex dynamical models will ultimately undergird, rather than undermine, the continuity of the Darwinian tradition. That is because the nonlinear dynamics of complex systems is extending and deepening the same probability revolution that has allowed Darwinism to be a fecund research tradition throughout the twentieth century. Nonlinear dynamics is extending the probability revolution by severing dynamics from its last links to classical physics. It is thereby offering new explanatory resources to the Darwinian tradition that may well enable it to remain as robust in the future is it has been in the past.

A most general argument for this projection is as follows. Boltzmannian systems share with their Newtonian predecessors commitment to equilibrium thinking. External forces are required to make a system change its inertial condition of motion or rest. Change is from one equilibrium state to another. Boltzmannian and Newtonian systems differ, however, insofar as the former are, while the latter are not, probabilistic. Boltzmannian systems share their probabilistic character with self-organizing systems. Both have an inherent arrow of time. They show what will spontaneously happen to an array of elements with various degrees of connectivity as time passes. But Boltzmannian and nonlinear systems differ because the latter do not assume that initial differences always average out. On the contrary, in self-organizing, nonlinear systems, outliers can unpredictably initiate large-scale spontaneous reorderings as systems move toward new attractors. From this perspective, self-organization advances probabilistic thinking by cutting it loose from the axioms and techniques of linearization that still tied Boltzmann to Newton, and in consequence tied Fisher to Darwin. This is likely to be significant for the Darwinian tradition because living systems seem intuitively to involve chance and self-organization as well as selection. In this respect, what seems odd is not that the new dynamics can be used to model the evolution of living things but that the evolution of living things could ever have been thought to fit particularly well with earlier dynamics.

We can peer a bit more deeply into the vision of nature that this conceptual shift portends by considering the various ways in which the idea of probability itself has been conceived. Historically, there have been three conceptions of the concept of probability. Probabilities can, in the first instance, be regarded as subjective estimates rather than as objective facts. This is the so-called "ignorance interpretation" assumed by Darwin and Herschel, as well as by nearly everyone else in their era. The roots of the ignorance interpretation of chance go back to late antiquity, when the Academic Skeptics, whose views were revived in the early modern period and were canonized by Hume, thought that being guided in uncertain matters by probabilities ("probabilism") was a substitute for real knowledge. (The assumption was that what is truly known must be certain, since the objects of knowledge were presumed to be unchanging or Platonic forms. It was the inaccessibility of these objects in a changing, sensory world that gave rise to the idea that we must content ourselves with probabilities.) Alternatively, probabilities can be regarded, from a slightly more object-centered perspective, as the long-run frequencies with which certain kinds of event occur. These frequencies are revealed as the number of tries and the range of available data go up. This "frequentist" view suggested itself naturally as the collection of statistical information turned in the nineteenth century into what Hacking calls an "avalanche of numbers" (Hacking 1990). Finally, probabilities can be regarded as real properties inherent in natural and social processes. What this means can be seen by reflecting on the fact that if, after many tries, one side of a die comes up more often than the other, it must be because of the die's underlying construction. Generalized, the point is that the world and the processes that make it up are full of dispositional properties or "propensities" that provide objective foundations for probability theory itself. Indeed, from the point of view of a processive rather than a statically substantial ontology, propensities are weighted or conditional probabilities that are inherent features of changing situations or occasions rather than absolute properties of isolated things. They are context-dependent properties of relational processes (Popper 1990). Even so, they are as real as can be, just as the indeterminacy of quantum systems and the probabilistically distributed energy levels of entropy are as real as can be, and not just measures of ignorance or frequency (Popper 1990; Denbigh and Denbigh 1985). Peirce was perhaps the first philosopher to grasp that we live in such a world. Not coincidentally, he adopted a propensity view of probabilities, as well as a philosophically realistic view about what science can learn about the world (Hacking 1990).

All three of these interpretations coexist and contest with each other even today.[1] Nonetheless, a slow but steady shift toward objectivist accounts of probability, and from frequentist to propensity interpretations, has taken place. It is easy to see why. The world that Laplace's Demon was supposed to be able to see was deterministic and predictable down to the last atom. Any appeal to probabilities would concern only what beings with lesser minds than the Demon fail to know. In the nineteenth century, determinism eroded but did not disappear, for while physicists like Maxwell and Boltzmann applied probabilistic reasoning to large arrays, they still clung to "in-principle-if-not-in-practice" determinism. Frequency interpretations of probability fit well with this halfway house. The propensity interpretation of probability could not easily flourish, accordingly, under either Newtonian or Boltzmannian conceptual regimes. Indeed, propensities, dispositions, tendencies, and capacities were still associated with the medieval obscurantism of decadent scholasticism. "Occult qualities" or "faculties," like the "dormitive virtue" of sleeping pills or the "nutritive virtue" of bread, says one of the characters in Hume's *Dialogues on Natural Religion*, are nothing but a "subterfuge" and "disguised ignorance" (Hume 1779). Ever since the quantum revolution began to write indeterminacy into the very warp of things, however, it has become increasingly easy to recognize that we do indeed live in a "world of propensities," full to the brim with all sort of "capacities" (Popper 1990; Cartwright 1983). Popper, for his part, regards this "world of propensities" as the outcome of a major intellectual shift away from explanations framed in terms of an atomistic and mechanistic ontology. "The world is no longer a causal machine," he writes. "It now can be seen as a world of propensities, as an unfolding process of realizing possibilities and of unfolding new possibilities" (Popper 1990, 18–19)[2] . A very deep arrow of time runs through this world as new regions of phase space are explored by the amplification of symmetry-breaking fluctuations followed by selective retention. Nor does every part of phase space have an equal chance of being explored. For the sensitivity to initial conditions that sends complex systems off toward new attractors is most readily explicable in terms of subtle propensities written into the most basic features of the world. We can view the emerging sciences of complexity as extending, or even completing, the probability revolution because we now recognize that, in spite of what Einstein believed, God not only plays with dice, but the dice are loaded.[3]

It is consistent with this new spirit that even before the emergence of complex systems dynamics, the concept of fitness, as evolutionary biologists use the term, had been given a powerful "propensity interpretation" (Mills and Beatty 1979; Brandon 1978; Sober 1984a). On this view, fitness is an organism's propensity to reproduce at a given level, or its chance of doing do. Counting actual numbers may be a good way to estimate or provide evidence of fitness in this sense, just as flipping a coin a number of times is a good way to establish its propensity for coming up heads or tails. The concept of fitness itself, however, is not equivalent to the way it is measured. It lies instead in the underlying construction of the things that are fit, just as the dispositional fragility of glass rests on the properties of silicon at certain temperatures, and as the chance of a coin falling heads or tails rests on how it was struck. Expected levels of fitness (or of adaptedness, when the same concept is looked at from the point of view of the "engineering" design of fit organisms [Burian 1983; Brandon 1990]) supervene upon a wide, and usually unknown, variety of underlying properties (Sober 1984a; Rosenberg 1985).

What this means is that nondeterministic propensities are accessible only through statistical reasoning and refer to statistically distributed properties. Admittedly, there is nothing about the propensity interpretation of the concept of *fitness* that absolutely requires it to be articulated in terms of a propensity interpretation of *probability*. But the explanatory virtues of the propensity interpretation of fitness become most salient when they are coupled with a propensity interpretation of the very probabilities that undergird those virtues. By taking this view of the matter, we can see why the propensity interpretation of fitness, far from reverting to the explanatory emptiness that early modern natural philosophers saw in capacities, potentialities, dispositions, and propensities, is, on the contrary, a way of freeing the notion of fitness from that very kind of emptiness, and so of freeing it from the false charge of tautology or circularity (Mills and Beatty 1979; Brandon 1978; Sober 1984a).

Yet the propensities that underlie fitness, and that inform the notion of natural selection, point to a far wider range of propensities that underlie the evolution of living systems. In nonlinear systems that operate far from equilibrium, the sensitivity to initial conditions that drives complex systems toward new attractors reflects subtle propensities that are pervasively at work throughout the world. These lead to self-organization, to emergent hierarchical structuring, and to complex adaptive behavior. Contemporary evolutionary theorists should feel at home in such a world. After all, Darwinism never really fit comfortably within Newtonian background assumptions. The dynamics of evolving biological systems are simply not those of Newtonian systems (Dyke 1988). The shift to Boltzmannian background assumptions created a liberating explanatory space more congruent with evolutionary and selective phenomena. That space was fully exploited by the modern synthesis. The rise of the sciences of complexity and self-organization now promises an even more robust set of background assumptions that is harmonious with the kinds and degrees of complexity that are at work in the evolution of living systems.

A world in which sensitivity to initial conditions can take a system unpredictably into new attractors, and in which spontaneous self-organization does much of the work classically assigned to impinging forces, is a world that does not fit well with received views about how science works any more than it fits with older evolutionary theories. At the outset of this book, we registered our conviction that the hypothetical-deductive or "covering law" view of scientific explanation reflects a world in which every system is presumed to be simple. We want to reiterate that point now that we are at the end of our journey. The idea that from a scientific law and a set of initial conditions one can predict successive states of deterministic systems is mirrored at the philosophical level in the idea that laws are universal statements from which, in conjunction with minor premises stating initial conditions, a conclusion deductively follows. The idea that particular conditions and universal laws jointly yield deterministic predictions of the subsequent state of a system is, however, a conceptual artifact of the Laplacean worldview. It is already at odds with deterministic chaos, and considerably more at odds with those portions of dynamics in which nonlinearities generate chaotic attractors in nondeterministic systems (Dyke 1988; Depew 1986). Similarly, the notion that simple systems can be fitted smoothly into even simpler ones stands tacitly behind every reductionistic research program since Descartes. The new sciences of organized complexity deny that the world, except for relatively small, highly constrained portions of it, can generally be presumed to be like that. Crossing the complexity barrier, accordingly, calls for equally radical revisions in how scientific theories are to be analyzed and in how they explain when they are applied to problems (Kellert 1993). This is why in writing this book we have consistently spoken in terms of matches between models and phenomena rather than of deductions of facts from laws.

Our point is not that there are no simple systems to which hypothetical-deductive models might apply. There clearly are. They have been at the center of scientific and philosophical attention ever since Galileo. In a presumptively complex world, however, whatever simple systems there may be are held in place by stable boundary conditions. In fact, once it has been recognized that what counts as a good explanation, or as explanatory at all, is context sensitive, it becomes clear that finding and citing the boundary conditions that hold both simple and complex systems in place is often more explanatory than finding transformation rules that govern state changes within them (Garfinkel 1980). Against a background in which it is assumed that the world is composed of nested simple systems, one does not have to worry much about boundary conditions or sufficient parameters. Finding initial conditions and laws that govern deterministic transitions, and trying to reduce these laws to still more general and parsimonious ones seems to be what does the explaining. In a presumptively complex world, on the other hand, explanatory illumination depends more on finding at what level of scale a system is operating, and what, if anything, is maintaining the boundary conditions that led us to generate qualitative predictions (Kellert 1993; Dyke 1988). It is here that we will find what Levins calls the "sufficient parameters" that enable us to explain and predict the behavior of complex systems (Levins 1966, 1968; Wimsatt 1980). From this perspective, the deepest problem with what have traditionally been called "universal laws of nature" is not so much their status as lawlike but the universality that is far too often claimed for them. It is becoming increasingly compelling, in fact, to reconceive even "laws of nature" such as Newton's gravitational law, the second law of thermodynamics as applying only under highly robust boundary conditions within which a certain range of dynamic transformations can reliably be expected to occur (Dyke 1988).

There was, of course, good reason for reductionist ideals in the early days of the scientific revolution. In a culture as dominated by the otherworldly as ours was until fairly recently, science required a spare form of materialism if it was to acquire any useful naturalistic breathing room at all. In an earlier era, in fact, natural science unaccompanied by reductionism usually led straight to dualism, vitalism, and spiritualism. We have seen, for example, how Lyell drew back from natural selection toward creationism. In an even more spectacular case, Wallace's inability to make natural selection account for mental phenomena eventually led him to out-and-out spiritualism (Richards 1987). Even today, overenthusiastic devotees of the Gaia hypothesis exhibit some of these tendencies (Lovelock 1979). To people like Huxley or Haeckel, or, in our own time, Crick and Watson, a little reductionistic materialism does not seem too high a price to pay if that will keep such unedifying regressions to dualism or idealism at bay.

The problem has been that when everything is antecedently considered to be "nothing but" atoms in the void, many real, important, and interesting phenomena tend to get explained away, brushed aside, eliminated, or, worse, crammed into the wrong explanatory box. While reductionistic and mechanistic approaches may have made it possible to explain some

Natural Selection and Self-Organization

things, they probably made it harder to explain others. They also made it harder for the scientific worldview to be received with equanimity by other sectors of culture. Indeed, since the reducing impulse undermines fairly huge tracts of experience, people like Wallace, who feel deeply about protecting phenomena they regard as existentially important, frequently conclude that they have no alternative except to embrace spiritualism, and sometimes even to attack the scientific worldview itself, if that is the only way to protect important spheres of experience that have been ejected from science's confining Eden.

In response, scientists and philosophers who feel strongly about the liberating potential of a spare, materialistic worldview began to patrol the borderlands between the high-grade knowledge scientists have of natural systems and the low-grade opinions that, in the view of science's most ardent defenders, dominate other spheres of culture and lead back toward the superstitious and authoritarian world of yesteryear. "Demarcating" science from other, less cognitively worthwhile forms of understanding was already a major feature of Darwin's world. A line beyond which the Newtonian paradigm could not apply was drawn at the boundary between physics and biology. We have seen how hesitant Darwin was to cross that line and what happened when he did. Twentieth-century people are sometimes prone to congratulate themselves for being above these quaint Victorian battles. They may have less reason to do so, however, than they think, for the fact is that throughout our own century, the same sorts of battles, with emotional overtones no less charged, have been waged at the contested line where biology meets psychology, and more generally where the natural sciences confront the human sciences. Dualisms between spirit and matter, and even between mind and body, may have been pushed to the margins of respectable intellectual discourse. But methodological dualisms between what is covered by laws and what is to be "hermeneutically appropriated" are still very much at the center of our cultural, or rather "two cultural," life. Cognitive psychologists and neurophysiologists are even now busy reducing mind-states to brain-states, while interpretive or humanistic psychologists are proclaiming how meaningless the world would be if mind is nothing but brain. Interpretive anthropologists are filled with horror at what would disappear from the world if the rich cultural practices that seem to give meaning to our lives were to be shown to be little more than extremely sophisticated calculations on the part of self-interested genes. Conflicts of this sort would have given Darwin stomachaches almost as bad as the ones he endured over earlier demarcation controversies.

The rhetorical pattern of these battles is still depressingly similar, in fact, to Huxley's confrontation with Wilberforce. Hermeneuts ridicule scientists like Hamilton, Dawkins, and Wilson when they suggest that nothing was ever known about social cooperation until biologists discov-

ered kin selection. Reductionists in turn criticize hermeneuts, now transformed largely into "culturists," for bringing back ghosts and gods, just
as their nineteenth-century predecessors were taxed with being "vitalists" every time they said something about the complexity of development. Humanists identify scientists with an outdated materialist
reductionism. Scientists insist that hermeneutical intentionality is little
more than disguised religion.

Perhaps a way out of this fruitless dialectic between the "two cultures"
can be found if each party could give up at least one of its cherished
preconceptions. It would be a good thing, for example, if heirs of the
Enlightenment would stop thinking that if cultural phenomena are not
reduced to some sort of mechanism, religious authoritarian will immediately flood into the breach. They should also stop assuming that nothing
is really known about human beings until the spirit of scientific reductionism gets to work. Students of the human sciences have, after all, been
learning things alongside scientists ever since modernity began. Among
the things they have learned are that humans are individuated as persons
within the bonds of cultures and cultural roles, and that as recipients and
transmitters of cultural meanings, they are bound together with others
in ways no less meaningful and valuable than the ways promoted by
strongly dualistic religions. By the same token, it would be helpful if
advocates of the interpretive disciplines would abandon a tacit assumption sometimes found among them to the effect that nature is so constituted that it can never accommodate the rich and meaningful cultural
phenomena humanists are dedicated to protecting, and that therefore
cultural phenomena *ought never* to be allowed to slip comfortably into
naturalism. Humanists seem to have internalized this belief from their
reductionist enemies, whose commitment to materialism is generally
inseparable from their resolve to show up large parts of culture, especially religion, as illusions. These opponents, we may safely say, take in
each other's laundry.

Thoughtful biologists like Gould, Eldredge, Lewontin, and Levins
seem to us to have been working toward rapprochement along these
lines. They have appealed to the irreducible complexity and hierarchical
structure of the biological world, and the no less irreducible status of
organisms as agents in worlds on which they confer meaning and value,
in order to resist sociobiological calls to integrate the human and the
natural sciences around reductionistic, and perhaps overly competitive,
notions like genic selectionism. Still, what is perhaps most striking about
much of this work is the rather stark contrast it draws between narrative
explanations of what is unique, irreversible, complex, particular, or contingent, and what is to be explained as instantiating general laws. This
contrast, as we have noted before, was first propounded by people who
were trying to defend the autonomy of the human sciences, especially
history, and to resist reductionistic calls for their absorption into the

natural sciences. A crucial component of their argument was that explanation by subsumption under general laws cannot respect the uniqueness of individual events or the complexity of the ways they are related. Only narrative explanation can do that. It is surprising, accordingly, to see a view that is usually used to draw a boundary between the natural and the human sciences now being brought into the natural sciences themselves as a way of defending the autonomy of evolutionary biology and of explaining its unique capacity to confront the complex, the particular, and the contingent. This is indeed one way of crossing the complexity barrier, for it recognizes that living things are essentially complex and that our methods of studying them must respect that fact. We are not sanguine enough to believe, however, that the old dialectic will be ended merely by inscribing an essentially humanistic viewpoint within evolutionary biology. On the contrary, an ingression of the old enemy into the home team's territory is much more likely to set off another round of even more intense boundary disputes.

One point of this book is to suggest that evolutionary biology will help end fruitless oppositions between the human and the natural not when it goes narrativist but when, in producing its narratives, it begins to use complex dynamical models to provide theoretical backing for understanding the complexity, uniqueness, contingency, and irreversibility of living systems. What makes it possible for us to believe that we may be on the verge of just such a liberation is the fact that science's view of nature is not by a long shot as reductionistic, mechanistic, and simplistic as it once was. Indeed, for a long time the explanatory power of science has been expanding in direct proportion as it has moved away from the reductionistic biases that made science possible in the first place. What may be surprising to humanists is that the most powerful agent and medium in opening up nature and the natural to complexity has been the one science that has for so long been blamed for "Newton's sleep," namely, dynamics.

Even in Maxwell's and Boltzmann's time, the probability revolution was expanding what counted as natural, and hence scientific, to accommodate phenomena previously banished as mere higgledy-piggledy in the age of Laplacean science. It was on these statistical terms that it first became possible to connect social phenomena to the natural order. That is the ultimate significance of the controversy surrounding Quetelet, into which Darwin's ideas were swept up. We see the complexity revolution as a potentially huge advance in realizing this dream without undermining what is uniquely, and uniquely valuable, about human experience. As our ability to grasp the dynamics of nonlinear systems has proceeded apace, it has become possible, perhaps for the first time, to articulate conceptions of nature that are expansive enough to accommodate many of the distinctively value-laden experiences, practices, and institutions that humanists and hermeneuts have been trying to protect without dismissing, betraying, or thinning them out. We are even willing to say that in our time, the confidence, maturity, and promise of a science should be measured not by its power to reduce the complex to the simple, or to throw away most of the interesting phenomena into an eliminative dump, but instead by its willingness to use complexity to countenance and understand phenomena that earlier advocates of science failed to respect.

One small piece of evidence that scientific naturalism is expanding to deal with complexity, and to make rapprochement between the natural and human sciences possible, is that some very large and vague notions that were banned from the discourse of science at the dawn of modernity seem of late to have been making a major comeback. In the previous chapter, the concept of teleology did not seem as hopelessly vitalistic as mechanistic materialists once believed. In this chapter, dispositional properties and propensities, ridiculed by the first scientific revolution as medieval obscurantism, have proved useful in explicating the ontology on which our best theories about the world depend. "Hierarchy" is another such notion. Outlawed by mechanists as redolent of the great chain of being that supposedly stretched between God, angels, humans, animals, plants, and dead matter, "hierarchy talk" seems now to occasion no scandal. On the contrary, advocates of an expanded synthesis speak freely of a "biological hierarchy," in which natural selection operates at various levels and on various units. There are ecological hierarchies and genealogical hierarchies. The world is fairly bursting, in fact, with entities that emerge spontaneously at different levels of scale (Allen and Starr 1982; Salthe 1985, 1993).

There are some people who believe that the recrudescence of such notions portends the reenchantment of the world after a long and depressing bout of mechanism (Toulmin 1990). That is not our view of what postmodernity will be like. On the contrary, we believe that old ideas like teleology, hierarchy, and propensity, having irreversibly lost their earlier associations under the stern discipline of modernity, are now available for new uses, including the uses which the new dynamics can make of them, and may give us a world that is at once as many-storied as the medieval and as naturalistic as the modern.

These reflections suggest that it should be at least a mild constraint on any evolutionary theory that claims to explain human phenomena that it should throw light on, rather than eliminate or reduce away, the interactional, relational, intentional, and symbolic features that interpretive social scientists have already discovered about social reality. Perhaps it is not too much to say that what we need is an evolutionary theory worthy of our best social theory, not a social theory trimmed to fit a rapidly receding, overly simplistic, evolutionary theory. For our part, we look forward to an ecologically grounded evolutionary theory whose point is the protection of individuals, communities, and their traditions in a natural world that is our true and only home.

Reading Guide to Part III

The story of the discovery of the structure of DNA and the development of molecular biology has been told in detail in Horace Freeman Judson's *The Eighth Day of Creation*. Robert Olby's *The Path to the Double Helix* is a more focused history of the events leading to Watson's and Crick's work. Watson's memoir, *The Double Helix*, is a breezy, personal account. An antidote to Watson's chauvinism is Anne Sayre's *Rosalind Franklin and DNA*. Crick's highly reductionistic vision of nature is on display in his *Of Molecules and Men*.

The biochemical background to the molecular revolution is described in Fruton's *Molecules and Life*, as well as in Needham, *The Chemistry of Life*; Leicester, *Development of Biochemical Concepts*; and Kohler, *From Medical Chemistry to Biochemistry*. The rise of biochemistry at Cambridge has been chronicled by Needham and Baldwin, *Hopkins and Biochemstry*, and by Weatherall and Kamminga, *Dynamic Science*. For its development at Caltech, see Fischer and Lipson, *Thinking about Science*, and more recently in Lily Kay's *The Molecular Vision of Life: Caltech, the Rockefeller Foundation and the Rise of the New Biology*. Additional information can be found in Garland Allen's *Life Science in the Twentieth Century*, as well as Allen's biography, *Thomas Hunt Morgan*. Judson's discussion of Monod's and Jacob's work in *The Eighth Day of Creation* is a good introduction to the topic of regulatory genes. Jacob's *The Possible and the Actual*, offers a fascinating insight into its author's peculiarly French version of Darwinism. On molecular processes that violate Mendel's laws, see Dover "Molecular Drive"; Doolittle, "The Origin and Function of Intervening Sequences in DNA"; and Doolittle and Sapienza, "Selfish Genes, the Phenotypic Paradigm and Genomic Evolution." On Barbara McClintock, and on how her intimate way of working on corn genetics allowed her to discover transposable elements, or jumping genes, long before the molecular revolution backed her up, see Evelyn Fox Keller's *A Feeling for the Organism*. Good summaries of the molecular revolution in genetics can be found in Christopher Wills's *The Wisdom of the Genes* and *Exons, Introns, and Talking Genes*. More speculative treatments are found in

Campbell, "An Organizational Interpretation of Evolution" and "The New Gene and Its Evolution."

The most readable account of the neutral theory of molecular evolution is Gould's "Through a Lens Darkly." More cautious, and only slightly less readable, are Ayala's "The Theory of Evolution: Recent Successes" and Stebbin's and Ayala's "The Evolution of Darwinism." At a more technical level, the reader should consult Lewontin's *The Genetic Basis of Evolutionary Change* and Kimura's *The Neutral Theory of Molecular Evolution*. To get the flavor of the debate about neutralism and balancing selection, see the special 1987 issue of the *Journal of Molecular Evolution* devoted to the subject.

On the notion of genic, rather than organism-level, selection, see G. Williams, *Adaptation and Natural Selection*, and Dawkins, *The Selfish Gene*, 1st and 2nd eds. The shift from selection solely at the level of organisms opened the way to considering whether selection might conceivably act at levels above as well as below the organism. This notion was pioneered by Lewontin, "The Units of Selection," but was quickly taken up and developed by philosophers of biology who were interested in distinguishing the conceptual structure of modern Darwinism from the separate, empirical question of what sorts of things (genes? proteins? cell lineages? organisms? colonies? demes? species?) might be selected and might evolve. Hull's "A Matter of Individuality," "Individuality and Selection," and "Units of Evolution" are seminal pieces and did much to ground Gould's contention that "a new and general theory of evolution," based on multilevel selection, is "emerging." After reading Gould's manifesto, "Is a New and General Theory of Evolution Emerging?" turn to the critical reply by Stebbins and Ayala: "Is a New Evolutionary Synthesis Necessary?" Then read Gould's response, "Darwinism and the Expansion of Evolutionary Theory." Next, go on to Eldredge's *Unfinished Synthesis* in which he argues that the synthesis will be unfinished until it is expanded to include group and species selection, as well as organismic and genic selection. For group selection, see Wade "Group Selection among Laboratory Populations of *Tribolium*," "An Experimental Study of Group Selection," and "A Critical Review of the Models of Group Selection"; D. S. Wilson, "A Theory of Group Selection" and "The Group Selection Controversy"; and a number of essays in Brandon and Burian, *Genes, Organisms, Populations*. For the way in which an expanded synthesis, including species selection, can accommodate Gould's and Eldredge's thesis of punctuated equilibrium, see Eldredge and Gould, "Punctuated Equilibria"; Gould and Eldredge, "Punctuated Equilibrium Comes of Age"; Stanley, *Macroevolution*; Vrba and Eldredge, "Individuals, Hierarchies, and Processes"; and Vrba and Gould, "The Hierarchical Expansion of Sorting and Selection." For philosophical defenses of the very idea of an expanded synthesis, consult Wimsatt, "The Units of Selection and the Structure of the Multilevel Genome"; Sober, *The Nature of Selection;* and

Brandon, *Adaptation and Environment.* For a vigorous argument that a truly expanded synthesis will not be selectionist, see Salthe's *Evolving Hierarchical Systems.* Much of the controversy about expanding the synthesis is nicely summarized by Ruse in his *Philosophy of Biology Today.* The journal *Biology and Philosophy,* edited by Ruse, is a continuing source of important and interesting articles on these issues.

The best introduction of Lewontin's and Levins's way of respecting biological complexity is Lewontin's remarkable little *Biology as Ideology* (1992). The essays in Levins and Lewontin, *The Dialectical Biologist,* are worth careful study and in fact constitute a good introduction to the notion that biological reality is essentially complex and that biological analysis must, and can, respect that complexity by abandoning optimization thinking for what Herbert Simons calls "satisficing," by rejecting reductionistic ideals and looking instead for what Levins calls "robust theorems," and by watching out for ideological biases of our culture that run roughshod over good, and usually highly concrete, science. For Lewontin's quarrel with E. O. Wilson on these points, see Segerstrale, "Colleagues in Conflict."

That Lewontin's and Levin's vision of biological reality already constitutes something of a complexity revolution can be seen by comparing their views to those of Herbert Simon, *The Sciences of the Artificial;* Wimsatt, "Reductionist Research Strategies and Their Biases in the Units of Selection Controversy" and "Developmental Constraints, Generative Entrenchment, and The Innate-Acquired Distinction"; and, recently, Bechtel and Richardson, *Discovering Complexity.* What we are recommending in this book is that this line of work (and the notion of an hierarchically expanded synthesis) be more explicitly connected to nonlinear dynamics and chaos theory, and, more generally still, to the complexity revolution in dynamics that many people now believe is upon us. For an explicit defense of this proposal, read Charles Dyke, *The Evolutionary Dynamics of Complex Systems.* For the more general notion that science is crossing the complexity barrier, see Heinz Pagels's *The Dreams of Reason.* Chaos and nonlinear dynamics are summarized for the general reader in Gleick's well-known *Chaos: Making a New Science* and in Stewart's more informative *Does God Play Dice?* Books requiring a more serious mathematical bent but that are helpful are Devaney's *Chaotic Dynamical Systems;* Sandefur's *Discrete Dynamical Systems;* Ott's *Chaos in Dynamical Systems;* and Jackson's *Perspectives of Nonlinear Dynamics.* For the broader philosophical implications of chaotic dynamic systems, and an introduction to the field that can be digested appreciatively by philosophers, see Stephen Kellert, *In the Wake of Chaos.*

The complexity revolution has from the start been connected to the self-organizational properties of complex systems. These in turn have been linked to what Ilya Prigogine calls "dissipative systems," and thus to the notion that modern dynamics, and computational capacity, allow us to see that the second law of thermodynamics is not at cross purposes with complexification. Prigogine and Stengers, *Order Out of Chaos*, is the best introduction to this aspect of the case. Nicolis and Prigogine also provide a thermodynamic perspective on complexity in *Exploring Complexity*. A more rigorous thermodynamic treatment is given by Arthur Peacocke in *The Physical Chemistry of Biological Organization*. Wider implications of nonequilibrium thermodynamics for complex systems dynamics are found in Swenson, "Emergent Attractions and the Law of Maximum Entropy Production," *Spontaneous Order, Evolution and Natural Law*, and "End-Directed Physics and Evolutionary Ordering," and Swenson and Turvey, "Thermodynamic Reasons for Perception-Action Cycles." Some implications for art and the humanities (as well as good summaries of some of the science) are found in Hayles, *Chaos Bound*.

The expectability of life from nonlife in origin-of-life research is best defended by bearing closely in mind the role of thermodynamics. Whether thermodynamics does the work, however, or simply provides boundary conditions for kinetic pathways, is a disputed question. Compare, for example, Morowitz, *Beginnings of Cellular Life*, with Wicken, *Evolution, Information, and Thermodynamics*. The thermodynamic aspects of complexifying, and dissipative, ecologies are dealt with by Lotka, *Elements of Physical Biology* and "The Law of Evolution as a Maximal Principle"; E. Odum, "The Strategy of Ecosystem Development" and *Environment, Power and Society*; H. Odum, "Self-organization, Transformity, and Information"; Odum and Odum, *Energy Basis for Man and Nature*; and, most fully, Ulanowicz, *Growth and Development*. For the application of thermodynamic complexity to evolutionary theory explicitly, see the essays in Weber, Depew, and Smith, *Entropy, Information, and Evolution*. and, less technically, Ronald Fox's *Energy and the Evolution of Life*. One way of making the link among evolution, complexification, and thermodynamics has been proposed in Wicken's stimulating *Evolution, Information, and Thermodynamics*. Another nonequilibrium approach, based on turning the second law into a higher-order law of information dissipation, has been defended by Brooks and Wiley, *Evolution as Entropy*, 1st and 2nd eds. They have their critics. Morowitz, "Entropy and Nonsense," in particular, is fairly hard hitting. See also Hariri, Weber, and Olmsted, "On the Validity of Shannon." On the other hand, Brooks's and Wiley's program has articulate and knowledgeable defenders (Collier, "Entropy in Evolution"; Salthe, *Development and Evolution*). Salthe's *Development and Evolution*, in fact, is a summary of how various ideas in nonequilibrium thermodynamics add up to a modern defense of the developmentalist tradition in evolutionary theory. It puts thermodynamic fire, as it were, under the structural or "laws of form" view of evolutionary change that has been espoused most prominently by Brian Goodwin. (The best summary of this program to date is Goodwin and Saunders, *Theoretical Biology*. See also the essays in Ho and Fox, *Process and Meta-*

phors in Evolution.) The consequences of this view for macroevolutionary pattern, direction, and progress are subject to dispute but generally contrast fairly vividly with Darwinian approaches. That even Darwinians are taking a less Lyellian view than a few decades ago can be seen by consulting many of the papers in Nitecki's anthology, *Evolutionary Progress*. That part of this sea change is due to the explanatory resources of the expanded synthesis idea can be seen in Bonner's *The Evolution of Complexity*. But the contrast between, for example, Salthe's *Development and Evolution* and Gould's *Wonderful Life* is still quite stark. The contrast with Dawkins, "Progress," is uncompromising. One can only conclude that the old rivalry between the Darwinian and developmentalist traditions is still alive.

The exploration of how complex systems adapt at the edge of chaos has been recently summarized for the general reader by Waldrop's *Complexity*, which also provides an institutional history of the Santa Fe Institute, where much of this line of analysis has been developed. Roger Lewin's *Complexity* focuses more explicitly on biological issues. In both of these books, the figure of Stuart Kauffman looms large. The more ambitious and mathematically literate reader will want to tackle Kauffman's *The Origins of Order*, as well as the publications of the Santa Fe Institute, such as the volumes on artificial life (Langton, *Artificial Life*; Langton, Taylor, Farmer, and Rasmussen, *Artificial Life II*). Kauffman, "Antichaos and Adaptation," is a more accessible treatment of cellular autonoma but a bit too silent on the consequences for evolutionary theory.

The realistic view of probabilities as underlying propensities that we defend in the final chapter can be found in Popper's *A World of Propensities*. Ulanowicz, "The Propensities of Evolving Systems," makes the connection to evolving systems. Philosophers of physics who are making the shift from the primacy of laws, epistemologically considered, to the primary of real capacities, propensities, dispositions, and powers include Nancy Cartwright, whose *How the Laws of Physics Lie* and *Nature's Capacities and Their Measurement* are highly recommended. One way of reconciling a realistic view of the entities referred to in scientific theories with what Kuhn and others have said about the relativity of scientific theories has been expressed by Hacking in *Representing and Intervening*, which is also an accessible introduction to the central issues in contemporary philosophy of science. More detailed studies of how we learn more about real entities, such as genes, as they are treated in successive theories have been undertaken by Burian, "On Conceptual Change in Biology," and Kitcher, *Abusing Science*.

Notes

CHAPTER 1

1. The distinction between *concepts* and *conceptions* is used by contemporary philosophers, notably John Rawls (1970). The concept of happiness, for example, can be conceptualized in terms of pleasure, yielding an Epicurean conception of happiness as pleasure; or as "activity in accord with virtue," as Aristotle says, or indeed as "a warm puppy." Similarly, Rawls has a conception of justice conceived in terms of fairness (Rawls 1970).

2. The heritability of an advantageous trait is what turns the environmental selection of variants in a single generation into the transgenerational process of natural selection, and over longer periods of time into (part of the explanation of) the evolution of species. So deeply is adaptation tied to what natural selection builds in this way that some contemporary Darwinians refuse to call a trait that has been built for one purpose but subsequently put to another an adaptation. Stephen Jay Gould has proposed calling these hijacked traits "exaptations," and the general category that covers useful traits without specific reference to their origin "aptations" (Gould and Verba 1982).

3. The term *chance* in "chance variation" does not mean that variations are uncaused. It means that the causes of variation, whatever they are, operate independently of what an organism needs. Thus, the traits whose fates Darwin tracks are not there because organisms recognize their utility, hatch them up, and pass them on. Organisms either have the right traits or they do not, and it is the environment rather than the organism itself that decides that.

A good way to grasp the role of chance variation in natural selection is to compare it with ideas that it resembles but from which it also crucially differs. It is a matter of *mere chance* whether lightning will hit and kill one member of a hiking party rather than another. Lightning presumably does not discriminate the traits of individuals, at least in a world where Jupiter Tonans or Wotan are presumed not to be at work. Darwin *denies* that what traits survive in a certain competitive environment is a matter of chance in this sense. That is not natural selection because it is not selection at all except in a very weak sense. Those traits that survive a genuine selective process do so for a reason.

The Darwinian notion of adaptation is not, moreover, a matter of what an organism does *to adapt*. "Adaptation," as it is used by Darwinians, does not mean "adaptability," the capacity to adjust to changing circumstances (although that capacity may itself be *an* adaptation), nor is the term used to describe a characteristic that just happens to perform some function or serve some purpose, although it had a completely different origin.

In addition to ambiguities about the role of chance, there are also differences between Darwin's notion of "the survival of the fittest," as it came to be called, and closely related conceptions. Those of us who are from California have heard since childhood the grisly story of the Donner party, whose wagon train was stranded in a blizzard in the High Sierra in gold rush days. There clearly was competition for scarce resources. Cannibalism broke out. There were differences between people, some of whom possessed traits capable of meeting these challenges better than others. In some sense, it was the "fit" who survived. Those who had the most moral scruples, for example, a trait that in palmier days and different environments would have been advantageous (in securing leadership roles, for example), now found themselves at a distinct disadvantage in this grimmer and changed set of circumstances. The same traits, it seems, may be differentially useful in different competitive environments. There is a reason why one person survived and another did not. Nonetheless, in spite of confusions on this point introduced by social Darwinism, the kind of competition and selection that took place among the Donner party does *not* illustrate natural selection or what Darwin meant when he talked about the "survival of the fittest," a phrase that he was prevailed upon by others to use in later editions of the *On the Origin of Species*. What happened to the Donner party occurred only at one time in the life of one generation. Those who survived did not continue to operate under these conditions, nor was the trait that enabled them to do so related to their general ability to have and rear human offspring. Nor was it necessarily heritable. "Survival of the fittest," as Darwin uses the phrase, *means* relative success in living long enough to have viable and fertile offspring.

4. Mayr (1988, 1991) is right to assert that Darwin's Darwinism involves more than one theory. Here we count three: a theory of adaptation, a theory of transmutation (speciation), and a theory of phylogenesis (descent of species from a common ancestor). Mayr counts five, adding Darwin's philosophical naturalism and his commitment to causal pluralism, according to which Darwin admits evolutionary causes other than natural selection in secondary roles, especially the inheritance of acquired characteristics ("Lamarckism") and sexual selection. In addition, Darwin had theories about inheritance ("pangenesis.").

5. In this connection, we agree wholeheartedly with the following: "It is surely arbitrary, as well as half-hearted, for a sociology of knowledge to presuppose that scientific theories are conditioned by their social context principally insofar as they are indebted to overtly social theories. More consistent and confident presuppositions for a whole-hearted sociological historiography would . . . take physical science sources too as suitable subjects for social condition and social construction" (Hodge 1987, 247). The influence of Newtonian thinking on Darwin is a case in point.

6. Schweber (1979) argues that Darwin himself was already party to the probability revolution. While we think that Darwin's theory fairly cries out for statistical treatment, we argue that the required treatment was provided first by Galton, Pearson, and Weldon and was completed when Fisher, Wright, and Haldane gave a statistical account of gene frequency changes. See chapter 6 for Schweber's argument. Chapters 8 through 13 deal with this sequence of ideas.

7. It appears, we hasten to note, in phase or state space, and not in three-dimensional space. In phase space, each dimension corresponds to a single variable of a system. Successive points in phase space give a "phase portrait" of how a system evolves over time. Cf. chapter 16.

8. "Chaos theory bears a certain resemblance to statistical science of physical, biological, and social systems. The invention of techniques for statistical analysis revealed orderly patterns . . . in the apparently random behavior of heated gases, animal populations, and undeliverable letters. . . . But while statistical techniques analyse averages over a large ensemble of systems, the techniques of nonlinear dynamics work on single systems or families of related systems. The "order" found in a system with chaotic dynamics has little in common within the "orderly" distribution of molecular velocities in a gas, for instance" (Kellert 1993, 84). Kellert stresses that nonlinearity does not imply giving up on predictions in science, even though it does mean giving up on the *determinism* that has so far undergirded our conception of prediction. Quantitative predictions based on the temporal in-

variance of laws and their deductive hold on cases are indeed ruled out. But *qualitative* predictions, based on modeling the parameter and boundary conditions within which a system will stay stable or change, are enhanced by nonlinear dynamics, particularly by their fractal properties.

9. Historians of science now recognize that Darwinism could not initially triumph over developmentalist theories of evolution in part because even Darwin still had one foot in the developmentalist views about inheritance (Hodge 1985; Bowler 1988). That gave evolutionists after Darwin two choices: devise theories of evolution that would pull Darwin's other foot back into developmentalism by envisioning phylogeny as ontogeny writ large, making natural selection at best a subordinate part of evolutionary causality, or maintain the Darwinian stress on external causation and natural selection by finding a new, nondevelopmentalist theory about how information is passed from generation to generation. The first alternative dominated evolutionary theory in the later nineteenth century. The latter course was not possible until the emergence of a new theory of heredity, conventionally called "Mendelian," which broke this link with the past. The severing of transgenerational information transmission from development and the consolidation of the probability revolution jointly made possible the emergence of modern genetic Darwinism. Even then, large chunks of the old developmentalist paradigm, including developmental biology and ecology, as we will see in chapter 15, were never fully integrated into the modern synthesis.

10. Another is the philosopher of biology William Wimsatt, whose use of complex models to explicate development is driven by long-standing objections to reductionistic biases in evolutionary biology. See Wimsatt (1980, 1986), Wimsatt and Schank (1988).

11. Kauffman's work presupposes what has been learned about deterministic chaos but does not hold that deterministic chaos is by itself a cause of evolution. Rather, in freeing dynamics of linearizing assumptions, the discovery of chaos in the phase space trajectory of deterministic equations allows us recognize that large arrays can, under the right conditions, *spontaneously* produce the kind of order that Kauffman calls "antichaos." Order, that is, does not have to be explained in terms of an external agent or force.

12. Among contemporary writers on evolutionary biology, none is quite so Kuhnian in spirit as Stephen Jay Gould. Gould takes a highly personal view about how individuals and research communities woven together by personal ties interpret their world. Individuals and groups are very sensitive to the cultural milieu in which they live. Gould's work on later nineteenth-century evolutionists, *Ontogeny and Phylogeny* (1977), and especially his more recent study of two generations of interpreters of the Cambian fossils found in the Burgess shale, *Wonderful Life* (1988), illustrate these Kuhnian tendencies.

13. Readers should not be misled by the title of this book into thinking that we are accounting for the evolution of the Darwinian tradition by applying the model of natural selection to selection of "ideas," or what Richard Dawkins calls "memes" (as opposed to genes, Dawkins 1976, 1989). There are books and papers that do that (Hull 1988; Campbell 1988). Ours, however, is not among them. Given our holistic and hermeneutical view of cultural change, we doubt whether cultural units stay put long enough to be objects of this kind of selection. Nor does Darwinism evolve on our account by "culture red in tooth and claw," that is, intense competition among scientists. It evolves as other cultural phenomena do. It follows that we do not posit or seek a clean demarcation line between science and other forms of inquiry, as, for example, Popper does (Popper, 1959). We think that that line is always being contested in local contexts.

14. In *Progress and Its Problems* (1977), Laudan seems to argue that research traditions are defined and bounded by their ontologies. When the ontology goes down, so does the tradition. In *Science and Values* (1984) Laudan seems to soften this view. The "reticular" theory of scientific change defended there holds that a tradition can maintain itself by

changing any of its elements, including, presumably, its ontology. We are sympathetic to the second account.

15. The example was first mentioned by Ernan McMullin.

16. The "semantic view of theories," according to which theories are definitions of kinds of systems, itself reflects the deepening influence of probability revolution on conceptual, as well as empirical, issues. Degrees of isomorphism between model and phenomena are judged by statistical and probabalistic methods.

17. In fact, Occam's razor, which bids us to prefer, *ceteris paribus*, the simpler of two hypotheses on the ground that we should not "multiply entities unnecessarily," was first advanced by a theologian trying as hard as he could to make the world conform better to what Christian theology required than Aristotelians like Thomas Aquinas. The entities in question were, in this original contest, categories rather than scientific theories. Aristotle admitted ten categories. Occam reduced them to three. Galileo and Descartes completed this process by reducing the categories to two: substance and quantity. Substance and quantity then became the pared-down ontology of modern science. Cf. Grant (1971).

CHAPTER 2

1. Mayr's (1982) well-known contrast between "typological essentialism" and "population thinking" contains a double conflation and a double inadequacy. It fails to recognize that there is a difference between the probabilistic thinking of Galton, Fisher, and Wright, and whatever approximation to it Darwin could achieve in a prestatistical framework, and also that not all essentialisms are typological. Characterizing all essentialisms as typological, and hence crypto-Platonic, may make it rhetorically easy to dismiss essentialism altogether but makes it difficult to see that Aristotelian or constitutive essentialism, far from being ripe for collapse, presented evolutionists with vigorous opposition at least until the middle of the nineteenth century, and so that Darwin's version of evolution was intended to address Aristotelians like Owen.

2. Constitutive essentialism is consistent with the idea that what makes something what it is its genealogy. Suppose that a being with all and only the marks of a horse could be hatched up from off-the-shelf chemicals and a recipe of some kind. Aristotle might be able to agree with David Hull that such a being is not really a horse, because it does not have the right genealogy. This is a good way of reinforcing the point that typological essentialism is not the same as constitutive essentialism and that Aristotle, unlike Plato, is not a typological essentialist. Cf. Furth (1988) for a defense of this view.

3. Sloan argues that Buffon's realism shows that he cannot have been a follower of Locke but of Leibniz (Sloan 1976). We agree but note that, in this respect as in many others, Leibniz was, and tried to be, a modern Aristotle.

4. In this respect the *Naturphilosophen*, along with other absolute idealists like Hegel, were expressing their discontent with Immanuel Kant's merely "critical" idealism, according to which Aristotle's old teleological maxim was to be used purely as a "regulative principle" or heuristic device for discovering the mechanical means by which functions and adaptations are carried out. We shall have more to say about Kant's philosophy of biology in chapters 4 and 7.

5. Hall, for example, claims that "von Baer developed the theory of recapitulation" (Hall 1990, 21), while Richards speaks of "von Baer's critique of recapitulation theory" (Richards 1992, 5). The issue can be at least semantically resolved by speaking, as we do in this book, of "strong" and "weak" recapitulationism and using the unmarked term *recapitulationism* to refer to the strong form. Strong recapitulationists believe that organisms pass through the adult stages of definitionally lower forms (for antievolutionists of this persuasion) or of

ancestors (for evolutionists). Weak recapitulationists hold that they pass only through undeveloped stages of (evolutionarily) ancestral or (definitionally) lower forms. There is a historical correlation (through idealism) between typological thinking and strong recapitulationism and an equally strong correlation (through Aristotelian developmentalism) between constitutive essentialism and weak recapitulationism. The claim in this paragraph is that Owen was substantially in the latter tradition but rhetorically in the former. Richards fails to keep all these distinctions in mind when he seemingly argues from the fact that Darwin criticized the antitransformist *implications* that von Baer and Owen drew from their weak recapitulationism to the conclusion that Darwin himself must have been a strong recapitulationist, and even a bit of a typologist in his talk about "archtypes" (Richards 1992). Darwin was in fact an evolutionary weak recapitulationist.

6. Sloan (1986) sees a certain vitalism in Darwin as well.

CHAPTER 3

1. The issues discussed in this paragraph are tangled. We do not want to take sides too firmly about the extent to which Darwin was a recapitulationist, except to assert that this element of Darwin's thought has been systematically underestimated by modern Darwinians until nearly the present; that whatever version of recapitulationism Darwin held was guided or constrained by his generally von Baerian views about ontogeny; and that Richards' proclamation that "the rib around which Darwinism fleshed his idea of progressive development was the principle of recapitulation" is tilted too far in the direction of strong recapitulationism (Richards 1992, 114).

2. This was a different use of the argument from that made by Malthus himself, at least when he first published the idea. For the dour parson, the law was first used to describe an economic mechanism for the Fall of Man and Original Sin, devised in a time of utopian revolutionary hopes to confute the secular perfectionism of William Godwin. Later revisions and reissues of the *Essay on Population* were less downbeat, regarding the law as stating what would happen unless something stopped it. This shift prepared the way for the distinctly upbeat interpretation of the *Westminster Review* circle, according to which Malthusian pressure motivates people to behave in a way that increases general utility.

3. The drive of organisms to maximize reproduction can itself be explained in terms of natural selection. Early life-forms that lacked this trait would not survive at all. Those that have survived have done so by acquiring competitive, and hence, in a Malthusian world, adaptive traits.

4. Why, we may well ask, did Malthus and Lyell, when yoked together like the major and minor premises of a syllogism, seem to lead to conclusions that both of these eminent thinkers would have, and actually, did resist? The answer, as Mayr says, is that both were essentialists. For both Malthus and Lyell there were limits to individual variations within kinds. Hence external pressures will lead to extinction rather than to transmutation. Darwin and Wallace both hit on indefinite variation and subsequent selection as a way out, although in slightly different ways.

5. The argument can also be schematized in terms of a famous flowchart produced by Ernst Mayr (Mayr 1977):

Malthusian superfecundity + Observed steady state of populations + limitations on resources = Struggle for existence

Struggle for existence + Individual differences + Heritability = differential survival or natural selection.

Natural selection + Many generations = Evolution.

CHAPTER 4

1. "The mere internal form of a mere blade of grass is sufficient to show that for our faculty of judgment, its origin is possible only according to the rule of purposes. . . . We can say boldly . . . that it is absurd for men to . . . hope that another Newton will arise in the future who shall make comprehensible to us the production of a blade of grass according to natural laws that no design has ordered" (Kant, *Critique of Judgment*, sec. 67). It is significant that Kant, unlike Paley, coupled this proclamation with an assertion that organisms cannot be regarded as mechanisms or machines (sec. 65). This made his view more coherent than Paley's.

2. Kant's German disciples postulated irreducible vital forces as a result of imbibing these lessons. Cf. Lenoir (1982).

CHAPTER 5

1. For some cautions against accepting this analogy, cf. Hodge (1987).

2. It is here that the socialist Wallace and, as we will see in chapter 11, Darwin's Russian disciples seem to have drawn back a bit, stressing the cooperative struggle of groups against the common external environment. To the extent that they do so, they are, as they often admit themselves, less Malthusian than Darwin (Kottler 1985; Todes 1989).

3. If there is anything wrong with Schweber's (1977) reconstruction of Darwin's early thought, it is his tendency to load too much of the individualist ontology of natural selection onto Darwin prior to his encounter with Malthus. What we take to be the slow absorption of the ontology of political economy into the theory of natural selection Schweber takes to be the framework that leads Darwin to Malthus and guides his reading. For a corrective, cf. Hodge and Kohn (1985, 192–93).

4. For more on what Darwin did and did not think about species, and on how far the views of modern Darwinians differ from his, see chapter 12.

CHAPTER 6

1. The proclamation of conservative prime minister Benjamin Disraeli, that he was "on the side of the angels," may have echoed Huxley's remark. Since then this expression has entered into the language.

2. Boltzmann said that the nineteenth was "Darwin's Century" in an address to the (Austrian) Imperial Academy of Science, May 29, 1886. The address, entitled "The Second Law of Thermodynamics," was published in L. Boltzmann, *Populäre Schriften* (Leipzig: J. A. Barth, 1905), and has been translated by Brian McGuiness in *Ludwig Boltzmann: Theoretical Physics and Philosophical Problems* (Boston: Dordrecht, Reidel, 1979).

3. Peirce, who in his work measuring gravity for the U.S. Coast and Geodetic Survey was intimately acquainted with the use of statistics to measure errors, as well as with recent developments in statistical mechanics, proposed an evolutionary theory that depended on the self-ordering properties of statistical arrays to drive evolution. On Peirce and statistics, cf. Hacking (1990, 200–15). Peirce also ascribed this view—wrongly in our opinion—to Darwin:

Mr. Darwin proposed to apply the statistical method of biology. The same thing has been done in a widely different branch of science, the theory of gases. Though unable to say what the movement of any particular molecule of gas would be on a certain hypothesis regarding the constitution of this class of bodies, Clausius and Maxwell were yet able, by the application of the doctrine of probabilities to predict that in the long run such and such a proportion of the molecules would, under given circumstances, acquire such and such velocities; that there would take place, every second, such and such a number of collisions, etc.; and from these propositions they were able to deduce certain properties of gases, especially in regard to their heat relations. In like manner, Darwin, while unable to say what the operation of variation and natural selection in every individual case will be, demonstrates that in the long run they will adapt animals to their circumstances. ("The Fixation of Belief," *Popular Science Monthly* 12 (1877), 3.)

Since this is substantially the view argued by Schweber, we propose to call it the Schweber-Peirce hypothesis. Note that on Peirce's view selection is less important than statistical sorting, and in any case cannot be brought to bear on particular cases. This in our opinion contrasts with Darwin's reasoning.

CHAPTER 7

1. Whatever matters of detail need to be amended, we owe it to Lenoir (1982) to have successfully disentangled the various strata that have been assimilated into the contemporary, and largely useless, term *vitalism*.

2. The tendency of Russian intellectuals to exaggerate and personalize these issues shows itself in the enthusiasm of young self-proclaimed Russian nihilists in the 1860s for mechanistic reductionism. It was a weapon that could be thrown in the face of their authoritarian elders. This spirit is beautifully and parsimoniously captured by Ivan Turgenev in his classic novel, *Fathers and Sons*. In 1858, an earnest young graduate, Arkady, just out of college, brings home to his provincial gentry family a brilliant, arrogant schoolmate, Bazarov, a medical student who promptly identifies himself as a nihilist, one who believes in nothing (from Latin *nihil*, "nothing"). It turns out that what Bazarov opposes is "romanticism" and that in fact he believes in quite a lot, especially the competence of scientific materialism to remake society completely if power is given to, or seized by, scientists. In the following passages Bazarov has a strained conversation with Pavel Kirsanov, Arkady's bachelor uncle, who, as a deeply convinced Romantic organicist, opposes the liberalization of Russia, and especially the liberation of the serfs, because he believes the bulk of its population is inherently, and rightly, dependent. "They hold tradition sacred," he says. "They are a patriarchal people, who cannot live without faith." Pavel takes Bazarov to be a cold, ruthless harbinger of the tough-minded sort who in 1917 actually took the country over and bashed heads in pursuit of a rationally managed society. Here is a sample of their conversation:

"Physics is your special subject, is it not?" asked Pavel.

"Physics, yes, and natural science in general."

"I am told that the Teutons have made great strides in that department lately."

"Yes, the Germans are our masters there," Bazarov replied.

"For my part I plead guilty to holding no brief for Germans. . . . Once upon a time there were a few Germans here and there—Schiller, for instance, and Goethe. But nowadays they only seem to churn out chemists and materialists."

"A decent chemist is twenty times more useful than any poet," interrupted Bazarov (Turgenev 1862 [1965], 96–97).

Arkady's good-hearted but weak-headed father, Nikolai, intervenes in this strained conversation by remarking, "I have heard that Liebig has made some astonishing discoveries having to do with improving the soil. Perhaps you can help me in my agricultural labors by giving me some useful advice.

"I am at your service, Nikolai Petrovich. But Liebig is miles above our heads. One must learn the alphabet before beginning to read, and we don't know the first letter yet" (Turgenev 1862 [1965], 98).

Here is a sudden revelation of the sense of inferiority that haunted the Russian intelligentsia. Here too is testimony to the wide influence of German science, under its materialist dispensation, which had excited Bazarov's generation and annoyed their Romantic elders. Russia in the 1850s and 1860s was full of young men like this. What they had in common was a belief in science as liberation and in strict reductionistic materialism, the view that organisms are "nothing but" arrangements of matter and energy, that minds are "nothing but" brains, and that all transcendent sources of meaning and ethical value are Romantic illusions. At one point in the book, Arkady, prompted by Bazarov, takes a book of Pushkin's poems from his father's hands and replaces it with Ludwig Buchner's popular materialist tract, *Energy and Matter*. Bazarov spends most of his time at the farm experimenting on frogs. "He has no faith in principles," remarks Pavel testily, "only in frogs." Bazarov says to the puzzled peasant who finds the frogs for him, "I shall cut the frog open to see what goes on inside him, and then, since you and I are much the same as frogs except that we walk about on our hind legs, I shall know what's going on inside us to." The peasant is suspicious. He is just gentry, the peasant says of Bazarov to himself. "And everyone knows they don't have much sense." (Turgenev 1862 [1965], 90)

3. The culmination of classical pragmatism, and its connection to Progressive politics and to Darwinism, can be found in the work of John Dewey. Cf. chapter 14 for more on this research program and its opposition to social Darwinism.

4. We will note the effect of this background on Sewall Wright in chapter 12.

5. An interesting, if belated, case in point is the directed evolutionism of the Jesuit paleontologist Pierre Teilhard de Chardin (Teilhard de Chardin 1959). Teilhard's theory, which has more than a little Bergsonian philosophy behind it, enjoyed great vogue among liberal American Catholics in the 1960s, just when large numbers of Catholics were being assimilated to the American middle class mainstream, and when their church was undergoing a profound internal upheaval. It is difficult not to think that Teilhard's teleological evolutionism provided educated American Catholics of that time with the same kind of spiritual balm that similar theories had offered to liberal Protestants, whose spiritual stuggles were similar, half a century earlier.

6. Leo Buss has shown that Weismann greatly exaggerated sequestation of the germ line. In a large array of plant and animal taxa, nothing of the sort happens (Buss 1987).

7. Gayon (1992) makes much of this point, arguing that more than anything else it was the inadequacy of Darwin's theory of inheritance that caused his *hypothesis* of natural selection to go into decline. When an adequate theory of inheritance became available (roughly a Mendelian version of hard inheritance), Darwin's hypothesis could finally be transformed into an adequate *theory* of natural selection, and of evolution by natural selection. This thesis accords fairly well with Mayr's reconstruction (1982) and Hodge's (1985) insistence that what held Darwin back was his residual fidelity to the long epigenetic tradition in which inheritance was seen as part of a theory of growth. Weismann's crucial contribution was to sever questions about somatic growth from questions about inheritance.

8. We say "allegedly" because, as Gould shows, the antlers of the Irish elk are not in fact outsized but allometrically in scale with general body growth (Gould 1977).

CHAPTER 8

1. We owe these textual comparisons to James Lennox.

2. Wallace was interestingly caught in the middle. He had never been a causal pluralist. He was a strong selectionist. However, his increasing despair of getting psychological traits

and social progress out of selection made a dualist of him in the end. He ascribed moral behavior to spiritual forces. On Spencer, Romanes, and Wallace, see Richards (1987).

3. This remark is intended as a corrective to Provine (1971), who, in a paradigmatic display of "internalist history of science," carefully screens off these theoretical issues from the social and ideological issues that surrounded and informed them.

4. There is a good deal of overlap between Comte's phenomenalist positivism and Mill's. Mill was a self-defined importer of French thought into England. Mill, however, did not countenance the French idea that a social science would be a "sociology," based on social laws that are distinct from and irreducible to the psychological laws that govern the acts of individuals, such as those to which the rational actions of economic agents conform. This difference between British and continental thought has deep roots and persists to this day.

CHAPTER 9

1. This is often referred to as a particulate theory of inheritance. We sometimes employ this term but warn readers that it can be misleading. Darwin and Weismann thought of heredity as coming in little particles, which Darwin called "gemmules," changing proportions of which are blended in offspring. They did not, however, think of these as subject to a combinatorial logic, as Mendel and Mendelians did. The combinatorial particles of the early Mendelians, on the other hand, were fairly abstract entities. Until Morgan's work, the idea was seldom broached that "genes" might in fact be physical entities.

2. It is commonly claimed that Darwin had a copy of Mendel's paper but that it lay uncut and unread on his shelf. It is usually added that Darwin hated to read German. It appears to be a myth that Darwin owned Mendel's paper. Even more mythical is the assumption that Darwin would have seen that in Mendel's theory of inheritance something that would complement natural selection better than his own blending inheritance. Darwin, as we have seen in part I, had his own ideas about these matters. He was not waiting around for genetic Darwinism to justify him.

3. See Bowler (1990) for a review of the literature.

4. Conceptions of what a gene is would continue to change both before and after the molecular revolution. Philosophers have written informatively about how referential stability was preserved through all these changing conceptions. See in particular Kitcher (1982b) and especially Burian (1985).

5. It is because of this that Sober calls the modern synthesis not only an equilibrium theory but *for that reason* a Newtonian theory, in which forces act on an inertial state of a system. Without denying the general truth of what Sober says, we distinguish between statistical and nonstatistical equilibrium theories, reserving the term *Newtonian* for the latter, for reasons set forth in part I of this book (Sober 1984a).

6. Haldane looked at data on the frequency of dark, or melanic, and light forms of *Biston betularia* that had been collected, mostly by botanizing parsons, since the eighteenth century. Over the course of the century, the relative proportions had shifted from 1 percent to 90 percent in some areas. These shifts correlate with the effects of industrial pollution. To this observation Haldane added experiment: 154 melanic (*DD* and *Dd*) and 64 light phenotypes (*dd*) were released near Birmingham. A few days later 82 (53 percent) dark and 16 (25 percent) light phenotypes were recaptured. Assuming that the recapture was unbiased, we can calculate the relative fitness (*w*) of the dark and light forms as $53/53 = 1.00$ and $25/53 = .47$ respectively. We now define a parameter, s, to measure the selection pressure against light forms: $s = 1 - w$. In this case the selection pressure against light forms is .53.

Further calculations will show that it would take nearly twenty generations (not long by human time scales) to obtain a 10 percent frequency of light moths. It would take another ninety generations, however, to get dd down to .01 percent of the total population, even if it is assumed that *dd* is lethal. This shows that it is very hard to purge even lethal recessives from a gene pool. Equilibrium will almost always be reached before then, and the recessive form will almost always be around to reassert itself in the environment, and hence to respond to new selection pressure.

CHAPTER 10

1. Gene linkage results when genes fail to abide by Mendel's law of independent segregation. Morgan showed that linkage is proportional to distance on the chromosome. Epistasis is the synergistic effect of two or more gene loci on a single phenotypic trait. (In this case the trait is said to be polygenic.) Pleiotropy is the counterpart of epistasis. It occurs when a single gene locus affects more than one phenotypic trait.

2. Determinism here, and unless noted henceforth, refers directly to the idea that in *deterministic equations* the state of a system at any one time is uniquely determined by its state at an immediately prior time, together with the relevant laws governing state transformations. Laplacean determinism adds the thought that if a deterministic equation applies to one instantaneous point in time, then all prior and subsequent states of the system are determined, and predictable, as well. That is what Laplace's Demon is supposed to be able to calculate. This generalization overlaps with ontological theses, and hence with philosophical or metaphysical determinism, and with controversies about the possibility of free will. For a useful typology of these ascending degrees of determinism, see Kellert (1993, 49–67).

   Additive, as we use it in this passage, and as Fisher uses it, names another property of deterministic sytems. It means that the solution of two equations dictates the solution of a third, and so forth, and therefore, when applied to actual systems, there is a constant proportion between causes and effects, or more generally between inputs and outputs. As applied to genetic variation, Fisher means to imply that the more of an allele there is in a population, the more there will be of the trait it codes for. For important failures of additivity in evolutionary dynamics, see chapter 14.

3. See chapter 6, note 2 above.

4. Bear in mind that after World War II, an expanded middle class, incorporating much of the working class, was in fact subsidized in this way in the United States. Birthrates in the 1950s soared even as opportunities for those children increased. That is what is meant by the baby boom. It is the looming collapse of those subsidies that has produced a new crisis in American politics as the century comes to close, and promises to do so for the foreseeable future.

5. The spread of the probability revolution from physics to evolutionary biology reached a point of intense self-consciousness in Fisher's assimilation of his research program to Maxwell's and Boltzmann's statistical models of dynamics. Similar attempts to reconceptualize classical economics had been taking place for some time. In the later nineteenth century, Francis Edgeworth had speculated that Maxwell's and Boltzmann's new physics might strengthen the tie between the principles of political economy and utilitarian philosophy. (According to utilitarians, such as Jeremy Bentham and John Stuart Mill, the good is defined in terms of the maximization of both personal and general utility. For the most part, utilitarians thought of units of pleasure as the good into which the utility of commodities can be cashed.) Edgeworth's aim is plainly stated in the title of his book: *Mathematical Psychics: An Essay on the Application of Mathematics to the Moral Sciences* (1882). (Note that

economics was *still* considered part of moral science, as Adam Smith considered it.) The utilitarian moral worldview is articulated by Edgeworth in a passage of vulgar lyricism: "As the movements of each particle, constrained or loose, in a material cosmos are continually subordinated to one maximum sum-total of accumulated energy, so the movements of each soul, whether selfishly isolated or linked sympathetically, may continually be realizing the maximum energy of pleasure, the divine love of the universe" (quoted in Porter 1986, 257).

What Edgeworth added to his predecessors' work in political economy was an explicit appeal to Maxwell's and Boltzmann's new paradigm to show that each economic actor is always converting utilities to pleasure, as potential energies are converted into kinetic energy, with the result that a *sum* of individuals, each seeking to maximize his or her own utility, would produce a cosmic orgy of pleasure. Marginal utility, the fundamental principle of neoclassical economists, is strictly analogous to gradient reduction by converting potential to kinetic energies. These billions of preferences could be summed by using price, what people were willing to pay at the margin, as an index, as Maxwell was able to appeal to average kinetic energies.

Later thinkers in this tradition have repudiated the grandiosity, if not the vulgarity, of Edgeworth's thought. They have contented themselves for the most part with taking a personal rather than a cosmic view of utility gradients, a turn that reflects the primacy of individual psychology over collective phenomena or "methodological socialism" in Anglo-American utilitarian thought (Dyke 1981). Each rational economic agent is supposed at any time to be maximizing a *personal* utility function, as he or she exploits slight differences in a personal field of utilities ("a basket of goods"). The metaphorical structure of such models is not statistical mechanics and thermodynamics, and far less anything that makes more extensive use of stochastic processes, but rather the physics of fields that preceded statistical mechanics (Mirowski 1989). In this respect, the subsequent trajectory of neoclassical economics has not paralleled evolutionary biology.

6. Kelvin threw this idea in Darwin's face, saying that the rate at which the sun's energy was dissipating showed that the universe would have come to a cold end long before enough time had elapsed for natural selection to produce what we know of the fossil record. Darwin not only reeled from the blow but was personally depressed that human achievement would necessarily end this way. This shows that he was perhaps more attached than he realized to the cozy purposive universe he had done so much to eliminate from biology. It is often thought that Darwin allowed an increased role for use inheritance, à la Lamarck, in later editions of the *On the Origin of Species* in order to answer to Kelvin. It would speed evolution up (Eiseley 1958). Whatever truth there may be in this claim, Darwin would have had the last laugh if he had lived long enough. Kelvin new nothing of thermonuclear fusion or fission, which increases the life expectancy of the universe many orders of magnitude above what is required for phylogenetic evolution. See chapter 17 for more on this topic.

7. Sober has argued that neo-Darwinism is itself a "Newtonian science" because it is a "theory of forces" impinging on gene frequencies in ways that can be tracked by differential equations. Sober means exogenous forces, in which an inertial "zero state" of the system is changed by something that makes genes frequencies depart from expected Hardy-Weinberg Equilibrium. "If I push a billiard ball north and you simultaneously shove it west, science and common sense predict that the ball will move northwest. Newtonian mechanics has made vector addition a familiar paradigm for computing the net effect of forces acting in concert. . . . The zero-force state in evolutionary theory is specified by the Hardy-Weinberg Law of population genetics. . . . Mendelism is the background against which evolutionary forces are described. It is not itself treated as a force." (Sober 1984a, 31)

This is fair enough, since we too insist that what stays the same in the change from Newtonian to Boltzmannian Darwinism is the notion of exogenous forces impinging on closed systems. We have allowed ourselves sometimes to talk of evolutionary processes as forces. But "Newtonian" acquires too lofty a general sense in Sober's use of it. It refers to any system whose natural or inertial state is changed at a mathematically computable rate by the imposition of a force. Sober's use of "Newtonian" for Darwinism, both old and new, thus blurs what are to us important differences between the classical Newtonianism that influenced Darwin, and the probabilistic model of systems dynamics pioneered by Maxwell and Boltzmann, which influenced Fisher, Wright, and Haldane. Everything Sober says about Darwinism makes it clear that probability thinking is essential to its explanatory power. In order to avoid whiggishly projecting this recognition onto Darwin, Sober might have distinguished more fully between Darwin's Darwinism, a theory of forces modeled on something like classical Newtonianism, and the theory Sober describes and analyzes with great acuity. That is a probabilistic theory, in which natural selection will occur whenever there is a fitness gradient, even if the population is not anywhere near its carrying capacity. The liberation of Darwinism from crude Malthusianism, as well as the other conceptual problems with which Sober wrestles, such as the role of frequency-dependent selection of genetic drift, and other nonselectionist forces, can arise, we think, only in the new paradigm.

8. Attempts to verify this empirically generated some ingenious experiments a few decades later. Dobzhansky's student Francisco Ayala, for example, put two strains of flies into identical laboratory conditions. One strain derived from crossbreeding two different strains, and so contained more variation than a second strain, which had been inbred to reduce variation. Various selection pressures were inflicted on both populations. Sure enough, the population containing more variation produced offspring at a higher rate than the other, suggesting its ability to utilize variation to fix adaptations to the new circumstances (Ayala 1965a, 1965b, 1968; Dobzansky et al. 1977, 32–34).

9. It is very easy to sense question begging here. Fisher's theorem, according to Marjorie Grene, is a bookkeeping device that says nothing about the causes of changes in gene frequencies. How does Fisher know that the causes of additive variance in fitness are so exclusively related to the ecological process of adaptation (Grene 1961, in Grene 1974, 154–63)? Freeing Fisher from this charge may be difficult, but it will be impossible unless Fisher's statistical dynamical models are borne in mind. Whatever it may turn out to be, the cause of changing gene frequencies is, on Fisher's physics-soaked mathematical model, a function of whatever entities are most numerous and most independent (Hodge 1992a, 248). These, Fisher thinks, are genes, whose multiplicity and independence allows them to respond quickly to environmental changes and so disproportionately to propel themselves, and the traits they code for, across generations. In other words, given Fisher's model, the burden of proof lies on causes other than adaptation on the basis of independent multiple genes.

10. We explain a few more points about variance and how to measure it here. Intuitively, there is more variability between 1 and 10 than between 4 and 6. In either case, the data are clustered around the mean, 5. What is important is that, in general, as variability increases, so does the likelihood that the scatter of data renders them statistically insig-nificant in making correlations or ascribing causes. That sort of insignificance is called a null hypothesis. You need to prove that a certain distribution of a range of data is *not* due to chance. Otherwise you assume that it is. Can you measure this likelihood precisely, or do you have to eyeball it? The mean value of the square of the deviations from the mean value gives a more sensitive measure of the variation in a population than does the range between the highest and lowest value, or even the mean deviation (for instance, between 1 and 10). This is because the further the deviation is from the mean, the more it contributes to the variance. The outliers give a more important contribution to the overall measure than variability. However, the value of the square of something is not easily visualized, so the square root of the variance is employed. This is called the standard deviation. For a normal distribution curve, 68 percent of the values will fall within standard deviation; 96 percent

will fall within two standard deviations. Thus, if a given range of data has a standard deviation of 3.4, while another has one of 2.0, this means that the first one has a calculably greater spread-out-ness, and hence greater variance.

11. British naturalists, especially butterfly fanciers, kept discovering, both in England and in the wilder places of the empire, a capacity some species have acquired to avoid being eaten by taking on the coloration and patterns of related species not so tasty to the local predators. Early in the century, mimicry had seemed to be a good confirmation of Mendelism. How could it have evolved if in its earliest stages it provided no protection from predators? It must have arisen all at once by a genetic "sport." This was, for example, what Punnett thought. Fisher gave a plausible account of how genetic gradualism could handle the problem. Explaining mimicry was regarded by the opposed Mendelians and Darwinians as a sort of *experimentum crucis*, which is why Fisher spends time using his theory to solve the problem (Turner 1983, 1985).

12. Most recently, Wynne Edwards, who in doing so raised the wrath of Fisher's modern disciples. Cf. chapter 14 below.

CHAPTER 11

1. Michael J. Wade and Charles Goodnight have recently reported experimental validation of Wright's mechanism in the laboratory, using populations of the flour beetle *Tribolium*, a species Wade had earlier used to confirm the possibility of group selection (Wade and Goodnight 1991).

2. In the course of debating among themselves how deep the so-called turn-of-the-century revolt against morphology was, or whether it even happened, historians of biology have been investigating how such a rift might map onto differences in intellectual style between naturalists, experimentalists, and theorists (Allen 1978; Maienshein 1991). We assume that there was something of a revolt from morphology. We think, however, that differences between naturalists, experimentalists, and theorists can be overdone. In comparison to the theoretical Fisher, Wright and Dobzhansky were both naturalists. Dobzhansky was arguably more of a naturalist than the experimentalist Wright. But Mayr is more of naturalist than either Wright or Dobzhansky without being less of a theorist. What he is not is an experimentalist. Conversations with David Magnus have helped us clarify these issues.

3. Wright remained a faithful Unitarian throughout his life, even to the point of preaching lay sermons. The contrast between Wright's religious views, or at least sensibility, and Fisher's Anglicanism is a topic that merits further investigation.

4. Wright, as a biologist, was interested more in causality than in elegant prediction. He applied the statistical technique of regression analysis to find causal pathways through which a population would move under the influence of various factors and parameters. Path analysis, pioneered by Wright, eventually became part of the arsenal of social scientists (Crow 1990).

5. For a lucid explanation of the difference between selection *of* traits and selection *for* traits, see Sober (1984a). Sober illustrates the difference by producing a toy, repeated shaking of which will propel balls of various sizes and colors through a series of holes of different diameters, so that at the bottom will remain only, say, green-colored balls of a certain size. The point is that there was selection *of* green balls, but selection *for* balls of the size that would make it to the bottom of the toy, all of which just happen to be colored green. See chapter 14.

6. The following remarks of Hodge are worth bearing in mind: "It would be fallacious to think that because drift is a corollary of a mathematical property of a population, its finite

size, while selection is a consequence of physical differences among individuals, it follows that drift is somehow a mathematical rather than physical processs. . . . Drift and selection are not to be contrasted as sampling with and without error, but as causally discriminate rather than causally indiscriminate erroneous sampling. An explanation that invokes drift invokes causation no less than a selection explanation does, but it invokes indiscriminate causation and so no causes of discrimination" (Hodge 1987, 253). Hodge believes that Beatty's worries (Beatty 1985) about how to partition natural selection from drift reflect too mathematical, and epistemological, a conception of the relevant phenomena. Hodge thinks the *vera causa* ideal is just as relevant to modern Darwinism as to Darwin's own. (Hodge 1987).

7. This is a research program still alive and well, living under names like "systems ecology," as can be seen in the work of the North Americans Howard and Eugene Odum, Robert Ulanowicz, and Lionel Johnson, among others. What it meant in its first phase of articulation was that Russian ecological evolutionists picked up from Darwin the principle of the division of labor, according to which in a "tangled bank" or "crowded heath" there is selection pressure to diversify life per unit of land to take advantage of new energy sources. It is not considered an orthodox view by most Darwinians because it requires that fitness be a function of energetic considerations alone. Most Darwinians are much more pluralistic than that about the components of fitness. We return to this issue in chapter 17.

8. Dobzhansky's insouciance about mathematics, and hence his total trust in Wright's judgment, is admirable and characteristic of the man. "My way of reading Sewall Wright's paper," he later said, "which I still think is perfectly defensible, is to examine the biological assumptions the man is making, to read the conclusions he arrives at, and hope to goodness that what comes in between is correct. "Papa knows best" is a reasonable assumption, because if the mathematics were incorrect, some mathematician would have found it out" (Dobzhansky 1962 in Provine 1986, 346).

9. This idea was made even more vivid by Leigh Van Valen, whose Red Queen hypothesis postulates that species have to run faster and faster, like the Red Queen in *Alice in Wonderland,* just to stay where they are as the environment degrades beneath them.

10. The internalization of this idea into the modern synthesis fostered a switch from an older conception of "niche," in which the term referred to the place of a species in a given community, to a view in which the niche was defined in terms of the total range of resources required by a single species. The first conception is that of Elton; the second of Hutchinson. See chapter 15.

11. As time went on, Dobzhansky developed a slightly different view about heterozygote superiority. A series of experiments in the 1940s showed that heterozygotes are *generally* advantageous (Burian, personal communication; Lewontin, unpublished). It may not be necessary, therefore, to be able to tell a story about the *particular* adaptive value of *particular* genes in *particular* environments. That is a good thing, since stories as plausible as the one about sickle cell anemia are hard to come by. This suggested that heterosis is inherently is a highly general evolutionary product, and an adaptive one.

CHAPTER 12

1. If another book deserves mention, it would be Bernard Rensch, *Neuere Probleme der Abstammungslehre* (1947).

2. Before looking into these topics more closely, we note that phrases like *modern synthesis* and its synonyms have a narrower scope than *neo-Darwinism.* That term was first applied to Weismann and to Darwinian naturalists who became hard selectionists after Weismann's

rejection of Lamarckism, and so abandoned Darwin's causal pluralism. The biometricians Galton, Pearson, and Weldon, were neo-Darwinians in the sense that they used statistical analysis to improve the explanatory power of Darwinian adaptationism. Fisher was a neo-Darwinian insofar as he transferred the biometricians" statistical approach from phenotypes to genotypes. The main charge of the modern synthesis was to apply the fundamental principles of population genetics to actual cases and problems. In applying these principles to specific problems, those who forged the modern synthesis agreed from the outset that in addition to natural selection, forces, processes, and boundary conditions like genetic drift, mutation pressure, migration, and population structure, would probably have to be, and could legitimately be, appealed to. This diverse series of uses should make one think twice before using the term as a synonym for the modern synthesis. Hence, we generally avoid using the term.

3. They are, however, inconsistent with Platonic or typological essentialism, for while the essential classes are real, in that view, the populations are not. For Platonists, populations are not spatially and temporally connected groups, and a fortiori are not individuated entities, but simply instances of kinds.

4. It is fascinating to speculate what role in giving birth to this idea was played by the fact that in developing it Dobzhansky was following populations of fruit flies and other organisms that were distributed into discrete populations in the real mountains, valleys, and deserts of California. The point is purely a matter of the "logic of discovery," however, and of creative psychology, rather than of conceptual importance.

5. The term *physiological isolation* was first used by Romanes in 1886. He was trying to show how speciation could be sympatric: "Some individuals living on the same geographical areas . . . are absolutely sterile with all other members of their species. . . . The barrier, instead of being geographical, is physiological" (Romanes 1886). By *physiological isolation* Dobzhanksy means genetic isolation. Note that Dobzhansky is by this token in principle more open to the idea that speciation can occur in the absence of external geographic isolation than Mayr, whose stress on allopatric speciation seems to imply a dim view of sympatric speciation of any kind. Note too that Dobzhansky's theory seems closer than Mayr's to Darwin's mature theory.

6. In later formulations Mayr removes *potentially* as redundant.

7. See chapter 13 for more on "the Eve hypothesis."

8. Among the developments for which Mayr takes retrospective credit is anticipating the discovery of a much more holistic genome, which produces true "genetic revolutions." Peripatric speciation, moreover, facilitated by such genetic revolutions, plays a significant role in Gould's and Eldredge's theory of punctuated equilibrium, which we will encounter in chapters 14 and 15. For Mayr, modern genetics does not threaten, but on the contrary serves to confirm, his version of the evolutionary synthesis, and his methodological recommendations to let natural history play the leading role in suggesting genetic hypotheses (Mayr 1988).

9. For more on "species as individuals" and its relation to "species selection," see chapter 14.

10. If it is true that the hardening of the synthesis should not be taken as tantamount to Fisher-and-Ford-ism, neither is it correct for Gould to imply that the pluralism of the early synthesis was anything remotely like what passes for pluralism today, or to imply that had it been left to its own devices the early synthesis would have evolved in that direction (Gould 1983). None of the architects of the modern synthesis ever allowed drift and other evolutionary processes to do much work on their own, or even in combination, apart from the jewel in the crown, adaptive natural selection working exclusively on organisms.

Today's "pluralists" about evolutionary forces advocate an "expanded Darwinism," which explicitly demotes the role of natural selection in evolution by acknowledging that genetic drift and constraints are in some cases more powerful than selection, and by allowing selection, or analogues of it, to range over a whole variety of units, from genes to species, rather than simply over organisms. It seems to us wrong-headed for contemporary pluralists to establish their genealogical roots by placing their own thoughts, as interesting as they are, in their intellectual fathers' heads.

11. V. B. Smocovitis seems to hold a stronger thesis than ours: that positivism, and in particular its demand for a unity of science, helped *form* rather than merely *harden* the synthesis (Smocovitis 1992). We have large reservations about this hypothesis; see chapter 14.

12. Ghiselin (1969) and Ruse (1979) also attempted to show that Darwin himself was a verificationist *avant la lettre*. Hodge has shown that Darwin's Herschelian *vera causa* ideals are sufficient to demonstrate that respect for empirical data and for testing hypotheses is not enough to make one a verificationist in anything like the positivist sense (Hodge 1977).

13. When Popper and others attacked the notion of fitness as an empty and meaningless tautology, a number of alarmed philosophers rose to its defense. One good solution emerged from attempts to think through a thought experiment first proposed by Michael Scriven (Scriven 1959). Scriven imagined two identical twins who together climb a mountain. One is struck with lightning and dies. The other goes on to raise a fecund brood. Which is more fit? Actual survivorship seems inadequate to answer this question. It is more compelling to say that both twins had the same *expected* fitness because they were made the same way, and so, all other things being equal, they should be presumed to have the same reproductive success. Darwinism on this view relies on *expected*, rather than actual or realized, fitness. Expected fitness can be defeated. Actual fitness cannot. Hence, expected fitness seems to be empirically meaningless on positivist criteria. That is a problem either for positivism or for the notion of fitness. Positivists blamed it, of course, on fitness.

The idea of expected fitness was a way out. It led in turn to the idea that fitness is a *dispositional property*. It does not refer to what an entity actually does but to a propensity it has for doing it.

The fact that some propensities are more deterministic than others does not undermine the fact they they are both dispositional. Fragility, for example, said of plate glass, is a deterministic disposition. Under given conditions, a ball thrown at a plate glass window at a certain speed will break it close to 100 percent of the time. By contrast, fitness, said of Scriven's twins, is a probabilistic disposition. It requires a comparative and populational context if the concept is to be applied to cases and assessed, precisely because what is expected does not always happen. Unlike the case of the window, you cannot infer directly from a generalization to a particular case. (From this perspective it becomes harder to assess the fitness of Scriven's twins without knowing a lot more than the thought experiment presents.) Fitness, accordingly, is the propensity of an organism that is a member of a given population in a particular environment to have an expected number of offspring (Mills and Beatty 1979). Put otherwise, the fitness of an organism is the *chance* that a member of a population has of surviving to reproduce at an expected level (Sober 1984a, 43). So construed, it is entirely possible that "in any single run organisms of the relatively fittest type may not out-reproduce their competitors; indeed, there are occasionally cases in which none of the fittest organisms survive" (Burian 1983, 301).

The propensity interpretation of fitness implies that expected fitness causally rests on, and tacitly refers to, the structural properties of fit organisms. Just as the fragility of glass rests on the properties of silicon at various temperatures, and the chance of a coin's falling heads or tails ultimately relies on what it is made of and how it was struck, so the fitness of an organism refers to a conjunction of real physical structures and properties of organisms. When fitness is looked at from this structural and causal point of view, it names what Darwinians have always called adaptedness. "Adaptedness," says Sober, "as I have used the term is simply fitness by another name" (Sober 1984a, 196; Brandon 1978, 1990; Burian 1983). One can speak of a certain level of relative adaptedness just as easily as one can refer to a certain level of fitness. In doing so, attention is drawn to the construction of an organism more than to its actual reproductive output. Fitness *qua* adaptedness is thus closely related to what is called "engineering fitness" (Burian 1983). Referring to fitness as adaptedness shows why experiments in which the engineering efficiency of organisms is compared to experimental setups are important in establishing whether organisms are working at peak efficiency, and hence are as optimally fit as they can be, or whether something is constraining them. Counting actual numbers may be a good way to estimate or provide evidence of fitness, just as flipping a coin a number of times is a good way to establish its propensity for coming up heads or tails. That does not mean, however, that the empirical meaning of the concepts of fitness or fairness resides in the ways these quantities are measured. It lies in the construction of the things that are fit and the coins that are flipped. So much for the positivist's tautology of fitness.

Unlike coins and glass, however, expected levels of fitness cannot be reduced to a small number of configuration of physical properties. A given level of fitness or adaptedness can be attained by a very wide, and usually unknown, variety of underlying physical substrates. Philosophers call such a relationship *supervenience* (Kim 1978). A property is supervenient when a given underlying physical condition is sufficient to produce it, but when at the same time many other such conditions could produce it as well. Supervenient properties are not reducible to a particular underlying configuration. However, they are not one whit less materially grounded than reductionists have always claimed. We may conclude, then, that expected fitness is not only a probabilistic disposition but a supervenient property of organisms (Sober 1984a). Readers of this book will recognize that the notion of supervenience is the philosophical or conceptual correlate of the microstate-macrostate distinction first introduced into statistical mechanics and thermodynamics by Maxwell and Boltzmann, and so is another product of the probability revolution. The beauty of this analysis is that in making *fitness* or *relative adaptedness* explanatory terms, philosophers are relying on and pointing to the probabilistic nature of modern Darwinism itself to solve a problem about its conceptual structure.

Some of this analysis can be achieved by interpreting probability theory itself in terms of frequencies of events in the long run or at the limit. However, when fitness is regarded as a propensity, it is more fertile to treat probabilities themselves as propensities as well, properties that supervene on the underlying structures of probabilistically distributed entities or properties. Under this more objective interpretation, the probability revolution enters more deeply into the complex structure of reality (Popper 1990). The progress of propensities in evolutionary biology thus serves as a particularly telling instance of the still unfolding consequences of the probability revolution in science. The problems of creationists and defenders of common sense with evolutionary theory often reflect failure to see how thinking in terms of probabilities makes it easier to understand both what is being explained and what does the explaining.

For more on this subject see chapter 18.

CHAPTER 13

1. Wilson and his former students used DNA restriction mapping and DNA sequencing techniques on human DNA from various populations to claim that present humans are descended from a single woman who lived about 150,000 years ago, when a "bottleneck" in our evolution occurred (Cann, Stoneking, and Wilson, 1987). That we are all descended from a very small group, perhaps even from a single female ("Eve"), should not be entirely

surprising in the light of Mayr's founder theory of allopatric speciation. Wilson and his colleagues also attempted to situate this event geographically and to substantiate current anthropological speculation about an African origin for our species. The interpretation of the data as supporting an African origin of humans has been challenged (Thorne and Wolpoff 1992; Gee 1992). However, there has been a spirited defense (Wilson and Cann 1992). This controversial claim is still being debated (Gibbons 1992).

2. While the evolutionary implications of editing of RNA are as yet unexplored, Walter Gilbert has pointed out that the "split gene" organization of eukaryotes (from the Greek for "good kernal," that is, a true or well-formed nucleus, as opposed to prokaryotes, which lack the nuclear structure in their cells) could allow genetic rearrangements that could produce novel enzyme activities (Gilbert 1978; see also Doolittle 1987). The capacity of such rearrangements to produce mosaic genes has been demonstrated for the proteins involved in blood clotting from vertebrates (Holland, Harlos, and Blake 1987).

3. John Campbell has argued that such multigene families appear to exhibit traits and capacities that are not described by the models of classical transmission genetics (Campbell 1985). These include creating not only variant genes but also variations in gene number and expression. Some multigene families contain dozens or hundreds of copies of a gene with the same function. In such a case, any one copy is expandable if it loses its function through mutation. Such mutations will accumulate. But there are times, under metabolic control, when there is a rectification of the genes to the normal or "wild type" or to one of the mutant forms, with the other variant alleles purged. The exact mechanisms by which this occurs are under study. Aside from neutral protein mutations, there can be point mutations that are selected for and are adaptations. In addition, there are types of mutations (transpositions, crossing over such that domains are shuffled, or point mutations that significantly alter the protein structural properties) that make possible new kinds of structure and behavior and thus represent marked evolutionary advance without disturbing existing functions. The emergent properties of hemoglobin or of a supergene complex or of multigene families discussed above are examples of such advance. This raises interesting questions about the relationship between classical and molecular genetics. Ultimately this complexity has important implications for evolutionary theory, as we will explore in the following chapters.

4. Waters (1990, 1994) argues that classical transmission genetics, as distinct from textbook caricatures of it, does not in the least retain a one-gene–one-trait assumption. Where people like Morgan, Bridges, and Sturtevant talk about one gene doing something, they are explaining not the *cause* of a trait but the *difference* in the making of a trait caused by one gene. These are not the same thing. Waters then goes on to argue that molecular biology explains the complexity already posited by classical transmission genetics. In this case, the *explanandum* of molecular genetics remains sufficiently the same as that of transmission genetics to justify the notion of reduction on the model of statistical theory of gases and heat. This argument represents a challenge to the consensus view we report here. The point about what classical transmission genetics explains is incontrovertibly true. Whether what Waters makes of it is also true is harder to say.

CHAPTER 14

1. In gel electrophoresis, a workhorse experimental technique of genetics during the last quarter of a century, cell-free extracts from individuals in a population are subjected to an electrical gradient across a plastic gel, which separates proteins based upon their electric charge and size. The enzyme of interest is visualized by using a chemical reaction that depends on the catalytic properties of the enzyme. Such studies show extensive polymorphism throughout all taxa, even allowing for the fact that only roughly a third of mutations

give rise to a change in the electric charge on a protein. To count as a polymorphism, Lewontin stipulated that at least 1 percent of a population has to possess a mutant enzyme. The level of polymorphism is the number of gene loci for which there are mutant alternative alleles divided by the total of gene loci studied for the whole population. In general, vertebrates have less polymorphism than insects, and both substantially less than plants. (The high degree of polymorphism for plants, along with polyploidy, probably compensates for the fact that plants cannot migrate with the ease of animals.)

2. Kimura, it seems reasonable to assume, had moral concerns too. As a Japanese he was intensely aware of the effects of radiation in producing harmful mutations, effects first demonstrated and worried over by "the great Muller." It is also intriguing that Kimura's Japan is a far less diverse society than Lewontin's, and Dobzhansky's, America. One hopes these issues will be pursued in William Provine's forthcoming work on Kimura.

3. The use of game theory has been one of the most fertile and powerful research programs in recent evolutionary theorizing. Game theory is particularly effective in showing how phenotypes get the best overall deal possible (optimize) by trading off one good against another. It has been particularly useful in analyzing life-cycle traits, such as number of offspring, timing of sexual maturation, and aging and death. (For a survey of results in primates, including humans, see Diamond 1992.) Game theory has also proved helpful in analyzing strategies of cooperation and competition in ways that do not necessarily require or imply taking a gene's-eye perspective. John Maynard Smith, for example, a population geneticist at Sussex University, has used mathematical game theory to show that any population with a mix of selfish and altruistic individuals will be more evolutionarily stable than one composed exclusively of "hawks" or "doves": A population composed exclusively of hawks will destroy itself; a population composed solely of doves will expose itself to the ravages of cheaters and invaders (Maynard Smith 1978). Robert Alexrod showed that if competitors play the same game long enough and respond to each other's moves in a tit-for-tat fashion, cooperation will generally arise even among self-interested individuals (Axelrod 1984).

Game theory may prove the most memorable contribution of the Cold War to science, for it was in the context of nuclear bluffing and brinksmanship that the notion was first put to serious use. Perhaps never before in human history have people's lives been so dependent on the truth of mathematical propositions.

4. This does not mean, however, that mothers have 100 percent of the genes of their sons, for they also carry around genes from the male that mated with them. That is because *Hymenopteran* queens mate only once, or, as in the case of the honey bee, a few times at most (Page et al. 1987). The sperm they acquire at that time is slowly doled out over their lifetime to many eggs. Not all eggs are fertilized. Those that are not become males. Those that are become females. One hundred percent of the genes of sons thus come from their mothers. They have no father. Sisters, meanwhile, who have both a mother and a father, acquire their genes in the usual 50 percent distribution from each parent. Sisters, then, share more of their genes with each other than with their brothers. In fact they share 75 percent of their genes because any given gene has a 50 percent chance of coming from the mother but a 100 percent chance of coming from the father. Thus it is in the interest of the sisters to make their mother produce more sisters.

5. Lewontin was especially upset when Wilson, an entomologist, went over to what Lewontin regarded as the side of the reactionaries. Wilson had been among those responsible for bringing Lewontin to Harvard from Chicago when their research group in population ecology began to drift eastward. Wilson's version of genic selectionism differs, however, from Dawkins's in ways that perhaps make him a less deserving target of Lewontin's scorn. It should be noted in this context that kin selection does not logically presuppose or entail genic selection of Dawkins's sort. It does not require, that is, that genes must be treated as

the beneficiaries of selective processes just because they are units of selection. See note 11 below

Wilson had been trained at Chicago by W. M. Wheeler to see insect colonies as highly cooperative "superorganisms." He was deeply impressed, therefore, with the results about *Hymenoptera*, which derived insect cooperation from the gamesmanship of insect genetics. Ironically, it was because Wilson could see competitive machinations behind the cooperation of insect societies that he was inclined to see how intense, almost socialistic forms of cooperation approximating those of insect societies might arise among competitive primates, including humans. Wilson is no apologist of capitalist individualism but an advocate of ecological communitarianism. (This is insufficiently selfish for Dawkins.)

Wilson's *Sociobiology: The New Synthesis* (1975) was designed to pull the social sciences, which since the fall of eugenics had shifted away from nature and toward nurture, back into the Darwinian fold (Degler 1991). It would do so, however, not by cleaving to Darwinism's traditional bias toward selfish behavior but, on the contrary, because Darwinism was now able to account for self-sacrificing altruism! In the last chapter of *Sociobiology*, and in a subsequent tract entitled *On Human Nature*, Wilson claimed that our selfish "genes have us on a leash," but a very long one (Wilson 1978, 167). Our resistance to their pull is both possible and desirable. It enables us to act more like ants and bees than like wolves.

One may well wonder why views like these would incense Lewontin more than those of other sociobiologists. Perhaps Lewontin saw in Wilson's model of cooperation something even more dangerous than the idea that selflessness always reduces to calculative selfishness: the notion that cooperation involves the surrender of free agency to group interest. Altruism like that is preferable only if the remaining viable alternative is a "war of all against all," as in Hobbes's *Leviathan*. That is a choice Lewontin, with his stress on the activity of organisms, does not feel we have to make, since humans are free agents, who might very well design a society in which their freedom and individuality is given full scope along with, and through, cooperation.

6. Lewontin and Levins take the liberal insistence on politically ensuring equality of opportunity not as a constraint on capitalism but as a way of justifying the otherwise doubtful assumption that competition occurs on a level playing field. On this assumption depends the very legitimacy of regimes that protect free-market capitalism. Meritocratic societies such as the United States, with much political interference in the distribution of social and economic goods, are for Levins and Lewontin forms of advanced capitalism, not harbingers of socialism (Lewontin 1992, 22).

7. Levins's connection to Dobzhansky's legacy, combined with his extreme sensitivity to the interactive nature of complex biological and social systems, is evident in his most famous work, *Evolution in Changing Environments* (1968), in which an effort is made to find "sufficient parameters" that can guide researchers through a maze of interacting variables.

8. It is intriguing that Dewey was still preaching these ideas in the same Columbia University in which Dobzhansky worked, where they had entered into the fabric of local culture. Whether there was any closer connection is a matter that still requires investigation. Dewey, in any case, remained unfortunately innocent of population genetics. If there is any connective tissue between Dewey and his disciples and Dobzhansky and his disciples, it will support the doubts we have already expressed about V. B. Smocovitis's argument about the influence of positivism in forming, rather than merely hardening, the modern synthesis (Smocovitis 1992; cf. chapter 12, note 11). Smocovitis's argument depends, in the first instance, on seeing reductionistic ideals in the attempt to unify biology through genetic Darwinism. That may be somewhat true of British versions of the synthesis, on which Smocovitis understandably concentrates, and of the hardened forms that became prominent in America under the influence of logical positivism. It is far less true, however, of prehardened American versions. These seem to have more affinities with naturalistic pragmatism

than with reductionistic positivism. In particular, pragmatism's "naturalized epistemology" is of a piece with its insistence on the role of organisms as active problem solvers, an insistence that Dewey shares with Dobzhansky. In any case, it is wise to remember that the unity of science does not depend on reductionism (Kitcher 1981; Darden 1991).

9. This does not mean that Lewontin and Levins do not bear the scars of the positivist ascendancy themselves. They even say they do (Levins and Lewontin 1985, 265). In explicitly denying that organisms are "problem solvers," for example, Lewontin *assumes* that problem solving is matter of adjusting one's behavior to a preexistent environment (Lewontin 1978, 1980, 1983, 1992). Rather than reconstructing adaptationism in the way Progressive Darwinians did, therefore, and as his own stress on the agency of organisms seemingly invites, Lewontin seems sometimes eager to surrender the very concept of adaptation, or at least adaptation*ism*, to his opponents as inseparable from the passive, mechanical conception of that process, and of problem solving, that has dogged the Darwinian tradition from the outset, and that came to the fore again under the influence of the hardened synthesis. So deeply is Lewontin under the spell of this passive conception of adaptation (which we regard as an artifact of the dynamic models in which Darwinism has encoded itself) that he sets out instead to define natural selection in a way that separates it from adaptation so construed.

10. Philosophers of biology have concerned themselves with epistemological as well as with ontological aspects of the expanded synthesis program. Here the pressing problem is to determine criteria by which units of selection are discriminated. One school of thought holds that a new unit should not be introduced until additivity of fitness fails at a lower level (Wimsatt 1981; Lloyd 1988; Thompson 1989; Godfrey-Smith 1992). Others think that this criterion still resonates too much with the Fisherian biases from which the idea of multilevel selection is seeking to free itself, and undermines the context sensitivity of fitness (Sober and Lewontin 1982; Brandon 1990). Advocates of and skeptics about a hierarchically expanded synthesis can be found on both sides of this still intensely debated, and highly technical, issue.

11. The question between Dawkins's selfish-gene hypothesis and those who oppose it is about whether causal primacy is to be ascribed to replicators or interactors. In probing this issue, it helps to distinguish, with Sober, between "selection for" and "selection of" (Sober 1984a, 97–102). Given this distinction, there can be selection *for* interactors alone, for they alone seem to be involved in causal processes (Sober 1984a). (Cf. chapter 11, note 5.)

Does Dawkins deny this, or does he think instead that replicators (selfish genes) are in fact interactors? In order to clarify this question, we must first see that the notion of "selection for" can itself be ambiguous. It can refer to the property or entity that is the environmental target or object of the selective process, or it can mean (or be confused with) the entity that *benefits* from natural selection. Elizabeth Lloyd shrewdly remarks that most of the bad tangles in the unit of selection controversy come from failing to recognize this distinction (Lloyd 1988). The consensus of the modern synthesis, especially on its American side, has it that there is selection *of* genes, selection *for* the phenotypic traits of individual organisms, and that evolving populations and lineages are *beneficiaries* of the selection process. From Dawkins's point of view organismic selectionists tacitly exaggerate the replicative prowess of organisms, which alone will turn the success of particular organisms into the transgenerational success of populations. Since the information that makes organisms what they are is scrambled and reassembled in meiosis, Dawkins, following Fisher, Hamilton, and Williams, claims that organisms do not make very good replicators. Genes, as he defines them, do. Because genotypes have the greatest longevity of any of the other entities we are considering, Dawkins treats them as the beneficiaries of selection. He then tacitly concludes, *for this reason*, that there is selection *for* genes. There is a sequence of slides in this argument. Defenders of the orthodox view can claim with some justice that Dawkins exaggerates the interactive prowess of replicators in order to get his conclusion. Dawkins's selfish genes are at one and the same time replicators and very powerful interactors. They are described not only as influencing other parts of the replicative machinery, or their cellular milieu, but as having power to manipulate whole organisms, which serve as their "vehicles." This level skipping seems magical. Thus, most philosophizing biologists and philosophers of biology have concluded that entities operating at the level of organismic interaction have a better claim to causal efficacy than anything the genes do (assuming they "do" anything) (Hull 1980; Brandon 1985, 1990; Wimsatt 1980; Sober and Lewontin 1982; Lewontin 1992).

12. One of Hull's philosophical reasons for pointing out such prejudices is to free evolutionary biology from the commonsense ontology from which physics and other advanced sciences have long since fled (Hull 1980, 1981, 1988). Hull intimates that commonsense ontologies encode oppressive prejudices that are harmful to the freedom of individuals.

13. Where entities from the ecological and genealogical hierarchies are systematically connected by means other than natural selection, or something analogous to it, we no longer remain within the boundaries of the Darwinian tradition. In versions of the expanded synthesis that do remain within these bounds, autonomous processes at work on both sides of the evolutionary hierarchy will certainly be acknowledged. Drift and its analogues, for example, will be at play at levels protected from the scrutiny of natural selection. On the genealogical side, internal connectivity and self-organization within genomes will constrain selection, even if they been created or sustained by it (Wimsatt 1986), and deflect it from one level to another (Gould 1982b). Ecological processes will also be allowed to exhibit forms of self-organization that cause evolutionary change (Eldredge 1985; Eldredge and Grene 1992). What gives such theories the right to call themselves Darwinian is the *stipulation* that a selection process modeled on organismic natural selection will knit together most of the interactions between entities on each side of the evolutionary synthesis. It is possible to produce theories of evolution in which this stipulation is abandoned, and causal primacy is accorded instead to processes other than natural selection on either the genealogical or ecological side of the hierarchy or both. Such theories abound and often make use of the dual-hierarchy notion (Brooks and Wiley 1986, 1988; Salthe 1993). Such proposals are no longer Darwinian, however, even by the generous standards offered by Gould. Nor, in most cases, do they want to be. See chapters 15 and 17 for details.

14. Lewontin's three conditions for natural selection make it difficult to differentiate between natural selection and genetic drift (Beatty 1984). Hodge thus thinks of Lewontin's three conditions for selection as "tautological selection," and not really natural selection at all, but a schema that must be filled in by genuine causal, if also statistical, processes, if natural selection or any of its analogues is to be ascribed (Hodge 1987). Hodge's reasoning is that just as Darwin's own theory cannot even be formulated without reference to the *vera causa* ideal, so the Darwinian tradition cannot retain its integrity without fidelity to such an ideal.

15. Biologists who have used the expanded synthesis idea to solve empirical problems about optimization differ from more orthodox Darwinians by appealing to conflicts and trade-offs *between* levels and units rather than to trade-offs at the *same* level. Bonner, for example, explains the evolution of increasing complexity in terms of such multilevel trade-offs (Bonner 1988). Buss explains the emergence, nature, and diversity of biological individuality in terms of conflicts between cell lineages and what is required if whole organisms are to be adapted to their environments (Buss 1987). Explaining how this happens requires Buss to reject the universality, even the normalcy, of Weismann's early segregation of the germ and soma cell lines. Buss is at the same time reviving and developing Weismann's rather ill-received appeal to conflicts between organismic and germinal selection (see chapter 7).

CHAPTER 15

1. As if to confirm this view, Brandon argues that biological hierarchies are not theory-neutral facts about the world. What one regards as interactors and replicators depends on the theory one is using. For his part, Brandon's lists, like those of Hull, includes only entities that can plausibly be objects or products of a process of genuine natural selection. "According to Hull's definition, which I have adopted, interactors *imply* selection," Brandon writes. "But there are many forms of interaction with the environment that do not necessarily lead to selection. Perhaps my dual hierarchy is a special case of Salthe's and Eldredge's" (Brandon 1990, 98, italics added).

2. If the general reader has a hard time getting a clear view of the history of ecology, that is because historians of ecology are themselves oriented toward one or another of these three traditions and construct their histories accordingly. Thus, Sharon Kingsley's *Modeling Nature* is rightly subtitled "Episodes in the History of Population Ecology," because she looks at systems ecology and community ecology with an eye to what would later prove useful to mathematical population ecologists like MacArthur and Wilson. Donald Worster's *Nature's Economy,* on the other hand, is entirely innocent of mathematical population ecology. One would never know from his book that any such thing had ever happened, or even that reformed systems ecology had continued to develop on its own. That is because Worster traces community ecology back to romanticism and forward to the green revolution. He writes in the tradition of Thoreauian pastoralism, in which Darwin turns out to be a mechanistic and capitalistic bad guy who helped expel us from Eden and despoil it. Ecology and evolutionary theory will not be unified until a comprehensive history of ecology is written.

3. E. O. Wilson shows himself to be Wheeler's student when he revives the idea that colonies, insect communities, and perhaps larger integrated biological systems are superorganisms (Wilson 1975). Revival of the group selection idea makes modest versions of the "superorganisms concept" defensible once again (D. Wilson and Sober 1989). At a higher level in the ecological hierarchy, advocates of the Gaia hypothesis explicitly claim that the earth itself is a superorganism because its atmospheric composition is regulated by expansions and contractions of its total biomass, with termites and other cellulose-digesting organisms doing most of the regulative work (Lovelock 1979; Margulis 1981).

4. Lotka avoided the term *autocatalytic,* with its specific chemical connotation, preferring instead *autocatakinetic* ("to self-transform") to describe organic autocatalysis, including the self-limiting tendency of a population to increase in numbers through reproduction.

5. The term *developmental biology* is anachronistic when applied to nineteenth-century embryology, although we will use it. It was coined by N. J. Berrill to name a field of study that would include colonial articulation as well as embryogenesis.

6. In these respects there are a good many Waddingtonian themes in Lewontin's work. See chapter 14.

7. In "Punctuated Equilibrium Comes of Age," Gould and Eldredge review what twenty years of research and argument have done to support punctuated patterns and species selection in macroevolution (Gould and Eldredge 1993). They profess to be modestly proud of the mettle their brain child has shown, although they caution that as a product of the contemporary *Zeitgeist* that produced Kuhn's punctuated model of scientific progress, it may go down with the *Zeitgeist* as well. Such is the modesty of Kuhnians.

8. Not all concerned parties, including some of Gould's own heroes, have agreed that the disparity is as wide as Gould makes it out to be (Morris 1993).

9. The case is exactly parallel to Scriven's case of two identical twins, one of whom is hit by lightning while his brother lives to see many healthy grandchildren. Since the fitness of the twins is the same, it cannot, by definition, be natural selection that explains the survival of the one and the death of the other. Lightning being indiscriminate about its targets, and not correlated to this or that organic trait, who survives is a matter of dumb luck. (See chapter 1, n. 3 and chapter 12, n. 13.) That is what Gould is saying, albeit on an enormously expanded scale, about why body plans survive catastrophes.

10. Gould seems to be writing a new kind of theodicy for a postmodern age. Hitherto, we have assumed that religious awe is keyed to the perception of purposive or functional order in the universe. The decline of that Judeo-Christian creationist belief, and its adaptationist afterglow, need not, Gould seems to imply, lead to a corresponding evisceration of the sense of wonder that lies at the core of religious sensibility. The sheer miracle that we are here at all in a universe that has planned or guaranteed nothing can do the trick.

CHAPTER 16

1. The periodic motion of the pendulum is the basis of clocks and became part of the metaphor of a "clocklike" universe. Consider an idealized and "fake" pendulum for which there is no friction and for which the displacements are small compared to the length, so that we can utilize the simple mathematics of a harmonic oscillator. For a harmonic oscillator the displacement in the x direction produces a restoring force $F = -kx$, from which we can derive a simple differential equation that allows us to calculate the angle of the displacement if we know the length of the pendulum, the acceleration due to gravity and the initial displacement. The resulting motion of the pendulum swinging back and forth in the x direction can be plotted in a phase space of the velocity (v) of the pendulum versus its position (x). A point in this phase space describes the pendulum's state at the instant of time t; at a later time there will be a point in a different position reflecting that the pendulum has moved. This process is repeated over and over, plotting each point until one swing of the pendulum back and forth in real space has been completed: The result in phase space will be a trajectory in the shape of a circle. A different set of initial conditions—for example, a larger displacement—will result in a different trajectory, here a larger circle. Working out all the possible trajectories for our ideal pendulum will give a concentric set of circular trajectories. If the pendulum is at rest, the trajectory will just be a point in the middle of the concentric circles. The sum of two possible trajectories is also a trajectory; this is a characteristic of linear differential equations.

As we have seen, a system is said to be linear if the differential equation defining its dynamics has the property that the sum of two solutions is also a solution. Thus our ideal and fake pendulum exhibits linear dynamics. A mathematical description of an idealized but "genuine" pendulum would still ignore friction but would require nonlinear terms to handle large displacements of the pendulum that are like those we would encounter with a real pendulum, for example, in a grandfather clock. The picture in phase space is of a series of ellipses that are not additive. This picture has more "structure" to it—in fact, it looks rather like an eye. The problem with nonlinear differential equations, prior to the advent of modern computers that allow their solution by numerical approximation, is that they are very difficult to solve. Generations of scientists have learned to take the nonlinear differential equations of genuine systems and "linearize" them by ignoring the inconvenient terms or approximating them. Thus pendulums were presented in elementary physics books as harmonic oscillators. Even with a genuine pendulum, that is, a model described by a nonlinear differential equation, we are still inhabiting an idealized world, for we have ignored friction. If friction, which dissipates the gravitational energy as heat, is taken into account, we find that the phase-space picture is not that of concentric circles or of ellipses but rather of a spiral into the central point of no motion. Regardless of the specific starting point, that is, the initial conditions, the system will always spiral to the same final point, rather like a marble spiraling down the mouth of a funnel and coming to rest in the spout.

2. It is precisely on this point that Kauffman's theory differs from the otherwise comparable work of Wimsatt. Wimsatt tends to think that natural selection itself is responsible for bringing about the genetic properties that make further adaptive natural selection possible. (Wimsatt 1986). See chapter 18.

3. The model also predicts that in these 317 attractors, 70 percent of the genes would be active identically in all the attractors. This roughly corresponds to the number of constitutive enzymes (enzymes synthesized in fixed amounts regardless of the metabolic state or rate of growth of the cell) expressed in all cells in a human body.

4. Kauffman's modeling of ecosystems explores only phenomena within the purview of population ecology. He has not attempted to deal with community ecology or energy-flow systems ecology. One can, however, imagine that his NK model can be so employed, as we will employ it in the next chapter. For example, to model the predator-prey relations in a community, N would represent the number of species and K would represent the connections between species—that is, who eats whom. The basic adaptive landscape would be similar to that employed for protein sequences.

5. He has explored the generation of complexity from simple dynamics (Kauffman 1985). More recently he has studied the production of simple patterns from a system with complex interactions (Kauffman 1991, 1993).

6. From the perspective of the new dynamics, and the essentially complex world it prefigures, we can see that the notion of law has acquired at least three different meanings, all of which are easily conflated in a presumptively simple world. *Law* refers, most simply, to observable regularities. Call this the Baconian sense. The concept acquires a new meaning when a rule for generating regularities is found. The law, in this case, is not an inductively grounded regularity itself but the rule that mathematically generates it. Call this the Galilean sense. Finally, rules that must to be obeyed by every possible system become "laws of nature." Call this the Newtonian (or perhaps Kantian) sense. In a presumptively simple world, Newton's second law is both a law of nature and a transformation rule.

CHAPTER 17

1. It is not strange that creationists should still be entangled in the problems of nineteenth-century science. Their own convictions come from a research tradition that stopped growing in that century.

2. In an earlier publication on this subject, writing in collaboration with others, we too closely assimilated natural selection to chemical selection of the efficient under thermodynamic imperatives (Weber et al. 1989). We wish to correct that impression now and to declare our belief that fitness measures cannot be reduced to the language and mathematics of energetics. From this fact it does not follow, however, that energetics is causally or explanatorily irrelevant to evolution by means of natural selection.

3. We know that even proteins would have folded up into compact globular structures, for even random sequences of amino acids will fold up into globular structures similar to those that now serve as enzymes in biological activity (Shakhnovich and Gutin 1990; Chan and Dill 1991). Even generic proteins, moreover, would have a higher-level structure possessing the potential of acting as a catalyst. In an aqueous environment, these peptide chains would fold up into ordered patterns of helix and extended chain to constitute the secondary structure of proteins (Dill 1985). Patterned modules consisting of several segments of

secondary structure fold to produce supersecondary structure. These in turn fold up in such a way that hydrophobic amino acids are buried and hydrophilic ones are exposed to water to produce the overall globular or tertiary structure with which biological enzymatic activity is associated.

As long as mutations in the amino acid sequence do not affect the tertiary structure-function, the lower-level structures are free to explore the surrounding sequence space. This phenomenon gives rise to neutral molecular evolution and to molecular clocks (Wilson, Carlson, and White 1977; Kimura 1983). There is an even greater degree of neutrality in the DNA that codes for proteins because the genetic code itself has redundancies, thereby giving nucleotide sequences even more plasticity in evolution, illustrated nicely in the term *hierarchical embeddedness* (Morowitz and Smith 1982).

Over 95 percent of the enzymes of contemporary organisms are composed of several polypeptide chains, usually identical, which are held together by chemical attraction, such as hydrophobic interactions and hydrogen bonding. This assemblage of subunits is termed the *quaternary structure*. Quaternary structure may have arisen originally to make more efficient use of DNA or to reduce the chance of deleterious somatic mutations. In any case, the emergence of quaternary structure made possible a *new* phenomenon. Some enzymes with quaternary structure exhibit a cooperativity of kinetics not seen in single subunit enzymes, which in certain types of graphic plots of the relationship of the rate of enzyme activity and substrate concentration show curvature rather than linearity. These emergent kinetics allow for a regulation of enzyme activity by feedback from other metabolites. These feedback loops self-regulate energy and other metabolism and introduce further nonlinearities to cellular metabolism. For example, the enzyme phosphofructokinase catalyzes the rate-limiting reaction in glycolysis (the metabolic pathway that breaks the six-carbon glucose molecule to the three-carbon pyruvate in ten enzyme-catalyzed reactions) and uses ATP as one of its substrates. Yet this enzyme is inhibited by high levels of ATP, one of the net ultimate products of the glycolytic pathway of which phosphofructokinase is a constituent part. This regulation by ATP gives rise to an oscillation of the flow of glucose through the glycolytic pathway that is mathematically analogous to the chemical clock reactions, such as the BZ reaction. Further, it allows the cell an instant-by-instant control of the flux of glucose through the glycolytic pathway in response to the cell's metabolic needs for ATP. Biological regulation, as exemplified by phosphofructokinase, it appears, evolves for the higher-level function of the pathway in the cell in which the enzyme is embedded rather than for the specific chemical function of the reaction that the enzyme catalyzes.

Beyond the emergence of the self-regulation of metabolism, what is important is that the modular aspect of the molecular design of proteins allows for rapid evolution of new enzyme activity by mixing and fusing the different genetic elements that code for these modules, especially domains, and generates a novel structure, which can be fine-tuned by natural selection. We may presume this ability to have evolved rapidly because it has great selective value. At the level of complex organization of genes into multigene families for traits that are especially valuable for survival in some organisms, observations have been made that mechanisms exist that similarly allow for rapid (in a geological time sense) deployment of variants. This has been termed by Christopher Wills as the evolution of evolvability (Wills 1989). This is the same conclusion that Kauffman reached.

4. We do not wish to be misunderstood on this point. We are not saying that thermodynamics drives life directly. Rather, while the condition of being away from equilibrium sets the thermodynamic precondition for the energy-flow description of embryonic development, it is the specific nonlinear *kinetic* pathways, which conform to these conditions, that can give rise to the self-organization. The equations for chemical self-organization turn out to describe these pathways. Even before the work of Belousov, the English mathematician Alan Turing, the great mid-twentieth-century genius who was also instrumental in developing the theoretical basis of the computer, had worked out the basic concepts of chemical

reactions that could cause patterns. Turing coined the phrase *self-organization* to describe this process (Turing 1952). Turing used self-organizing pattern formation in chemical reactions to model how pattern could form and symmetry be broken in the biological process of morphogenesis, the process by which a fertilized egg develops into an embryo and onward to its adult form. He derived a system of equations that described how chemical reactions would change in space and time. Over one range of possible parameters, the equations have the usual homogeneous solutions. But just past the "critical point" there is observed instability, and bifurcation to a solution that has less homogeneity and more structure. The previously obtaining symmetry of space and/or time is thereby broken. Harrison shows how Turing's ideas still are vital guides to modeling the phenomena of developmental biology (Harrison 1988). It is worth noting how well these thermodynamic and kinetic descriptions of the morphogenesis of embryos fit with Kauffman's treatment of the self-organization of genetic regulatory systems. Kauffman's computer simulations, based on Boolean networks and cellular automata, reveal systems dynamics that are consistent with Harrison's analyses, based on Turing's chemical kinetics, and both are provocatively consistent with observed biological phenomena.

5. A different view has been suggested by Brooks and Wiley, whose *Evolution as Entropy* caused a flurry of discussion in the agitated 1980s (Brooks and Wiley 1986, 1988; Collier 1986; Weber et al. 1989; Hariri, Weber, and Olmsted 1990; Morowitz 1992). They claim that informational entropy (Shannon entropy, see Shannon and Weaver 1949) is a more general notion of which thermodynamic entropy is a specific example. In Brooks and Wiley's view, the second law of thermodynamics itself ensures self-organization in informational systems just as in energy-processing ones. They have made some converts (Salthe 1993). On the whole, the argument fails, if only because thermodynamics of any sort says very little about how it is to be obeyed. More fundamentally, there is no conservation principle for information comparable to that provided for energy by the first law of thermodynamics. The explanation for the dynamics that Brooks and Wiley model is more likely to be that the chemical basis of the genetic code, the translation and regulation of genetic information, *along with the action of selection*, constrains genetic regulatory systems to the fecund edge of order and chaos. Also, it should be recalled that the genetic information space and its regulation occurs within the context of cellular metabolism (Weber et al. 1989); without being embedded in this context, genetic information makes no sense. Furthermore, it codes for the catalysis of metabolic transformations that occur far from equilibrium (Weber 1991a). It is only at this metabolic interface that the genotype is expressed as a phenotype, which through morphogenesis becomes an organism that interacts with its environment.

CHAPTER 18

1. Subjective probabilities still survive, indeed thrive, among those who take a "Bayesian" approach to confirming hypotheses and to the problem of induction generally. This is, however, more an epistemological preoccupation than a scientific one.

2. Contingency and chance are not as opposed to causality in natural processes and human affairs as we might think, for propensities can causally skew outcomes. Gould cites the novel *A Fatal Inversion* by Ruth Rendell, writing as Barbara Vine (Vine 1987), as an elegant example of multiple contingencies leading to extraordinary consequences in human action (Gould 1989a). While it is absolutely true that the details of the trajectories of the individual characters' lives reflect the contingencies and contingent cascades of events in their situations, one cannot escape the sense that, despite these factors, not all possible trajectories are available to each character, but rather each character has propensities due to innate or acquired limitations and characteristics. Certain characters are prone to panic or make thoughtlessly rash decisions, others are apt to misjudge situations or to need acceptance,

and so forth. Hence, one can conceive a number of possible alternative outcomes for the characters, but their overall success or failure in life will reflect in part how their propensities interacted with their contingencies. This view of propensity in human action, assumed by generations of novelists and explicitly explored by George Eliot, suggests a role for internal dispositional factors that is stronger than the new "higgledy piggledy" of Gould but weaker than the old proposition that character is destiny. To understand what novelists apparently knew long before scientists did, we should consider the possibility that physical systems are not only lawlike, but that the physical evolution of the universe and of smaller-scale self-organizing systems reflects chance fluctuations that are amplified. In the evolution of biological systems, where there is a tension between lawlike generic tendencies and the contingencies upon which natural selection acts, we see an ever greater role for the effects of attractors, propensities, and contingencies. Naturally, this idea suggests that, just as Maxwell's statistics helped legitimate social science, so the new dynamics adds weight to the tacit epistemology and ontology of the humanities. (Cf. Hayles 1990)

3. The analysis in this paragraph should not be taken to imply that Maxwell was unaware of the sensitivity of statistical arrays to initial conditions and their consequent tendency to depart from the effects of averaging. When Maxwell says that "we may perhaps say that the observable regularities of nature belong to statistical molecular phenomena that have settled down into permanent stable condition," he is aware that not everything is like that (Maxwell 1920, 14). It was perhaps their suspicion that there are many phenomena that do not settle down in the way Maxwell describes that led Poincaré and Liapunov to study nonlinear dynamics. In any case, the computational capacities we have only recently acquired should not lead anyone to think that nonlinear dynamics itself was born yesterday. It was not. (See Kellert 1993; Collier 1993.)

References

Adams, M. B. 1968. The founding of population genetics: Contributions of the Chetverikov school, 1924–1934. *Journal of the History of Biology* 1:23–49.

———. 1980. Sergei Chetverikov: The Kol'tsov Institute and the evolutionary synthesis. In *The Evolutionary Synthesis*, pp. 242–278. E. Mayr and W. B. Provine, eds. Cambridge, MA: Harvard University Press.

———, ed. 1990. *The Wellborn Science: Eugenics in Germany, France, Brazil and Russia*. New York: Oxford University Press.

———, ed. 1994. *The Evolution of Theodosius Dobzhansky*. Princeton NJ: Princeton University Press.

Agassiz, L. 1857. *Contributions to the Natural History of the United States*. Boston: Little, Brown.

Allen, G. A. 1975. *Life Science in the Twentieth Century*. New York: Wiley.

———. 1978. *Thomas Hunt Morgan: The Man and His Science*. Princeton, N. J.: Princeton University Press.

Allen, T. F. H., and T. B. Starr. 1982. *Hierarchy: Perspectives for Ecological Complexity*. Chicago: University of Chicago Press.

Alvarez, L. W., W. Alvarez, F. Asaro, and H. V. Michel. 1980. Extraterrestrial cause for the cretaceous-tertiary extinction. *Science* 208:1095–1108.

Alvarez, W., and F. Asaro. 1990. What caused the mass extinction? An extraterrestrial impact. *Scientific American* 263 (4):78–84.

Appel, T. A. 1987. *The Cuvier-Geoffrey Debate: French Biology in the Decades before Darwin*. Oxford: Oxford University Press.

Avery, O. T., C. M. MacLeod, and M. McCarthy. 1944. Studies on the chemical nature of the substance inducing transformation of pneumoccocal types. Induction of transformation by a deoxyribosenucleic acid fraction isolated from pneumoccocus type III. *Journal of Experimental Medicine* 79:137–158.

Axelrod, R. 1984. *The Evolution of Cooperation*. New York: Basic Books.

Ayala, F. J. 1965a. Relative fitness of populations of *Drosophila serrata* and *Drosophila birchii*. *Genetics* 51:527–544

———. 1965b. Evolution of fitness in experimental populations of *Drosophila serrata*. *Science* 150:903–905.

———. 1968. Genotype, environment, and population numbers. *Science* 162:1453–1459.

————. 1974. Biological evolution: Natural selection or random walk? *American Scientist* 62:692–701.

————. 1982. The genetic structure of species. In *Perspectives on Evolution*, pp. 60–82. R. Milkman, ed. Sunderland, MA: Sinauer.

————. 1985. Reduction in biology: A recent challenge. In *Evolution at a Crossroads: The New Biology and the New Philosophy of Science*, pp. 65–80. D. J. Depew and B. H. Weber, eds. Cambridge, MA: MIT Press.

Baer, K. E. von. 1828. *Entwicklungsgeschichte der Thiere: Beobactung und Reflexion*. Königsberg: Bornträger.

Bak, P., and K. Chen. 1991. Self-organized criticality. *Scientific American* 264 (1):46–53.

Balme, D. M. 1962. Genos and eidos in Aristotle's biology. *Classical Quarterly* 12:81–98.

Bartel, D. P., and J. W. Szostak. 1993. Isolation of new ribozymes from a large pool of random sequences. *Science* 261:1411–1418.

Bateson, W. 1928. *Scientific Papers*. R. C. Punnett, ed. Cambridge: Cambridge University Press.

Beatty, J. 1981. What's wrong with the received view of evolutionary theory? In P. D. Asquith and R. M. Giere, eds. *PSA 1980*, 2:341–55. East Lansing, Mich.: Philosophy of Science Association.

————. 1984. Chance and natural selection. *Philosophy of Science* 51:183–211.

————. 1985. Speaking of species: Darwin's strategy. In *The Darwinian Heritage*, pp. 265–281. D. Kohn, ed. Princeton, NJ: Princeton University Press.

————. 1986. The synthesis and the synthetic theory. In *Integrating Scientific Disciplines*, pp. 125–136. W. Bechtel, ed. Dordrecht: Nijhoff.

————. 1987. Dobzhansky and drift: Facts, values and chance in evolutionary biology. In *The Probabilistic Revolution*, 2:271–311. L. Krüger, G. Gigerenzer, and M. Morgan, eds. Cambridge, MA: MIT Press.

Bechtel, W. 1986. *Integrating Scientific Disciplines*. Dordrecht: Nijhoff.

————. 1993. Integrating sciences by creating new disciplines: The case of cell biology. *Biology and Philosophy* 8:277–299.

Bechtel, W., and R. C. Richardson. 1993. *Discovering Complexity: Decomposition and Localization as Strategies in Scientific Research*. Princeton, NJ: Princeton University Press.

Beer, G. 1983. *Darwin's Plots: Evolutionary Narrative in Darwin, George Eliot and Nineteenth-Century Fiction*. London: Routledge and Kegan Paul.

Bell, C. 1833. *The Hand: Its Mechanism and Vital Endowments as Evincing Design*. London: W. Pickering.

Belousov, B. P. 1958. Oscillation reaction and its mechanisms [in Russian]. *Sborn Referat. Radiat. Meditsin Za*. (Collection of abstracts on radiation medicine). Moscow: Medgiz.

Benzer, S. 1955. Fine structure of a genetic region in bacteriophage. *Proceedings of the National Academy of Science* (USA) 41:344–354.

————. 1962. The fine structure of the gene. *Scientific American* 206(1):70–84.

Berrill, N. J. 1961. *Growth, Development and Pattern*. San Francisco: Freeman.

Berrill, N. J., and C. K. Liu. 1948. Germplasm, Weismann, and Hydrozoa. *Quarterly Review of Biology* 23:124–132.

Black, M. 1962. *Models and Metaphors*. Ithaca, NY: Cornell University Press.

Bogen, J., and J. Woodward. 1988. Saving the phenomena. *Philosophical Review* 97:303–352.

Bonner, J. T. 1988. *The Evolution of Complexity by Means of Natural Selection*. Princeton, NJ: Princeton University Press.

Bowlby, J. 1991. *Charles Darwin: A New Life*. New York: Norton.

Bowler, P. J. 1983. *Evolution: The History of an Idea*. Berkeley and Los Angeles: University of California Press. Rev. ed., 1989.

———. 1988. *The Non-Darwinian Revolution*. Baltimore: Johns Hopkins University Press.

———. 1989. *The Mendelian Revolution: The Emergence of Hereditarian Concepts in Modern Science and Society*. Baltimore: Johns Hopkins University Press.

———. 1990. *Charles Darwin: The Man and His Influence*. Oxford: Basil Blackwell.

Bowring, S. A., J. P. Grotzinger, C. E. Isachsen, A. H. Knoll, S. M. Pelechaty, and P. Kolosor. 1993. Calibrating rates of early Cambrian evolution. *Science* 261:1293–1298.

Box, J. F. 1978. *R. A. Fisher: The Life of a Scientist*. New York: Wiley.

Boyd, R., P. Gasper, and J. D. Trout. 1991. *The Philosophy of Science*. Cambridge, Mass.: MIT Press.

Bradie, M. 1980. Models, metaphors, and scientific realism. *Nature and System* 2:3–20.

———. 1984. The metaphorical character of science. *Philosophica Naturalis* 21:229–243.

———. 1986. Assessing evolutionary epistemologies. *Biology and Philosophy* 1:401–459.

Brandon, R. N. 1978. Adaptation and evolutionary theory. *Studies in History and Philosophy of Science* 9:181–206.

———. 1981. Biological teleology: Questions and explanations. *Studies in History and Philosophy of Science* 12:91–105.

———. 1985. Adaptive explanations: Are adaptations for the good of replicators or interactors? In *Evolution at a Crossroads: The New Biology and the New Philosophy of Science*, pp. 81–96. D. J. Depew and B. H. Weber, eds. Cambridge, MA: MIT Press.

———. 1990. *Adaptation and Environment*. Princeton, NJ: Princeton University Press.

Brandon, R. N., and R. M. Burian. 1984. *Genes, Organisms, Populations: Controversies over the Units of Selection*. Cambridge, MA: MIT Press.

Brannigan, A. 1981. *The Social Basis of Scientific Discoveries*. Cambridge: Cambridge University Press.

Brewster, D. 1831. *The Life of Sir Isaac Newton*. London: John Murray.

Brooks, D. R. and E. O. Wiley. 1986. *Evolution as Entropy: Toward a Unified Theory of Biology*. Chicago: University of Chicago Press.

———. 1988. *Evolution as Entropy: Toward a Unified Theory of Biology*. 2d ed. Chicago: University of Chicago Press.

Brown, H. 1977. *Perception, Theory and Commitment: The New Philosophy of Science*. Chicago: University of Chicago Press.

Brown, R. H. 1986. Rhetoric and the science of history: The debate between evolutionism and empiricism as a conflict of metaphors. *Quarterly Journal of Speech* 72:148–161.

Brush, S. G. 1983. *Statistical Physics and the Atomic Theory of Matter, from Boyle and Newton to Landau and Onsager*. Princeton, NJ: Princeton University Press.

Buckle, H. T. 1857. *History of Civilization in England*. London: Longmans.

Buffon, G. L. 1954. *Oeuvres philosophiques*. J. Piveteau, ed. Paris: Presses universitaires de France.

Burian, R. M. 1977. More than a marriage of convenience: On the inextricability of history and philosophy of science. *Philosophy of Science* 44:1–42.

———. 1983. Adaptation. In *Dimensions of Darwinism*, pp. 287–314. M. Grene, ed. Cambridge: Cambridge University Press.

———. 1985. On conceptual change in biology: The case of the gene. In *Evolution at a Crossroads: The New Biology and the New Philosophy of Science*, pp. 21–42. D. J. Depew and B. H. Weber, eds. Cambridge, MA: MIT Press.

———. 1988. Challenges to the evolutionary synthesis. *Evolutionary Biology* 23:247–269.

———. 1992. How the choice of experimental organism matters: Biological practices and discipline boundaries. *Synthese* 92:151–166.

———. 1993. Unification and coherence as methodological objectives in the biological sciences. *Biology and Philosophy* 8:301–318.

———. 1994. Dobzhansky on evolutionary dynamics. In *The Evolution of Theodosius Dobzhansky*. M. Adams, ed. Princeton, N.J.: Princeton University Press.

Burian, R. M., J. Gayon, and D. Zallen. 1988. The singular fate of genetics in the history of French biology. *Journal of the History of Biology* 21:357–402.

Burian, R. M., and R. C. Richardson. 1991. Form and order in evolutionary biology: Stuart Kauffman's transformation of theoretical biology. *PSA 1990*, 2:267–87.

Bush, S. G. 1983. *Statistical Physics and the Atomic Theory of Matter from Boyle and Newton to Landau and Onsager*. Princeton, NJ: Princeton University Press.

Buss, L. W. 1987. *The Evolution of Individuality*. Princeton, NJ: Princeton University Press.

Butler, S. 1879. *Evolution, Old and New*. London: Hardwick and Bogue.

Cain, A. J., and P. M. Shepard. 1954. Natural selection in *Cepaea*. *Genetics* 39:89–116.

Cairns, J., J. Overbaugh, and S. Miller. 1988. The origin of mutants. *Nature* 335:142–145.

Callebaut, W. 1993. *Taking the Naturalistic Turn: Or How Real Philosophy of Science Is Done*. Chicago: University of Chicago Press.

Callendar, L. A. 1988. Gregor Mendel—an opponent of descent with modification. *History of Science* 26:41–75.

Campbell, D. T. 1974. "Downward causation" in hierarchically organised biological systems. In *Studies in the Philosophy of Biology*, pp. 179–186. F. J. Ayala and Th. Dobzhansky, eds. Berkeley and Los Angeles: University of California Press.

———. 1988. A general "selection theory" as implemented in biological evolution and social belief-transmission-with-modification in science. *Biology and Philosophy* 3:171–177.

Campbell, J. 1982. *Grammatical Man*. New York: Simon and Schuster.

Campbell, J. H. 1985. An organizational interpretation of evolution. In *Evolution at a Crossroads: The New Biology and the New Philosophy of Science*, pp. 133–167. D. J. Depew and B. H. Weber, eds. Cambridge, MA: MIT Press.

———. 1987. The new gene and its evolution. In *Rates of Evolution*, pp. 283–309. K. Campbell and M. F. Day, eds. London: Allen and Unwin.

Cann, R. L., M. Stoneking, and A. C. Wilson. 1987. Mitochondrial DNA and human evolution. *Nature* 325:31–36.

Carlson, E. A. 1966. *The Gene: A Critical History.* Philadelphia: Saunders.

Carter, C. W., Jr., and J. Kraut. 1974. A proposed model for interaction of polypeptides with RNA. *Proceedings of the National Academy of Science USA* 71:283–287.

Cartwright, N. 1983. *How the Laws of Physics Lie.* Oxford: Oxford University Press.

Cech, T. R. 1990. Self-splicing of group 1 introns. *Annual Reviews of Biochemistry* 59:543–568.

Chambers, R. 1844. *Vestiges of the Natural History of Creation.* London: John Churchill.

Chan, H. S., and Dill, K. A. 1991. Compact polymers. In *Conformations and Forces in Protein Folding,* pp. 43–66. B. T. Nall and K. A. Dill, eds. Washington, DC: American Association for the Advancement of Science.

Charlesworth, B. R., R. Lande, and M. Slatkin. 1982. A neo-Darwinian commentary on macroevolution. *Evolution* 36:474–498.

Chetverikov, S. S. 1926. On certain aspects of evolutionary processes from the standpoint of genetics. *Zhurnad Exp. Biol.* 1:3–54 (Russian); English translation in *Proceedings of the American Philosophical Society* 105:167–195 (1959).

Child, C. M. 1915. *Individuality in Organisms.* Chicago: University of Chicago.

Churchill, F. R. 1974. William Johannsen and the genotype concept. *Journal of the History of Biology* 7:5–30.

Clark, R. W. 1968. *JBS: The Life and Work of J. B. S. Haldane.* New York: Coward-McCann.

Clements, F. E. 1916. *Plant Succession: An Analysis of the Development of Vegetation.* Publication 242. Washington, DC: Carnegie Institute.

Cohen, I. B. 1960. *The Birth of a New Physics.* New York: Norton.

Coleman, W. 1971. *Biology in the Nineteenth Century: Problems of Form, Function, and Transformation.* New York: Wiley.

———. 1986. Evolution into ecology: The strategy of Warming's ecological plant geography. *Journal of the History of Biology* 19:181–196.

Collier, J. 1986. Entropy in evolution. *Biology and Philosophy* 1:5–24.

———. 1993. Holism and the new physics. *Descant* 79–80:135–153.

Colp, R., Jr. 1977. *To Be an Invalid: The Illness of Charles Darwin.* Chicago: University of Chicago Press.

———. 1987. Charles Darwin's "insufferable grief." *Free Associations* 9:7–24.

Conklin, E. G. 1915. *Heredity and Environment in the Development of Man.* Princeton, NJ: Princeton University Press.

Conry, Y. 1974. *Introduction de darwinisme en France au XIX^{e}me siècle.* Paris: VRIN.

Cooper, J. 1990. Metaphysics in Aristotle's embryology. In *Biologie, logique et métaphysique chez Aristote,* pp. 55–84. D. Devereux and P. Pelligrin, eds. Paris: CNRS.

Corsi, P. 1988. *The Age of Lamarck: Evolutionary Theories in France, 1790–1830.* Berkeley and Los Angeles: University of California Press.

Corsi, P., and Weindling, P. J. 1985. Darwinism in Germany, France, and Italy. In *The Darwinian Heritage,* pp. 683–729. D. Kohn, ed. Princeton, NJ: Princeton University Press.

Courtillot, V. E. 1990. What caused the mass extinction? A volcanic eruption. *Scientific American* 263(4):85–92.

Coveney, P., and R. Highfield. 1991. *The Arrow of Time.* New York: Fawcett Columbine.

Crick, F. 1957. On protein synthesis. *Symposia of the Society for Experimental Biology* 12:138–163.

———. 1966. *Of Molecules and Men.* Seattle: University of Washington Press.

———. 1988. *What Mad Pursuit: A Personal View of Scientific Discovery.* New York: Basic Books.

Crow, J. F. 1990. Sewall Wright's place in twentieth-century biology. *Journal of the History of Biology* 23:57–89.

Culotta, E. 1991. How many genes had to change to produce corn? *Science* 252:1792–1793.

Cunningham, A., and N. Jardine, eds. 1990. *Romanticism and the Sciences.* Cambridge: Cambridge University Press.

Darden, L. 1986. Relations among fields in the evolutionary synthesis. In *Integrating Scientific Disciplines,* pp. 113–123. W. Bechtel, ed. Dordrecht: Martinus Nijhoff.

———. 1991. *Theory Change in Science: Strategies from Mendelian Genetics.* New York: Oxford University Press.

Darden, L., and N. Maull, 1977. Interfield theories. *Philosophy of Science* 44:43–64.

Darwin, C. R. 1839. *Journal of Researches into the Geology and Natural History of the Various Countries Visited by H. M. S. Beagle under the Command of Captain Fitzroy, R. N. from 1832 to 1836.* London: Henry Colburn.

———. 1859. *On the Origin of Species by Means of Natural Selection or the Preservation of Favored Races in the Struggle for Life.* London: John Murray, and New York: D. Appleton, 1860. A facsimile reprint, with an introduction by E. Mayr, was published in 1964, Cambridge, MA: Harvard University Press.

———. 1868. *The Variation of Animals and Plants under Domestication.* London: John Murray.

———. 1871. *The Descent of Man.* London: John Murray.

———. 1872. *On the Origin of Species by Means of Natural Selection, or the Preservation of Favored Races in the Struggle for Life.* 6th ed. London: John Murray. New York: D. Appleton and Co. 1873.

———. 1958. *The Autobiography of Charles Darwin.* N. Barlow, ed. London: Collins.

———. 1987. *Charles Darwin's Notebooks, 1836–1844: Geology, Transmutation of Species, Metaphysical Enquiries.* Transcribed and edited by P. H. Barrett, P. J. Gautrey, S. Herbert, D. Kohn, and S. Smith. Ithaca, NY: Cornell University Press.

Darwin, F. 1887. *The Life and Letters of Charles Darwin.* Vols. 1–3. London: John Murray; New York: D. Appleton and Co., 2 vols. Quotations are from the American edition.

Darwin, F., and A. C. Seward. 1903. *More Letters of Charles Darwin.* 2 vols. London: John Murray; New York: D. Appleton, 2 vols..

Dawkins, R. 1976. *The Selfish Gene.* Oxford: Oxford University Press.

———. 1982. *The Extended Phenotype: The Long Reach of the Gene.* Oxford: Oxford University Press.

———. 1986. *The Blind Watchmaker.* New York: Norton.

———. 1989. *The Selfish Gene*. 2nd ed. Oxford: Oxford University Press.

———. 1992. Progress. In *Keywords in Evolutionary Biology*, pp. 263–272. E. F. Keller and E. A. Lloyd, eds. Cambridge, Mass.: Harvard University Press.

Deamer, D. W. 1985. Boundary structures are formed by organic components of the Murchison carbonaceous chondrite. *Nature* 317:792–794.

———. 1986. Role of amphiphilic compounds in the evolution of membrane structure on the early earth. *Origins of Life* 17:3–25.

———. 1992. Polycyclic aromaic hydrocarbons: primitive pigment systems in the prebiotic environment. *Advances in Space Research* 12 (4):183–189.

Deamer, D. W., and E. Harang. 1990. Light-dependent pH gradients are generated in liposomes containing ferricyanide. *BioSystems* 24:1–4.

Deamer, D. W., and R. M. Pashley. 1989. Amphiphilic components of the Murchison carbonaceous chondrite: Surface properties and membrane formation. *Origin of Life and Evolution of the Biosphere* 19:21–38.

de Beer, G., ed. 1958. *Evolution by Natural Selection*. Cambridge: Cambridge University Press.

de Chardin, T. 1959. *The Phenomenon of Man*. B. Wall, trans. New York: Harper & Row.

Degler, C. N. 1991. *In Search of Human Nature: The Decline and Revival of Darwinism in American Social Thought*. New York: Oxford University Press.

Delbrück, M. 1940a. Absorption of bacteriophages under various physiological conditions in the host. *Journal of General Physiology* 23:631–642.

———. 1940b. The growth of bacteriophage and lysis of the host. *Journal of General Physiology* 23:643–660.

———. 1971. Aristotle-totle-totle. In *Of Microbes and Life*. J. Monod and E. Borek, eds. New York: Columbia University Press.

Delbrück, M., N. W. Timofeeff-Ressovsky, and K. G. Zimmer. 1935. Über die Natur der Genmutation und Genstruktur. *Nachr. Ges. Wiss. Göttingen Math.-Phys. Kl.* 6(13):190–245.

Denbigh, K. G., and J. S. Denbigh. 1985. *Entropy in Relation to Incomplete Knowledge*. Cambridge: Cambridge University Press.

Depew, D. J. 1985. Narrativism, cosmopolitanism, and historical epistemology. *Clio* 14:357–378.

———. 1986. Nonequilibrium thermodynamics and evolution: A philosophical perspective. *Philosophica* 37:27–58.

———. 1988. Music, politics and philosophy in Aristotle's ideal state. In *A Companion to Aristotle's Politics*. D. Keyt and F. Miller, eds. London: Blackwell.

Depew, D. J., and B. H. Weber. 1985. Innovation and tradition in evolutionary theory: An interpretive afterword. In *Evolution at a Crossroads: The New Biology and the New Philosophy of Science*, pp. 227–260. D. J. Depew and B. H. Weber. Cambridge, MA: MIT Press.

———. 1988. Consequences of nonequilibrium thermodynamics for the Darwinian tradition. In *Entropy, Information and Evolution: New Perspectives on Physical and Biological Evolution*, pp. 317–354. Cambridge, MA: MIT Press.

———. 1989. The evolution of the Darwinian research tradition. *Systems Research* 6:255–263.

Desmond, A. 1989. *The Politics of Evolution*. Chicago: University of Chicago Press.

Desmond, A., and J. Moore. 1991. *Darwin.* London: Michael Joseph.

Devany, R. L. 1989. *An Introduction to Chaotic Dynamical Systems.* Redwood City, CA: Addison-Wesley.

Devereux, D., and P. Pellegrin. 1990. *Biologie, logique et métaphysique chez Aristote.* Paris: CNRS.

Dewey, J. 1894. Review of Ward's *Psychic Factors in Civilization.* In *John Dewey: Early Works 4.* J. A. Bydston, ed. Carbondale: Southern Illinois University Press, 1967–1972.

———. 1910. The influence of Darwinism on philosophy. In *The Influence of Darwin on Philosophy and Other Essays.* New York: Holt.

Diamond, J. 1992. *The Third Chimpanzee: The Evolution and Future of the Human Animal.* New York: Harper Collins.

Dickerson, R. E. 1971. Sequence and structure homologies in bacterial and mammalian-type cytochromes. *Journal of Molecular Biology* 57:1–15.

———. 1980. The cytochromes c: An exercise in scientific serendipity. In *The Evolution of Protein Structure and Function: A Symposium in Honor of Professor Emil L. Smith,* pp. 173–202. D. S. Sigman and M. A. B. Bazier, eds. New York: Academic Press.

Dickerson, R. E., R. Timkovich, and R. J. Almassy. 1976. The cytochrome fold and the evolution of bacterial energy metabolism. *Journal of Molecular Biology* 100:473–491.

Dill, K. A. 1985. Theory for the folding and stability of globular proteins. *Biochemistry* 24:1501–1509.

Di Trocchio, F. 1991. Mendel's experiments: A reinterpretation. *Journal of the History of Biology* 24:485–519.

Dobzhansky, Th. 1937. *Genetics and the Origin of the Species.* New York: Columbia University Press.

———. 1941. *Genetics and the Origin of the Species,* 2d ed. New York: Columbia University Press.

———. 1951. *Genetics and the Origin of the Species.* 3d ed. New York: Columbia University Press.

———. 1962. *Mankind Evolving.* New Haven: Yale University Press.

———. 1973. *Genetic Diversity and Human Equality.* New York: Basic Books.

———. 1980. The birth of the genetic theory of evolution in the Soviet Union in the 1920's. In *The Evolutionary Synthesis,* pp. 229–242. E. Mayr and W. Provine, eds. Cambridge, MA: Harvard University Press.

Dobzhansky, Th., F. J. Ayala, G. L. Stebbins, and J. W. Valentine. 1977. *Evolution.* San Francisco: Freeman.

Doolittle, W. F. 1987. The origin and function of intervening sequences in DNA: A review. *American Naturalist* 130:915–928.

Doolittle, W. F., and C. Sapienza. 1980. Selfish genes, the phenotypic paradigm and genomic evolution. *Nature* 284:601–603.

Dover, G. 1982. Molecular drive: A cohesive mode of species evolution. *Nature* 299:111–117.

Dunn, L. C. *A Short History of Genetics.* New York: McGraw-Hill. Reprint ed., Ames: Iowa State University Press, 1991.

Dyke, C. 1981. *The Philosophy of Economics.* Englewood Cliffs, NJ: Prentice-Hall.

———. 1988. *The Evolutionary Dynamics of Complex Systems*. Oxford: Oxford University Press.

Dyke, C., and D. J. Depew. 1988. Natural selection as explanation of last resort. *Revista di Biologia* 81(1):115–129.

Eigen, M. and P. Schuster. 1979. *The Hypercycle*. Berlin: Springer-Verlag.

———. 1982. Stages of emerging life: Five principles of early organization. *Journal of Molecular Biology* 19:47–61.

Eisley, L. 1958. *Darwin's Century: Evolution and the Men Who Discovered It*. New York: Doubleday.

Eldredge, N. 1985. *Unfinished Synthesis: Biological Hierarchies and Modern Evolutionary Thought*. New York: Oxford University Press.

———. 1989. *Macroevolutionary Dynamics: Species, Niches, and Adaptive Peaks*. New York: McGraw-Hill.

Eldredge, N., and S. J. Gould. 1972. Punctuated equilibria: An alternative to phyletic gradualism. In *Models in Paleobiology*, pp. 82–115. T. J. M. Schopf, ed. San Francisco: Freeman Cooper.

Eldredge, N., and M. Grene. 1992. *Interactions: The Biological Context of Social Systems*. New York: Columbia University Press.

Ellis, E. L., and M. Delbrück. 1939. The growth of bacteriophage. *Journal of General Physiology* 22:365–384.

Elton, C. S. 1927. *Animal Ecology*. London: Sidgwick and Jackson.

Endler, J. 1986a. *Natural Selection of the Wild*. Princeton, NJ: Princeton University Press.

———. 1986b. The newer synthesis? Some conceptual problems in evolutionary biology. *Oxford Surveys in Evolutionary Biology* 3:224–243.

Ereshefsky, M. 1992. *The Units of Evolution: Essays on the Nature of Species*. Cambridge, MA: MIT Press.

Estabrook, A. H. 1916. *The Jukes in 1916*. Washington, DC: Carnegie Institution.

Field, R. J. and L. György 1993. *Chaos in Chemistry and Biochemistry*. Singapore: World Scientific.

Fine, A. 1986. *The Shaky Game: Einstein, Realism and Quantum Theory*. Chicago: University of Chicago Press.

Finley, M. I. 1973. *The Ancient Economy*. Berkeley and Los Angeles: University of California Press.

Fisch, M., and S. Schaffer, eds. 1991. *William Whewell: A Composite Portrait*. Cambridge: Cambridge University Press.

Fischer, E. P., and C. Lipson. 1988. *Thinking about Science: Max Delbrück and the Origin of Molecular Biology*. New York: Norton.

Fisher, R. A. 1918. The correlation between relations on the supposition of Mendelian inheritance. *Transactions of the Royal Society of Edinburgh* 52:399–433.

———. 1922. On the dominance ratio. *Proceedings of the Royal Society of Edinburgh* 42:321–341.

———. 1925. *Statistical Methods for Research Workers*. Edinburgh: Oliver and Boyd.

———. 1930. *The Genetical Theory of Natural Selection*. Oxford: Oxford University Press. 2nd rev. ed., New York: Dover, 1958.

———. 1936. Has Mendel's work been rediscovered? *Annals of Science* 1:115–137.

Fitch, W. M., and E. Margoliash. 1967. Construction of phylogenetic trees: A method based on mutation distances as estimated from cytochrome c sequences is of general applicability. *Science* 155:279–284.

Ford, E. B. 1980. Some recollections pertaining to the evolutionary synthesis. In *The Evolutionary Synthesis*, pp. 334–342. E. Mayr and W. B. Provine, eds. Cambridge, MA: Harvard University Press.

Foucault, M. 1971. *The Order of Things: An Archaeology of the Human Sciences*. New York: Pantheon.

Fox, R. F. 1988. *Energy and the Evolution of Life*. New York: Freeman.

Fox, S. W. 1965. Simulated natural experiments in spontaneous organization of morphological units from protenoid. In *The Origins of Prebiological Systems and Their Molecular Matrices*, pp. 361–382. S. W. Fox, ed. New York: Academic Press.

———. 1980. Metabolic microspheres. *Naturwissenschaften* 67: 378–383.

———. 1984. Proteinoid experiments and evolutionary theory. In *Beyond Neo-Darwinism: An Introduction to the New Evolutionary Paradigm*, pp. 15–60. M.-W. Ho and P. T. Sanders, eds. London: Academic Press.

———. 1988. *The Emergence of Life: Darwinian Evolution from the Inside*. New York: Basic Books.

Fruton, J. S. 1972. *Molecules and Life: Historical Essays on the Interplay of Chemistry and Biology*. New York: Wiley-Interscience.

Furth, M. 1988. *Substance, Form and Psyche: An Aristotelian Metaphysics*. Cambridge: Cambridge University Press.

Gaissinovitch, A. E. 1980. The origins of Soviet genetics and the struggle with Lamarckism, 1922–29. *Journal of the History of Biology* 13:1–51.

Gallie, W. B. 1964. *Philosophy and the Historical Understanding*. New York: Schocken.

Galton, F. 1869. *Hereditary Genesis*. London: Macmillan.

———. 1875–1876. Typical laws of heredity. *Journal of the Royal Institution* 8:282–301.

———. 1877. Typical laws of inheritance. *Nature* 15:512–514.

———. 1888. Correlations and their measurement. *Proceedings of the Royal Society*, London 45:135–145.

———. 1889. *Natural Inheritance*. London: Macmillan.

Garfinkel, A. 1980. *Forms of Explanation*. New Haven, CT: Yale University Press.

———. 1987. The slime mold *Dictyostelium* as a model of self-organization in social systems. In *Self-Organizing Systems: The Emergence of Order*, pp. 181–212. F. E. Bates, ed. New York: Plenum.

Gasper, P. 1992. Reduction and instrumentalism in genetics. *Philosophy of Science* 59:655–670.

Gatlin, L. 1972. *Information Theory and the Living System*. New York: Columbia University Press.

Gayon, J. 1992. *Darwin et l'apres-Darwin: Une Historie de l'hypothèse de selection naturelle*. Paris: Editions Kine.

Gee, H. 1992. Statistical cloud over African Eden. *Nature* 355:583.

Ghiselin, M. T. 1969. *The Triumph of the Darwinian Method*. Berkeley: University of California Press.

———. 1974. A radical solution to the species problem. *Systematic Zoology* 23:536–544.

Gibbons, A. 1992. Mitochondrial Eve: wounded, but not dead yet. *Science* 257:873–875.

Gibson, J. J. 1966. *The Senses Considered as Perceptual Systems*. Boston: Houghton Mifflin.

Gibson, J. J. 1986. *The Ecological Approach to Visual Perception*. Hillsdale, NJ: Laurence Erlbaum.

Giere, R. N. 1979a. Propensity and necessity. *Synthese* 40:439–451.

———. 1979b. *Understanding Scientific Reasoning*. New York: Holt, Rinehart and Winston.

———. 1984. Toward a unified theory of science. In *Science and Reality*, pp. 5–31. J. T. Cusing, C. F. Delaney, and G. M. Gutting, eds. Notre Dame, IN: Notre Dame University Press.

———. 1988. *Explaining Science: A Cognitive Approach*. Chicago: University of Chicago Press.

Gigerenzer, G., Z. Swijtink, T. Porter, L. Daston, J. Beatty, and L. Krüger. 1989. *The Empire of Chance: How Probability Changed Science and Everyday Life*. Cambridge: Cambridge University Press.

Gilbert, S. F. 1978. Embryological origins of gene theory. *Journal of the History of Biology* 11:307–351.

———. 1988. Cellular politics: Ernest Everett Just, Richard B. Goldschmidt, and the attempt to reconcile embryology and genetics. *The American Development of Biology*, pp. 311–346. in R. Rainger, K. R. Benson, and J. Maienschein eds. Philadelphia: University of Pennsylvania Press.

———. 1991. Epigenetic landscaping. Waddington's use of cell fate bifurcation diagrams. *Biology and Philosophy* 6:135–154.

———. 1992. Cells in search of community: Critiques of Weismannism and selectable units in ontogeny. *Biology and Philosophy* 8:473–487.

Gilbert, W. 1978. Why genes in pieces? *Nature* 271:501.

Gillespie, J. H. 1991. *The Causes of Molecular Evolution*. New York: Oxford University Press.

Gilmour, J. S. L., and J. W. Gregor. 1939. Demes: A suggested new terminology. *Nature* 144:333–334.

Gleick, J. 1987. *Chaos: Making a New Science*. New York: Viking.

Godfrey-Smith, P. 1992. Additivity and the units of selection. *PSA 1992* 1:315–328.

Goldschmidt, R. B. 1938. *Physiological Genetics*. New York: McGraw-Hill.

———. 1940. *The Material Basis of Evolution*. New Haven: Yale University Press. Reprint ed. with an introduction by S. J. Gould, 1982.

———. 1960. *In and Out of the Ivory Tower: The Autobiography of Richard B. Goldschmidt*. Seattle: University of Washington Press.

Goodwin, B. C. 1984. A relational or field theory of reproduction and its evolutionary implications. In *Beyond Neo-Darwinism: An Introduction to the New Evolutionary Paradigm*, pp. 219–241. M.-W. Ho and P. T. Saunders, eds. London: Academic Press.

———. 1988. Morphogenesis and heredity. In *Evolutionary Processes and Metaphors*, pp. 145–162. M.-W. Ho and S. Fox, eds. Chichester: Wiley-Interscience.

————. 1989a. A structuralist research programme in developmental biology. In *Dynamic Structures in Biology*, pp. 49–61. B. Goodwin, A. Sibatani, and G. Webster, eds. Edinburgh: Edinburgh University Press.

————. 1989b. Evolution and the generative order. In *Theoretical Biology*, pp. 89–100. B. C. Goodwin and P. Saunders, eds. Edinburgh: Edinburgh University Press.

Goodwin, B. C., and P. Saunders, eds. 1989. *Theoretical Biology: Epigenetic and Evolutionary Order from Complex Systems*. Edinburgh: Edinburgh University Press. Reprint ed. Baltimore, MD: Johns Hopkins University Press, 1982.

Goodwin, B. C., and L. E. H. Trainor. 1980. A field description of the cleavage process in embryogenesis. *Journal of Theoretical Biology* 85:757–770.

Gotthelf, A. 1976. Aristotle's conception of final causality. *Review of Metaphysics* 30:226–254.

Gotthelf, A., and J. G. Lennox. 1987. *Philosophical Issues in Aristotle's Biology*. Cambridge: Cambridge University Press.

Gould, S. J. 1977a. *Ontogeny and Phylogeny*. Cambridge: Harvard University Press.

————. 1977b. *Ever Since Darwin: Reflections in Natural History*. New York: Norton.

————. 1980. Is a new and general theory of evolution emerging? *Paleobiology* 6:119–130.

————. 1982a. The meaning of punctuated equilibrium and its role in validating a hierarchical approach to macroevolution. In *Perspectives on Evolution*, pp. 83–104. R. Milkman, ed. Sunderland, MA: Sinauer.

————. 1982b. Darwinism and the expansion of evolutionary theory. *Science* 216:380–387.

————. 1982c. The uses of heresy: An introduction to Richard Goldschmidt's *The Material Basis of Evolution*. In R. Goldschmidt. *The Material Basis of Evolution*, pp. xiii–xlii. Reprint ed. New Haven, CT: Yale University Press.

————. 1983. The hardening of the modern synthesis. In *Dimensions of Darwinism*, pp. 71–93. M. Grene, ed. Cambridge: Cambridge University Press.

————. 1987. *Time's Arrow, Time's Cycle*. Cambridge, MA: Harvard University Press.

————. 1988. On replacing the idea of progress with an operational notion of directionality. In *Evolutionary Progress*, pp. 319–338. M. H. Nitecki, ed. Chicago: University of Chicago Press.

————. 1989a. *Wonderful Life: The Burgess Shale and the Nature of History*. New York: Norton.

————. 1989b. Through a lens, darkly: Do species change by random molecular shifts or natural selection? *Natural History* 98 (9):16–24.

————. 1990. Darwin and Paley meet the invisible hand. *Natural History* 99(11): 8–12.

————. J. 1991. Knight takes bishop. In *Bully for Brontosaurus: Reflections in Natural History*, pp. 385–401. New York: Norton.

Gould, S., J. and N. Eldredge. 1977. Punctuated equilibria: The tempo and mode of evolution reconsidered. *Paleobiology* 3:115–151.

————. 1993. Punctuated equilibrium comes of age. *Nature* 366:223–227.

Gould, S. J., N. L. Gilinsky, and R. Z. German. 1987. Asymmetry of lineages and the direction of evolutionary time. *Science* 236:1437–1441.

Gould, S. J., and R. C. Lewontin. 1979. The spandrels of San Marco and the panglossian paradigm: A critique of the adaptationist programme. *Proceedings of the Royal Society London B* 205:581–598.

Gould, S. J., and E. S. Vrba. 1982. Exaptation—a missing term in the science of form. *Paleobiology* 8:4–15.

Grant, E. 1971. *Physical Science in the Middle Ages.* New York: Wiley.

Grant, V. 1971. *Plant Speciation.* New York: Columbia University Press.

Graunt, John. 1662. *Observations upon the Bills of Mortality.* London. Reprint of the second edition, 1676, in *The Economic Writings of Sir William Petty.* C. H. Hull, ed. Cambridge: Cambridge University Press.

Gray, A. 1860a. Design versus necessity: Discussion between two readers of Darwin's treatise on the origin of species, upon its natural theology. *American Journal of Science and Arts* 30:226–239.

———. 1860b. Darwin on the origin of species. *Atlantic Monthly* 6:109–116, 229–239.

———. 1874. Charles Darwin: A sketch. *Nature.*

———. 1876. *Darwiniana.* New York: Appleton. Reprint ed., Cambridge, MA: Harvard University Press, 1963.

———. 1889. *Scientific Papers of Asa Gray.* C. S. Sargent, ed. Boston.

Grene, M. 1961. Statistics and selection. *British Journal for the Philosophy of Science* 12:25–42.

———. 1974. *The Understanding of Nature.* Dordrecht: Reidel.

———. 1983. *Dimensions of Darwinism: Themes and Counterthemes in Twentieth-Century Evolutionary Theory.* Cambridge: Cambridge University Press.

———. 1985. Perception, interpretation and the sciences: Toward a new philosophy of science. In *Evolution at a Crossroads: The New Biology and the New Philosophy of Science,* pp. 1–20. D. J. Depew and B. H. Weber, eds. Cambridge, MA: MIT Press.

———. 1990a. Is evolution at a crossroads? *Evolutionary Biology* 23:247–269.

———. 1990b. Evolution, "typology" and "population thinking." *American Philosophical Quarterly* 27:237–244.

Gruber, H. E. 1974. *Darwin on Man.* New York: Dutton.

———. 1985. Going the limit: Toward the construction of Darwin's theory. In *The Darwinian Heritage,* pp. 9–34. D. Kohn, ed. Princeton, NJ: Princeton University Press.

Gutting, G. 1980. *Paradigms and Revolutions: Appraisals and Applications of Thomas Kuhn's Philosophy of Science.* Notre Dame, IN: Notre Dame University Press.

Hacking, I. 1975. *The Emergence of Probability: A Philosophical Study of Early Ideas about Probability and Statistical Inference.* Cambridge: Cambridge University Press.

———. 1983. *Representing and Intervening: Introductory Topics in the Philosophy of Natural Science.* Cambridge: Cambridge University Press.

———. 1987. Was there a probabilistic revolution 1800–1930? In *The Probabilistic Revolution,* 1:45–55. L. Krüger, L. J. Datson, and M. Heidelberger, eds. Cambridge, MA: MIT Press.

———. 1990. *The Taming of Chance.* Cambridge: Cambridge University Press.

Haeckel, E. 1866. *Generelle Morphologie der Organismen.* Berlin: Reimer.

———. 1879. *The Evolution of Man: A Popular Exposition of the Principal Points of Human Ontogeny and Phylogeny.* New York: Appleton.

———. 1905. *The Evolution of Man.* 2 vols. J. McCabe, trans. London: Watts.

Haldane, J. B. S. 1924. A mathematical theory of natural and artificial selection. *Transactions of the Cambridge Philosophical Society* 23:19–41.

———. 1932. *The Causes of Evolution.* London: Longmans, Green. Reprint ed., Princeton, NJ: Princeton University Press, 1990.

Hall, B. 1989. Adaptive evolution that requires multiple spontaneous mutations, I: Mutations involving an insertion sequence. *Genetics* 120:887–897.

———. 1990. *Evolutionary Developmental Biology.* London: Chapman and Hall.

Hamilton, W. D. 1964. The genetical evolution of social behavior, I and II. *Journal of Theoretical Biology* 7:1–52.

Haraway, D. J. 1976. *Crystals, Fabrics, and Fields: Metaphors of Organicism in Twentieth-Century Biology.* New Haven, CT: Yale University Press.

Hariri, A., B. H. Weber, and J. Olmsted III. 1990. On the validity of Shannon-information calculations for molecular biological sequences. *Journal of Theoretical Biology* 147:235–254.

Harmon, P. M. 1982. *Energy, Force and Matter: The Conceptual Development of Nineteenth-Century Physics.* Cambridge: Cambridge University Press.

Harold, F. M. 1990. To shape a cell: An inquiry into the causes of morphogenesis in microorganisms. *Microbiology Reviews* 54:381–431.

Harrison, L. G. 1988. Kinetic theory of living pattern and form and its possible relationship to evolution. In *Entropy, Information and Evolution: New Perspectives on Physical and Biological Evolution,* pp. 53–74. B. H. Weber, D. J. Depew, and J. D. Smith, eds. Cambridge, MA: MIT Press.

Hayles, N. K. 1990. *Chaos Bound: Orderly Disorder in Contemporary Literature and Science.* Ithaca, NY: Cornell University Press.

Hempel, C. G. 1966. *Philosophy of Natural Science.* Englewood Cliffs, NJ: Prentice-Hall.

Hennig, W. 1966. *Phylogenetic Systematics.* Urbana: University of Illinois Press.

Herbert, S. 1991. Charles Darwin as a prospective geological author. *Journal of the History of Biology* 24:159–192.

Herschel, J. F. W. 1830. *A Preliminary Discourse on the Study of Natural Philosophy.* London: Longman, Rees, Orme, Brown, and Green. Reprint ed., Chicago: University of Chicago Press, 1987.

———. 1850. Quetelet on probabilities. *Edinburgh Review* 92:1–57.

———. 1861. *Physical Geography of the Globe.* London: Longmans, Green.

Hershey, A., and M. Chase. 1952. Independent functions of viral protein and nucleic acid in growth of bacteriophage. *Journal of General Physiology* 36:39–56.

Hesse, M. 1966. *Models and Analogies in Science.* Notre Dame, IN: University of Notre Dame Press.

———. 1980. *Revolutions and Reconstructions in the Philosophy of Science.* Bloomington: Indiana University Press.

Hexter, J. H. 1971. *Doing History.* Bloomington: Indiana University Press.

Hilts, V. L. 1978. *Aliis exterendum,* or, the origins of the Statistical Society of London. *Isis* 69:21–43.

Himmelfarb, G. 1959. *Darwin and the Darwinian Revolution.* Garden City, NY: Doubleday.

————. 1968. Varieties of social Darwinism. In *Victorian Minds,* pp. 314–332. New York: Knopf.

Hirshmann, A. O. 1977. *The Passions and the Interests: Political Arguments for Capitalism before Its Triumph.* Princeton, NJ: Princeton University Press.

Ho, M.-W., and S. W. Fox. 1988. Process and metaphors in evolution. In *Evolutionary Process and Metaphors,* pp. 1–17. Chichester: Wiley-Interscience.

Ho, M.-W., and P. T. Saunders. 1984. *Beyond Neo-Darwinism: An Introduction to the New Evolutionary Paradigm.* London: Academic Press.

Hodge, M. J. S. 1972. The universal gestation of nature: Chamber's *Vestiges* and *Explanations. Journal of the History of Biology* 5:127–151.

————. 1977. The structure and strategy of Darwin's "long argument." *British Journal for the History of Science* 10:237–246.

————. 1982. Darwin and the laws of the animate part of the terrestrial system (1835–1837): On the Lyellian origins of his zoonomical explanatory program. *Studies in the History of Biology* 6:1–106.

————. 1983. The development of Darwin's general biological theorizing. In *Evolution from Molecules to Men,* pp. 43–62. Cambridge: Cambridge University Press.

————. 1985. Darwin as a lifelong generation theorist. In *The Darwinian Heritage,* pp. 207–244. D. Kohn, ed. Princeton, NJ: Princeton University Press.

————. 1987. Natural selection as a causal, empirical and probabilistic theory. In *The Probabilistic Revolution,* 2:233–270. L. Krüger, G. Gigerenzer, and M. S. Morgan, eds. Cambridge, MA: MIT Press.

————. 1989a. Generation and the origin of species, 1837–1839: A historiographical suggestion. *British Journal for the History of Science* 22:267–281.

————. 1989b. Darwin's theory and Darwin's argument. In *What the Philosophy of Biology Is: Essays Dedicated to David Hull,* pp. 163–182. Dordrecht: Kluwer.

————. 1992a. Biology and philosophy (including ideology): A study of Fisher and Wright. In *The Founders of Evolutionary Genetics,* pp. 231–293. S. Sarkar, ed. Dordrecht: Kluwer.

————. 1992b. Darwin's argument in the *Origin. Philosophy of Science* 59:461–464.

————. 1992c. Natural selection: Historical perspectives. In *Keywords in Evolutionary Biology,* pp. 212–219. E. F. Keller and E. A. Lloyd, eds. Cambridge, MA: Harvard University Press.

Hodge, M. J. S., and P. Kohn. 1985. The immediate origins of natural selection. In *The Darwinian Heritage,* pp. 185–206. P. Kohn, ed. Princeton: Princeton University Press.

Hogben, L. T. 1931. *Genetic Principles in Medicine and Social Science.* London: Allen and Unwin.

Holland, S. K., K. Harlos, and C. C. F. Blake. 1987. Deriving the generic structure of fibronectin type II domains from the prothrombin Kringle I crystal structure. *EMBO* 6:1875–1880.

Holmes, F. L. 1964. Introduction, in J. Liebig, *Animal Chemistry,* facsimile of 1842 Cambridge translation. New York: Johnson Reprint Corporation.

Hopkins, F. G. 1913. The dynamic side of biochemistry. *Report of the British Association* 1913:652. Reprinted in *Hopkins and Biochemistry,* J. Needham and E. Baldwin, eds. 1949, Cambridge: Heffer, pp. 136–159.

Horan, B. 1994. The statistical character of evolutionary theory. *Philosophy of Science* 61:76–95.

Hull, D. L. 1965. The effect of essentialism on taxonomy: Two thousand years of stasis. *British Journal for the Philosophy of Science* 15:314–26,16:1–18.

———. 1973. *Darwin and His Critics*. Cambridge, MA : Harvard University Press.

———. 1974. *Philosophy of Biological Science*. Englewood Cliffs, NJ: Prentice-Hall.

———. 1976. Are species really individuals? *Systematic Zoology* 25:174–191.

———. 1978. A matter of individuality. *Philosophy of Science* 45:335–360.

———. 1980. Individuality and selection. *Annual Reviews of Ecology and Systematics* 11:311–332.

———. 1981. Units of evolution: A metaphysical essay. In *The Philosophy of Evolution*, pp. 23–44. U. L. Jensen and R. Harré, eds. Brighton: Harvester Press. Reprinted in *Genes, Organisms, Population: Controversies over the Units of Selection*, R. N. Brandon and R. M. Burian, eds. Cambridge, MA: MIT Press, pp. 142–160.

———. 1985. Darwinism as an historical entity: A historiographic proposal. In *The Darwinian Heritage*. D. Kohn, ed. Princeton, NJ: Princeton University Press.

———. 1988a. *Science as a Process: An Evolutionary Account of the Social and Conceptual Development of Science*. Chicago: University of Chicago Press.

———. 1988b. A mechanism and its metaphysics: An evolutionary account of the social and conceptual development of science. *Biology and Philosophy* 3:123–155.

Hume, D. 1779. *Dialogues Concerning Natural Religion*. Edinburgh. Edited by J. V. Price, Oxford: Oxford University Press, 1977.

Hutchinson, G. E. 1965. *The Ecological Theater and the Evolutionary Play*. New Haven, CT: Yale University Press.

Hutton, J. 1788. Theory of the earth or an investigation of the laws observable in the composition, dissolution, and restoration of land upon the globe. *Transactions of the Royal Society* (Edinburgh) 1:209–304.

———. 1795. *Theory of the Earth with Proofs and Illustrations*. 2 vols. Edinburgh.

Huxley, J. S. 1942. *Evolution, the Modern Synthesis*. London: Allen and Unwin.

Huxley, T. H. 1894. *Evolution and Ethics*. London: Macmillan.

Jackson, E. A. 1991. *Perspectives on Nonlinear Dynamics*. Cambridge: Cambridge University Press.

Jacob, F. 1973. *The Logic of Life: A History of Heredity*. New York: Pantheon.

———. 1977. Evolution and tinkering. *Science* 196:1161–1166.

———. 1982. *The Possible and the Actual*. Seattle: University of Washington Press.

Jacob, F., and J. Monod. 1961. Genetic regulatory mechanisms in the synthesis of proteins. *Journal of Molecular Biology* 3:318–356.

Jenkin, F. 1867. The origin of species. *North British Review* 45:277–318.

Jensen, A. R. 1969. How much can we boost IQ and scholastic achievement? *Harvard Educational Review* 39:1–123.

Johannsen, W. 1903. *Über Erblichkeit in Populationen und in reinen Linien*. Jena: Fischer.

Johnson, L. 1981. The thermodynamic origin of ecosystems. *Canadian Journal of Fisheries and Aquatic Science* 38:571–590.

————. 1988. The thermodynamic origin of ecosystems: A tale of broken symmetry. In *Entropy, Information, and Evolution: New Perspectives on Physical and Biological Evolution*, pp. 75–105. B. H. Weber, D. J. Depew, and J. D. Smith, eds. Cambridge, MA: MIT Press.

————. 1990. The thermodynamics of ecosystems. In *Handbook of Environmental Chemistry*, vol. 1, pt. E: *The Natural Environment and Biogeochemical Cycles*, pp. 1–47. U. Hutzinger, ed. Heidelberg: Springer-Verlag.

————. 1992. An ecological approach to biosystem thermodynamics. *Biology and Philosophy* 7:35–60.

Johnsson, K., R. K. Allemann, H. Widmer, and S. A. Brenner. 1993. Synthesis, structure and activity of artificial, rationally designed catalytic polypeptides. *Nature* 365:530–532.

Jordan, D. S. 1898. *Footnotes to Evolution*. New York: Appleton.

————. 1905. The origin of species through isolation. *Science* 22:545–562.

————. 1925. Isolation with segregation in organic evolution. In *Smithsonian Report for 1925*, Washington, DC: Smithsonian Institute, pp. 321–326.

Judson, H. F. 1979. *The Eighth Day of Creation: The Makers of the Revolution in Biology*. New York: Simon and Schuster.

Jukes, T. H. 1987. Transitions, transversions and the molecular evolutionary clock. *Journal of Molecular Evolution* 26:87–98.

Jungck, J. R. 1983. Is the neo-Darwinian synthesis robust enough to withstand the challenge of recent discoveries in molecular biology and molecular evolution? *PSA* 2:322–330.

Kalmus, H. 1983. The scholastic origins of Mendel's concepts. *History of Science* 21:61–83.

Kant, I. 1790. *Critique of Judgment*. Berlin. Trans. J. H. Bernard, New York: Hafner, 1951.

Kauffman, S. A. 1985. Self-organization, selective adaptation, and its limits: A new pattern of inference in evolution and development. In *Evolution at a Crossroads: The New Biology and the New Philosophy of Science*, pp. 169–207. D. J. Depew and B. H. Weber, eds. Cambridge, MA: MIT Press.

————. 1989. Origins of order in evolution: Self-organization and selection. In *Theoretical Biology*, pp. 67–88. Edinburgh: Edinburgh University Press.

————. 1991. Antichaos and adaptation. *Scientific American* 265 (2):78–84.

————. 1993. *The Origins of Order: Self-Organization and Selection in Evolution*. New York: Oxford University Press.

Kay, L. E. 1993. *The Molecular Vision of Life: Caltech, the Rockefeller Foundation and the Rise of the New Biology*. New York: Oxford University Press.

Keller, E. F. 1983. *A Feeling for the Organism: The Life and Work of Barbara McClintock*. San Francisco: Freeman.

Keller, E. F., and E. A. Lloyd. 1992. *Keywords in Evolutionary Biology*. Cambridge, MA: Harvard University Press.

Kellert, S. H. 1993. *In the Wake of Chaos: Unpredictable Order in Dynamical Systems*. Chicago: University of Chicago Press.

Kellogg, V. 1907. *Darwinism Today: A Discussion of Present-Day Scientific Criticism of the Darwinian Selection Theories*. New York: Holt.

Kerr, R. A. 1991. Dinosaurs and friends snuffed out? *Science* 251:160–162.

————. 1992a. Extinction by a one-two comet punch? *Science* 255:160–161.

———. 1992b. Huge impact tied to mass extinction. *Science* 257:878–880.

Kettlewell, H. B. D. 1955. Selection experiments on industrial melanism in the *Lepidoptera*. *Heredity* 9:323–342.

———. 1956. Further selection experiments on industrial melanism in the *Lepidoptera*. *Heredity* 10:287–301.

———. 1973. *The Evolution of Melanism: The Study of a Recurring Necessity*. Oxford: Oxford University Press.

Kevles, D. 1985. *In the Name of Eugenics: Genetics and the Uses of Human Heredity*. New York: Knopf.

Kim, J. 1978. Supervenience and nomological incommensurables. *American Philosophical Quarterly*

Kimura, M. 1968. Evolutionary rate at the molecular level. *Nature* 217:624–626.

———. 1983. *The Neutral Theory of Molecular Evolution*. Cambridge: Cambridge University Press.

———. 1987. Molecular evolutionary clock and the neutral theory. *Journal of Molecular Evolution* 26:24–33.

King, J. L., and T. H. Jukes. 1969. Non-Darwinian evolution. *Science* 164:788–798.

King, M. -C., and A. C. Wilson. 1975. Evolution at two levels: Molecular similarities and biological differences between humans and chimpanzees. *Science* 188:107–116.

Kingsland, S. E. 1985. *Modeling Nature: Episodes in the History of Population Ecology*. Chicago: University of Chicago Press.

Kitcher, P. 1981. Explanatory unification. *Philosophy of Science* 48:507–531.

———. 1982a. *Abusing Science*. Cambridge, MA: MIT Press.

———. 1982b. Genes. *British Journal for the Philosophy of Science* 33:337–359.

———. 1984. 1953 and all that: A tale of two sciences. *The Philosophical Review* 93:335–373.

Kohler, R. E. 1982. *From Medical Chemistry to Biochemistry: The Making of a Biomedical Discipline*. Cambridge: Cambridge University Press.

Kohn, D., ed. 1985. *The Darwinian Heritage*. Princeton, NJ: Princeton University Press.

Kottler, M. J. 1985. Charles Darwin and Alfred Russel Wallace: Two decades of debate over natural selection. In *The Darwinian Heritage*, pp. 367–432. D. Kohn, ed. Princeton, NJ: Princeton University Press.

Krüger, L. 1987. The slow rise of probabilism: Philosophical arguments in the nineteenth century. In *The Probabilistic Revolution*. L. Krüger, L. Dasten, and M. Heidelberger, eds. Cambridge, MA: MIT Press.

Krüger, L., L. Daston, and M. Heidelberger. 1987. *The Probabilistic Revolution*. 2 vols. Cambridge, MA: MIT Press.

Kuhn, T. S. 1962. *The Structures of Scientific Revolutions*. Chicago: University of Chicago Press.

———. 1970. *The Structures of Scientific Revolutions*. 2d ed. Chicago: University of Chicago Press.

———. 1987. What are scientific revolutions? In *The Probabilistic Revolution*, 1:7–22. L. Krüger, L. J. Daston, and M. Heidelberger, eds. Cambridge, MA: MIT Press.

Lack, D. L. 1947. *Darwin's Finches*. Cambridge: Cambridge University Press.

Lakatos, I. 1970. Falsification and the methodology of scientific research programmes. In *Criticism and the Growth of Knowledge*, pp. 91–195. I. Lakatos and A. Musgrave, eds. Cambridge: Cambridge University Press.

———. 1978. *The Methodology of Scientific Research Programs*. Cambridge: Cambridge University Press.

Lakatos, I., and A. Musgrave. 1970. *Criticism and the Growth of Knowledge*. Cambridge: Cambridge University Press.

Lamarck, J. B. 1809. *Philosophie zoologique, ou Exposition des considérations relatives à l'histoire naturelle des animaux*. Paris. English trans. H. Elliot. 1914. *The Zoological Philosophy*. London: Macmillan.

Lamarck, J. B. 1815. *Histoire naturelle des animaux sans vertèbres*. Paris.

Langton, C. G. 1986. Studying artificial life with cellular automata. *Physica (D)*. 22:120–149.

———. 1989. *Artificial Life*. Reading, MA: Addison-Wesley.

———. 1992. Life at the edge of chaos. In *Artificial Life II*, pp. 41–91. C. G. Langton, C. Taylor, J. D. Farmer, and S. Rasmussen, eds. Reading, MA: Addison-Wesley.

Langton, C. G., C. Taylor, J. D. Farmer, and S. Rasmussen. 1992. *Artificial Life II. Reading, MA: Addison-Wesley*.

Laporte, L. F. 1990. The world into which Darwin led Simpson. *Journal of the History of Biology* 23:499–516.

Lauden, L. 1977. *Progress and Its Problems: Toward a Theory of Scientific Growth*. Berkeley and Los Angeles: University of California Press.

———. 1984. *Science and Values: The Aims of Science and Their Role in Scientific Debate*. Berkeley: University of California Press.

———. 1990. *Science and Relativism*. Chicago: University of Chicago Press.

Lederberg, J., and E. L. Tatum. 1946. Novel genotypes in mixed cultures of biochemical mutants of bacteria. *Cold Spring Harbor Symposia on Quantitative Biology* 11:113–114.

Leicester, H. M. 1974. *Development of Biochemical Concepts*. Cambridge, MA: Harvard University Press.

Lennox, J. 1982. Teleology, chance, and Aristotle's theory of spontaneous generation. *Journal of the History of Philosophy* 20 : 219–38.

———. 1987. Kinds, forms of kinds, and the more or less in Aristotle's biology. In *Issues in Aristotle's Biology*, pp. 339–359. A. Gotthelf and G. Lennox, eds. Cambridge: Cambridge University Press.

Lenoir, T. 1982. *The Strategy of Life: Teleology and Mechanics in Nineteenth Century German Biology*. Dordrecht and Boston: Reidel.

Levine, G. 1988. *Darwin and the Novelists*. Cambridge, MA: Harvard University Press.

Levins, R. 1966. The strategy of model building in population biology. *American Scientist* 54:421–431.

———. 1968. *Evolution in Changing Environments*. Princeton, NJ: Princeton University Press.

Levins, R., and R. Lewontin. 1985. *The Dialectical Biologist*. Cambridge, MA: Harvard University Press.

Lewin, R. 1992. *Complexity: Life at the Edge of Chaos*. New York: Macmillan.

Lewontin, R. C. 1970. The units of selection. *Annual Review of Ecology and Systematics* 1:1–18.

———. 1974. *The Genetic Basis of Evolutionary Change*. New York: Columbia University Press.

———. 1978. Adaptation. *Scientific American* 239:212–230.

———. 1980. Adattamento. *The Encyclopedia Einaudi*. Milan: Einaudi; reprinted as Adaptation in Sober 1984b, pp. 234–251.

———. 1983. Gene, organism and environment. In *Evolution from Molecules to Men*, pp. 273–285. D. S. Bendall, ed. Cambridge: Cambridge University Press.

———. 1991. Facts and the factitious in natural science. *Critical Inquiry* 18 (1):140–153.

———. 1992. *Biology as Ideology: The Doctrine of DNA*. New York: Harper Collins.

Lewontin, R. C., and J. L. Hubby. 1966. A molecular approach to the study of genic heterozygosity in natural populations, II: amount of variation and degree of heterozygosity in natural populations of *Drosophila pseudoobscura*. *Genetics* 54:595–609.

Lewontin, R. C., J. A. Moore, W. B. Provine, and B. Wallace, eds. 1981. *Dobzhansky's Genetics of Natural Populations I–XLIII*. New York: Columbia University Press.

Lewontin, R. C., S. Rose, and L. J. Kamin. 1984. *Not in Our Genes: Biology, Ideology, and Human Nature*. New York: Pantheon.

Liebig, J. 1842. *Die Organische Chemie in ihrer Anwendung auf Physiologie und Pathologie*. Braunschweig: Vieweg.

Lloyd, E. A. 1983. The nature of Darwin's support for the theory of natural selection. *Philosophy of Science* 50:112–129.

———. 1988. *The Structure and Confirmation of Evolutionary Theory*. Westport, CT: Greenwood Press.

Loomis, W. F. 1988. *Four Billion Years: An Essay on the Evolution of Genes and Organisms*. Sunderland, MA: Sinauer.

Lorenz, E. N. 1963. Deterministic nonperiodic flow. *Journal of Atmospheric Science* 20:282–293.

———. 1993. *The Essence of Chaos*. Seattle, WA: University of Washington Press.

Lotka, A. J. 1922. Contribution to the energetics of evolution. *Proceedings of the National Academy of Science: USA* 51:148–154.

———. 1924. *Elements of Physical Biology*. Baltimore: Williams and Wilkins. Reprinted as *Elements of Mathematical Biology*. New York: Dover, 1956 (page numbers are from this edition).

———. 1945. The law of evolution as a maximal principle. *Human Biology* 17:167–194.

Lovejoy, A. O. 1936. *The Great Chain of Being*. Cambridge, MA: Harvard University Press.

Lovelock, J. E. 1979. *Gaia: A New Look at Life on Earth*. Oxford: Oxford University Press.

Luria, S. E., and M. Delbrück. 1943. Mutations of bacteria from virus sensitivity to virus resistance. *Genetics* 28:491–511.

Luria, S. E., M. Delbrück, and T. F. Anderson. 1943. Electron microscope studies of bacterial viruses. *Journal of Bacteriology* 46:57–77.

Lyell, C. 1830–1833. *Principles of Geology, Being an Attempt to Explain the Former Changes of the Earth's Surface, by Reference to Causes Now in Operation*. 3 vols. London: John Murray. Facsimile reprint, Chicago: University of Chicago Press, 1990.

———. 1881. *Life, Letters and Journals of Sir Charles Lyell*. London: John Murray.

———. 1970. *Sir Charles Lyell's Scientific Journals on the Species Question*. New Haven, CT: Yale University Press.

MacIntyre, A. 1977. Epistemological crisis, dramatic narrative and the philosophy of science. *Monist* 60:453–471.

———. 1981. *After Virtue*. Notre Dame, IN: Notre Dame University Press.

Magnus, D. In Defense of Natural History: David Starr Jordan and the Role of Isolation in Evolution. Ph.D. dissertation, Stanford University.

Maienschein, J. 1991. *Transforming Traditions in American Biology*. Baltimore: Johns Hopkins University Press.

Maienschein, J., R. Rainger, and K. R. Benson. 1980. Were American morphologists in revolt? *Journal of the History of Biology* 14:83–87.

Malthus, T. R. 1798. *An Essay on the Principle of Population, as It Affects the Future Improvement of Society, with Remarks on the Speculations of Mr. Goodwin, M. Condorcet, and Other Writers*. London: J. Johnson.

Manier, E. 1978. *The Young Darwin and His Cultural Circle*. Dordrecht: Reidel.

Manuel, F. E. 1968. *A Portrait of Isaac Newton*. Cambridge, MA: Harvard University Press.

Margulis, L. 1981. *Symbiosis in Cell Evolution: Life and Its Environment on the Early Earth*. San Francisco: Freeman.

Marx, K. and F. Engels. 1937 (1965). *Selected Correspondence*. I. Lasker, trans. Moscow.

Maxwell, J. C. 1890. *Scientific Papers*. 2 vols. W. D. Niven, ed. Cambridge: Cambridge University Press.

———. 1920. *Matter and Motion*. New York: Macmillan

May, R. M. 1974. Biological populations with nonoverlapping generations: Stable points, stable cycles, and chaos. *Science* 186:645–647.

———. 1987. Chaos and the dynamics of biological populations. In *Dynamical Chaos*, M. V. Berry, I. C. Percival, and N. O. Weiss, eds. Princeton, NJ: Princeton University Press, pp. 27–43.

Maynard Smith, J. 1978. Optimization theory in evolution. *Annual Review of Ecology and Systematics* 9:31–56.

———. 1983. *Evolution and the Theory of Games*. Cambridge: Cambridge University Press.

———. 1988. *Games, Sex, and Evolution*. New York: Harvester Wheatsheaf.

Maynard Smith, J., R. Burian, S. Kauffman, P. Alberch, J. Campbell, B. Goodwin, R. Laude, D. Raup, and L. Wolpert. 1985. Developmental constraints and evolution: A perspective from the Mountain Lake conference on development and evolution. *Quarterly Reviews of Biology* 60:265–287.

Mayr, E. 1942. *Systematics and the Origin of Species*. New York: Columbia University Press.

———. 1954. Change of genetic environment and evolution. In *Evolution as a Process*, pp. 157–188. J. Huxley, A. C. Hardy, and E. B. Ford, eds. London: Allen and Unwin.

———. 1963. *Animal Species and Evolution*. Cambridge, MA: Harvard University Press.

———. 1977. Darwin and natural selection: How Darwin may have discovered his highly unconventional theory. *American Scientist* 65:321–327.

———. 1978. Evolution. *Scientific American* 239:46–55.

———. 1980. Prologue: Some thoughts on the history of the evolutionary synthesis. In *The Evolutionary Synthesis: Perspectives on the Unification of Biology*, pp. 1–48. E. Mayr and W. B. Provine, eds. Cambridge, MA: Harvard University Press.

————. 1982. *The Growth of Biological Thought: Diversity, Evolution, and Inheritance.* Cambridge, MA: Harvard University Press.

————. 1984. The unity of the genotype. In *Genes, Organisms, Populations: Controversies over the Units of Selection,* pp. 69–84. R. N. Brandon and R. M. Burian, eds. Cambridge, MA: MIT Press.

————. 1985. How does biology differ from the physical sciences? In *Evolution at a Crossroads: The New Biology and the New Philosophy of Science,* pp. 43–63. D. J. Depew and B. H. Weber, eds. Cambridge, MA: MIT Press.

————. 1988. *Toward a New Philosophy of Biology: Observations of an Evolutionist.* Cambridge, MA: Harvard University Press.

————. 1991. *One Long Argument: Charles Darwin and the Genesis of Evolutionary Thought.* Cambridge, MA: Harvard University Press.

Mayr, E., and W. Provine. 1980. *Perspectives on the Unification of Biology.* Cambridge, MA: Harvard University Press.

Mendel, J. [G.]. 1865. Versuche über Pflanzen-hybriden. *Verhandlungen des naturforschenden Vereines in Brünn* 4:3–47.

Midgley, M. 1985. *Evolution as Religion: Strange Hopes and Stranger Fears.* London: Methuen.

Mikulecky, D. C. 1983. Network thermodynamics: A candidate for a common language for theoretical and experimental biology. *American Journal of Physiology* 245:R1–R9.

————. 1990. Network thermodynamics: A unifying approach to dynamic nonlinear living systems. In *Theoretical Ecosystems Ecology: The Network Perspective.* T. P. Burns and M. Higashi, eds. Cambridge: Cambridge University Press.

Mill, J. S. 1843. *A System of Logic.* London: Longmans, Green.

Miller, D., ed. 1985. *Popper Selections.* Princeton, NJ: Princeton University Press.

Miller, S., and L. E. Orgel. 1974. *The Origins of Life on Earth.* Englewood Cliffs, NJ: Prentice-Hall.

Mills, S. K., and J. H. Beatty. 1979. The propensity interpretation of fitness. *Philosophy of Science* 46:263–286.

Milne-Edwards, H. 1844. Considérations sur quelques principes relatifs à la classification naturelle des animaux. *Annals des Science Naturelle* 3d ser. 1:65–99.

————. 1851. *Introduction à la zoologie générale ou considérations sur les tendence de la nature dans la constitution du regue animal.* Paris: Masson.

Mink, L. 1979. Narrative form as an instrument of cognition. In *The Writing of History,* R. H. Canary and H. Kozicke, eds., Madison: University of Wisconsin Press, pp. 130–150.

Mirowski, P. 1989. *More Heat Than Light.* Cambridge: Cambridge University Press.

Mishler, B. D., and M. J. Donoghue. 1992. Species concepts: A case for pluralism. In *The Units of Evolution: Essays on the Nature of Species,* pp. 121–137. M. Ereshefsky, ed. Cambridge, MA: MIT Press.

Mitchell, M., P. T. Hraber, and J. P. Crutchfield 1993. Revisiting the edge of chaos: cellular automata to perform computations. *Santa Fe Institute Working Paper 93-03-014.*

Mitchell, P. D. 1961. Coupling of phosphorylation to electron and hydrogen transfer by a chemiosmotic type of mechanism. *Nature* 191:144–148.

————. 1962. Metabolism, transport and morphogenesis: Which drives which? *Journal of General Microbiology* 29:25–37.

Monod, J. 1974. On the molecular theory of evolution. In *Problems of Scientific Revolutions: Progress and Obstacles to Progress in Science*. R. Harré, ed. Oxford: Oxford University Press.

Moore, J. R. 1977. Could Darwinism be introduced in France? *British Journal of the History of Science* 5:474–476.

———. 1979. *The Post-Darwinian Controversies: A Study of the Protestant Struggle to Come to Terms with Darwin in Great Britain and America, 1870–1900*. Cambridge: Cambridge University Press.

———. 1985. Darwin of Down: The evolutionist as Squarson-naturalist. In *The Darwinian Heritage*. D. Kohn, ed. Princeton, NJ: Princeton University Press.

Morgan, T. H. 1919. *The Physical Basis of Heredity*. New York: Lippincott.

Morowitz, H. 1986. Entropy and nonsense. *Biology and Philosophy*. 1:473–476.

———. 1992. *Beginnings of Cellular Life: Metabolism Recapitulates Biogenesis*. New Haven: Yale University Press.

Morowitz, H. J., D. W. Deamer, and T. Smith. 1991. Biogenesis as an evolutionary process. *Journal of Molecular Evolution* 33:207–208.

Morowitz, H. J., B. Heinz, and D. W. Deamer. 1988. The chemical logic of a minimum protocell. *Origins of Life and Evolution of the Biosphere* 18:281–287.

Morris, S. C. 1993. The fossil record and the early evolution of the Metazoa. *Nature* 361:215–225.

Muller, H. J. 1935. *Out of the Night: A Biologist's View of the Future*. New York: Vanguard.

Murray, J. D. 1989. *Mathematical Biology*. Berlin: Springer-Verlag.

Naess, A. 1983. The deep ecology movement: some philosophical aspects. *Philosophical Enquiry* 8:10–31.

Nagel, E. 1961. *The Structure of Science*. New York: Harcourt, Brace and World.

Needham, J. 1970. *The Chemistry of Life: Lectures on the History of Biochemistry*. Cambridge: Cambridge University Press.

———. 1984. Foreword to *Beyond Neo-Darwinism: An Introduction to the New Evolutionary Paradigm*, pp. vii–viii. M.-W. Ho and P. T. Saunders, eds. London: Academic Press.

Needham, J., and E. Baldwin. 1949. *Hopkins and Biochemistry*. Cambridge: Heffer.

Newton, I. 1687. *Philosophiae Naturalis Principia Mathematica*. London: Streater. English translation 1729, reprint ed., Berkeley and Los Angeles: University of California Press, 1934.

Nicolis, G., and I. Prigogine. 1989. *Exploring Complexity: An Introduction*. New York: Freeman.

Nitecki, M. H. 1988. *Evolutionary Progress*. Chicago: University of Chicago Press.

Norton, B. 1983. Fisher's entrance into evolutionary science: The role of eugenics. In *Dimensions of Darwinism: Themes and Counterthemes in Twentieth-Century Evolutionary Theory*. M. Grene, ed. Cambridge: Cambridge University Press.

Odum, E. 1953. *Fundamentals of Ecology*. Philadelphia: Saunders.

———. 1969. The strategy of ecosystem development. *Science* 164:262–270.

Odum, H. T. 1971. *Environment, Power and Society*. New York: Wiley Interscience.

———. 1988. Self-organization, transformity, and information. *Science* 242:1132–1139.

Odum, H. T., and E. Odum. 1982. *Energy Basis for Man and Nature*. 2d ed. New York: McGraw-Hill.

Officer, C. B., A. Hallam, C. L. Drake, and J. D. Devine. 1987. Late Cretaceous and paroxysmal Cretateous/Tertiary extinctions. *Nature* 326:143–149.

O'Grady, R. T., and D. R. Brooks. 1988. Teleology and biology. In *Entropy, Information, and Evolution: New Perspectives on Physical and Biological Evolution*, pp. 285–316. B. H. Weber, D. J. Depew, and J. D. Smith, eds. Cambridge, MA: MIT Press.

Olby, R. 1985. *Origins of Mendelism.* 2nd ed. Chicago: University of Chicago Press.

———. 1987. William Bateson's introduction of Mendelism to England: A reassessment. *British Journal for the History of Science* 20:399–420.

Olmsted, J. 1988. Observations on evolution. In *Entropy, Information and Evolution: New Perspectives on Physical and Biological Evolution*, pp. 243–261. B. H. Weber, D. J. Depew, and J. D. Smith, eds. Cambridge, MA: MIT Press.

Oparin, A. I. 1938. *The Origin of Life.* London: Macmillan. Reprint ed,, New York: Dover, 1953.

Ospovat, D. 1976. The influence of Karl Ernst von Baer's embryology, 1828–1859. *Journal of the History of Biology* 9:1–28.

———. 1977. Lyell's theory of climate. *Journal of the History of Biology* 10:317–339.

———. 1981. *The Development of Darwin's Theory: Natural History, Natural Theology and Natural Selection, 1838–1859.* Cambridge: Cambridge University Press.

Ott, E. 1993. *Chaos in Dynamical Systems.* Cambridge: Cambridge University Press.

Owen, R. 1992. *The Hunterian Lectures in Comparative Anatomy, May–June, 1837.* P. R. S. Sloan, ed. Chicago: University of Chicago Press.

Oyama, S. 1985. *The Ontogeny of Information.* Cambridge: Cambridge University Press.

———. 1988. Stasis, development and heredity. In *Evolutionary Processes and Metaphors,* pp. 255–274. M. W. Ho and S. W. Fox, eds. Chichester: Wiley-Interscience.

Page, R., G. E. Robinson, N. W. Calderone, and W. C. Rothenbuhler. 1987. Genetic structure, division of labor and the evolution of insect societies. In *The Genetics of Social Evolution.* M. D. Breed and R. E. Page, eds. Boulder, CO: Westview Press.

Pagels, H. R. 1988. *The Dreams of Reason: The Computer and the Rise of the Sciences of Complexity.* New York: Simon and Schuster.

Paley, W. 1802. *Natural Theology, or Evidences of the Existence and Attributes of the Deity Collected from the Appearances of Nature.* London: Fauldner.

Palladino, P. 1991. Defining ecology: Ecological theories, mathematical models and applied biology in the 1960's and 70's. *Journal of the History of Biology* 24 (2):223–243.

Paradis, J. and G. C. Williams. 1989. *Evolution and Ethics.* Princeton NJ: Princeton University Press.

Parshall, K. H. 1982. Varieties as incipient species: Darwin's numerical analysis. *Journal of the History of Biology* 15:191–214.

Patterson, H. E. H. 1992. The recognition concept of species. In *The Units of Evolution: Essays on the Nature of Species*, pp. 139–158. M. Ereshefsky, ed. Cambridge, MA: MIT Press.

Paul, D. B. 1984. Eugenics and the Left. *Journal of the History of Ideas* 45:567–590.

Pauling, L., H. A. Itaus, S. J. Singer, and I. C. Wells. 1949. Sickle cell anemia, a molecular disease. *Science* 110:543.

Peacocke, A. R. 1983. *An Introduction to the Physical Chemistry of Biological Organization.* Oxford: Oxford University Press.

Pearson, K. 1892. *The Grammar of Science*. London: Scott.

———. 1896. Mathematical contributions and the theory of evolution III: Regression, heredity and panmixia. London: *Philosophical Transactions of the Royal Society A* 187:253–318.

Pellegrin, P. 1986. *Aristotle's Classification of Animals*. A. Preus, trans. Berkeley: University of California Press.

Pierce, C. S. 1877. The fixation of belief. *Popular Science Monthly* 12:1–15.

Playfair, J. 1802. *Illustrations of the Huttonian Theory of the Earth*. Edinburgh.

Plotkin, H. C. 1988. Behavior and evolution. In *The Role of Behavior in Evolution*, pp. 1–17. H. C. Plotkin, ed. Cambridge, MA: MIT Press.

Popper, K. R. 1945. *The Open Society and Its Enemies*. 2 vols. London: Routledge.

———. 1959. *The Logic of Scientific Discovery*. London:. Hutchinson.

———. 1972. *Objective Knowledge*. Oxford: Oxford University Press.

———. 1974. Intellectual autobiography. In *The Philosophy of Karl Popper,* 1:3–181. P. A. Schilpp, ed. La Salle, IL: Open Court.

———. 1978. Natural selection and the emergence of mind. *Dialectica* 32:339–355.

———. 1980. Letter to the editor. *New Scientist* 87:611.

———. 1984. Evolutionary epistemology. In *Evolutionary Theory: Paths into the Future*. J. W. Polland, ed. Chichester: Wiley-Interscience.

———. 1990. *A World of Propensities*. Bristol: Thoemmes.

Porter, T. M. 1986. *The Rise of Statistical Thinking*. Princeton, NJ: Princeton University Press.

Price, R. 1972. Fisher's fundamental theorem made clear. *Annals of Human Genetics* 36:129–140.

Prigogine, I., and I. Stengers. 1984. *Order Out of Chaos: Man's New Dialogue with Nature*. New York: Bantam.

Prout, W. 1834. *Chemistry, Meteorology and the Function of Digestion Considered with Reference to Natural Theology*. London: William Pickering.

Provine, W. B. 1971. *The Origins of Theoretical Population Genetics*. Chicago: University of Chicago Press.

———. 1983. The development of Wright's theory of evolution. In *Dimensions of Darwinism*, pp. 43–70. M. Grene, ed. Cambridge: Cambridge University Press.

———. 1985. The Wright-Fisher controversy. *Oxford Surveys in Evolutionary Biology* 2:197–219.

———. 1986. *Sewall Wright and Evolutionary Biology*. Chicago: University of Chicago Press.

Punnett, R. C. 1915. *Mimicry in Butterflies*. Cambridge: Cambridge University Press.

———. 1950. Early days of genetics. *Heredity* 4:1–10.

Quetelet, L. A. J. 1835, translated 1842. *A Treatise on Man*. Facsimile edition of English translation 1969. Gainesville, FL: Scholars Facsimiles and Reprints.

———. 1847. De l'Influence de libre arbitre de l'hommes sur les faits sociaux. *Bulletin de la Commission Central de Statistique* 3:135–155.

———. 1848. *Du Système social et des loís quí le réqissent*. Paris: Guillaumin.

Rasmussen, N. 1991. The decline of recapitulationism in early 20th c. biology. *Journal of the History of Biology* 24 (1):51–89.

Rawls, J. 1970. *A Theory of Justice.* Cambridge, MA: Harvard University Press.

Recker, D. A. 1987. Causal efficacy: The structure of Darwin's argument strategy in the *"Origin of Species." Philosophy of Science* 54:147–175.

Rennie, J. 1993. DNA's new twists. *Scientific American* 266 (3):122–132.

Rescher, N., ed. 1985. *Reason and Rationality in Natural Science.* Pittsburgh, PA: University of Pittsburgh Press.

Ricardo, D. 1817. *On the Principles of Political Economy and Taxation.* London: John Murray.

Richards, R. J. 1987. *Darwin and the Emergence of Evolutionary Theories of Mind and Behavior.* Chicago: University of Chicago Press.

———. 1988. The moral foundations of the idea of evolutionary process: Darwin, Spencer and the neo-Darwinians. In *Evolutionary Progress,* pp. 129–148. M. H. Nitecki, ed. Chicago: University of Chicago Press.

———. 1992. *The Meaning of Evolution: The Morphological Construction and Ideological Reconstruction of Darwin's Theory.* Chicago: University of Chicago Press.

Richardson, R. C., and R. M. Burian. 1992. A defense of the propensity interpretation of fitness. *PSA 1992.* 1:349–362.

Robson, G. C., and O. W. Richards. 1936. *The Variation of Animals in Nature.* London: Longmans.

Romanes, G. 1886. Physiological selection: An additional suggestion on the origin of species. *Journal of the Linnean Society: Zoology* 19:337–411.

Rosen, R. 1991. *Life Itself: A Comprehensive Inquiry into the Nature, Origin, and Fabrication of Life.* New York: Columbia University Press.

Rosenberg, A. 1983. Fitness. *Journal of Philosophy* 80:457–473.

———. 1985. *The Structure of Biological Science.* Cambridge: Cambridge University Press.

———. 1989. From reductionism to instrumentalism. In *What the Philosophy of Biology Is: Essays Dedicated to David Hull,* pp. 245–262. M. Ruse, ed. Dordrecht: Kluwer.

Roux, W. 1894. The problem, method and scope of developmental mechanics. W. M. Wheeler, trans. *Biological Lectures of the MBL* 149–190.

Rudwick, M. J. S. 1970. The strategy of Lyell's *Principles of Geology. Isis* 61:4–33.

———. 1985. *The Great Devonian Controversy: The Shaping of Scientific Knowledge among Gentlemanly Specialists.* Chicago: University of Chicago Press.

Runnegar, B. 1982. The Cambrian explosion: Animals or fossils? *Journal of the Geological Society of Australia* 29:395–411.

Ruse, M. 1975. Darwin's debt to philosophy: An examination of the influence of the philosophical ideas of John F. W. Herschel and William Whewell on the development of Charles Darwin's theory of evolution. *Studies in the History and Philosophy of Science* 6:159–181.

———. 1979. *The Darwinian Revolution: Science Red in Tooth and Claw.* Chicago: University of Chicago Press.

———. 1984a. Is there a limit to our knowledge of evolution? *BioScience* 34 (2):100–104.

———. 1984b. The morality of the gene. *Monist* 67:167–199.

———. 1986. *Taking Darwin Seriously.* London: Basil Blackwell.

————. 1988a. *Philosophy of Biology Today.* Albany: State University of New York Press.

————. 1988b. Molecules to men: Evolutionary biology and thoughts of progress. In *Evolutionary Progress*, pp. 97–126. M. H. Nitecki, ed. Chicago: University of Chicago Press.

Sahlins, M. 1976. *The Uses and Abuses of Biology.* Ann Arbor: University of Michigan Press.

Salthe, S. N. 1985. *Evolutionary Hierarchical Systems: Their Structure and Representation.* New York: Columbia University Press.

————. 1991. Formal considerations on the origin of life. *Uroboros* 1:45–56.

————. 1992. Hierarchical non-equilibrium: self-organization as the new post-cybernetic perspective. In *New Perspectives on Cybernetics*, pp. 49–58. G. van der Vijver, ed. Dordrecht: Kluwer.

————. 1993. *Development in Evolution: Complexity and Change in Biology.* Cambridge, MA: MIT Press.

Sandefur, J. T. 1990. *Discrete Dynamical Systems: Theory and Applications.* Oxford: Clarendon Press, University of Oxford Press.

Sarich, V. M., and A. C. Wilson. 1966. Quantitative immunochemistry and the evolution of primate albumins: Microcomplement fixation. *Science* 154:1563–1566.

Sarkar, S. 1992. *The Founders of Evolutionary Genetics.* Dordrecht: Kluwer.

Sauve-Meyer, S. 1992. Aristotle, teleology and reduction. *Philosophical Reviews* 101 (4):791–825.

Sayer, A. 1975. *Rosalind Franklin and DNA.* New York: Norton.

Schaffer, W. M., and M. Kot. 1985a. Nearly one-dimensional dynamics in an epidemic. *Journal of Theoretical Biology* 112:403–427.

————. 1985b. Do strange attractors govern ecological systems? *Bioscience* 35:342–350.

Scherer, S. 1990. The protein molecular clock: Time for a reevaluation. *Evolutionary Biology* 24:83–106.

Schneider, E. D. 1988. Thermodynamics, ecological succession, and natural selection: A common thread. In *Entropy, Information, and Evolution: New Perspectives on Physical and Biological Evolution*, pp. 107–138. B. H. Weber, D. J. Depew, and J. D. Smith, eds. Cambridge, MA: MIT Press.

Schneider, E. D., and J. J. Kay. In press. Life as a manifestation of the second law of thermodynamics. *Computing and Mathematics.*

Schopf, J. W. 1983. *Earth's Earliest Biosphere.* Princeton, NJ: Princeton University Press.

Schrödinger, E. 1944. *What Is Life? The Physical Aspect of the Living Cell.* Cambridge: Cambridge University Press.

Schweber, S. S. 1977. The origin of the *Origin* revisited. *Journal of the History of Biology* 10:229–316.

————. 1979. The young Darwin. *Journal of the History of Biology* 12:175–192.

————. 1980. Darwin and the political economists. *Journal of the History of Biology* 13:195–289.

————. 1985. The wider British context of Darwin's theorizing. In *The Darwinian Heritage*, pp. 35–69. D. Kohn, ed. Princeton, NJ: Princeton University Press.

————. 1988. The correspondence of the young Darwin. *Journal of the History of Biology* 21:501–519.

———. 1989. John Herschel and Charles Darwin: A study in parallel lives. *Journal of the History of Biology* 22:1–71.

Scriven, M. 1959. Explanation and prediction in evolutionary theory. *Science* 130:477–482.

Scudo, F. M., and M. Acanfora. 1985. Darwin and Russian evolutionary biology. In *The Darwinian Heritage*, pp. 731–754. D. Kohn, ed. Princeton, NJ: Princeton University Press.

Secord, J. A. 1991. Discovery of a vocation: Darwin's early geology. *British Journal for the History of Science* 24:133–157.

Segerstrale, U. 1986. Colleagues in conflict: An "in vivo" analysis of the sociobiology controversy. *Biology and Philosophy* 1:53–88.

Selzer, J. 1993. *Understanding Scientific Prose*. Madison: University of Wisconsin Press.

Servos, J. W. 1990. *Physical Chemistry from Ostwald to Pauling: The Making of a Science in America*. Princeton, NJ: Princeton University Press.

Shakhnovich, E. I., and A. M. Gutin. 1990. Implications of thermodynamics of protein folding for evolution of primary sequences. *Nature* 346:773–775.

Shannon, C. E., and W. Weaver. 1949. *The Mathematical Theory of Communication*. Urbana: University of Illinois Press.

Shaw, R. 1984. *The Dripping Faucet as a Model Chaotic System*. Santa Cruz, CA: Aerial Press.

Simon, H. A. 1962. The architecture of complexity. *Proceedings of the American Philosophical Society* 106:467–482.

———. 1969. *The Sciences of the Artificial*. Cambridge, MA: MIT Press.

Simpson, G. G. 1944. *Tempo and Mode in Evolution*. New York: Columbia University Press.

———. 1961. *Principles of Animal Taxonomy*. New York: Columbia University Press.

———. 1964a. *This View of Life*. New York: Harcourt, Brace and World.

———. 1964b. Organisms and molecules in evolution. *Science* 146:1535–1538.

Skinner, B. F. 1971. *Beyond Freedom and Dignity*. New York: Knopf.

———. 1978. *Reflections on Behaviorism and Society*. Englewood Cliffs, NJ: Prentice-Hall.

Sloan, P. R. 1976. The Buffon-Linneaus controversy. *Isis* 67:356–375.

———. 1979. Buffon, German biology, and the historical interpretation of biological species. *British Journal for the History of Science* 12:109–153.

———. 1985. Darwin's invertebrate program, 1826–1836: preconditions for transformationism. In *The Darwinian Heritage*, pp. 71–120. D. Kohn, ed. Princeton, NJ: Princeton University Press.

———. 1986. Darwin, vital matter, and the transformation of species. *Journal of the History of Biology* 19:369–445.

Smart, J. J. C. 1963. *Philosophy and Scientific Realism*. London: Routledge and Kegan Paul.

Smith, A. 1776. *An Inquiry into the Nature and Causes of the Wealth of Nations*. London: Codell.

Smith, C., and M. N. Wise. 1989. *Energy and Empire: A Biographical Study of Lord Kelvin*. Cambridge: Cambridge University Press.

Smith, E. L. 1968. The evolution of proteins. *Harvey Lectures* 62:231–291.

Smith, T. F., and H. J. Morowitz. 1982. Between history and physics. *Journal of Molecular Evolution* 18:265–282.

Smocovitis, V. B. 1992. Unifying biology: The evolutionary synthesis and evolutionary biology. *Journal of the History of Biology* 25:1–65.

Sober, E. 1984a. *The Nature of Selection: Evolutionary Theory in Philosophical Focus.* Cambridge, MA: MIT Press.

———. 1984b. *Conceptual Issues in Evolutionary Biology: An Anthology.* Cambridge, MA: MIT Press.

———. Darwin on natural selection: A philosophical perspective. In *The Darwinian Heritage,* pp. 867–899. D. Kohn, ed. Princeton, NJ: Princeton University Press.

———. 1988. *Reconstructing the Past: Parsimony, Evolution, and Inference.* Cambridge, MA: MIT Press.

———. 1993. *Conceptual Issues in Evolutionary Biology: An Anthology.* 2d ed. Cambridge, MA: MIT Press.

Sober, E., and R. Lewontin. 1982. Artifact, cause and genic selection. *Philosophy of Science* 49:157–162.

Spencer, H. 1864. *The Principles of Biology.* London: Williams and Norgate.

Stanley, S. M. 1979. *Macroevolution: Pattern and Process.* San Francisco: Freeman.

Stebbins, G. L., and F. J. Ayala. 1981. Is a new evolutionary synthesis necessary? *Science* 213:267–271.

———. 1985. The evolution of Darwinism. *Scientific American* 253 (1):72–82.

Stein, C., and E. R. Sherwood. 1966. *The Origin of Genetics: A Mendel Sourcebook.* San Francisco: Freeman.

Stewart, I. 1989. *Does God Play Dice? The Mathematics of Chaos.* Oxford: Blackwell.

Stigler, S. M. 1986. *The History of Statistics: The Measurement of Uncertainty before 1900.* Cambridge, MA: Harvard University Press.

Suppe, F. 1977. *The Structure of Scientific Theories.* Urbana: University of Illinois Press.

Suppes, P. 1967. What is a scientific theory? In *Philosophy of Science Today,* pp. 55–67. G. S. Morgenbesser, ed. New York: Basic Books.

Swenson, R. 1989. Emergent attractions and the law of maximum entropy production: Foundations to a theory of general evolution. *Systems Research* 6:187–197.

———. 1991a. End-directed physics and evolutionary ordering: Obviating the problem of the population of one. In *The Cybernetics of Complex Systems: Self-Organization, Evolution, and Social Change,* pp. 41–60. F. Geyer, ed. Salinas, CA: Intersystems.

———. 1991b. Order, evolution and natural law: Fundamental relations in complex systems theory. In *Handbook of Systems and Cybernetics,* C. Negotia, ed. New York: Marcel Decker.

———. 1992. Autocatakinetics, yes—autopoesis, no: Steps toward a unified theory of evolutionary ordering. *International Journal of General Systems Research* 21:207-228.

———. 1995. *Spontaneous Order, Evolution and Natural Law: An Introduction to the Physical Basis for an Ecological Psychology.* Hillsdale, NJ: Lawrence Erlbaum Associates.

Swenson, R., and M. T. Turvey. 1991. Thermodynamic reasons for perception-action cycles. *Ecological Psychology* 3:317–348.

Symons, R. H. 1992. Small catalytic RNAs. *Annual Reviews of Biochemistry* 61:641–671.

Tabony, J., and D. Job. 1990. Spatial structures in microtubular solutions requiring a sustained energy source. *Nature* 346:448–450.

Thom, R. 1972. *Structural Stability and Morphogenesis*. New York: Benjamin.

———. 1989. An inventory of Waddington concepts. In *Theoretical Biology: Epigenetic and Evolutionary Order from Complex Systems*, pp. 1–7. B. Goodwin and P. Saunders, eds. Edinburgh: Edinburgh University Press.

Thompson, D. W. 1917. *Growth and Form*. Cambridge: Cambridge University Press.

———. 1942. *Growth and Form*. 2nd ed. Cambridge: Cambridge University Press.

Thompson, P. 1989. *The Structure of Biological Theories*. Albany: State University of New York Press.

Thorne, A. G., and M. H. Wolpoff. 1992. The multiregional evolution of humans. *Scientific American* 226 (4):76–83.

Todes, D. P. 1989. *Darwin without Malthus: The Struggle for Existence in Russian Evolutionary Thought*. New York: Oxford University Press.

Toulmin, S. 1990. *Cosmopolis: The Hidden Agenda of Modernity*. New York: Free Press.

Tritton, D. 1986. Chaos in the swing of a pendulum. *New Scientist* 111(1518):37–40.

Turgenev, I. S. 1862 [1965]. *Fathers and Sons*. R. Edmonds, tr., London: Penguin.

Turing, A. M. 1952. The chemical basis of morphogenesis. *Philosophical Transactions of the Royal Society London*. B237:37–72.

Turner, J. R. G. 1983. "The hypothesis that explains mimetic resemblance explains evolution": The gradualist-saltationist schism. In *Dimentions of Darwinism*, pp. 129–169. M. Grene, ed. Cambridge: Cambridge University Press.

———. 1985. Fisher's evolutionary faith and the challenge of mimicry. *Oxford Surveys in Evolutionary Biology* 2:159–196.

———. 1987. Random genetic drift: R. A. Fisher and the Oxford school of ecological genetics. In *The Probabilistic Revolution*, 2:313–354. L. Krüger, L. Dasten, and M. Heidelberger, eds. Cambridge, MA: MIT Press.

Tyson, J. L. 1976. *The Belousov-Zhabotinski Reaction: Lecture Notes in Biomathematics 10*. Berlin: Springer-Verlag.

Ulanowicz, R. E. 1986. *Growth and Development: Ecosystems Phenomenology*. New York: Springer-Verlag.

———. 1989. A phenomenology of evolving networks. *Systems Research* 6:209–217.

———. 1994. The propensities of evolving systems. In *Social and Natural Complexity*. E. L. Kahlil and K. E. Boulding, eds. Cambridge: Cambridge University Press.

van Fraassen, B. C. 1980. *The Scientific Image*. Oxford: Oxford University Press.

van Holde, K. E. 1980. The origin of life: A thermodynamic critique. In *The Origins of Life and Evolution*, pp. 31–46. H. O. Halvorson and K. E. van Holde, eds. New York: Liss.

van der Steen, W. J. 1993. Towards disciplinary disintegration in biology. *Biology and Philosophy* 8:259–275.

van Valen, L. 1992. Ecological species, multispecies, and oaks. In *The Units of Evolution: Essays on the Nature of Species*, pp. 69–77. M. Ereshefsky, ed. Cambridge, MA: MIT Press.

Vine, B., [Ruth Rendell]. 1987. *A Fatal Inversion*. New York: Bantam.

Volterra, V. 1926. Fluctuations in the abundance of species considered mathematically. *Nature* 118:558–560.

Vrba, E. S., and N. Eldredge. 1984. Individuals, hierarchies and processes: Towards a more complete evolutionary theory. *Paleobiology* 10:146–171.

Vrba, E. S., and S. J. Gould. 1986. The hierarchical expansion of sorting and selection: Sorting and selection cannot be equated. *Paleobiology* 12:217–228.

Vucinich, A. 1989. *Darwin in Russian Thought.* Berkeley: University of California Press.

Waddington, C. H. 1940. *Organizers and Genes.* Cambridge: Cambridge University Press.

———. 1956. *Principles of Embryology.* New York: Macmillan.

———. 1962. *New Patterns in Genetics and Development.* New York: Columbia University Press.

———. 1977. *Tools for Thought: How to Understand and Apply the Latest Scientific Techniques of Problem Solving.* New York: Basic Books.

Wade, M. J. 1976. Group selection among laboratory populations of *Tribolium. Proceedings of the National Academy of Science USA* 73:4604–4607.

———. 1977. An experimental study of group selection. *Evolution* 31:134–153.

———. 1978. A critical review of the models of group selection. *Quarterly Review of Biology* 53:101–114.

Wade, M. J., and C. J. Goodnight. 1991. Wright's shifting balance theory: An experimental study. *Science* 253:1015–1018.

Wagner, M. 1873. *The Darwinian Theory and the Law of the Migration of Organisms.* J. L. Laird, trans. London.

Waldrop, M. M. 1990. Spontaneous order, evolution and life. *Science* 247:1543–1545.

———. 1992. *Complexity: The Emerging Science at the Edge of Order and Chaos.* New York: Simon and Schuster.

Wallace, A. R. 1855. On the law which has regulated the introduction of new species. *The Annals and Magazine of Natural History.* ser. 2, 16:184–196.

———. 1858. On the tendency of varieties to depart indefinitely from the original type. *Journal of the Proceedings of the Linnean Society (Zoology)* 3:53–62.

———. 1870. *Contributions to the Theory of Natural Selection.* London: Macmillan.

———. 1876. *The Geographical Distribution of Animals.* London: Macmillan.

———. 1889. *Darwinism: An Exposition of the Theory of Natural Selection.* London: Macmillan.

Wallace, B. 1992. *The Search for the Gene.* Ithaca, NY: Cornell University Press.

Wallace, D. G., L. R. Maxson, and A. C. Wilson. 1971. Albumin evolution in frogs: A test of the evolutionary clock hypothesis. *Proceedings of the National Academy of Science USA* 68:3127–3129.

Waters, C. K. 1990. Why the anti-reductionist consensus will not survive. *PSA 1990* 1:125–140.

———. 1991. Tempered realism about the force of selection. *Philosophy of Science* 58:553–573.

———. 1994. Genes made molecular. *Philosophy of Science* 61:163–185.

Watson, J. D. 1968. *The Double Helix: A Personal Account of the Discovery of the Structure of DNA.* New York: Atheneum.

Watson, J. D., and F. C. Crick. 1953. Molecular structure of nucleic acids: A structure for deoxyribose nucleic acid. *Nature* 171:737–738.

Weatherall, M., and H. Kamminga. 1992. *Dynamic Science: Biochemistry in Cambridge, 1898–1949*. Cambridge: Wellcome Unit for the History of Medicine.

Weber, B. H. 1991a. Complex systems dynamics and the evolution of biological hierarchies. In *The Cybernetics of Complex Systems: Self-Organization, Evolution, Social Change*, pp. 31–40. F. Geyer, ed. Salinas, CA: Intersystems.

———. 1991b. Glynn and the conceptual development of the chemiosmotic theory: A retrospective and a prospective view. *Bioscience Reports* 11:577–617.

Weber, B. H., D. J. Depew, and J. D. Smith. 1988. *Entropy, Information, and Evolution: New Perspectives on Physical and Biological Evolution*. Cambridge, MA: MIT Press.

Weber, B. H., D. J. Depew, C. Dyke, S. N. Salthe, E. D. Schneider, R. E. Ulanowicz, and J. S. Wicken. 1989. Evolution in thermodynamic perspective: An ecological approach. *Biology and Philosophy* 4:373–405.

Webster, G. 1984. The relations of natural forms. In *Beyond Neo-Darwinism: An Introduction to the New Evolutionary Paradigm*, pp. 195–217. M.-W. Ho and R. T. Sanders, eds. London: Academic Press.

———. 1989. Structuralism and Darwinism: Concepts for the study of form. In *Dynamic Structures in Biology*, pp. 1–15. B. Goodwin, A. Sibatini and G. Webster, eds. Edinburgh: Edinburgh University Press.

Webster, G., and B. C. Goodwin. 1982. The origin of species: A structuralist approach. *Journal of Social and Biological Structures* 5:15–47.

Weldon, W. F. R. 1893. On certain correlated variations in *Carcinus moenus*. *Proceedings of the Royal Society* 54:329.

Wesson, R. 1991. *Beyond Natural Selection*. Cambridge, MA: MIT Press.

West, B. J. 1985. *An Essay on the Importance of Being Nonlinear. Lecture Notes in Biomathematics 62*. Berlin: Springer-Verlag.

Westheimer, F. H. 1987. Why nature chose phosphates. *Science* 235:1173–1178.

Whewell, W. 1833. *Astronomy and General Physics Considered in Reference to Natural Theology*. London: William Pickering.

———. 1837. *History of the Inductive Sciences*. 3 vols. London: Parker.

Wicken, J. S. 1985. An organismic critique of molecular darwinism. *Journal of Theoretical Biology* 117:545–561.

———. 1987. *Evolution, Information and Thermodynamics: Extending the Darwinian Program*. New York: Oxford University Press.

Wiley, E. O. 1981. *Phylogenetics: The Theory and Practice of Systematics*. New York: Wiley-Interscience.

Williams, G. C. 1966. *Adaptation and Natural Selection*. Princeton, NJ: Princeton University Press.

———. 1989. A sociological expansion of *Evolution and Ethics*. In *Evolution and Ethics*. J. Paradis and G. C. Williams, eds. Princeton, NJ: Princeton University Press.

———. 1992. *Natural Selection: Domains, Levels, and Challenges*. New York: Oxford University Press.

Williams, M. B. 1970. Deducing the consequences of evolution: A mathematical model. *Journal of Theoretical Biology* 29:343–385.

Williams, R. J. P. 1961. Possible functions of chains of catalysts. *Journal of Theoretical Biology* 1:1–17.

———. 1991. The chemical elements of life. *Journal of the Chemical Society: Dalton Translations,* 1991:539–546.

Wills, C. 1989. *The Wisdom of the Genes: New Pathways in Evolution.* New York: Basic Books.

———. 1991. *Exons, Introns, and Talking Genes: The Science behind the Human Genome Project.* New York: Basic Books.

Wilson, A. C., and R. L. Caun. 1992. The recent African genesis of humans. *Scientific American* 266(4):68–73.

Wilson, A. C., S. S. Carlson, and T. J. White. 1977. Biochemical evolution. *Annual Reviews of Biochemistry* 46:573–639.

Wilson, D. S. 1975. A theory of group selection. *Proceedings of the National Academy of Science USA.* 72:143–146.

———. 1983. The group selection controversy: History and current status. *Annual Reviews of Ecology and Systematics* 14:159–188.

Wilson, D. S., and E. Sober. 1989. Reviving the superorganism. *Journal of Theoretical Biology* 136:337–356.

Wilson, E. B. 1986. *The Cell in Development and Inheritance.* New York: Macmillan.

Wilson, E. O. 1975. *Sociobiology: The New Synthesis.* Cambridge, MA: Harvard University Press.

———. 1978. *On Human Nature.* Cambridge, MA: Harvard University Press.

Wimsatt, W. C. 1980. Reductionist research strategies and their biases in the units of selection controversy. In *Scientific Discovery,* vol. 2: *Case Studies,* pp. 213–259. T. Nickles, ed. Dordrecht: Reidel.

———. 1981. The units of selection and the structure of the multilevel genome. In *PSA 1980,* 2:122–183. P. Asquith and R. Giere, eds. East Lansing, MI: Philosophy of Science Association.

———. 1986. Developmental constraints, generative entrenchment, and the innate-acquired distinction. In *Integrating Scientific Disciplines,* pp. 185–208. W. Bechtel, ed. Dordrecht: Martinus-Nijhoff.

Wimsatt, W. C., and J. C. Schank. 1988. Two constraints on the evolution of complex adaptations and the means for their avoidance. In *Evolutionary Progress,* pp. 231–273. M. H. Nitecki, ed. Chicago: University of Chicago Press.

Wolpoff, M., and A. Thorne. 1991. The case against Eve. *New Scientist* 130 (1774):37–41.

Worster, D. 1977. *Nature's Economy: The Roots of Ecology.* San Francisco: Sierra Club.

Wright, L. 1973. Functions. *Philosophical Review* 82:139–168.

———. 1976. *Teleological Explanations.* Berkeley: University of California Press.

Wright, S. 1930. Review of the *Genetical Theory of Natural Selection* by R. A. Fisher. *Journal of Heredity* 21:349–356.

———. 1931. Evolution in Mendelian populations. *Genetics* 16:97–159.

———. 1932. The roles of mutation, inbreeding, crossbreeding and selection in evolution. *Proceedings of the Sixth International Congress of Genetics* 1:356–366.

———. 1964. Biology and the philosophy of science. *Monist* 48:265:290.

———. 1977. *Evolution and the Genetics of Populations*, vol. 3: *Experimental Results and Evolutionary Deductions*. Chicago: University of Chicago Press.

———. 1986. *Evolution*. Chicago: University of Chicago Press.

Wynne-Edwards, V. C. 1962. *Animal Dispersion in Relation to Social Behavior*. Edinburgh: Oliver and Boyd.

Young, R. M. 1985. Darwinism *is* social. In *The Darwinian Heritage*, pp. 609–640. D. Kohn, ed. Princeton, NJ: Princeton University Press.

Zhabotinskii, A. M. 1964. Periodic movement in oxidation of malonic acid in solution [in Russian]. *Biofizka* 9:306–311.

Zotin, A. I. 1972. *Thermodynamic Aspects of Developmental Biology*. Basel: Karger.

Zuckerkandel, E. 1987. On the molecular evolutionary clock. *Journal of Molecular Evolution* 26:34–46.

Index

for Fisher, 250–251, 252
Jenkin's critique of, 193–195
Boerhave, Hermann, 41
Bohr, Niels, 254, 342–343
Boltzmann, Ludwig, 3–4, 11, 17, 93
and "deep time," 459–460
and statistical law, 154–155, 246, 253–256,
260–266
Boltzmannian, Newtonian, and nonlinear
systems, 15–18, 24, 486–487, 488, 490.
See also Boltzmann, Ludwig; Newtonian
dynamical models; Nonlinear dynamics
Boolean networks, 431–434
Boundary conditions, for laws, 455–456,
490–491
Boveri, Theodore, 230
Bowen, Francis, 182
Bowler, Peter, 169–170
Box, Joan Fisher, 249
Bragg, Sir Lawrence, 345
Brandon, Robert, 315, 477, 525n1
Breeders, 280. *See also* Natural selection:
breeders as models for
Brewster, David, 142
Bridge laws, 26, 328
Bridges, Robert, 230
Bridgewater Treatises, The, 102–106, 163
Bright, John, 62
British Association for the Advancement
of Science (BAAS) 142–144, 151, 159, 160
Brooks, D. R. and E. O. Wiley: *Evolution as
Entropy,* 529n5
Brooks, W. R., 224
Brougham, Lord Henry, 51
Brownian motion, 264
Bryan, William Jennings, 379
BSC. *See* Biological species concept
Buchner, Ludwig, 178
Buckle, Thomas Henry, 150, 210, 258
History of Civilization in England, 151, 207
Buffon, Georges Louis de, 33, 34, 39, 41,
303
catastrophism of, 44
Natural History, 34
Burgess Shale fossils, 421
Buss, Leo: *The Evolution of Individuality,* 414
"Butterfly effect," 16
BZ reaction, 465–466

California Institute of Technology. (Caltech), 341–345, 355, 416
Cambrian explosion, 318, 421, 424, 425,
453

Cambridge University, 339–341, 345, 346,
355, 416
Campbell, D. T., 471
Campbell, John, 353, 485, 520n3
Canalization, Waddington's, 417
Capitalism, 61, 63, 120–121, 158, 195. *See
also* Free market
American, 182
and Dawkins, 375
and eugenics, 239
and the exploitative view of nature, 404
and Levins and Lewontin, 522n6
and social Darwinism, 379–380
and sociobiology, 376
and Wilson, 522n5
Carlyle, Thomas, 67
Carnot, Sadi, 11, 256, 260
Carnot's heat machine, 256, 266
Carus, Carl Gustav, 48
Castle, William, 229, 278
Casuistry, 204
"Catalytic task space," Kauffman's, 445
Catastrophism, 44, 58, 95–96
Lyell's attack on, 97
modern, 422, 424
"Catastrophe theory," René Thom's, 418
Cause, and causality. *See also* Chance
and chance, 529–530n2
Darwin's view of external, 110–111, 124–
125, 126, 136–137
and genes, 377–380 (*see also* Gene)
ignorance of, and probability, 113
and probability in biology, 286
secondary, 73, 95, 97, 99, 102, 126, 145
Cell, 176, 339–340
Benard, 464–465
convection, 463
division, and generation of organisms,
413–414
membranes, 403
and the origin of life, 400
as self-replicating, dissipative autocatalytic cycles, 467
types as attractors, 449–450
Weismann's germ and somatic, 188–189
Cellular automata, 440, 442, 456
Cellular respiration, 340
Chambers, Robert: *Vestiges of the Natural
History of Creation,* 73, 75, 84, 110–111
Chance, 361. *See also* Cause, and causality
and causality, 529–530n2
and the conditions for natural selection
(Kauffman), 454

Chance (cont.)
in Darwin's natural selection, 81, 113, 149–150, 155
in the expanded synthesis, 390
in the fixation of genes, 337, 435
ignorance interpretation of, 113, 206, 487
laws of, 254, 264
logic of, 423–424
and genetic drift (Wright), 283, 286
on the hierarchical view, 382
vs. natural selection, 429
in the neutral theory, 364, 382
and phylogeny, 421–422, 426
and regulatory genes, 378
and self-organization, 481
and variation, 503n3
Chaos, 16, 20, 499. *See also* Antichaos; Complex systems; Nonlinear dynamics
at or near the edge of, 440–444, 452, 501
and natural populations, 451
and statistical analysis, 504–505n8
Chase, Martha, 344
Chemical selection, 402, 403, 411
and competition, 469–470
and dissipative pathways, 465
and natural selection, 470–471
"Chemiosmotic theory," Mitchell's, 340
Chernesevsky, N., 289
Chetverikov, Sergei, 13, 287, 288, 291, 292, 307, 332
Childs, Charles Manning, 414
Chromasomal theory of heredity, 355
Chromasomes, 230–231, 341
haploid, 342
Circle, as Aristotelian natural motion, 8, 87, 88–89, 90, 186
Clades, and cladism, 304–305, 314, 381, 421
Clausius, Rudolf, 11, 94, 253, 257–258, 261
Clements, Frederick, 406–407
The Phytogeography of Nebraska, 406
Plant Succession, 406
"Climax community," 393–394, 406–407
Codons, 346
Coevolution, 425–426, 451–453, 473
Coleridge, Samuel Taylor, 54–55, 62, 105
Colonies, Wright's, 280
Common descent
and Darwin, 2, 6, 110–111
and the "Eve" hypothesis, 519–520n1
and Geoffroy, 49
and Haeckel, 170
and Lamarck, 46
and von Baer, 413

Competition
and chemical selection, 469–470, 485
ecological, 405, 411, 473, 485
Malthusian, 79, 81, 82, 120–121, 134–136, 155, 269
Complementarity thesis, 342–343
Complexity, and complexification. *See also* Complex systems; Nonlinear dynamics
adaptationist misconstrual of, 386
catastrophe, 434–435
and the Darwinian tradition, 429–430
ecological, 393–394, 395, 472
and the expanded synthesis, 361
Geoffroy's view of, 49
Haeckel's view of, 179
Lamarck's view of, 5, 46, 136
Lewontin and Levins's view of, 378–379, 499
and narrative explanation, 390
and natural selection, 18, 420, 426–427
neo-Lamarckian, 186
in phylogeny, 420, 424, 426, 427, 475–478
and self-organizaton, 431
Complexity barrier, 25, 430, 490
Complexity revolution, 499–500
Complex systems, 437–440. *See also* Auto-catalytic cycles; Chaos; Complexity; Nonlinear dynamics; Self-organization
and continuity of the Darwinian tradition, 486
dynamics of, 438–440
nonlinearities in, 438
and opposition between the human and natural, 494–495
and phase space, 438–439
and prediction, 436, 438
and the probability revolution, 486–487
and self-organization, 432–433, 437
Computational baselines, 363. *See also* Null hypothesis
Computer modeling, 337, 432, 436
Comte, Auguste, 62, 208–209
Constitutive essentialism, 37–38, 506n1, 506n2, 507n5
Constraints
design, 434
developmental, on natural selection, 481–482
evolution explained by, 387–388, 389–390, 429
"local," 482

race to determine the structure of, 345–346

"selfish," 348, 373–374

Dobzhansky, Theodosius, 13, 287, 291–297, 332–333, 484

and Dewey, 522–523n8

and diversity of populations, 296

and eugenics, 297

and Fisher, 294, 296

and genetic drift, 294–295

Genetics and the Origin of Species, 292, 294, 295, 299, 307, 316–317, 367

and Kauffman, 443

and Lewontin, 366–368, 379

and Mayr, 311–312, 313

and the modern synthesis, 299–302

moral dimension of, 296–297

and natural selection, 276, 293–296

and Simpson, 319

and speciation, 306–310

and Wright, 276–277, 292–296, 307–308, 516n8

Dominant traits, 221

Donne, John, 88

Doppler, Christian, 221

Dostoyevsky, Fyodor: *Notes from Underground*, 210

Dreisch, Hans, 230, 413–414

Drosophila. See Fruit fly

Dualism, 491, 492

Dugdale, Robert, 240

Durkheim, Emile, 209, 211

Dynamics, and dynamic systems, 3, 25–26, 437. *See also* Complex systems; Newtonian dynamical model; Nonlinear dynamics

trajectories of, 26, 438

East, Edward, 240

"Ecological species concept," 320

Ecology, 301, 393–394, 403–412

and biogeography, 405

and the "climax community," 406–407, 474

and community ecology, 407, 472

competition and cooperation in, 405–406, 411, 485

"deep," 407

and developmentalism, 405

at or near the edge of chaos, 452

and evolution, 472–473

history of, 104–412, 525n2

and Kauffman's model, 451–453

and ontogeny, 393–394, 405, 412–420, 473–475

and phylogeny, 393–394, 405, 420–427, 475–478

and reformed systems ecology, 407–412

and the reunion with evolutionary theory, 404, 412

and "superorganisms," 405, 407, 475

Edgeworth, Francis: *Mathematical Psychics: An Essay on the Application of Mathematics to the Moral Sciences*, 512–513n5

Edinburgh Review, 142, 152

Egerton, Francis Henry, 102

Eigen, Manfred, 402

Einstein, Albert, 254, 264

Eldredge, Niles, 19–20, 381, 421–422

Eliot, George, 67

Middlemarch, 159

The Mill on the Floss, 159

Ellis, Emory, 343

Elton, Charles: *Animal Ecology*, 407

Embranchements, Cuvier's, 44–45, 48

Embryogenesis, 42

Embryology, 190. *See also* Developmentalism; Ontogeny and Phylogeny; Recapitulationism

Aristotle's, 41–42

and Morgan, 230

and mutation, on Kauffman's model, 450

von Baer's, 42–43, 170–172, 412–413

Empedocles, 36, 46

Empiricism. *See also* Logical positivism

positivist, 212

Victorian, 148

Energetics, 290–291. *See also* Ecology

and evolution, 472–473

and natural selection, 472, 485–486

revolution in mid-nineteenth century, 255–257

Energy, 154, 255–257

cellular, 340

and fitness, 472–473

Engels, Friedrich: *Condition of the Working Class in England*, 151

Enlightenment, 87, 91, 94, 119–120

Counter-, 87

and the probability revolution, 204

Scottish, 51, 94, 102, 115, 118

Entropy, 11, 12, 93–94. *See also* Equilibrium; Second law of thermodynamics

and autocatalytic kinetic systems, 426

Entropy (cont.)

Boltzmann's statistical approach to, 261–264

in Fisher's theory, 12, 253, 270–271, 284

"informational," 483

and living things, 460–462

negative, ("negentropy"), 461

and time, 263, 460

"Environmental selection," 315

Enzymes, 340, 346

Epigenesis, 41, 42, 189

Epigenetic landscape, Waddington's, 417, 434

Epistasis, 512n1

Equilibrium, dynamic, 9, 90, 102, 265. *See also* Hardy-Weinberg Equilibrium formula; Punctuated equilibrium; Second law of thermodynamics

Maxwellian-Boltzmannian, 154, 265

in Darwin's theory, 126–127, 130, 131

Fisher's conception of, 251–253

and heat flow, 256

in Newtonian and Boltzmannian systems, 486–487

systems far from, 461–462, 463–464

thermodynamic, 261, 262–263

Ergodic systems, 464

"Error curve," 207, 222

Essentialism, 39–40, 128–129, 303, 310. *See also* Constitutive essentialism; Typological Essentialism

Estabrook, Arthur, 240

Eugenics, 13, 194, 199–202, 212, 227, 380

demise of, 297

and Dobzhansky, 297

Fisher's, 244, 248–250, 269–270, 272–273

and genetics, 238–242

and Kimura's neutral theory, 367

negative, 200, 239

positive, 200, 244, 270

and Wilson's sociobiology, 375, 376

Wright's, 279

Eugenics Education Society of London, 199

"Eve" hypothesis, 519–520n1

"Evolutionary molecular clock," 348

Evolutionary progress, 158, 159–160

Darwin's view of, 136–138

Haeckel's view of, 179, 181

neo-Lamarckian view of, 186–187

Evolutionary species concept, 319

Evolutionary theories

emergence of, 43

post-Darwinian, 169–170

Evolvability, 21, 485

"Exaptation," 388, 503n2

"Exons," 350–351, 351–352, 353

Expanded synthesis, 19–20, 360–362, 381–390, 498–499

and adaptationism, 429

and complexity, 361

and Dawkins, 384–385

and developmentalism, 361, 429

and genetic drift, 383

and laws, 388–389

and Lewontin, 384, 385–390, 499

natural selection in, 360, 381–388, 402, 518n10, 524n13

narrative in, 389–390

and new developmentalists, 397–399

and philosophy of science, 384, 385–386, 388–389, 398–399

and reductionism, 384, 388

self-organizing properties in, 381, 382

Explanation, 26–30, 129–130, 315. *See also* Law

as context sensitive, 388, 491

and "just-so stories," 388, 389, 390

"law-covered," 389

logical positivist, 326 (*see also* Logical positivism)

narrative, 389–390 (*see also* Narrative)

statistics, and predictions, 152 (*see also* Probability revolution; Statistical method)

Extinction, 77, 124

for Cuvier, 45

for Darwin, 70, 81

of demes, 281

for Gould, 422–424, 425

for Kauffman, 452–453

for Lyell, 107–108

for neo-Lamarckians, 186

uniformitarian, 96

Falsification, empirical, 325

First Reform Bill, 61, 102, 121, 143

Fisher, Eileen, 250

Fisher, R. A., 10, 11–12, 17, 332, 360, 372

background of, 243

"The Correlation between Relatives on the Supposition of Mendelian Inheritance," 244–245

and Darwin, 251–252, 269–270

and Dobzhansky, 294, 296

and eugenics, 241, 245–246, 248–250, 269–270, 272–273

Gene (cont.)

"jumping," 338, 351, 497

Kauffman's arrays of symbols as, 432

linkage, 231, 235, 512n1

as nonatomistic, 281

as physical entity, 511n1

"pseudo-", 352

recessive, 237, 281

structural, 349, 350–351, 467

as unit and beneficiary of selection, 369–370, 373

"Genetic assimilation," Waddington's, 416, 417

Genetic code, working out of, 337

Genetic drift, 13, 19. *See also* "Sewall Wright effect"

developed by Kimura, 364

difficulty of distinguishing, from selection, 286–287

Dobzhansky's rejection of, 295

English Darwinian misinterpretation of, 321–324

and the expanded synthesis, 383

Fisher's view of, 275

Wright's view of, 275, 280–281, 307–308

"Genetic load," 367–368

Genetic maps, 231–232

"Genetic revolution," 314, 319

Genetics, 1, 173, 228

developmental, 14

and eugenics, 238–242

in France, 349

molecular, 1, 14

Genic selectionism, 282, 368–381, 387, 483, 498

Dawkins's, 372–375, 521n5

and Fisher, 368–369

Hamilton's, 370–372

Lewontin and Levins's critique of, 375–380

Williams's, 369–370

Wilson's, 375, 521–522n5

Genome, 14, 337

Genotype and phenotype, 226, 228, 232, 234

for Dobzhansky, 294

for Fisher, 244, 246

for Johannsen, 228

for Mayr, 310–315

for Waddington, 416

Geoffroy Saint-Hilaire, Etienne, 34, 35, 161, 418, 427, 483. *See also* Geoffroyism

debate of with Cuvier, 48–50, 52, 161–162

and Lamarck (*see* Geoffroyian Lamarckism)

and the *Naturphilosophen*, 48–49

Geoffroyian Lamarckism, 49–50, 162, 171, 172

and Darwin, 69, 72, 136, 170–171

and Haeckel, 178–179

and "theoretical biology," 418

Geoffroyism, 52, 362, 433, 483. *See also* Geoffroy Sainte-Hilaire, Etienne; Geoffroyian Lamarckism

Geographic isolation, 278, 305. *See also* Physiological isolation; Reproductive isolation; Speciation: allopatric

and Darwin's variation, 130–131, 133

for Dobzhansky, 276–277, 292–293

for Wright, 276–277

"Germinal selection," Weismann's, 190–191

Gibbs, Josiah Willard, 339

Gilbert, Charles, 278

Gilbert, Walter, 347

Glycolysis, 340

God. *See also* Deism; Theism

American transcendentalist, 184

Cuvier's, 45

and the design argument, 51, 100–102, 103–105, 146, 162, 476–478

Dissenting Whigs', 105–106

Haldane's view of, 236

Herschel's, 66

Lyell's, 106, 107–108

neoclassical biology's, 33

Newton's, 68–69, 85–86, 92, 103

Prout's, 103

and secondary causes, 94–95, 99, 102

Tory intellectual conception of, 105

Goddard, Herbert, 240

Goethe, Johann Wolfgang von, 49, 87, 173. *See also Naturphilosophie*

Goldschmidt, Richard, 301, 317, 350, 395, 414–415

The Material Basis of Evolution, 415

Physiological Genetics, 415

and saltationist speciation, 415

Goodwin, Brian, 418–420, 433–434, 483, 500

Gould, John, 60

Gould, Stephen Jay, 19–20, 381. *See also* Punctuated equilibrium

and adaptationism, 387–388

developmentalist rebuttal of, 424–427

and the hierarchical theory of selection, 382–383

as Kuhnian, 505n12

Lotka, Alfred, 407–412, 474
and autocatalytic cycling, 409–411
background of, 407–408
Elements of Physical Biology, 408
"law of evolution" of, 409
and physics, 408–409
Lotka-Volterra equation, 411–412
Lucretius: *De Rerum Natura*, 69, 271
Luria, Salvadore, 344, 345
Lwoff, André, 349
Lyell, Charles, 10
and Darwin, 60, 67, 73–75, 83–84, 109–110, 111, 124
and the locus of competition, 127
objection of, to Lamarckism, 108–109, 124
Principles of Geology, 58, 67, 73, 78, 83, 95, 102, 124
and tension between external forces and fixed kinds, 106–108
as uniformitarian and Newtonian, 97–98, 108–109
on variability and adaptability, 106–109
Lysenko, T. D., 242, 287

MacArthur, Robert, 301, 375, 412
McClintock, Barbara, 338, 351, 497–498
McCullough, Warren, 433
Mach, Ernst, 211, 264
MacIntyre, Alasdair, 24
Macroevolution and microevolution, 14–15, 317, 319, 381, 415, 483
"Magic molecule," Dreisch's and Morgan's, 414
Maimonides, Moses, 38
Malthus, Thomas, 7–8, 62, 120
Essay on Population, 7, 124, 507n2
influence of, on Darwin, 7–8, 71–72, 76–80, 115, 121–124, 126- 128, 132–136, 155, 269–270
Russian view of, 289–290
Malthusian limits, 120–121, 269
Margoliash, Emmanuel, 348
Market mechanism, 116–118. *See also* Capitalism; Free market
Martineau, Harriet, 67, 71, 121, 156
Marx, Karl, 82, 120–121, 376
Capital, 151
Marxism, 327, 379, 380
Mass, 90–91, 154
mass selection, 280
Materialism
ancient Greek, 36
of biochemists, 338–339

dialectical, 287
Dissenting Whig, 95
Haldane's, 236
Lamarck's, 47–48, 68–69
as passive, 68, 103
reductionistic, 174–175, 177, 178, 236
Russian, 1860's, 289–290
Simpson's humanistic, 318
Maupertuis, Pierre de, 93
Maximization (and minimization) principles, 93–94, 119, 284. *See also* Entropy
Fisher's, 251
Maxwell, James Clerk, 3–4, 11, 17, 92, 93
and Darwinism, 266
and statistical law, 153–154, 253–256, 257–260, 264–266, 530n3
and "switching mechanisms," 271
Maxwell's Demon, 154, 259
May, Robert, 451
Maynard Smith, John, 301, 434, 521n3
Mayr, Ernst, 2, 13, 20, 301–302, 309–316
background of, 309–310
and the biological species concept, 311–314
and Darwin, 310–311, 314
and Dobzhansky, 310, 311–312, 313
and Fisher, 316
and the historical conception of evolutionary biology, 310- 311, 314–315
and natural selection, 315–316
and reductionism, 315, 316, 354
and Simpson, 319
stress of, on phenotypic level, 310–316
Systematics and the Origin of Species, 299, 310
systematics of, 314–315
"Mean kinetic energy," 257, 258
Meckel, Johann, 48
Meckel-Serres law, 52
Meiosis, 230
Mendel, Father Gregor, 1, 215, 217–224. *See also* Mendelism; Mendel's laws
as Aristotelian, 221–222
and Darwin, 220
education of, 221
experimental method of, 221–223
"Experiments on Plant Hybridization," 217–220
monastic life of, 220–221
and the myth of simultaneous independent rediscovery, 223–224
and physics, 221–222

Mendelism, 28, 331–332, 331–332. *See also* Mendel's laws
Bateson's 224–228
vs. biometry, 224–228, 233–234, 238, 305–306
combinatorial particles of, 511n1
and eugenics, 238–242
and the preformationalist tradition, 189, 228–229, 396
reduction of, to molecular genetics, 354–358, 369, 374
Weismannian, 395, 396
Mendel's laws, 28–29, 217–218, 229–230, 328, 352. *See also* Mendelism
Messenger RNA, 346, 350–351
Metabolism, 468–470, 486–487
Metaphor, as explanation, 29, 30, 374, 456
Migration, 306, 312
Mill, John Stuart, 65, 67, 121, 152, 156
System of Logic, 156
Miller, Stanley, 400
Millikin, R. A., 341
Milne-Edwards, Henri, 82, 133, 171
Introduction à la zoologie général, 135
Milton, John, 59
Mimics, 387, 515n11
"Missing links," 52, 183, 383
Mitchell, Peter, 340
Modern evolutionary synthesis, 13–14, 234, 299, 516–517n2
American, 276–277, 299–321, 324–325, 332
British, 321–325
and Chetverikov, 291, 332
and Dobzhansky, 276–277, 307
expansion of (*see* Expanded synthesis)
and gene expression, 414
"hardening" of, 324–325, 388–389, 517n10
as marriage of natural selection and genetics, 1
and molecular biology, 337–338, 356–357, 359
and physics, 302–303
and population thinking, 300–301
as positivist reduction, 333
and the probability revolution, 11, 173, 302–303
as solution to "the species problem," 300
as synthesis among fields, 299–300
as a theory, 299, 307, 333
Molecular biology, 189–190, 497–498. *See also* Biology
and developmentalism, 395, 396

and the modern synthesis, 337–338, 356–357, 359
and the origin of life, 400–403
natural selection in, 357
philosophical implications of, 353–357
popularity of, 346–347
as reductive model, 328
as threat to Darwinism, 1, 395
"Molecular clock," of protein evolution, 348, 357, 528n3
and the neutral theory, 362–368
"Molecular drive," 353
Monod, Jacques, 14, 349, 401, 418, 431
Morgan, Thomas Hunt, 221, 240, 279, 291
background of, 230
and Driesch, 414
and gene linkage, 231–232
and the Morgan school, 232–235
and natural selection, 235
The Physical Basis of Heredity, 235
sensitivity of, to embryology, 230–231
Morphogenesis, Turing's model of, 418, 528–529n4
"Morphogenetic field," Goodwin and Webster's, 418, 419
"Mosaic development," 414
Morowitz, Harold, 365, 402–403, 403–404, 468–469
Muller, Hermann, 230, 235, 241–242, 287, 296
and the Kimura-Lewontin controversy, 360, 366–367, 381
Out of the Night, 241, 242
Müller, Johannes, 54
Multigene family, 351–352, 520n3
Mutations, 224, 225, 226, 229. *See also* "Sports," and "single variations"
and "genetic assimilation," 416
Haldane's view of, 237
in Kauffman's model, 435
Morgan's redefinition of, 234–235
and multigene families, 520n3
in the neutral theory, 362, 364, 382
radiation-induced, 343
in Wimsatt's model, 484–485

Narrative
approaches to complexity and contingency, 423–424, 436, 457, 493–494
and catastrophism, 95, 96, 98
in the expanded synthesis, 389–390, 398, 423–424

Narrative (cont.)

Kauffman's resistance to Darwinian, 455

and Mayr's evolutionary natural history, 316

Natural selection. *See also* Adaptationism; Adaptedness and adaptation; Balancing selection; Chemical selection; Fitness; Genic selectionism; Group selection; Interdemic selection; Physical selection

and autocatalytic cycles, 466, 470–473, 475, 485

Bateson's view of, 224–228

breeders as models for, 5, 29, 71, 78, 124–125, 148, 149

and complexification, 18, 420, 426–427

as conceptual core of Darwinian tradition, 2

Darwin's view of, 6, 71–72, 80–82, 113–115, 123, 124–125, 133

Dawkins's view of, 374–375

developmental constraints and, 481–482

Dobzhansky's view of, 276, 293–296

emergent from chemical selection, 470, 485

and energetics, 472, 485–486

and evolvability, 20–21

in the expanded synthesis, 360, 381–388, 402, 524n13

as explanation of first resort, 481

as explanation of last resort, 481

and explanatory models from physics, 2–3

Fisher's view of, 244, 245, 247–248, 250–253, 266–273

genetic theory of, 172–173, 357

Haeckel's view of, 179

Haldane's view of, 236, 237, 238, 243

Kauffman's view of, 429–430, 432, 434–436, 443–444, 446, 451, 454

Lewontin's view of, 386, 524n14

and "life-cycle" traits, 475

Lotka's view of, 408–411

Mayr's view of, 315–316

in molecular biology, 357

Morgan's view of, 235

neo-Lamarckian view of, 181–182, 186

as Newtonian external force, 71, 124–126, 153

and the neutralist theory, 359–360, 362–364, 366–368, 481

on the edge of chaos, 471

and the Oxford School, 321–323

and post-Darwinian evolutionary theory, 170–171

"Progressive Darwinist," 379–380

and protein synthesis, 347

Russian, anti-Malthusian, 290–291

and self-organization, 429–430, 443–444, 454, 479–486

Simpson's view of, 366

and statistical law, 153–155, 201–202

as tautology, 327

and teleology, 477

Waddington's view of, 418

Weismann's view of, 188, 189, 190–191

Weldon's view of, 214–215

Wright's view of, 280–287

"Natural theologians," 99, 100

Naturphilosophie, 48–49, 52, 55, 161, 174–176, 506n4

and vitalism, 175, 177

Needham, John, 41

Needham, Joseph, 415, 416

Negative heuristic, 22

"Negentropy," 461

"Neo-Darwinism," 188, 198, 513–514n7, 516–517n2

Neo-Lamarckism, 183–184, 185–187, 201, 317, 330

and natural selection, 181–182, 186

Neotenous retention of juvenile characters, 187

Neurospora, 342

Neutral theory of protein evolution, 14, 362–368, 381, 498, 528n3. *See also* Kimura, Motoo

and the modern synthesis, 365–366

and natural selection, 359–360, 362–364, 366–368, 481

Newton, Sir Isaac, 3, 8–10, 64, 85–86. *See also* Newtonian dynamic models

laws of motion of, 88–91

materialism of, 68–69

Opticks, 92

physics of, 86–87

Principia Mathematica, 86

reconstruction of, 94–95

Newtonian dynamical models, 8–10, 28, 71, 86–92, 265. *See also* Darwin, Charles, and Malthusian political economy; Newton, Sir Isaac

and biology, 100–102, 109–111, 114–115, 125, 132–133, 265–266, 490

and Boltzmannian models, 15–18, 24, 486–487, 488, 490

and *The Bridgewater Treatises*, 102–106

Ostwald, Friedrich William, 408
Owen, Richard, 33, 35, 39, 62, 105
  and Aristotle, 55
  and Darwin, 55–56, 60, 70, 142
  and Grant, 52, 54
  Hunterian Lectures of, 53–54, 161
  Neoplatonism of, 54
  and recapitulationism, 52–54
  and typological essentialism, 55
  and von Baer, 52–54, 70
Oxford school of population genetics, 321,
  323

Packard, Alphaeus, 181
Paleontology, 52
  Agassiz's, 183–184
  Gould's, 421–424
  neo-Lamarckian, 185–187
  Simpson's, 317–318
  and the stasis recorded in the fossil re-
  cord, 482
Paley, William, 4–5, 6, 54, 57. See also De-
  sign argument
  Natural Theology, 67, 99–102
Pangene, 228
Pangenesis, 131, 228
"Panglossian paradigm, the," 105, 106
Paradigm. See "Scientific paradigm"
Particulate inheritance, 221, 225, 226, 511n1
Pascal, Blaise, 204
"Pascal's Wager," 204
Path analysis, 515n4
Patterson, Hugh, 320
Pauling, Linus, 341, 342, 345
Pearl, Raymond, 408
Pearson, Karl, 211–212, 213–215, 226, 227–
  228, 238
  and Fisher, 245, 247–248, 270
  The Grammar of Science, 211–212, 285
Peas, Mendel's, 217, 218–220
Peel, Sir Robert, 61
Peelites, 62, 67
Peirce, Charles Sanders, 155, 182, 286, 484,
  488
  evolutionary theory of, 508–509n3
Pendulum, 439, 526n1
Pepper moth (Biston betularia), Haldane's,
  237–238
Permian extinction, 453
Permo-Triassic extinction, 423
Perutz, Max, 340, 345
Petty, William: Political Arithmetic, 203
"Phage." See Bacteriophage

Phase space, 438–439, 456, 488–489, 504n7
Phenomena, as explananda, 27–28
Phenomenalism, 212, 263–264, 326
Philipchenko, I. A., 287, 292, 317, 332
Philosophical Radicals, 62–63, 65, 121
Philosophy of science, 26, 148, 165, 263,
  325, 501
  and the expanded synthesis, 384, 385–
  386, 388–389, 398–399
  Victorian, 64–67
"Phyletic evolution," 314
"Phylogenetic systematics," 304–305
Phylogenetic trees, calculated from protein
  sequences, 348, 350
Phylogeny, 6, 393, 420–427. See also Ontogeny and phylogeny
  complexification and simplification in,
  420, 424, 426, 427
  contingency in, 421–422, 423–424
  developmentalist rebuttal of Gould's
  analysis of, 424–427
  and mass extinction, 422–423
  and narrative, 423–424
  and punctuated equilibrium, 420–421
"Physical biology," Lotka's, 408
Physical selection, 402, 464–465, 469, 470,
  483–484
Physics. See Boltzmann, Ludwig; Law;
  Maxwell, James Clerk; Newtonian dy-
  namic models; Nonlinear dynamics; Phi-
  losophy of science; Probability
  revolution; Reductionism
Physiocrats, 119, 120
Physiological isolation, 309, 311, 517n5
Pikaia, 423
Planck, Max, 264
Plato, and (Neo-) Platonism, 36, 38–39, 49,
  54, 87–88, 229
Playfair, John: Illustrations of the Huttonian
  Theory of the Earth, 93, 97
Pleiotropy, 232, 512n1
Political economy. See Darwin, Charles:
  and Malthusian political
  economy
Polypeptide, 346
Polyphosphate, 469
Pope, Alexander, 88
Popper, Karl, 325, 488
Population genetics, 10–13, 243, 284, 288,
  299, 301
Population pressure, Malthusian, 76, 78,
  120–126, 127, 155
  vs. usable energy sources, 290

Populations, natural vs. laboratory, 288
"Population thinking," 233, 300–301
 Mayr's, 310–311
 and species as "individuals," 321
 and typological essentialism, 506n1
Porter, T. M., 151, 259
Positive heuristic, 22
Positivism, 325–326. *See also* Logical positivism; Phenomenolism
Poulton, E. B., 226
"Power-law diagnostic," 440–442
Prediction, 26, 65, 152, 315, 326
 and complex systems, 436, 438
 and nonlinearity, 504–505n8
Preformationism, 41–42, 189, 228–229, 396
Priestly, Joseph, 338, 339
Prigogine, Ilya, 461–462
"Principle of competitive exclusion," 412
"Principle of plenitude," 38
Probabilism, 204
Probability, 487–490. *See also* Probability revolution
 frequency interpretation of, 487, 488
 ignorance interpretation of, 113, 206, 258, 487
 propensity interpretation of, 487–490, 519n13
 as real properties, 487–488
 and speciation, 306–307
Probability revolution, 11, 202–206, 257–273, 430. *See also* Probability; Statistical method
 and Boltzmann, 260–264
 and Darwin, 113–114, 302, 330–331, 504n6
 and Darwinism, 264–266, 330–331
 and Fisher's use of Maxwellian-Boltzmannian physics, 255, 266- 273
 as Kuhnian scientific paradigm, 330
 and Maxwell, 257–260
 and modern Darwinism, 302–303
 and nonlinear dynamics of complex systems, 486
 and the semantic view of theories, 506n16
 and supervenience, 519n13
 and teleology, 476–477
 utilitarianism, and political economy, 512–513n5
 and Wright, 285–286
Prokaryotes and eukaryotes, 424–425, 520n2
Progress. *See* Evolutionary progress
Propensities. *See* Probability: propensity interpretation of

Protein, 340, 346
 evolution, 347–350
 human and chimpanzee, 349–350
 and the origin of life, 400–401, 445–446, 467–470
 polymorphisms, 338, 367–368, 520–521n1
Prout, William: *Chemistry, Meteorology, and the Function of Digestion Considered with Reference to Natural Theology*, 102–103
"Pseudo-genes," 352
"Punctuated equilibrium," Gould and Eldredge's, 15, 318, 320, 383, 420–421, 476
 in Kauffman's model, 450–451, 452, 476
Punnett, Reginald W., 218, 227, 232, 233, 243
Punnett square, 218
"Pure lines," 228–229, 234
"Purifying selection," Muller's, 235, 241, 288
 and the neutral theory, 360, 363, 366–367, 368, 381

Qualitative predictions, 438
"Quantum evolution," 319–320
Quantum mechanics, 254, 342–343, 345, 488
Quaternary structure of protein, 528n3
Quesnay, François, 119
Quetelet, Adolphe, 150–153, 194
 Letters on the Theory of Probabilities, 209
 "On the Influence of Man's Free Will on Social Facts," 209
 and the probability revolution, 207–208, 209, 210
 "The Social System and the Laws That Govern It," 209
Queteletismus, 210, 258

Racism, 195, 197, 368, 376
"Rational economic agent," 120, 157
"Rational morphology," Geoffroy's, 48
Realism, scientific, 29–30
Recapitulationism, 161, 170–173, 190, 506–507n5
 and Darwin, 138–139, 170–172
 decline of, 331
 Haeckel's, 170, 176–177, 329–330
 and Owen, 52–53, 170–171
Recessive traits, 218, 221
Red Queen hypothesis, 516n9
Reductionism, 25, 174–175, 177, 178
 Dawkins's radical materialist, 374–375
 and the expanded synthesis, 384, 388